CONTROLLED REPRODUCTION
in Sheep and Goats

Controlled Reproduction in Farm Animals Series

Ian Gordon, *Emeritus Professor of Animal Husbandry, University College, Dublin, Ireland*

1: Controlled Reproduction in Cattle and Buffaloes

2: Controlled Reproduction in Sheep and Goats

3: Controlled Reproduction in Pigs

4: Controlled Reproduction in Horses, Deer and Camelids

CABI Publishing is a division of CAB International

CABI Publishing
CAB International
Wallingford
Oxon OX10 8DE
UK

Tel: +44 (0)1491 832111
Fax: +44 (0)1491 833508
Email: cabi@cabi.org
Web site: www.cabi-publishing.org

CABI Publishing
875 Massachusetts Avenue
7th Floor
Cambridge, MA 02139
USA

Tel: +1 617 395 4056
Fax: +1 617 354 6875
Email: cabi-nao@cabi.org

A catalogue record for this book is available from the British Library, London, UK

ISBN 0 85199 115 7
This book comprises Volume 2 in a four-part series on
Controlled Reproduction in Farm Animals
Four volume set: 0 85199 118 1

First published 1997
Reprinted 1999
Transferred to print on demand 2004

Printed and bound in the UK by Antony Rowe Limited, Eastbourne.

CONTROLLED REPRODUCTION
in Sheep and Goats

IAN GORDON
BSc, MA, PhD, DipAgric (Cantab), NDA

Emeritus Professor of Animal Husbandry
University College
Dublin
Ireland

CAB INTERNATIONAL

Contents

Preface

More than a decade has passed since the appearance of the book entitled *Controlled Breeding in Farm Animals* (Pergamon Press), on which this new four-volume series is based. The aim of this second volume of the series is to provide a fairly detailed and up-to-date view of the literature dealing with the many different ways in which reproduction in sheep and goats may be controlled and manipulated. The hope is that this book will prove to be of value and interest not only to students of animal science and veterinary medicine but also to those concerned with the practical aspects of reproduction control, whether in research, in an advisory capacity or in applying the techniques directly on the farm. Although the book does not concern itself with reproductive disorders or infertility problems, parts of the text will be of interest to veterinary practitioners. For advanced undergraduates in animal science and veterinary medicine contemplating research in reproductive physiology, the work may provide some insight into the nature and scope of current reproductive technology and of the problems that await solution.

It would be foolish to claim that a work such as this can be other than incomplete, in view of the vastness of the literature, but an attempt has been made to ensure that most statements of substance are backed by an appropriate reference. It should be emphasized that the text covers areas such as embryo transfer technology in which there is considerable research activity; for many readers, the chapters may serve as nothing more than a starting point in seeking information on their particular interests. A major objective of the work is to draw attention to information that may be used directly to increase the worldwide efficiency of sheep and goat production systems.

This volume has been distilled from research and teaching interests in the UK, the USA and Ireland, spanning a period of more than 40 years. As a graduate in Agricultural Science from Nottingham University in the early 1950s, the author had the opportunity of working as a graduate student under the late Sir John Hammond at the School of Agriculture and Animal Research Station in Cambridge. At that time, in the 1950s, difficulties facing a researcher in reproductive physiology in Cambridge included lack of pasture and space to keep farm animals. For that reason, the author was to spend much of the 1950s out of the laboratory working directly with farmers and their sheep and cattle in many counties of England and Wales. Later, in Ireland, ably supported by an enthusiastic band of graduate students, the author continued to work happily with farmers in developing controlled reproduction techniques. In more recent times, I have been confined to the laboratory, mainly in an effort to see progress in the laboratory production of cattle and sheep embryos. Nevertheless, it is with farmers and the animals they tend that the author feels most at home. It is hoped that this book will be of some value to all those who have the best interests of animal agriculture at heart.

Acknowledgements

The author wishes to express his gratitude to the many editors and publishers who have given permission to reproduce illustrations and tables from books and original articles. Sincere thanks must also be extended to the staff of CAB INTERNATIONAL for their excellent help during the preparation and production of this volume.

The author greatly appreciates the provision of office and facilities at Lyons Research Farm by University College Dublin to allow the writing to be done in the peace and quiet of the countryside.

Finally, thanks are due to my wife, who cheerfully tolerated the many hours of absence involved in writing the book.

List of Abbreviations

ACTH	adrenocorticotrophic hormone
AI	artificial insemination
BSA	bovine serum albumin
CIDR	controlled internal drug release
CL	corpus luteum
DES	diethyl-stilboestrol
DF	dominant follicle
DMSO	dimethyl sulphoxide
EGF	epidermal growth factor
EIA	enzyme immunoassay
ET	embryo transfer
FGA	fluorogestone acetate
FGF	fibroblast growth factor
FSH	follicle stimulating hormone
GnRH	gonadotrophin releasing hormone
GST-AI	Guelph system for transcervical AI
HAP	horse anterior pituitary
hCG	human chorionic gonadotrophin
3β-HSD	hydroxysteroid dehydrogenase (Epostane)
HMG	human menopausal gonadotrophin
ICSI	intracytoplasmic sperm injection
IFN	interferon
IGF-I	insulin-like growth factor-I
INRA	National Institute of Agricultural Research (France)
IU	international units
IVF	*in vitro* fertilization
IVM	*in vitro* maturation
LH	luteinizing hormone
MAP	medroxyprogesterone acetate
MGA	melengestrol acetate
MOET	multiple ovulation and embryo transfer

MOPS	3-(*N*-morpholino)propanesulphonic acid
ODB	oestradiol benzoate
oPL	ovine placental lactogen
PBS	phosphate-buffered saline
pFSH	purified FSH
PG	prostaglandin
pLKH	purified LH
PMSG	pregnant mare serum gonadotrophin
PRL	prolactin
PSPB	pregnancy-specific protein B
rBST	recombinant bovine growth hormone
RIA	radioimmunoassay
RSD	ram semen diluent
SOF	synthetic oviduct fluid
SPA	sperm penetration assay
TGF-α/β	transforming growth factor-alpha/-beta

Introduction to Controlled Reproduction in Sheep

1

1.1. Introduction

Sheep are believed to have been one of the first mammals to be domesticated and are known to have been closely associated with man from a very early date. Shelton (1995) draws attention to the fact that sheep offer the potential of making an important and continuing contribution to providing food and fibre for a growing world population. The same author notes that the efficiency of meat production from sheep can be increased by exploiting some of the unique advantages offered by this species. One of the most important advantages of sheep lies in its reproductive rate. In that regard, it should be noted that for more than 60 years, research workers around the world have examined the possibility of employing hormones in the control of oestrus and ovulation in sheep. The fact that most ewes in the agriculturally productive countries are seasonal breeders and often produce smaller lamb crops than the farmer may actually wish has made sheep a rather obvious target for the reproductive physiologist's attention.

1.1.1. Seasonality of reproduction in sheep

It is well known that reproduction in sheep is seasonal, at least in breeds originating from temperate climates. Over the years, natural selection presumably has favoured sheep that give birth at the most appropriate time of year in terms of climate and food availability. Seasonal breeding patterns may even be found in farm species such as cattle, when they have returned to the feral state (Ortavant *et al.* 1985). It appears that natural selection has provided mammals, especially sheep, with signalling systems which couple certain forms of environmental variation with the appropriate neuroendocrine responses to ensure that reproductive activity occurs at the most favourable time of year, depending on the length of gestation. These neuroendocrine responses continue to operate in the sheep despite the selection practised by man over several thousand years of domestication.

1.1.2. Sheep at home and abroad

The breeding ewe flock in the Irish Republic currently stands at around 4.62 million head (June 1994 livestock census); the total flock number runs to some 8.5 million animals. Sheep numbers in the countries of the European Union (EU) show an estimated number of breeding ewes in 1995 around the 71 million mark, with the United Kingdom having the largest flock in the EU. Although sheep numbers in the EU countries showed some decline in 1994, in other parts of the world there have been marked increases in sheep populations. In China, for example, Cheng and Ma (1992) record a fourfold increase in the number of sheep in that country between 1949 and 1989, the number rising from 26.2 million to 113.5 million; several new breeds have been developed in China by crossing with imported breeds from the former USSR, New Zealand, Australia and the UK.

It should also be mentioned that in the post-communist restructuring of agriculture in Eastern European countries, there are examples of marked declines in sheep populations. Following the introduction of a freemarket economy in Poland in 1989, for example, sheep numbers declined from about five million to less than one million in 1994 (Martyniuk and Rzepecki, 1995).

1.1.3. Sheep as efficient meat producers

Authors have drawn attention to the fact that low lamb output per ewe is a major factor limiting the energetic efficiency of sheep meat production (Blaxter, 1964); it has also been observed that with an estimated 'biological ceiling' of five lambs per ewe and a lambing interval of six months, the ewe has further to go than most of the farm species in realizing its full reproductive potential.

From the farmer's viewpoint, it is probably fortunate that over the past three decades the ewe has become one of the animals preferred by the biologist for research in understanding more of the details of the physiological and endocrinological mechanisms which mammals use to turn their reproductive systems on and off (Karsh, 1980, 1995; Deveso *et al.*, 1992); there are also many areas of interest in human medicine (e.g. fetal physiology) which have been advanced using the ewe as the experimental animal. Progress towards commercially acceptable controlled reproduction techniques was greatly accelerated in the 1950s with the elucidation of the role of progesterone in facilitating the induction of coincident oestrus and ovulation (Dauzier and Wintenberger, 1952; Dutt, 1952; Robinson, 1952) and with the availability of the highly potent progesterone analogues that had been developed for use in human fertility control.

Early days in controlled reproduction

An event of major significance in the development of controlled reproduction in sheep was the report of Robinson (1965) showing that progesterone and progestagens can be administered in physiologically effective doses over a period of some two weeks by the intravaginal route. Without recourse to needless historical detail, most attempts in controlled reproduction up to that time had centred around a series of progesterone injections or oral administrations of certain progestagens, with or without pregnant mare serum gonadotrophin (PMSG); quite apart from the fact that oestrous response and conception rate left much to be desired, the time and labour involved remained a major factor militating against their acceptance in commercial sheep farming.

Much of the work in controlled reproduction in sheep in the UK and Ireland in the 1950s and early 1960s was concerned with attempts to simplify the progesterone/progestagen administration procedures down to the point at which they might be used on the farm (Gordon, 1958a, 1963; Crowley, 1964).

Advent of the intravaginal device

In 1964, came the development by Robinson and associates in Sydney of a simple method of controlling the time of mating and ovulation in the ewe (Robinson, 1965). The key to controlled reproduction developed at that time was a simple device used for the long-term administration of progesterone or one of its more active analogues. This was the polyurethane sponge, with a drawstring attached, impregnated with a suitable progestagen, which was inserted into the vagina and left in place for 12 days (Fig. 1.1). The progestagen was absorbed through the vaginal wall and entered the circulation. In the cyclic ewe, the device took over the role of the normal corpus luteum after it regressed, thus preventing ovulation and at the same time conditioning the central nervous system to respond to the events associated with the subsequent controlled ovulation. In the anoestrous sheep, it conditioned the reproductive tract and receptors in the brain to respond in a normal physiological manner to stimulation by a gonadotrophin such as PMSG given on cessation of treatment. Whether in anoestrus or in the normal breeding season, when the sponge was withdrawn, the progestagen inhibition of pituitary function was removed, and ovulation occurred some 72 h later.

1.1.4. Sheep in intensive and extensive farming

Sheep are among these farm species whose production methods can still be profitably and substantially intensified, right from the point of breeding the ewe through until the time of dispatching the lamb at the abattoir; it is also a species that is capable of coping with difficult terrains and harsh environments. Although in theory it is possible to talk about sheep producing two litters of five lambs in the space of one year, the more realistic target to keep in mind for intensive lamb production systems is increasing the output of the ewe to four

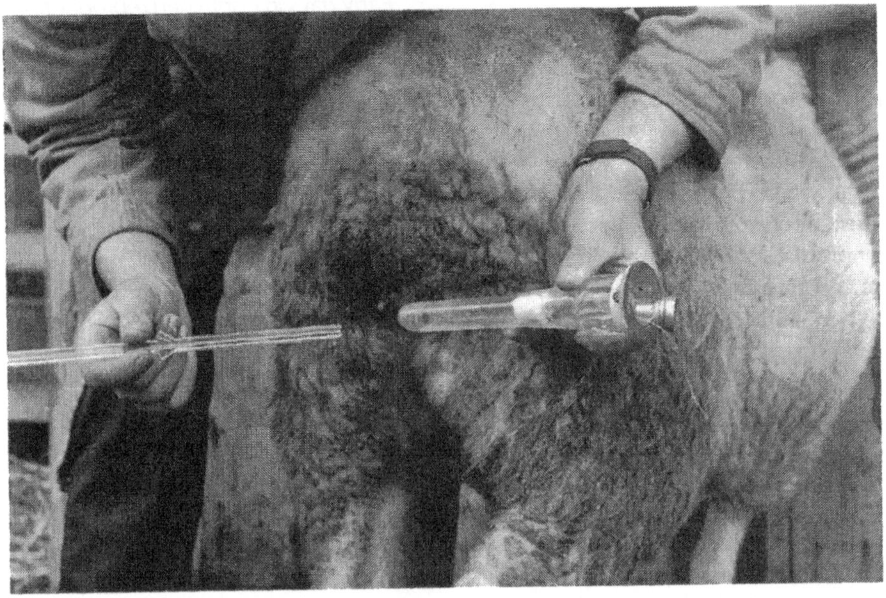

Fig. 1.1. The advent of the progestagen-impregnated intravaginal sponge. Of all the advances in controlled reproduction in sheep, that achieved by Professor Terry Robinson and his associates at Sydney was by far the most important. Irish sheep producers have used the 'sponge' method in sheep for more than 20 years and have found it to be a very acceptable method of producing 'early-lambs'. Of the countries using sponges, France is the one that makes the greatest use of this procedure.

lambs by improvements in litter size and lambing frequency; as noted by Robinson (1979), such a lamb output could result in a considerable improvement in the efficiency of feed utilization.

Sheep and the environment

Quite apart from using sheep in the production of meat, milk and fibre, the species has attributes which appeal to those concerned with the preservation of the environment. Glimp (1995) points out that sheep can produce meat and fibre from renewable natural resources and contribute to agricultural and ecosystem sustainability. Sheep are currently being used on rangelands in the USA to control various noxious plants and sheep grazing may be the preferred grazing system by wildlife biologists. In Ireland, on the other hand, there is some concern about the possible deterioration of the environment as a result of overstocking with sheep in some hill areas.

1.1.5. Dairy sheep

During the Middle Ages in the UK, many sheep were kept for their milk; reports from the fourteenth century talked of 6000 milking ewes on some of the large church estates. Wensleydale cheese, for example, was originally made from sheep's milk. With the rise of the dairy cow in popularity, however, dairy sheep all but disappeared. Currently, dairy sheep in the UK are becoming increasingly popular, not only with hobby farmers but also in larger dairy operations. It should be noted that certain husbandry practices employed in sheep farming are different from those involved in farming meat-producing sheep, particularly the early weaning of lambs in order to increase the amount of marketable milk and to facilitate machine milking.

As to the main products of sheep dairying, these include fresh milk, cheese and yoghurt. The milk is said to be easily digested and has proven to be a valuable alternative to cow's milk for those subject to some allergies. The high total solid levels in ewe's milk makes it particularly suitable for cheese making; 4 litres of sheep's milk is required for the production of 1 kg of cheese, versus 10 litres of cow's milk. The main breeds currently used in commercial sheep dairying in the UK are the British Friesland and the British Milksheep.

Milking sheep in France

In France, according to Barillet and Bocquier (1993) sheep dairying remains fairly traditional, with small farms producing milk primarily for cheese making. Most ewe milk production is concentrated in the three mountainous regions of the Massif Central (Roquefort cheese), the Pyrenees Atlantiques and Corsica. In the 1970s, a programme was introduced to increase ewe milk production by improving husbandry, management and milking techniques; over a 20-year period, the average size of flocks tripled and the production of ewe milk doubled. The most significant factors contributing to these increases were the widespread mechanization of milking and improved genetic selection techniques. By 1990, 75% of farms producing ewe milk in France were part of a milk recording system; artificial insemination has been a potent tool in such sheep improvement schemes in France, used in conjunction with oestrus control procedures (Cognie *et al.*, 1984).

Dairy sheep in Middle-East countries

Milk is also an important product of sheep systems in the Middle-East countries. A paper by Kassem *et al.* (1989) concluded that under semi-arid conditions, Awassi sheep can achieve acceptable levels of reproductive performance with good management. The Awassi breed dominates the sheep found in countries such as Jordan, Lebanon and Syria and about 50% of the 7.8 million sheep found in Iraq are Awassis. Nearer home, there are about one million Awassi sheep in Turkey, mainly along the border with Syria. Controlled reproduction techniques have been reported in Awassi ewes by Treacher *et al.* (1994) in Syria. It is believed that there is considerable scope for breeding improvement within the Awassi breed, using the type of nucleus breeding schemes discussed by Timon (1993).

1.1.6. Accurately identifying sheep

For breeding and animal health programmes, accurate identification of sheep is required. Various reports have come from work in the electronic identification of animals. As part of an EU supported project on electronic identification of farm animals, Caia *et al.* (1994) evaluated capsule formation and migration movements of transponders implanted in different body sites. The differences which they observed in migration movements, breakage and losses between body sites, indicated the convenience of subcutaneous implantation in the axilla and the base of the ear.

1.2. Areas of Controlled Reproduction in Sheep

Controlled reproduction in sheep, as the term might be applied to conditions in Ireland, could be expected to cover the full spectrum of lowland lamb production systems (Fig. 1.2). It may mean breeding sheep towards the end of the normal anoestrous period (early-lamb production); it may mean breeding ewes to permit an extremely compact spring lambing period in small flocks or even getting most of the lambs in a flock born on almost the one day; it may mean breeding ewes to top-quality rams by artificial insemination or it may even mean a rapid build-up of stocks of certain breeds of sheep by embryo transfer.

The most important consideration in quality lamb production is the breed genotype used. Clearly, there is the influence of the ewe breed on carcass quality, but the selection of the terminal ram breed can be one area in which controlled breeding may be able to make a useful contribution (e.g. by the use of appropriate sheep artificial insemination (AI) programmes). In Northern Ireland, Johnston and Steen (1995) recorded that Dutch Texel sired lambs produced a significantly leaner carcass than Suffolk sired lambs; carcasses of Texel sired lambs contained 40g kg^{-1} more lean and 30g kg^{-1} less inter-muscular fat. In the Irish Republic, a sheep AI service based on Dutch Texel semen was operated for some years, using controlled reproduction techniques (Fig. 1.3).

1.2.1. Flock size considerations

The scope for certain controlled reproduction applications is likely to vary with flock size and environment. A New Zealand farmer with an average flock size of 2600 breeding ewes, or even a British farm with an average of 200, would view compact lambings in quite a different light from an Irish farmer with an

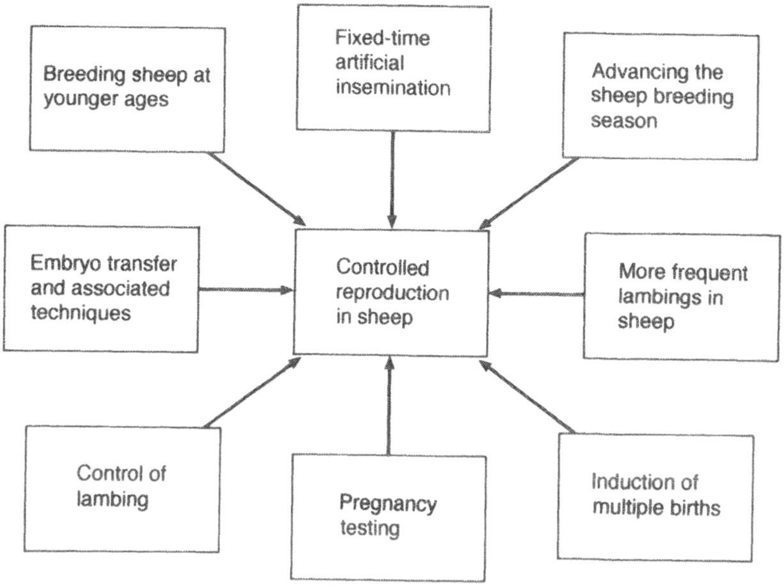

Fig. 1.2. Areas of controlled reproduction in sheep.

Fig. 1.3. The Texel ram and lean-lamb production. In the Irish Republic, sheep AI was seen as a method of making Texel rams available to farmers who would not otherwise have had an opportunity of using them. A sheep AI service was run for 15 years from the University farm.

average of less than 100 breeding ewes. In some countries, out-of-season breeding may hold no great interest, either because of the adverse effects of high summer temperatures on ram fertility and lamb viability or because autumn and winter climatic conditions are quite inappropriate for low-cost lamb rearing.

For the Irish sheep producer, and for farmers elsewhere, the technology for the control of reproduction, including control of oestrus, fixed-time AI, early pregnancy diagnosis and synchronization of lambings is now available in basic form; this offers possibilities in allowing lamb production to be planned from A to Z in a way which is just not feasible under nature. In many parts of the world, including Ireland, the approach to sheep farming is still very traditional; lowland sheep often suffer from the handicap that they are usually regarded as a sideline to the dairy or other enterprises and do not always command the physical and mental effort which is called for if they are to improve their comparative efficiency.

1.2.2. Advantages of controlled reproduction

Controlled reproduction can be important in saving time and labour, especially on the small farm at critical periods such as lambing, and in permitting the use of breeding methods (e.g. AI using rams of a particular breed and quality) which would otherwise be debarred on the grounds of the high costs involved. Using the new technology as an essential ingredient, sheep systems have now been developed which are capable of permitting the annual output per ewe to be doubled, even trebled, under appropriate farming conditions. Scaramuzzi and Martin (1984) noted at that time there had been a slow but continuing increase in the range of preparations available for the artificial control of reproduction in the ewe and a similar increase in the specificity and efficiency with which they act. In particular, it is noted that many advances in reproduction control stem from ideas and concepts developed during the course of basic research. In this regard, they draw attention to one of the most striking examples of controlled reproduction in sheep to emerge recently and record how the development of immunization against steroids went from an investigative laboratory technique to a commercially available product.

1.2.3. Factors influencing response to controlled reproduction

It should also be emphasized, however, that successful controlled reproduction in sheep is not only a matter of appropriate hormonal techniques but is also one of ensuring that they are employed in situations where they can give acceptable results. Difficulties in the past in some areas of controlled reproduction have undoubtedly arisen, not only from inadequacies in the hormone treatments themselves, but in trying to pursue unnecessarily ambitious objectives, such as two lamb crops within the calendar year. It is necessary to bear in mind that many external and internal factors (season, age, lactational status) are likely to influence the type of response achieved by a controlled reproduction programme. In the light of the new information emerging in the literature each year, the prospects for improving the reproductive performance of commercial sheep flocks by hormonal means becomes all the greater; certainly, compared with the 1960s, a more scientific and rational approach to manipulating reproduction in the ewe is now possible, based on a more complete understanding of the hormonal mechanisms involved.

Developments in reproduction control in sheep have been reviewed (Haresign, 1990, 1992), and attention drawn to various defects in existing techniques, in overcoming problems such as the induction of pregnancy in the anoestrous ewe. Although there is an obvious need for good mating management if acceptable results are to be achieved after the use of intravaginal progestagen devices and PMSG, the overall value of the methods developed by Australian researchers in the early 1960s should not be underestimated. Indeed, as noted by Robinson (1988), and his sentiments would certainly be echoed by farmers in France and the Irish Republic, the current 'sponging' technology for controlled breeding in sheep has stood the test of time, despite its acknowledged limitations; it is likely to continue in use for many years to come.

Advent of melatonin treatments

Haresign (1992) has discussed the role of melatonin in the regulation of seasonal breeding in sheep and to the development of slow-release formulations for use on the farm in inducing out-of-season breeding; such preparations can advance the breeding season by some 4–6 weeks, with lambing percentages in early (January) lambing flocks comparable to those in traditional (March) lambing flocks. Other topics covered by this author included the development of laparoscopic procedures for intrauterine insemination and for the collection and transfer of sheep embryos to assist in breeding programmes.

1.2.4. Embryo transfer and breeding improvement programmes

According to Brash (1994), dealing with advanced breeding techniques for sheep improvement under Australian conditions, multiple ovulation and embryo transfer (MOET) is a complex and expensive procedure which must be used wisely. MOET has the ability to increase the rate of genetic gain in parent stud flocks which will then flow down to industry. Embryo splitting and marker-assisted selection are newer techniques which may become common in parent studs during the next decade if costs are reasonable. Embryo cloning for mass production is likely to be available within 10 years, while the use of transgenic animals is at least 10 years away from practical application. However, each of these possibilities could have dramatic effects on sheep breeding if they become practicable.

Quite apart from the use of embryo transfer in MOET programmes and other breeding improvement programmes, there are likely to be many instances in which this technology may be used with more limited objectives in mind. There may be a case for some pedigree sheep breeders to employ embryo transfer (ET) technology as a means of increasing income from their flock. One suggestion is that the breeder recovers and freezes embryos from ewes in the summer period and markets them to commercial farmers in the autumn. By this means, the commercial farmer would produce high value pedigree stock from his ewes rather than commercially priced crossbred lambs.

1.3. Factors Affecting Ewe Fertility

As a background to a discussion of the techniques used in controlled reproduction in sheep, some mention is desirable on conception rates and breeding activity as they may normally be expected to occur in sheep.

1.3.1. Conceptions to first and later services

When the farmer introduces the ram among lowland sheep, of the breeds native to Ireland and the UK, it can be expected that some 80% of ewes conceive to first service (Averill, 1955; Gordon, 1955, 1975); this is talking of adult sheep exposed to a ram of good fertility in the autumn breeding season (see Table 1.1). The conception rate of 80% and better in the ewe is in marked contrast to the 55% or less in the dairy cow and presumably is the result, among other things, of the fact that sheep tend to shed two oocytes at ovulation rather than one.

Breed differences in conception rates

With exotic breeds, such as the Finn Landrace or the Russian Romanov, which are exceptionally prolific, conception rates of the order of 90% and better may be achieved; this would largely be the result of the greater number of ovulations occurring. Thus, a farmer may be able to apply controlled reproduction to

Table 1.1 Lambing outcome in 51 sheep flocks bred in the autumn breeding season (from Gordon, 1975).

	Ewes bred	Gave birth	Lambs born	Conception rate (%)*	Lambs per conception
First services	1991	1608	2266	80.8	1.41
First and second services		1845	2564	92.7	1.39

*Ewes that gave birth as a percentage of those bred.

breed a small flock of Finn or Finn-cross sheep simultaneously, and by application of induction treatment for controlled parturition at the other end of pregnancy get almost all the lambs born on a predetermined date; this would work less well with breeds or conditions in which 80% or less conceive at first service.

Barrenness in sheep

Between first and later services, it can be expected that most sheep become pregnant. Constitutional barrenness in sheep is rare, a fact recorded at the turn of the century by Heape (1899). In a comprehensive analysis of lowland sheep records at the time, Heape found only 8.6% to be barren; much the same was reported by Marshall (1905). At a later stage, a study of several breeds in the UK and Ireland recorded that 6–7% of adult breeding ewes exposed to the ram failed to give birth (Gordon, 1958b, 1967); elsewhere, Lees (1978), dealing with data for the UK, quotes a barren rate of 7.6% in lowland sheep.

Evaluating controlled reproduction techniques

To be commercially acceptable, as opposed to being technically possible, controlled reproduction methods need to be simple to apply, cheap (relative to the product involved) and highly effective. In assessing the effectiveness of controlled reproduction measures, conception rate to first service should be of the same order (i.e. 80% or better) as that expressed by the average cyclic ewe in its natural breeding season. The number of lambs produced per conception (litter size) is also of great practical interest; unless gonadotrophin treatment is part of the protocol, litter size can be expected to vary in the same way as it does in the untreated ewe in the autumn breeding season. Breed, age and body condition of the animal, as well as several well-recognized environmental influences (daylength, temperature, feed supplies), may all contribute towards the incidence of multiple births in the sheep under treatment.

1.3.2. New sheep breeds for reproductive performance

Much work in attempts to improve litter size and productivity in lowland sheep by crossbreeding and selection programmes has been carried out over the past 25 years. In the UK, the prolific Cambridge breed was developed and met with much farmer approval under suitable lowland management conditions. In Ireland, a new breed, the Belclare, was established. In that country, results of research on fertility and lamb production of crossbred ewes using Belclare and other breeds have been presented by Hanrahan (1994). Further afield, in The Netherlands, Suss and Strittmatter (1991) give an account of the prolific Swifter breed, developed from crossing Texel and Flemish sheep and the North Holland breed, based on the Texel and the Finn Landrace. The Coopworth and Perendale in New Zealand and the Polypay in the USA are other examples of new breeds that found widespread acceptance. In Germany, where the supply of home-produced lamb meat at the time only met 42% of the demand,

Wassmuth (1990) discussed methods of improving the competitiveness of sheep using appropriate crossbreeding and selection programmes.

The Booroola Merino

Sometimes, in the search for increased reproductive performance, it has been a matter of selecting strains within a particular sheep breed. In Australia, the Booroola provided breeders with an opportunity to make marked increases in reproductive efficiency. Problems in the controlled use of this Merino strain in the sheep industry have been discussed by Fogarty (1984); he noted that economically feasible management systems were required to improve the survival of the higher order multiple births under extensive conditions in that country.

1.3.3. Seasonal breeding activity

In the ewe, all native breeds of sheep in Ireland show a well-defined breeding and non-breeding season; for most lowland breeds, the breeding season would span the 6-month period running from September to February (Fig. 1.4). A similar story holds true for British breeds, as reported by Hafez (1952) from his work in Cambridge. Although some reports in the literature at one time suggested that rams as well as ewes are seasonal breeders, this would not seem to apply to any of the sheep breeds in Ireland (Smyth and Gordon, 1967); this is not to say that seasonal fluctuations in semen quality and libido do not occur, but simply that if they do, they do not materially reduce the efficiency of the mating process.

Daylight, environment and the breeding season

It is widely accepted that the breeding season of sheep is regulated by changes in daylength, the photoperiodic effect acting via the hypothalamic-pituitary axis and mediated by way of the pineal gland (Karsh *et al.*, 1984); much of this acceptance stemmed from the work of Yeates (1949), working with Suffolk-crossbreds in the UK, who reversed the seasonal breeding activity of the animals by artificially controlling the light environment. Later studies with ewes of the Merino breed in Australia showed that the breeding season of these sheep was regulated by seasonal changes in daylength, although breeding activity was less decisively separated from the anoestrous season compared with sheep of British origin (Yeates, 1956). It also became evident around this time that environmental temperature may genuinely influence the onset of the breeding season (Dutt and Bush, 1955; Lees, 1966; Neville and Neathery, 1970).

Temperature effects

The precise role of temperature as it may operate under natural, as opposed to contrived, experimental conditions is not clear. It is known, however, that on a long-term basis seasonal fluctuations in temperature do not override seasonal

Fig. 1.4. Irish sheep are seasonal breeders. In the early 1970s work on the University farm in Dublin with groups of adult Galway, Cheviot and other breeds showed that it was the end of August before Galways were coming into oestrus and into September before Cheviots (seen here) responded. Without controlled reproduction techniques, there would be no easy way of getting Cheviots pregnant during the summer months of June, July and August.

changes in photoperiodicity; ewes kept in a light environment six months out of phase with normal seasonal lighting showed their breeding season during the warm weather of late spring and early summer and a period of anoestrus during the colder autumn and winter months (Robertson, 1977).

As observed by Radford (1966), some years ago, no explanation of the sheep breeding season based solely on light can cover all the available facts. Even on the question of light itself, there may be questions of moonlight as opposed to sunlight; in the former Czechoslovakia, there was a suggestion, based on a 10-year study involving about 10,000 ewes, that the start of cyclical breeding was associated with the occurrence of the full moon (Horak and Potucek, 1978). In talking about tropical sheep, it is well-known that in equatorial regions of the world, where seasonal variations in daylength do not occur as they do in the more temperate regions, sheep breeds may have no distinct breeding season.

Breeding season and lengthening daylight

It certainly need not be a question of the sheep breeding season being determined by a change-over from lengthening daylight to decreasing daylength; Watson and Radford (1955) drew attention to instances in which Merino ewes started breeding while natural daylength was still increasing and similar evidence has been recorded with other breeds elsewhere (Thimonier and Mauleon, 1969). Other studies with Merinos in Australia, in which ewes were exposed for prolonged periods either to continuous or equinoctial lighting, clearly showed that seasonal variations in breeding activity developed in the absence of changes in the duration of light (Radford, 1961); the breeding of Merinos can also vary markedly according to the particular locality in which they are found (Watson, 1962).

In New South Wales, some Merino flocks can show maximum breeding activity in spring and may enter anoestrus in autumn, a complete reversal of the breeding pattern to be expected on the basis of daylength changes (Robinson *et al.*, 1970). In discussing the Merino, it should be remembered that the breed makes up about 75% of Australia's immense sheep population; a paper by Jeffries (1989) describes breeds of sheep, other than the Merino, which have been developed in that country in more recent times.

Breeding patterns in tropical hair sheep

Seasonal reproduction in sheep, as understood in the temperate regions of the world, may not be under the same influences in other countries and with other genetic strains of sheep. The tendency for sheep to exhibit marked variation in reproductive activity during the year is markedly reduced towards the equator, where ewes can show breeding activity throughout the year (Hombolu *et al.*, 1985). Tropical hair sheep exist in several countries in Latin America, having originated from West African sheep brought in by the Portuguese and the Spaniards in the sixteenth and seventeenth centuries. The Pelibuey hair sheep is the most widespread breed of sheep in Cuba, where it is very tolerant of the environmental conditions in that country.

Seasonal breeding in such tropical hair sheep is considered to be different from that in wool breeds (Reyna *et al.*, 1991); environmental cues other than light appear to be important influencing seasonal patterns in hair sheep. Instead, factors such as nutrition (forage availability and digestibility) and possibly relative humidity and/or precipitation and temperature may explain periods of reduced cyclicity at certain times of the year. It has been suggested that hair sheep may provide a unique model to study reproductive processes independent of environmental influences such as photoperiod, which affects seasonal breeding in wool sheep.

Fertility and fat-tailed sheep

There may be certain fertility problems that stem from the particular breed of sheep used in some farming systems. In the USA, for example, Shelton (1990) studied the reproductive performance of tail-docked and undocked Karakul ewes over a six-year period; the percentage of ewes that lambed was significantly higher for docked sheep (92.9%) than for those with tails (78.9%).

1.3.4. Effect of nutrition and body condition

It has been recognized for many years by commercial sheep farmers that the 'flushing' of ewes prior to the start of the breeding season may have a profound influence on the lamb crop produced by those ewes in the following spring. Flushing refers to the practice of having ewes on a rising plane of nutrition for some defined period to improve ovulation rates prior to the introduction of the ram. As noted by Scaramuzzi and Murray (1994), the practice of flushing is an excellent example of the interactions that exist between nutrition and reproduction. However, it is also clear that the precise mechanisms by which nutrition does influence reproduction are not well understood. It is clear from various reports that body fat content (body condition) directly affects hypothalamic activity and GnRH secretion and that effects on reproductive performance are mediated by way of changes in ovarian hormones or in hypothalamic-pituitary sensitivity to ovarian hormones (Rhind *et al.*, 1989).

Post-mating feeding effects

In Aberdeen, McKelvey *et al.* (1988) used reciprocal embryo transfer to separate the effects of pre- and post-mating nutrition on embryo survival and growth of the sheep fetus. They concluded that high-plane feeding during the post-mating period can adversely affect embryo survival, and that this was mediated through a decline in plasma progesterone concentrations; low-plane feeding at this time, in contrast, had little effect on embryo survival. In the USA, Pope *et al.* (1994), who studied the influence of short-term fasting on blastocyst morphology and survival, showed that fasting mated ewes for three days (day 10–day 13 after mating) tended to increase the percentage of sheep that lambed (92% vs. 80% in controls) and increased the concentrations of progesterone by day 13; on the other hand, blastocyst development was not altered by such short-term fasting.

However, not all studies have shown that a high plane of nutrition after ovulation influences embryo survival, although there may be a significant reduction in peripheral progesterone concentrations (Wallace *et al.*, 1994). When undernutrition is severe, there can be a significant decrease in pregnancy rate in ewes, although this does not appear to be attributable to inadequacy of corpus luteum function (Abecia *et al.*, 1994).

Applying new information on nutrition

Opportunities for applying new knowledge of the effect of nutrition on reproduction in the pregnant ewe have been discussed by Robinson (1990). The effect of nutrition on reproductive performance of ewes has also been reviewed by Rhind (1992). Research continues to provide new information. In arid environments, for example, where roughage feeds are often in short supply, it is not unusual to feed all-concentrate diets to sheep at mating time. Work at Aberdeen indicates that when concentrates are used to supplement a roughage diet, the proportion of energy contributed by concentrates should lie in the range 0.4–0.6 (Al-Khozam *et al.*, 1995). The effect of nutrition in prenatal and early postnatal life on subsequent lifetime reproductive performance of Scottish Blackface sheep has been examined by Gunn *et al.* (1995). Further studies along these lines by Borwick *et al.* (1995) demonstrated that undernutrition of the fetus in the first two months of pregnancy may influence oocyte number and could be a factor in explaining the known reduction in reproductive performance of ewes undernourished in early life.

Nutrition and the breeding season

One question sometimes asked is whether the breeding season of sheep can be materially affected by the level of nutrition that they enjoy. In temperate breeds and under normal conditions of feed and management, it appears that nutrition has little effect in this regard. Irish sheep farmers who wish their ewes to breed in summer (July) rather than autumn (September) cannot induce the animals to show oestrus by providing additional feed inputs. Those who have worked with tropical sheep appear to have reached similar conclusions. In Ethiopia, for example, Mukasa-Mugerwa *et al.* (1993) concluded that, although Menz ewes are year-round breeders, they experience an apparent reduction in sexual activity at certain periods, which appears to be independent of their level of nutrition.

The nutritional status of the sheep is reflected in various ways and generally it would be expected that a positive correlation exists between bodyweight and ovulation rate in the ewe, as shown by various authors (e.g. Morley *et al.*, 1978). As in cattle, various scoring systems have been devised for assessing the body condition of sheep, and these can be applied on the farm as a means of maintaining ewes in the most appropriate nutritional state during the breeding season and during gestation.

Protein levels and embryo survival

In cattle, there is some evidence that feeding a rumen-undegradable protein supplement may enhance embryo survival rate. On the other hand, work in the west of Ireland reported by Diskin and Hanrahan (1995) showed no such effect in sheep; in fact, there was evidence that energy or protein supplementation reduced rather than enhanced conception rate.

Clover sickness in sheep

Certain feeds are known to contain active constituents that may directly affect the fertility of sheep that eat them. Infertility of ewes that grazed predominantly subterranean clover pastures suddenly emerged as a serious problem in the mid-1940s in Australia (Bennetts *et al.*, 1946); it was found that the consumption of subterranean clovers that contain high levels of weak non-steroidal oestrogens was associated with various forms of reproductive failure in sheep. Subsequent work showed that other legumes, such as lucerne, which is known to contain the phyto-oestrogen coumesterol, can depress ovulation rate in the ewe (Adams, 1990).

In sheep, two forms of infertility are evident. Firstly, sheep grazing oestrogenic pasture may suffer 'temporary infertility', with a reduced twinning rate or anovulation; fertilization and embryo survival rates are slightly depressed. Secondly, permanent and cumulative infertility can occur in ewes exposed to oestrogen for more than six months; ewes become less able to conceive because of impaired transport of sperm through the cervix.

Attempts to control clover infertility

The primary method available to control clover infertility has been the replacement of pasture by less oestrogenic cultivars. On average, affected flocks have around a 10% increase in non-pregnant ewes, but most farmers can achieve an acceptable lambing rate by increasing their management and feed inputs to produce more twin lambs (Adams, 1990). However, even today, it should be noted that oestrogenic clovers make up a substantial proportion of pasture legumes in Western Australia and are likely to continue to be important for the foreseeable future. Subclinical permanent infertility is now recognized as the most important oestrogenic problem in sheep in Australia, with more than one million ewes having permanently damaged reproductive tracts because of grazing oestrogenic pastures (Adams, 1994). The Australian experience should be kept in mind by sheep farmers in all countries where clovers form a predominant component of the pasture. A report from New Zealand by McDonald *et al.* (1994), for example, has shown some evidence of impaired fertility even when ewes graze low-oestrogenic strains of red clovers.

There are also instances, on this occasion recorded in New Zealand, in which the ingestion of fusarium, a common fungal contaminant of some pastures, can result in reduced fertility (Smith *et al.*, 1986).

Parasite control and depilatory treatments

Sheep may be treated with a variety of drugs in routine disease and parasite control programmes; they may also be treated in some instances with agents for the removal of wool. Keisler *et al.* (1993) in the USA concluded from a study that their results did not support the hypothesis that ivermectin has a detrimental effect on the reproductive performance of ewes during the breeding season. In Australia, Brown *et al.* (1994) examined the effects of depilatory doses of epidermal growth factor (EGF) on the fertility of Merino ewes; they concluded that an interval of five weeks between EGF treatment and breeding was sufficient time for the sheep to resume normal cyclicity and fertility.

1.3.5. Environmental factors – internal and external

Internal factors

The presence of an adjacent male in the uterus during pregnancy has been known to influence the subsequent reproductive performance of females in some species (e.g. rodents and pigs). A study reported by Avdi and Driancourt (1995) investigated whether such an effect occurred in sheep, using Chios ewes as a model; in contrast to rodents and pigs, the presence of a male embryo(s) in sheep had no effect on subsequent reproductive performance of the female siblings. However, studies by Fitzgerald *et al.* (1993) in Idaho did lead them to conclude that the sex of the other fetus in a twin pregnancy and the number of offspring born were associated with differences in serving capacity of mature rams.

External factors

It is clear, from many reports, that high temperatures can adversely affect ewe fertility. According to Casu *et al.* (1991), heat stress has adverse effects on ovulation in the ewe, the greatest effect being observed in the first three days after ovulation. The same Italian authors found that exposure to heat in the second half of pregnancy resulted in damage to the embryo, and decreased birth weight and viability. These effects, which are most clearly observed in the tropical regions, also occur in the Mediterranean area. In India, Hooda and Naqvi (1990) concluded from their studies that nutritionally stressed sheep experience more heat stress than those on nutritionally adequate diets and that some breeds are more affected by heat than others.

1.3.6. Stress and ewe reproduction

A growing amount of evidence in the 1970s pointed to the adverse effects of stress on sheep reproduction; this may sometimes be a consideration in talking about the application of controlled reproduction methods. In the UK, Doney *et al.* (1976a, b) demonstrated that the stress of handling ewes in normal husbandry operations may influence ovulation rates adversely and increase the extent of embryo mortality; similar effects could be induced experimentally, using exogenous adrenocorticotrophic hormone (ACTH) doses. In New Zealand, the effect of shearing as a factor inducing stress was the subject of several reports. Shearing, widely practised throughout New Zealand at that time during or shortly after the mating season, could be regarded as one of the most stressful events that can happen to sheep (Kilgour and de Langen, 1970); it involved forcing ewes into strange situations, handling and isolation from other sheep, as well as the stresses associated with shearing itself and consequent readjustments in their body metabolism. Welch *et al.* (1979) provided evidence that shearing carried out a week or so after mating exerted a dramatic effect on the lambing pattern subsequently shown by the flock; shearing markedly reduced the proportion of ewes lambing at the expected time, although the sheep did become pregnant at a later stage.

There is ample evidence for the view that any form of stress should be avoided as far as possible during the mating period (Gunn and Doney, 1979); in early pregnancy in the ewe, the indications are that a sustained moderate degree of undernutrition, resulting in a 3–4% loss in liveweight during the first month of gestation is unlikely to have any significantly harmful effect in ewes that are in appropriate body condition at mating (Russel, 1979).

1.3.7. Hormonal approaches to enhanced fertility

Exogenous progesterone

The effect of post-mating hormonal supplementation on fertility was examined by Scaramuzzi *et al.* (1988) in Australia; they confirmed a number of earlier reports by showing that ewe fertility could be improved by supplementary progesterone treatment between days 10 and 25 post-mating. In the USA, Pope *et al.* (1995) examined dose-response relationships of exogenous progesterone given shortly after ovulation (days 2–4) on blastocyst development and fertility in two sheep breeds. Lambing rates of Targee ewes was not different following progesterone treatment, but in Polypay ewes treatment significantly increased lambing rate from 200 to 256%; the authors concluded that treatment of ewes having an ovulation rate > 2 with progesterone improved embryo survival.

Use of human chorionic gonadotrophin (hCG)

There have been reports of work in which hCG has been employed in early pregnancy in an effort to influence blastocyst growth and pregnancy rate. In Ohio, Nephew *et al.* (1994) reported on the effect of hCG administered prior to the time of maternal recognition of pregnancy; their results indicated that hCG treatment on day 11.5 of the oestrous cycle stimulated uterine secretion and blastocyst growth sufficiently to increase pregnancy rate.

Use of GnRH analogues

A number of studies in recent years have reported that GnRH treatment on day 12 post-insemination will improve the fertility of cattle and sheep. In Wales, Beck *et al.* (1994) reported on the effect of GnRH (buserelin) treatment on day 12 post-mating on the reproductive performance of sheep in different flocks; treatment led to a significant increase in litter size in one flock of yearling ewes. In other studies, Beck *et al.* (1995) investigated the effect of GnRH (Buserelin) injection on day 12 post-mating on ovarian function in ewes; they concluded that treatment induced ovulation in a majority of ewes and probably influenced ovarian hormone secretion.

Use of ovine interferon

In ewes, early embryo loss has been attributed largely to a failure of maternal recognition of pregnancy. There have been reports that the treatment of ewes with recombinant bovine interferon alpha could improve pregnancy rates (Nephew *et al.*, 1990; Martinod *et al.*, 1991; Schalue-Francis *et al.*, 1991). However, results reported by Imig *et al.* (1995) did not support the hypothesis that recombinant ovine interferon, administered from day 9 to day 19 post-breeding, enhanced the establishment of pregnancy in ewes; an alternative explanation for lack of success might relate to the dose, mode of administration or hyperthermic response associated with treatment. The latter explanation appears to be strengthened by the report of L'Haridon *et al.* (1995) who developed formulations permitting sustained release of ovine interferon, resulting in a more persistent antiluteolytic action; it remains for these formulations to be used in attempts to improve sheep fertility.

1.3.8. Suppressing reproductive function in ewes

Reports have been provided by some workers on the effect of immunizing ewe-lambs against GnRH soon after birth or around the time of puberty. In Australia, Brown *et al.* (1995) employed a prototype commercial preparation in Merino ewe-lambs and studied them over a 2-year period. They report that at least 60% of the GnRH-immunized ewes did not show oestrus and possessed small uteri and ovaries which lacked follicle development; growth rates of immunized and control ewes were similar. These workers suggest that the lack of GnRH stimulation and the consequent deprivation of gonadotrophins early in the life of ewes may result in permanent impairment of hypothalamic and/or pituitary function.

1.4. Ram Fertility and Breeding Activity

The success or otherwise of many controlled reproduction techniques is not alone a question of influencing the reproductive processes of the ewe; the outcome also depends on the capability of the ram (mating activity and semen quality) when natural service is the mode of breeding, and on semen quality and insemination procedures when AI is the chosen method (Fig. 1.5).

As described by Setchell (1984), the testes of rams are relatively large (each testis is about 0.5% of body weight) in comparison with many other mammals; the male gonads are concerned with sperm production and secretion of the male sex hormones. The sex hormones of the ram are responsible for sexual drive (libido) and the function of the male accessory glands, such as the prostate, seminal vesicles and Cowper's glands. The testis consists mainly of convoluted seminiferous tubules (each testis contains about 7000m of tubules) in which the sperm are formed; these tubules contain no blood vessels or nerves but between them is the interstitial tissue, with the blood and lymphatic vessels, the Leydig cells, which secrete the male sex hormones, and some other cells, including a substantial number of macrophages. Spermatogenesis is an exceedingly complex process involving the interactions between the somatic Sertoli cells on the one hand and the germinal cells at various stages of development on the other.

Estimates of daily sperm production

In the fetal ram-lamb, the proliferation of testicular somatic and germinal cells occurs throughout the growth of the testes but at a higher rate before day 100 of gestation than later (Hochereau-De Reviers *et al.*, 1995). The rate of sperm production can be determined in various ways and there is general agreement that this is about 20 million sperm g^{-1} day^{-1} for the ram. It is well accepted that

Fig. 1.5. Rams in Ireland are capable of breeding in all seasons. In the 1950s, it was commonly believed by sheep farmers in the UK that rams were often incapable of breeding during the spring and summer months of the year and that they experienced something akin to the ewe's non-breeding season. It is now clear that this is not the case.

an excellent index of sperm production in the ram is testicular size. Sperm formed in the testes are matured in the epididymes and held there awaiting ejaculation. Each epididymis is a convoluted duct, some 50m long; sperm take from 10 to 14 days to pass through this duct.

Seasonal changes in gonadotrophin secretion

Ram and ewe sexual activity depends on pituitary gonadotrophic function. Dealing with Ile-de-France rams in their native country, Ortavant *et al.* (1988) notes evidence of differences in luteinizing hormone (LH) concentrations according to time of year; LH pulse frequency was found to be low (3 pulses every 24h) by the start of winter and high (6 pulses every 24h) as early as June; each LH pulse induced a testosterone peak. Studies of follicle stimulating hormone (FSH) in the ram have also shown evidence of seasonal changes, with a maximum occurring in August–September (Ravault *et al.*, 1980). It is also clear that gonadotrophin secretion and release in sheep are mainly regulated by light and gonadal steroids. According to the views advanced by Ortavant *et al.* (1988), during decreasing daylength LH release is stimulated and the negative feedback effect of steroids is reduced; conversely, in periods of increasing daylength, LH release is less and the feedback effects of steroids is increased. However, there is evidence of the crucial role of FSH in spermatogenesis. A report by Kilgour *et al.* (1994) showed that the inhibition (by passive immunization) of FSH but not LH markedly affected sperm production in the ram; treatment with FSH antiserum significantly decreased daily production of B2 spermatogonia and of leptotene and pachytene spermatocytes.

Ram : ewe ratios

Where rams are joined with ewes for natural service, semen quality and the sperm doses that operate must be such as to ensure a high fertilization rate in the ewe. In normal commercial practice, two or three rams are usually joined for every 100 ewes in the flock; with one ram for every 30–50 ewes, the male has three ewes or so to work with each day, based on a 17-day oestrous cycle. However, the normal servicing ability of the ram is high and New Zealand workers have shown that the proportion of rams to ewes may be reduced to about 1% and still remain adequate (Allison, 1975, 1978); this means that a ram would be covering on average, some six ewes per day. Elsewhere, in Australia, there have been studies indicating that individual rams may not always be capable of that level of performance (Synott *et al.*, 1981). In view of the strong relationship between testicular size and sperm production, Gherardi *et al.* (1980) suggested that about 400g of testicular tissue per 100 ewes joined is required for successful joining at ram : ewe ratios of 1 : 100.

1.4.1. Evaluation of ram libido and fertility

Although the importance of the ram in determining the lambing performance of the flock has been widely accepted, with some exceptions, relatively few efforts have been made to quantify and evaluate ram mating performance. On the assumption that an average ram must be capable of two or more services for every ewe that comes in oestrus, a pen mating test was devised by Mattner *et al.* (1971); in this, rams were placed in a pen with five oestrous ewes for 20 min and the number of services recorded. It was found that a relationship existed between the total number of services in three such tests and the service performance of the ram in a flock situation; other workers, however, found no such relationship.

High and low serving rams

In other studies, Kilgour (1979) reported using service tests of longer duration, occupying periods of 1 or 3 h rather than 20 min; it was found that it was possible to arrange rams into high and low serving categories and that these could be related to the performance of males under flock conditions. An additional aspect of this ram testing work was in finding that high performance rams tended to have daughters of higher than average fertility (Wilkins and Kilgour, 1978).

Other work by these same authors showed that ram serving capability (measured by a 1-h pen test in the morning and another in the afternoon) affected ewe reproductive performance when the rams were employed in flock matings during the early weeks after joining (Kilgour and Wilkins, 1980);

service tests were regarded as important when the mating load was higher than usual or when the matings had to be completed in a short time span.

Low-cost serving capacity tests

According to Fowler (1984) in Australia, a reliable low cost test of ram serving capacity was desirable, but did not exist at that time. According to this author, the value of such a serving capacity test would lie in eliminating inactive rams and identifying those whose higher serving capacity enable them to make better use of larger testes. A reliable serving capacity test would enable rams to be used according to their capabilities. The development of a commercially usable serving capacity test is discussed by Blockey and Wilkins (1984); such a test must be simple to conduct, of short duration and, most important of all, should be an accurate predictor of mating performance. According to these authors, the potential benefits of serving capacity testing to the improvement of flock mating efficiency have been well demonstrated but some areas require further clarification.

Ram testing and flock fertility

The relationship between ram breeding capacity and flock fertility was the subject of further studies reported by Kilgour (1993) in Australia. As in earlier work, these studies were based on the premise that breeding frequency of the ram is an important determinant of flock fertility (i.e. ewes served more than once during their heat period have a greater chance of becoming pregnant than ewes served only once). The effect of breeding capacity in this more recent Australian work was studied by exposing each of 15 rams to 200 cyclic ewes for 17 days; five rams were designated as being of high, medium or low breeding capacity. Breeding capacity was measured in various tests (3-h or 1-h pen breeding tests). Results showed that breeding frequency (number of services per ram during flock mating) of rams was closely related to flock fertility, but breeding capacity, as measured in the tests, was only a moderately accurate indicator of breeding frequency. It was noted, however, that some rams were capable of impregnating more than 100 ewes during the course of one oestrous cycle and that the identification of such rams would increase flock breeding performance. This could be of economic benefit by virtue of reduced ram purchases and lower flock maintenance costs.

Age of ram

Age of sheep, whether talking of rams or ewes, can be a relevant consideration in assessing flock fertility. Several reports have emphasized the importance of age in rams and ewes. In Australia, dealing with Merino sheep, ewes and rams mated at the first time at 1 1/2 years of age consistently gave lower lambing percentages than mature animals (Dawe *et al.*, 1974). As with bulls, testicular development appears to be more closely related to liveweight rather than age of ram-lambs; by increasing the plane of nutrition of ram-lambs, the age at which puberty is reached can be reduced (Cameron *et al.*, 1984). It is known that ejaculates collected shortly after puberty in rams are likely to contain a high proportion of abnormal sperm which exhibit poor motility, although several weeks later these characteristics are similar to those of adult rams (Skinner and Rowson, 1968). In Ireland, the use of Suffolk ram-lambs in midsummer (July) may be associated with a fertility level markedly below that of the same animals used in October.

Development of sexual responsiveness of rams

It is also relevant to note certain behavioural aspects of ram reproductive performance in young animals. Studies by Price *et al.* (1995) in California led them to conclude that the sexual responsiveness of ram-lambs towards females was sufficiently undeveloped at 6 months of age (i.e. puberty) that extended exposure to sexually-receptive ewes is required for many ram-lambs to exhibit their inherent sexual (mating) potential. At 8 months of age, on the other hand, the sexual development of ram-lambs had matured sufficiently so that relatively brief encounters with oestrous ewes released the full expression of adult sexual behaviour. The authors suggest that such findings may be of benefit to sheep breeders who wish to use ram-lambs in their breeding programmes.

Responsiveness of rams to daylength changes

It should be noted that for most of the species studied, daylength influences the age of puberty equally in both males and females. In sheep, however, there appears to be strong evidence that the responsiveness to photoperiod is sexually differentiated (Herbosa and Foster, 1995); although the timing of the pubertal rise in LH in ewe-lambs is highly influenced by daylength, in ram-lambs it is not. These workers found that this sex difference was due to the organizing action of androgens on the brain during prenatal development.

Rams and the lambing percentage

Although sheep farmers have always recognized that rams differ in their mating capabilities and in their fertility, it only became evident some 30 years ago that rams may contribute to variation in the litter size of their mates by way of differences in the fertilizing ability of their semen or in the prenatal survival of their offspring (Turner, 1969; Bradford, 1972). There has also been some evidence that selection for high fertility in ewes has been accompanied by an enhanced ability of semen from related rams to fertilize ova; the implication here is that the ram can exert a direct effect on litter size.

Brucella ovis *and epididymitis*

Brucella ovis infection in sheep is the cause of a chronic epididymitis in rams which can result in reduced fertility. The infection in sheep was first reported in New Zealand and Australia in the early 1950s and according to FAO sources it is now known to exist in many sheep producing countries, although not in Ireland and the UK. Under some conditions, it has proved possible for veterinary measures to eliminate the problem from an entire sheep population. The eradication of this form of brucellosis from the Falkland Islands has been dealt with in a report by Reichel *et al.* (1994); the protocol adopted in this involved a combination of serological tests and culling. Rams that showed palpable signs of epididymitis were immediately removed from the ram flock and positive reactors to serological tests placed in isolation until retested two weeks later, and if seropositive culled.

1.4.2. Seasonal variations in ram activity

In the British Isles, despite one report suggesting markedly inferior ram performance in the ewe anoestrus (Yeates, 1949), the bulk of evidence for rams accumulated since that time would indicate that they are quite capable of maintaining high mating vigour and acceptable semen quality during the spring and summer months (Gordon, 1958b, 1963; Smyth and Gordon, 1967; Jennings, 1972). In Australia, workers reported that there was no distinct breeding season in the ram as in the ewe, although the rate of sperm production may be decreased in the spring months. Among environmental factors (daylength, feed, climate) likely to reduce the effectiveness of the ram markedly, elevated temperature would seem to be the most important. In Australia (Gunn *et al.*, 1942) and North America (McKenzie and Terrill, 1937; Dutt and Hamm, 1957), several studies have shown that high summer temperature is often responsible for temporary infertility.

Competent rams for controlled reproduction

With the development of more intensive lamb production systems, especially those calling for more frequent lambings and births throughout the year, it is clearly essential to assess the nature and extent of seasonal changes in ram fertility. The availability of new methods for overcoming the seasonal anoestrus in the ewe (e.g. use of sheep with extended seasons; induction of oestrus and ovulation by hormone treatment or even photoperiodic manipulation) does make it necessary that rams of high libido and good fertility should be available at all times of the year. There is also the question of developing practical routines that may assist in improving ram performance (e.g. light manipulations; melatonin treatment), where this might be considered necessary to meet the needs of natural service or artificial insemination.

Testicular size and ejaculation frequency

The effect of frequent ejaculation and of season on the semen characteristics of rams has been the subject of several reports. It was evident from such studies that successive ejaculations affected the quantity but not the quality of ram spermatozoa. In Israel, Amir *et al.* (1986) observed that the fertility of Finn-cross ram ejaculates was not affected by the frequency of collection, even when this involved five daily ejaculations during 17 consecutive days. Testicular volume and scrotal circumference can often decrease markedly during the mating period when rams are active (Thwaites and Hannan, 1989). Some reports have shown testicular volume dropping to almost half its pre-joining volume; there can also be a substantial decline in the bodyweight of the ram due to its mating activity.

Rams with small testes may be less fertile than those with large testes under conditions of field mating; Gherardi *et al.* (1980) considered that this may be due to rams with small testes producing less sperm than those with large testes. Cameron *et al.* (1986) found that semen quality of rams with large or small testes did not differ; this indicated that there is no general relationship between testicular weight in rams and semen quality.

1.4.3. Manipulating the light environment

There have been several reports drawing attention to the relationship between decreasing photoperiod, i.e. short days, and increased reproductive activity in rams; on this basis, peak fertility in the male occurs in the autumn (Yeates, 1949; Ortavant *et al.*, 1985). There is no doubt that considerable seasonal variation in testis size and semen production is evident in primitive breeds of ram such as the Soay, used in studies by Lincoln (1976); although a case could possibly be made for employing photoperiodic manipulations if a farmer was faced with using Soay rams to breed ewes throughout the year, the need for light treatment in dealing with the improved mutton breeds, such as the Suffolk, is much less evident.

Seasonal variations in Ile-de-France ram fertility

In France, although it has been noted that seasonal variations in gonadal activity are less pronounced in the ram than in the ewe (Ortavant *et al.*, 1988), there is a well-developed body of evidence recording seasonal changes in the Ile-de-France breed. The testis weight in the adult Ile-de-France ram has been reported as varying from 180–190 g in late winter to early spring to 300–320g in late summer and autumn; quality of semen and fertility is lower in spring than in autumn, as recorded by Colas (1979). Seasonal effects on semen quality in German Merino rams were investigated by Menger *et al.* (1988); season had a significant effect on ejaculate volume, sperm motility and sperm concentration, but there was no evidence that fluctuations in semen quality led to problems in achieving pregnancies. In Poland, there was some seasonal variation in the quality of fresh semen collected from Longwool sheep (Udala *et al.*, 1991), but data on its fertilizing ability were not provided.

Light manipulation and testicular volumes in Suffolk rams

There have been some reports from the USA that have maintained that considerable seasonal variations occur in testicular volume of Suffolk rams and that application of short-day (8h light : 16h darkness) light treatments over a period of 3–10 weeks resulted in increases of 40–50% in their volume (Schanbacher and Ford, 1979). However, as already mentioned, there appears to be ample evidence from work in Ireland and elsewhere that Suffolk rams are able to perform satisfactorily in all seasons. If it was a question of dealing with

Suffolks that were incompetent breeders in the sheep anoestrus, then light manipulations might deserve some consideration.

As it is, there may be more of a case for using light control in keeping rams in peak breeding condition when they are supplying semen for breeding ewes by AI in the spring and summer months of the year. Those who have employed light treatments reported that rams respond to light changes faster than do ewes (Evans and Robinson, 1980). This suggested that the neuroendocrine mechanisms that control the onset of the breeding season are sexually differentiated. Results from studies by Lubbers and Jackson (1993) showed that gonadectomized oestradiol- or testosterone-treated male sheep have much less pronounced seasonal changes of LH secretion than similarly treated female sheep.

Maintaining a high testis weight by light manipulation

In France, studies with Ile-de-France rams have shown that frequent and brief stimulations of LH release can be obtained using regimens of increasing and decreasing daylengths whose periods are reduced from 6 months to 4, 3 or even 2 months (Pelletier and Almeida, 1987); testicular weights of rams were found to follow the photoperiodic changes in the first three instances but it increased and stayed close to the maximum when changes were made at 2-month intervals. According to Ortavant *et al.* (1988), rams so treated continued showing a persistent high testis weight for more than 2.5 years (Fig. 1.6).

1.4.4. Identifying low fertility rams

Clearly, ram fertility is important in determining the proportion of ewes that conceive; methods of distinguishing rams of good fertility from those of poor fertility have been the subject of many studies. It is either a matter of examining or measuring the ram's reproductive organs, in particular the testes, taking semen samples for microscopic examination or employing a combination of clinical examination and semen testing. On the question of semen testing, Crowley and Walsh (1971), in work in Ireland, did find that semen testing could be useful for detecting the small proportion of rams of inherently very low fertility, but its general use as a screening procedure was not satisfactory. In south-west Scotland, MacLaren (1988) found that one-third of 547 rams examined had recognizable abnormalities of the reproductive tracts, other systems or both; the importance of careful physical examination of the scrotum and its contents is stressed. Semen examination of ram-lambs by this worker frequently revealed poor sperm density, as well as morphological abnormalities such as retained protoplasmic droplet, accepted by many as an index of immaturity.

Physical and semen examinations

Examining semen prior to using rams for breeding may not always detect those rams which subsequently suffer temporary infertility, nor does it necessarily

Fig. 1.6. Manipulating the light environment to influence the breeding capabilities of rams. Much research has been conducted over the years on the effect of different light regimes on the breeding response of ewes and rams. Light manipulation can be employed as a means of maintaining semen production at a high level, both in rams and bucks.

discriminate between the best and the moderate fertility males. The fertility examination of rams before sale or the start of mating has been a common practice in countries such as New Zealand (Bruere, 1971). It is perhaps noteworthy that New Zealanders take careful note of the size and tone of the testes rather than paying undue attention to the examination of semen samples.

Sperm output is proportional to testicular size, and a ram possessing large symmetrical testes free from defects is likely to produce semen of good quality (Kilgour, 1979). It has been shown that a good estimate of semen production can be obtained by measuring the size of the testes. Knight (1972) estimated that semen is produced at the rate of about 20 million sperm per gram per testis per day; in later work it was demonstrated that measurement of scrotal volume, scrotal circumference (scrotal wool removed) and mean diameter of the testis could all give equally good measure of semen production (Knight, 1977). In Ireland, a survey of lowland rams indicated that the incidence of constitutional infertility in rams is very low (Crowley and Walsh, 1971) as it is with ewes.

It is clear from many reports that mating can have a direct and adverse effect on the testicular size of rams which appears to be accelerated by the sudden decrease in liveweight often experienced by active rams (Raadsma and Edey, 1985). Clearly, rams that are busy mating may well lose out on time devoted to eating.

1.4.5. Nutritional and environmental effects

The testicular size of adult Merino rams has been shown to increase when animals are placed on a high plane of nutrition and to decrease when placed on a low plane (Thwaites and Hannan, 1989; Murray *et al.*, 1991). It is believed that the hypothalamic-pituitary axis, acting primarily by way of gonadotrophins, is involved in testicular responses to nutrition in much the same way as it is involved in responses to other environmental stimuli, such as daylength and social cues (Martin *et al.*, 1994); it is evident, however, that the effects of diet on testicular growth may also involve mechanisms that are independent of changes in gonadotrophin secretion. Results from work by Thwaites (1995a) indicated that undernutrition in Merino rams had a greater effect than exercise on testicular volume and that exercise alone was associated with increased testicular tone; the same author also showed that undernutrition caused a progressive decrease in the testicular tone and volume of Merino rams (Thwaites, 1995b).

Nitrogen supplements

The effects of feeding supplements containing different levels and sources of nitrogen on bodyweight and testicular volume have been studied by several workers. In Australia, Thwaites (1994) working with Merino rams recorded that during 4 weeks of mating, liveweight and testicular volume decreased in unsupplemented rams by 16% and 36%, respectively; loss in testicular volume was progressively reduced as the nitrogen content of the supplement increased from 1.7 to 5.0%.

Olfaction and the ram

In sheep, it is widely known that the ram's presence induces rapid changes in the pattern of pituitary hormone secretion in the female; the presentation of the ram to anoestrous ewes may induce an LH pulse in a matter of minutes. A similar effect has been described in the ram; the presence of sexually receptive ewes can induce an increase in LH and testosterone concentrations. In France, studies by Gonzalez *et al.* (1991) suggest that this female-induced secretion of LH and testosterone was not the result of olfactory cues.

1.4.6. Hormonal approaches to enhanced ram performance

Effect of L-tyrosine

The oral administration of L-tyrosine has been reported to significantly increase ejaculate volumes in Egyptian rams in a paper by El-Sayed *et al.* (1993).

Effect of prostaglandin

It is known that certain prostaglandins have their origin in the male accessory glands, largely in the seminal vesicles. The concentration of prostaglandin in

$F_{2\alpha}$ in semen is higher in the ram than in bulls, boars or stallions. The clinical administration of prostaglandin in farm animals has been reported in several papers dealing with cattle, horses and buffaloes. In sheep, the effect of $PGF_{2\alpha}$ and the synthetic PG analogue, cloprostenol, on sex drive and semen characteristics was examined in Irish rams by Mekonnen *et al.* (1989). In this study, PG was administered 30 min prior to attempting semen collection. Prostaglandin treatment resulted in significant increases in ejaculate volume and total sperm per ejaculate compared with control rams.

Melatonin implants

The role of the pineal gland and melatonin in reproduction in ruminants and rams has been reviewed by several workers (D'Occhio and Suttie, 1992; Yellon *et al.*, 1992). A role for the pineal gland was established by studies in which pinealectomy and superior cervical ganglionectomy caused disruption of reproductive responses to changes in natural and artificial photoperiods. Melatonin is known to be involved in photoperiodic time measurement and circulating concentrations of the hormone follow light–dark cycles, with significantly increased secretion during darkness. Treatment of rams with melatonin implants advanced the seasonal increase in LH pulse frequency and testicular size in studies reported from New Zealand by Webster *et al.* (1991); results demonstrated that the advance in reproductive development induced by melatonin was due to an effect on the hypothalamic pulse generator, which increased LH pulse frequency. In Uruguay, on the other hand, Fernandez Abella and Villegas (1992) studied the effect of melatonin implants on the semen quality of Corriedale and Polwarth rams in spring and summer, but concluded that such treatment did not influence the fertility of rams at the latitude of that country.

1.4.7. Suppressing reproductive function in rams

Castration by vaccine

Sheep farmers in many parts of the world have considered that the castration and tail docking of lambs is necessary for disease prevention and good flock management. Castration and tail docking, however, can result in behavioural and physiological changes indicative of considerable pain in lambs. A study by Kent *et al.* (1995) of different methods led them to conclude that a combined rubber ring and Burdizzo method of castration and tail docking of lambs less than seven days old caused least acute pain, as judged by the behavioural and physiological indices measured. None the less, this method still produced considerable acute pain and the authors saw the need to seek improvements or alternative treatments. In terms of castration, one alternative could eventually take the form of an anti-GnRH vaccine, administered soon after birth or before the onset of puberty. Research in this area was first reported some time ago.

There were several reports during the late 1970s and early 1980s showing that repeated immunization of adult male and female farm ruminants against

GnRH can inhibit the secretion of gonadotrophins and lead to the suppression of reproductive function (Jeffcoate *et al.*, 1978; Robertson *et al.*, 1982); such suppression in adult sheep and cattle is known to be reversible with time (Keeling and Crighton, 1984). A study by Brown *et al.* (1994) showed that similar immunization of rams very early in life or around puberty was capable of producing profound and long-acting effects on their reproductive capacity, with some rams having juvenile-like reproductive organs and a complete lack of libido two years after treatment. It appeared that immunization early in life may have induced some degree of impairment to hypothalamic function, resulting in long-term suppression of GnRH and of gonadotrophin release in the majority of the immunized rams; growth rates of immunized and untreated rams were similar. In the USA, Daley *et al.* (1995) also reported on the effect of immunization against GnRH of ram-lambs at a month of age on growth, carcass characteristics and reproductive development; their results indicated that immunization was an effective alternative to physical castration in lamb production.

Short-scrotum rams

Short-scrotum rams can be produced by forcing the testes tight against the abdominal wall of the young lamb and placing an elastrator ring immediately below the testes. The sterility and productivity of such rams was studied by Morcombe *et al.* (1990) in Australia. Histological examination of testes from short-scrotum rams after mating showed severe degeneration of tubular epithelium and almost without exception a lack of evidence of sperm production; growth rate was significantly greater than in castrated lambs. The use of such short-scrotum rams for detecting oestrus in sheep flocks has been reported by Zhao *et al.* (1992) in China; histological examination showed no evidence of spermatogonia in the testes. The rams did, however, possess strong libido and high mating capacity.

Vasectomy

One certain way of obtaining sterile teaser rams, whether for use in research or in commercial practice, is by way of vasectomy, the surgical procedure by which the vas deferens from each testis is cut. One question that occasionally arises with such rams is fertility in the days immediately following surgery. In Uruguay, Villegas *et al.* (1992) reported on the effects of vasectomy on semen traits in various breeds of rams; all rams showed very low fertility by day 4 post-operation and they were sterile within a week of vasectomy.

1.5. Place of Artificial Insemination

The first serious efforts with sheep AI were in the former USSR during the 1920s and the number of ewes bred by this method increased through the years to the point where 42–44 million were inseminated each year, representing 72–76% of all animals (Jheltobruch, 1979); in certain regions, 90–95% of ewes were bred by AI. Developments in AI technology and related topics have been discussed in various reviews (Maxwell, 1984; Evans, 1988; Gourley and Riese, 1990; Wallace, 1992). Experiences in the application of sheep AI in New Zealand (Dyer, 1990), Norway (Berg and Aamdal, 1991), Sweden (Lillo, 1989; Nilsson, 1989), Greece (Georgoudis *et al.*, 1995), Germany (Hollerrieder, 1991) and the former East Germany (Becker *et al.*, 1986; Strittmatter and Peter, 1991) have been described in various papers.

1.5.1. Early days in farm application of sheep AI

The development of sheep AI in the former USSR was part of a massive upgrading programme applied to native sheep which started after World War I, when Australian Merinos were imported into the USSR. The AI procedures adopted at that time were uncomplicated and have remained so over the years; ram semen is generally collected and used immediately, occasionally with some degree of dilution. Thus, although the USSR could claim to be the first in the application of sheep AI, in terms of new developments in ram semen technology, it has not been associated with any notable advances.

According to Ryder (1981), the sheep industry at that time in the former USSR had made considerable progress since World War II, not only in expanding sheep numbers, but in moving rapidly towards a predominance of Merino types by the extensive use of AI. Sheep were maintained on large farms carrying anything from 3000 to 60,000 sheep. The usual arrangement was for breeding ewes to be kept in flocks of 600 to 800; for artificial insemination, heat periods among ewes were checked by running 8–10 colour-marked 'aproned' rams with the flock and selecting out those marked, once daily in the morning. About 70% of ewes were inseminated using freshly undiluted semen, the insemination being carried out within 20–30 min of semen collection, using 0.05ml volumes and estimated sperm doses of 120–150 million. The 'work-load' on any one ram was usually on more than 400–600 ewes in a breeding season; conception rates to first service of 75–80% were claimed.

Self-drafting oestrous ewes

It might be mentioned that New Zealand workers in later times developed a self-drafting system for oestrous ewes destined for AI (Matthews *et al.*, 1991). This system was based on the principle that most ewes in oestrus seek out and remain near rams. The New Zealand workers designed various systems for trapping oestrous ewes attracted to two decoy rams. They concluded that the catching rate with their trapping system was sufficiently high for practical use in AI programmes.

Sheep AI used in few countries

Despite the extensive use of AI in the former USSR and several other eastern and central European countries and in some parts of Latin America, with the exception of France, sheep AI has certainly not become a common practice in the other sheep producing regions of the world, including Australia, New Zealand, Western Europe and North America. This results from several problems relating to the management of the ewe flock, the costs involved in AI and the handling procedures necessary for ram semen. The fact that most AI, as currently practised in Russia, is still based on fresh undiluted semen, as it was more than 70 years ago, serves to show that methods of dilution and frozen storage of semen applicable in cattle do not necessarily hold true for sheep.

Use of fresh semen

Research since 1930 on storage of semen at reduced temperatures (0°–15°C) and at ambient temperature has been reviewed by Maxwell and Salamon (1993). Diluents used have included buffers combined with sugars plus egg yolk or its constituents, milk from various sources, glycine and other substances. Other reports confirm that irrespective of the diluent, dilution rate, temperature and conditions of storage, ram sperm deteriorated during storage. Changes included reduced motility and morphological integrity of sperm, accompanied by a decline in their survival in the ewe reproductive tract. A rapid decline in fertility occurs when ram semen is stored for more than 24 h and used in cervical inseminations.

In this regard, it is of interest to note that in New Zealand, Upreti *et al.* (1995) have reported the development of a chemically defined ram semen diluent (RSD-1) which was able to maintain sperm motility in diluted semen incubated at 38°C for about 24 h; in contrast, a conventional milk-based diluent supported motility for less than 6 h at 38°C. The same workers reported that ram sperm motility was influenced by the buffering capacity, osmolarity and the presence or absence of macromolecules and calcium in the chemically defined diluent. Among the organic buffers tested, MOPS (3-(*N*-morpholino)propanesulphonic acid) had a marked effect on the maintenance of sperm motility.

Frozen semen

The potential influence of one ram on the sheep industry using technology current at the time is outlined in Fig. 1.7, taken from Maxwell (1984). This deals with the Australian scene and compares the number of lambs produced per year from natural mating, AI with fresh semen and AI with frozen semen. Based on the assumption that frozen semen may be collected and stored from an outstanding sire for most of the year, a conservative estimate of 8 months' semen collection from one ram at a rate of nine ejaculates per week should provide enough semen to leave some 12,000 lambs per year.

The effects of cervical insemination with frozen semen on fertility and litter size in Norwegian sheep was reported by Olesen (1993). The NRR% (defined as the percentage of ewes not returning to service within 20 days of AI) was about 60% and was significantly affected by various factors, including technician/veterinarian, ram, age of ewe, number of inseminations per oestrus; there was a reduction in litter size of about 0.2–0.4 lambs in inseminated ewes compared with those bred by natural service. The optimal insemination time

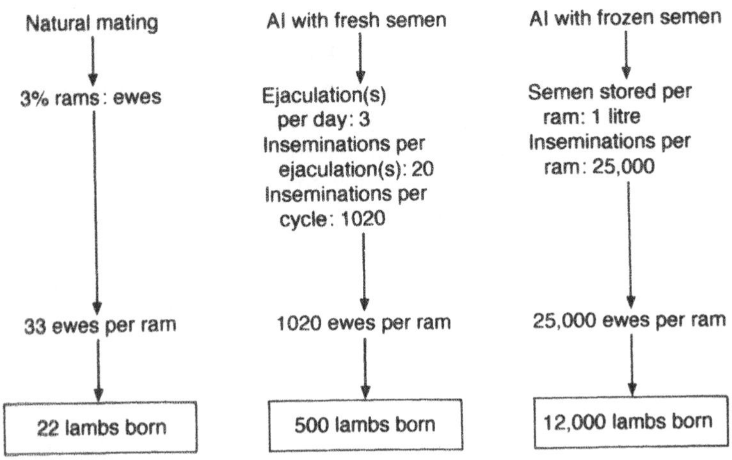

Fig. 1.7. The potential number of lambs sired by a ram each year using natural mating, AI with fresh semen and AI by laparoscopy with frozen semen (from Maxwell, 1984).

was found to be 15–20h from oestrus detection.

The cost and benefits associated with the use of AI in commercial flocks were evaluated by Abbott (1994) with reference to wool-producing sheep in Australia; the cost per lamb weaned after laparoscopic AI was estimated as $A100; benefits exceeded this cost for rams of very high merit when wool prices were moderate or high. AI with purchased semen also provided benefits for owners of commercial flocks who wished to breed their own replacement flock rams.

Sheep AI in small flocks

For any thought of applying AI economically under the small farm and flock conditions found in countries such as Ireland, the need for oestrus detection must be eliminated completely and all members of the selected group inseminated at a predetermined hour. This calls for precise control of oestrus and ovulation, using controlled reproduction techniques, not only among cyclic ewes in the autumn breeding season but also in sheep during other seasons of the year when they could normally be in anoestrus.

It was only during the 1970s that methods, both of controlled reproduction and ram semen processing, became available to enable the application of a new fixed-time approach to sheep AI; France is one country which has been in the forefront of such developments. In that country, in the 1979–86 period, the number of sheep inseminations doubled from 218,260 to 452,293, which accounted for about 27% of the national breeding flock. By 1990, the number of inseminated ewes had increased to 696,955, with 36% of dairy ewes and 4% of mutton ewes being bred by this method (Kupferschmied, 1992); in 1991, 740,000 ewes were inseminated in the country (Mesnil du Buisson, 1994). The French sheep AI industry has been based on distributing fresh semen, used on the day of collection, from more than 20 centres around the country; demand for AI in milk and mutton sheep in France is generally concentrated in the spring/early summer period.

1.5.2. Fresh and chilled semen

The main semen diluents used in sheep AI have either been based on cow's milk (whole or skim) or egg-yolk with Tris, citrate or phosphate buffers (Graham and Foote, 1985). The formulation of these diluents was often based on the Russian experience, as reviewed by Maxwell (1984).

Use of antifreeze proteins

Although most marine fish species occur in shallow, relatively warm waters, a small number of species live at least part of their life in the cold waters of the Arctic and Southern oceans. In the 1960s, it was discovered that the blood plasma of Antarctic fish demonstrated a phenomenon generally known as 'thermal hysteresis'; it did not freeze until about –2.5°C, but on rewarming, melted at about 0.8°C, indicating that unfrozen plasma contained some agent which reduced its freezing point. It is now known that antifreeze glycopeptides are universal in Antarctic fish (Davenport, 1992); these antifreeze proteins are now being used in various studies in the cryopreservation of embryos and gametes of farm animals, including sheep. In New Zealand, Payne *et al.* (1994) have studied the effects of antifreeze proteins from winter flounder and Antarctic cod in the motility of ram sperm after chilling (5°C) and after freeze-thawing; at a concentration of $10 \mu g\ ml^{-1}$, one antifreeze protein significantly increased the percentage of motile sperm after freeze-thawing.

1.5.3. Problems in freezing ram semen

Any consideration of sheep AI as a method of breeding would be incomplete without referring to the possibility of employing frozen semen. Although techniques for the insemination of sheep with freshly collected semen have been available for many years (reviewed by Emmens and Robinson, 1962), the development of freeze-thawing methods has been slow and procedures still leave much room for improvement. Although there is no scarcity of reports showing that the *in vitro* revival of ram spermatozoa after cryopreservation can be satisfactory, reports of acceptable conception rates after normal cervical insemination of such spermatozoa have been few. In France, Colas (1979) reported a 50% rejection rate of ejaculates after freezing and thawing, even with the best of his freezing treatments.

Fresh and frozen semen comparisons

In many comparative trials, the fertility of freeze-thawed ram semen has been much below that of fresh semen; this reduced fertility has been attributed to impaired sperm transport through the cervix, resulting in failure to establish adequate numbers in the sperm reservoirs that are known to exist in the ewe's reproductive tract (Fiser *et al.*, 1987). It is also now well recognized that the insemination of freeze-thawed semen directly into the uterus can result in a fertilization rate much the same as that of fresh semen; such evidence is usually taken as supporting the view that the basic problem is one of establishing adequate numbers in the sperm reservoirs (cervical and isthmic). According to Hunter and Nichol (1993), at least 30–60 min are required for adequate colonization of the sheep cervix with a fertilizing population of fresh sperm; how far this applies to frozen-thawed sperm is not known. It should be mentioned that variation in fertility between males is regarded as even more marked in rams than in bulls (Watson, 1979).

Evaluation of semen freezing techniques

One of the problems in the cryopreservation of ram semen is in being able to evaluate the success of the freeze-thaw methods employed. A technique for the evaluation of sheep-semen-freezing methodology was reported by Hattingh and Kay (1992) from South Africa. In this, semen was evaluated using a sperm penetration assay (SPA) on zona-free hamster oocytes. In Spain, Garde *et al.* (1993) evaluated the SPA before and after freezing ram sperm and suggested that this test would be suitable for the routine evaluation of frozen ram sperm. In California, Choudhry *et al.* (1995) also reported data suggesting that the SPA may be a useful test for predicting the relative *in vivo* fertility of freeze-thawed ram sperm after laparoscopic intrauterine insemination; these authors noted that the SPA may be more relevant to the outcome of laparoscopic AI than to other forms of sheep insemination (cervical) which do involve sperm transport.

The use of cattle oocytes for the evaluation of the fertility of ram semen has been reported by Smith and Murray (1995). This is based on the availability of cattle oocytes in most cities (from ovaries obtained at abattoirs) and the fact that heterologous *in vitro* fertilization (IVF) between cattle and sheep is possible. Such a sperm penetration assay may be valuable where it is not possible to use the hamster test.

1.5.4. Diluents and semen processing

Australian workers have favoured the pellet-freezing of ram semen and have been careful to observe the same principles as apply in fresh semen AI; namely, the need to employ a dense inseminate of highly motile sperm if an adequate population of sperm is to be established in the ewe tract. The Australian freezing method was based on the use of a Tris–glucose–citric acid egg-yolk diluent, added to semen at a low dilution rate (1 : 2 semen to diluent ratio); ewes inseminated once or twice during oestrus with sperm doses of 360 million motile sperm gave conception rates of 50–55% (Visser and Salamon, 1974). Work in the UK reported by Maxwell *et al.* (1980) showed a 52% lambing rate for semen frozen in this way in pellets as against 29% for semen frozen in straws.

Pellets vs. straws for frozen ram semen

Freezing in pellets has an advantage over freezing in straws by virtue of its simplicity and ease of operation. On the negative side, handling pellets on the farm is not easy because of the steps required before inseminations can be performed (thawing of pellets, filling of insemination pipettes or straws).

Buffers

An evaluation of zwitterion buffers in diluents for freezing ram sperm was reported by Molinia *et al.* (1994b). In this, diluents based on the zwitterion buffers TES, HEPES and PIPES, adjusted to pH 7.0 with NaOH or Tris, were compared with Tris citrate diluents by assessing post-thaw motility and acrosome integrity of frozen ram sperm. The authors concluded that the zwitterion buffers were superior to Tris-citrate; inclusion of egg-yolk (13.5%), centrifugation of the diluents, use of low dilution rates (three- or sixfold) and freezing in pellets rather than minitubes or straws improved motility and acrosome integrity. Whether the use of these novel diluents is reflected in improved fertility remains to be demonstrated. Other studies reported by Molinia *et al.* (1994a) from Sydney showed that a range of sugar concentrations can be successfully incorporated into ram freezing diluents.

Without doubt, frozen ram semen will eventually prove to be valuable in commercial sheep breeding and production programmes. Artificial insemination should prove to be extremely useful in speeding the transfer of genetic gains between stud flocks and commercial flocks. In sheep, there is obviously scope for accumulating stocks of frozen semen for use during periods of peak demand in the autumn breeding season at other times; this could be valuable for the operation of a commercial viable insemination service.

1.5.5. Season and semen quality

French workers have maintained that the fertility of frozen semen is highly dependent upon the season of collection and that best fertility is evident during the autumn breeding season. It has been reported by Colas (1979) that the incidence of morphologically abnormal sperm (distal cytoplasmic droplets and abnormal tails) can be much greater under increasing daylength conditions than under decreasing light (22.5 vs. 10.3%); fertilizing capacity of fresh sperm is also said to be affected, with the poorest semen being produced under conditions of increasing daylength. As noted earlier, such marked fertility differences with season have not been observed with the particular ram breeds found in Ireland (Fig. 1.8).

1.5.6. Insemination techniques

The efficacy of depositing semen into the vagina, cervix or uterus of the ewe has been reported by various workers, in terms of the conception rates achieved. Vaginal insemination is the simplest technique, and involves depositing fresh diluted semen deep into the vagina of the ewe without attempting to locate the cervix. The Australians refer to it as the 'shot-in-the-dark' or SID method, which sums up its efficiency rather well. Not surprisingly, conception rate after SID has been reported to be lower than that following cervical insemination (Kerton *et al.*, 1984; Rival *et al.*, 1984). Cameron *et al.* (1986) also found that vaginal insemination was inferior to cervical insemination, although the difference between the two methods was only about half that reported by the earlier workers; the difference appeared to be greatest when low numbers of sperm were inseminated. Contrary to some previous reports, Cameron *et al.* (1986) found that resting after insemination did not improve conception rates. The vaginal method is ineffective for frozen-thawed semen.

Cervical insemination

The success of cervical insemination in sheep, using chilled (+2–5°C) rather than fresh semen, appears to be dependent on dilution rate, sperm dose and the duration of semen storage before use. As with chilled ram semen, the major factor limiting the fertility of ewes inseminated with frozen ram semen is the inability of ram sperm to cross the cervix of the ewe after being subjected to the freeze-thaw process. The exact nature of the cellular changes responsible for this problem remain undefined. Unlike fresh semen, frozen semen must be deposited within the uterus to ensure adequate fertility.

Intrauterine AI

Intrauterine insemination by way of mid-ventral laparotomy with fresh-chilled ram semen resulted in high fertilization rates (Salamon *et al.*, 1979; Maxwell, 1984). Such results established two important facts: firstly, problems of poor conception rates, using chilled or frozen semen with cervical insemination, could be circumvented by depositing sperm directly in the uterus; secondly, fewer sperm are needed with intrauterine AI. The method of intrauterine AI was further developed by Killeen and Caffrey (1982), who reported a high fertilization rate, using both fresh and frozen semen, with the aid of laparoscopy. By the mid-1980s, several reports of promising fertility results after laparoscopic intrauterine AI had been published by workers in Australia, the UK, New Zealand and elsewhere (Armstrong and Evans, 1984; Maxwell *et al.*, 1984; Tervit, 1984; McKelvey *et al.*, 1985). The potential of using intrauterine AI and frozen semen rather than the ram in natural service or with cervical inseminations with fresh semen has been given in Fig. 1.7.

According to Maxwell (1984), in Western Australia alone more than 22,000 ewes were inseminated commercially with frozen semen by laparoscopy during the 1983/84 breeding season. Some reports suggest that intrauterine insemination into one uterine horn may be sufficient to obtain a high

Fig. 1.8. Performance of Dorset Horn rams during the ewe anoestrus. Dorset Horn sheep in the UK have a long history of lambing in the autumn (breeding season) – the breed was kept in the south of England and at one time supplied a trade for young lamb at Christmas in London. Why the Dorset Horn should differ from the other English Long-wool breeds is not entirely clear.

fertilization rate. A paper by Correa *et al.* (1994) in Chile recorded a significantly higher fertilization rate in ewes inseminated unilaterally with frozen semen (into the right horn of the uterus) as compared with cervical insemination (84% vs. 19%). The most widely practised form of uterine insemination in the ewe is laparoscopic AI, which is used within the Australian stud Merino industry (Evans, 1991).

Transcervical insemination in sheep

Although there is ample evidence showing that a single laparoscopic AI in sheep with frozen semen can result in acceptable conception rates, its wider application may be limited by its expense.

An alternative to laparoscopic AI is the passage of a pipette into the uterus from the vagina via the cervix; this procedure is known as transcervical AI. There have been several reports of transcervical AI (see review by Eppleston and Maxwell, 1993). A report by Halbert *et al.* (1990) in Guelph reported pregnancy rates of 50–68% in three flocks inseminated transcervically and 70% for ewes inseminated laparoscopically. One more recent study has evaluated the use of the Guelph system for transcervical AI (GST-AI) in Australian Merinos (Windsor *et al.*, 1994); although this technique could result in conception rates similar to those with laparoscopic AI, this was only in ewes in which semen was deposited within the uterus rather than the cervix (about 50% of sheep). The cervical penetration rate (proportion of ewes in which the AI pipette could be passed through the cervix) was identified as a major factor limiting the commercial use of this technique.

Further studies by Windsor (1995) led to a paper describing the influence of a number of factors on the cervical penetration rates achieved by the GST-AI system. Cervical penetration rates ranged from 21% to 80%. The author concluded that it may often be wise to exclude ewes that have lambed only once as well as maiden ewes. Penetration rates appeared to be higher in ewes showing natural rather than progestagen controlled heats and in ewes in the breeding season rather than at other times of the year. The author does not specify the time taken per sheep to perform the transcervical insemination, which, as well as the welfare aspects of the GST-AI system, may need much further evaluation.

Insemination pipettes

From Hungary, a report by Magyar (1994) deals with a new insemination pipette for the intrauterine laparoscopic insemination of sheep; this double-wall pipette was tested successfully on 350 ewes. The author concluded that its application could reduce the costs of insemination, allow easy and accurate dosage of semen, provide good heat insulation and prevent introduction of the needle deeper than 5 mm.

1.6. References

Abbott, K.A. (1994) Cost-benefit evaluation of artificial insemination for genetic improvement of wool-producing sheep. *Australian Veterinary Journal* 71, 353–360.

Abecia, J.A., Rhind, S.M. and McMillen, S.R. (1994) Effect of undernutrition on luteal function and the distribution of progesterone in endometrial tissue in ewes. *ITEA, Produccion Animal* 90A(2), 63–71.

Adams, N.R. (1990) Permanent infertility in ewes exposed to plant oestrogens. *Australian Veterinary Journal* 67, 197–201.

Adams, N.R. (1994) Phytoestrogens in legumes. *Journal of Animal Science* 72 (Suppl. 1)/ *Journal of Dairy Science* 77 (Suppl. 1), p. 56.

Al-Khozam, N.M., Robinson, J.J., McEvoy, T.G., Aitken, R.P., Findlay, P.A. and Robertson, I.S. (1995) Effect of dietary energy concentrations during the peri-ovulatory period on the in vivo and in vitro development of fertilized sheep ova. *Proceedings British Society of Animal Science* (Winter Meeting), Paper 74 (2 pp).

Allison, A.J. (1975) Ewe and ram fertility in commercial flocks mated with differing numbers of ewes per ram. *New Zealand Journal of Experimental Agriculture* 3, 161–167.

Allison, A.J. (1978) Flock mating in sheep. IV. Effect of number of ewes per ram on ejaculated characteristics of libido during the mating period. *New Zealand Journal of Agricultural Research* 21, 187–195.

Amir, D., Gacitua, H., Ron, M. and Lehrer, A.R. (1986) Seasonal variation in semen characteristics and the fertility of Finn cross rams subjected to frequent ejaculation. *Animal Reproduction Science* 10, 75–84.

Armstrong, C.S. and Evans, G. (1984) Intrauterine insemination enhances fertility of frozen semen in superovulated ewes. *Journal of Reproduction and Fertility* 71, 89–94.

Avdi, M. and Driancourt, M.A. (1995) Effects of litter sex composition on the subsequent reproductive performance of females in sheep. *Journal of Reproduction and Fertility* Abstract Series No. 15, p. 51.

Averill, R.L.W. (1955) Fertility of the ewe. *Proceedings of the Society for the Study of Fertility* 7, 139–148.

Barillet, F. and Bocquier, F. (1993) Dairy sheep production in France: the main objectives of research and development and conditions of implementation. *Production Animales* 6 (1), 17–24.

Beck, N.F.G., Peters, A.R. and Williams, S.P. (1994) The effect of GnRH agonist (buserelin) treatment on day 12 post mating on the reproductive performance of ewes. *Animal Production* 58, 243–247.

Beck, N.F.G., Jones, M., Davies, B., Mann, G.E. and Peters, A.R. (1995) The effect of GnRH analogue (buserelin) treatment on day 12 post mating on ovarian function in ewes. *Journal of Reproduction and Fertility* Abstract Series No. 15, p. 73.

Becker, K., Liebscher, D. and Hennig, H. (1986) 30 years of artificial insemination of sheep – a contribution to the intensification of sheep breeding in the Dresden. Tierzucht 40, 422-425.

Bennetts, H.W., Underwood, E.J. and Shier, F.L. (1946) A specific breeding problem of sheep on subterranean clover in Western Australia. *Australian Veterinary Journal* 22, 2–12.

Berg, K.A. and Aamdal, J. (1991) Artificial insemination with frozen semen in ewes at different times of the breeding season. *Reproduction in Domestic Animals* 26, 27–30.

Blaxter, K.L. (1964) Dietary factors affecting energy utilization. *Proceedings of the Nutrition Society* 23, 3–11.

Blockey, M.A. de B and Wilkins, J.F. (1984) Field application of the ram serving capacity test. In: Lindsay, D.R. and Pearce, D.T. (eds) *Reproduction in Sheep.* Cambridge University Press, Cambridge, pp. 53–58.

Borwick, S.C., Rhind, S.M. and McMillen, S.R. (1995) Effects of undernutrition from the time of mating on ovarian development in foetal sheep at 62 d of gestation. *Journal of Reproduction and Fertility* Abstract Series No. 15, p. 52.

Bradford, G.E. (1972) Genetic control of litter size in sheep. *Journal of Reproduction and Fertility* Supplement 15, 23–41.

Brash, L.D. (1994) Advanced breeding and techniques for wool sheep improvement. *Wool Technology and Sheep Breeding* 42, 327–337.

Brown, B.W., Stockwell, P.R. and Panaretto, B.A. (1994) Effects of depilatory doses of epidermal growth factor on subsequent fertility, pregnancy rate and lambing performance in Merino ewes. *Australian Journal of Agricultural Research* 45, 333–338.

Brown, B.W., Mattner, P.E., Carroll, P.A., Holland, E.J., Paull, D.R., Hoskinson, R.M. and Rigby, R.D.G. (1994) Immunization of sheep against GnRH early in life: effects on reproductive function and hormones in rams. *Journal of Reproduction and Fertility* 101, 15–21.

Brown, B.W., Mattner, P.E., Carroll, P.A., Hoskinson, R.M. and Rigby, R.D.G. (1995) Immunization of sheep against GnRH early in life: effects on reproductive function and hormones in ewes. *Journal of Reproduction and Fertility* 103, 131–135.

Bruere, A.N. (1971) Practical aspects of fertility in the ram. *Sheep Farming Annual.* Massey University, New Zealand, pp. 31–40.

Caia, C., Ribo, O. and Nehring, R. (1994) Formation of fibrous capsule and migration movements at different body sites of transponders implanted in sheep for electronic identification. *Journal of Animal Science* 72 (Suppl. 1)/*Journal of Dairy Science* 77 (Suppl. 1), p. 307.

Cameron, A.W.N., Fairnie, I.J. and Keogh, E.J. (1984) Semen quality, quantity and flock fertility. In: Lindsay, D.R. and Pearce, D.T. (eds) *Reproduction in Sheep*. Cambridge University Press, Cambridge, pp. 79–84.

Cameron, A.W.N., Tilbrook, A.J., Lindsay, D.R., Keogh, E.J. and Fairnie, I.J. (1986) The effect of testicular weight and insemination technique on fertility of sheep. *Animal Reproduction Science* 12, 189–194.

Casu, S., Cappai, P. and Naitana, S. (1991) Effects of high temperatures on reproduction in small ruminants. In: *Animal Husbandry in Warm Climates*. EAAP Publication No. 55, pp. 103–111.

Cheng, G. and Ma, N. (1992) Achievements of Chinese sheep and goat raising industries over the last forty years. *Animal Genetic Resources Information* No. 9, pp. 57–70.

Choudhry, T.M., Berger, T. and Dally, M. (1995) In vitro fertility evaluation of cryopreserved ram semen and its correlation with relative in vivo fertility. *Theriogenology* 43, 1195–1200.

Cognie, Y., Colas, G. and Thimonier, J. (1984) Control of reproduction in the ewe. In: *The Reproductive Potential of Cattle and Sheep*. INRA Publication, pp. 175–190.

Colas, G. (1979) Fertility in the ewe after artificial insemination with fresh and frozen semen at the induced estrus, and influence of the photoperiod on the semen quality of the ram. *Livestock Production Science* 6, 153–166.

Correa, J.E., Bergmann, B. and Gatica, R. (1994) Fertilization rate in sheep unilaterally inseminated with frozen semen. *Small Ruminant Research* 13, 99–101.

Crowley, J.P. (1964) The extension of the breeding season in sheep. *Proceedings of the 5th International Congress of Animal Reproduction and AI* (Trento) 2, 378–383.

Crowley, J.P. and Walsh, M.A. (1971) Infertility in rams in Ireland. *Irish Veterinary Journal* 25, 27–30.

Daley, D.A., Adams, T.E., Daley, C.A., Patton, W.R. and Evans, J.L. (1995) Effects of immunocastration on growth, carcass characteristics and reproductive development in ram lambs. *Sheep and Goat Research Journal* 11, 31–34.

Dauzier, L. and Wintenberger, S. (1952) Declanchement simultane de l'oestrus dans un lot de brebis avec possibilite de gestation ulteriece. *Annales Zootechnica* 49–52.

Davenport, J. (1992) The precarious life of high-latitude marine fish. *Biolgoist* 39(5), 218–221.

Dawe, S.T., Archer, W.R., Bennett, N.W., Brunskill, A., Cahill, J.R., Donnelly, F.B., Roberts, B.C. and Trimmer, B.I. (1974) The effect of ram percentage on the fertility of maiden ewes. *Proceedings Australian Society of Animal Production* 10, 274–278.

Deveson, S.L., Arendt, J. and Forsyth, I.A. (1992) The influence of the pineal gland and melatonin on the reproductive performance of domesticated female ungulates. *Animal Reproduction Science* 30, 113–134.

Diskin, M.G. and Hanrahan, J.P. (1995) The effect of energy or protein supplementation on embryo survival rate in ewes. *Proceedings British Society Animal Science* (Winter Meeting), Paper 55.

D'Occhio, M.J. and Suttie, J.M. (1992) The role of the pineal gland and melatonin in reproduction in male domestic ruminants. *Animal Reproduction Science* 30, 135–155.

Doney, J.M., Gunn, R.G. and Smith, W.F. (1976a) Effects of premating environmental stress, ACTH, cortisone acetate or metapone on oestrus and ovulation in sheep. *Journal of Agricultural Science, Cambridge* 87, 127–132.
Doney, J.M., Smith, W.F. and Gunn, R.G. (1976b) Effects of post-mating environmental stress or administration of ACTH on early embryonic loss in sheep. *Journal of Agricultural Science, Cambridge* 87, 133–136.
Dutt, R.H. (1952) Induction of estrus and ovulation in anestrual ewes by use of progesterone and pregnant mare serum. *Journal of Animal Science* 11, 792.
Dutt, R.H. and Bush, L.F. (1955) The effect of low environmental temperature on initiation of the breeding season in sheep. *Journal of Animal Science* 14, 885–889.
Dutt, R.H. and Hamm, P.T. (1957) The effect of high environmental temperature and shearing on semen quality of rams. *Journal of Animal Science* 16, 328–334.
Dyer, A. (1990) A practical look at the use of ovine cervical artificial insemination to upgrade commercial flocks. *Wool Technology and Sheep Breeding* 38, 112–113.
El-Sayed, A.I., El-Azab, A.I. and Nasr, M.T. (1993) Effect of oral administration of L-tyrosine on the quality of ram semen. *Annals of Agricultural Science, Moshotor* 30, 1207–1218.
Emmens, C.W. and Robinson, T.J. (1962) AI in sheep. In: Maule, J.P. (ed.) *The Semen of Animals and Artificial Insemination.* Commonwealth Agricultural Bureau, Slough, pp. 205–251.
Eppleston, J. and Maxwell, W.M.C. (1993) Recent attempts to improve the fertility of frozen ram semen inseminated into the cervix. *Wool Technology and Sheep Breeding* 41, 291–302.
Evans, G. (1988) Current topics in artificial insemination of sheep. *Australian Journal of Biological Science* 41, 103–116.
Evans, G. (1991) Application of reproductive technology to the Australian livestock industries. *Reproduction, Fertility and Development* 3, 627–650.
Evans, G. and Robinson, T.J. (1980) Reproductive potential and endocrinological responses of sheep kept under controlled lighting. II. Pituitary and gonadal responses of ewes and rams to a six-monthly light cycle. *Animal Reproduction Science* 3, 39–56.
Fernandez Abella, D. and Villegas, N. (1992) Effect of melatonin on semen quality in spring and summer. *Boletin Tecnico de Ciencias Biologigicas, Universidad de la Republica (Salto)* 2(1), 51–57.
Fiser, P.S., Ainsworth, L. and Fairfull, R.W. (1987) Evaluation of a new diluent and different processing procedures for cryopreservation of ram semen. *Theriogenology* 28, 599–607.
Fitzgerald, J.A., Perkins, A. and Hemenway, K. (1993) Relationship of sex and number of siblings in utero with sexual behaviour of mature rams. *Applied Animal Behaviour Science* 38, 283–290.
Fogarty, N.M. (1984) Breeding for reproductive performance. In: Lindsay, D.R. and Pearce, D.T. (eds) *Reproduction in Sheep.* Cambridge University Press, Cambridge, pp. 226–233.
Fowler, D.G. (1984) Reproductive behaviour of rams. In: Lindsay, D.R. and Pearce, D.T. (eds) *Reproduction in Sheep.* Cambridge University Press, Cambridge, pp. 39–46.
Garde, J., Gutierrez, A., Artiga, G., Perez Guzman, M., Montoro, V. and Vazquez, I. (1993) A comparison of the hamster test with other methods for the evaluation of the quality of frozen ram semen. *Proceedings of the 5th International Symposium on Animal Reproduction* (Luso, Portugal) 2, pp. 285–291.

Georgoudis, A., Hatziminaoglou, I. and Pappas, V. (1995) The breeding scheme of the Karagouniko sheep in Greece. *Cahiers Options Mediterraneennes* 11, 61–65.

Gherardi, P.B., Lindsay, D.R. and Oldham, C.M. (1980) Testicle size in rams and flock fertility. *Proceedings of the Australian Society of Animal Production* 13, pp. 48–50.

Glimp, H.A. (1995) Using sheep to enhance the environment. *Journal of Animal Science* 73 (Suppl. 1), p. 242.

Gonzalez, R., Levy, F., Orgeur, P., Poindron, P. and Signoret, J.P. (1991) Female effect in sheep. II. Role of volatile substances from the sexually receptive female; implication of the sense of smell. *Reproduction, Nutrition and Development* 31, 103–109.

Gordon, I. (1955) The hormonal augmentation of fertility in sheep. *Proceedings of the British Society of Animal Production* pp. 55–63.

Gordon, I. (1958a) The use of progesterone and PMS in the control of fertility in sheep: with special reference to the practical utility of such treatment. *Cambridge University Dissertation Abstracts* 1956/57, p. 3.

Gordon, I. (1958b) Hormonal augmentation of fertility in the ewe during the breeding season. *Journal of Agricultural Science, Cambridge* 50, 123–151.

Gordon, I. (1963) The induction of pregnancy in the anoestrous ewe by hormonal therapy. *Journal of Agricultural Science, Cambridge* 60, 31–79.

Gordon, I. (1967) Aspects of reproduction and neonatal mortality in ewe lambs and adult sheep. *Journal of the Irish Department of Agriculture* 64, 76–127.

Gordon, I. (1975) Hormonal control of reproduction in sheep. *Proceedings of the British Society of Animal Production* 4, pp. 79–93.

Gourley, D.D. and Riese, R.L. (1990) Laparoscopic artificial insemination in sheep. *Veterinary Clinics of North America, Food Animal Practice* 6(3), 615–633.

Graham, J. and Foote, R.H. (1985) Predicting fertility of fresh and frozen semen. *The Advanced Animal Breeder* 1 December issue, p. 6.

Gunn, R.G. and Doney, J.M. (1979) Fertility in Cheviot ewes. 1. The effect of body condition at mating on ovulation rate and early embryo mortality in North and South country Cheviot ewes. *Animal Production* 29, 11–16.

Gunn, R.G., Sim, D.A. and Hunter, E.A. (1995) Effects of nutrition in utero and early life on the subsequent lifetime reproductive performance of Scottish Blackface ewes in two management systems. *Animal Science* 60, 223–230.

Gunn, R.M.C., Sanders, R.N. and Granger, W. (1942) Studies in fertility in sheep. 2. Seminal changes affecting fertility in rams. *Bulletin of the Council Scientific and Industrial Research in Australia* No. 148, 140 pp.

Hafez, E.S.E. (1952) Studies on the breeding season and reproduction of the ewe. *Journal of Agricultural Science, Cambridge* 42, 189–265.

Halbert, G.W., Dobson, H., Walton, J.S., Sharpe, P. and Buckrell, B.C. (1990) Field evaluation of a technique for transcervical intrauterine insemination of ewes. *Theriogenology* 33, 1231–1243.

Hanrahan, J.P. (1994) Evaluation of crossbred ewe types. *Irish Grassland and Animal Production Association Journal* 28, 106–113.

Haresign, W. (1990) Controlled breeding in sheep. In: *New Developments in Sheep Production*. Occasional Publication No. 14, BSAP, 23–37.

Haresign, W. (1992) Manipulation of reproduction in sheep. *Journal of Reproduction and Fertility* Suppl. 45, 127–139.

Hattingh, H. and Kay, G.W. (1992) The sperm penetration assay: a useful technique for the evaluation of sheep semen freezing methodology. *Animal Technology* 43, 121–126.

Heape, W. (1899) Abortion, barrenness and fertility in sheep. *Journal Royal Agricultural Society* Series III, 10, 217.

Herbosa, C.G. and Foster, D.L. (1995) Masculinization of the reproductive response to photoperiod occurs early in prenatal development in the sheep. *Biology of Reproduction* 52 (Suppl. 1), p. 203.

Hochereau-De Reviers, M.T., Perreau, C., Pisselet, C., Locatelli, A. and Bosc, M. (1995) Ontogenesis of somatic and germ cells in sheep fetal testis. *Journal of Reproduction and Fertility* 103, 41–46.

Hollerrieder, J. (1991) Investigations on the freezing of ram semen, with special reference to acrosome morphology and video computer analysis. Thesis, Ludwig-Maximiliansuniversitat Munchen Germany, 108 pp.

Hombolu, J.O., Ojo, S.A., Jamdar, M.N. and Molokwu, E.C. (1985) Ovarian activity of Yankasa sheep using abattoir specimens. *Theriogenology* 23, 263–272.

Hooda, O.K. and Naqvi, S.M.K. (1990) Effect of thermal load and feed restriction on the relative adaptability of Malpura and Avikalin sheep in semi-arid region. *Indian Journal of Animal Sciences* 60, 608–611.

Horak, F. and Potucek, M. (1978) The effect of the lunar phase on the sexual activity of ewes. *Sbornik Vysoke Skoly Zemedelske* 23, 743–749.

Hunter, R.H.F. and Nichol, R. (1993) Rate of establishment of a fertilising population of spermatozoa in the sheep cervix after a single mating at the onset of oestrus. *Journal of Experimental Zoology* 266, 168–171.

Imig, J.L., Powell, M.R., Naivar, J., Roberts, R.M. and Keisler, D.H. (1995) Effect of recombinant ovine interferon-tau on pregnancy rate in the ewe. *Biology of Reproduction* 52 (Suppl. 1), p. 73.

Jeffcoate, I.A., Foster, J.P. and Crighton, D.B. (1978) Effect of active immunisation of ewes against synthetic luteinising hormone releasing hormone. *Theriogenology* 10, 323–335.

Jeffries, B.C. (1989) New sheep breeds of Australia. In: Erskine, K. (ed.) *Colored Sheep and Wool: Exploring their Beauty and Function*. Black Sheep Press, Oregon, USA.

Jennings, J.J. (1972) Some factors affecting the reproductive performance of rams. PhD Thesis, National University of Ireland, Dublin.

Jheltobruch, N.A. (1979) AI of sheep in the Soviet Union. In: *Sheep Breeding* 2nd edn. (revised by W. Haresign). Butterworths, London, pp. 565–570.

Johnston, S.D. and Steen, R.W.J. (1995) The effect of genotype on lamb carcass characteristics. *Proceedings of the British Society Animal Science* (Winter Meeting), Paper 45.

Karsh, F.J. (1980) Seasonal reproduction: a saga of reversible fertility. *The Physiologist* 33(6), 29–38.

Karsh, F.J. (1995) Neuroendocrine signals for ovulation – fitting pieces to an unsolved puzzle. *Journal of Reproduction and Fertility* Abstract Series No. 15, p. 1.

Karsh, F.J., Bittman, J.E., Foster, D.L., Goodman, R.L., Legan, S.J. and Robinson, J.E. (1984) Neuroendocrine basis of seasonal reproduction. *Recent Progress in Hormone Research* 40, 185–232.

Kassem, R., Owen, J.B., Fadel, I., Juha, H. and Whitaker, C.J. (1989) Aspects of fertility and lamb survival in Awassi sheep under semi-arid conditions. *Research and Development in Agriculture* 6, 161–168.

Keeling, B.J. and Crighton, D.B. (1984) Reversibility of the effects of active immunization against LH-RH. In: Crighton, D.B. (ed.) Immunological Aspects of Reproduction in Mammals. Butterworths, London, pp. 379–397.

Keisler, D.H., Bettencourt, C.M.V. and Moffat, R.J. (1993) Effects of invermectin on

reproductive functions in ewes. *Journal of Animal Science* 71, 2293–2296.
Kent, J.E., Molony, V. and Robertson, I.S. (1995) Comparison of the Burdizzo and rubber ring methods for castrating and tail docking lambs. *Veterinary Record* 136, 192–196.
Kerton, D.J., McPhee, S.R., Davies, I.F., White, M.B., Banfield, J.C. and Cahill, L.P. (1984) A comparison of insemination techniques in Corriedale ewes. *Proceedings of the Australian Society of Animal Production* 15, 701.
Kilgour, R.J. (1979) The importance of the ram on flock fertility. *Wool Technology and Sheep Breeding* 27, 41–44.
Kilgour, R.J. (1993) The relationship between ram breeding capacity and flock fertility. *Theriogenology* 40, 277–285.
Kilgour, R.J. and de Langen, H. (1970) Stress in sheep resulting from management practices. *Proceedings of the New Zealand Society of Animal Production* 30, 65–76.
Kilgour, R.J. and Wilkins, J.F. (1980) The effect of serving capacity of the ram on flock fertility. *Australian Journal of Experimental Agriculture and Animal Husbandry* 20, 662–666.
Kilgour, R.J., Courot, M., Pisselet, C., Dubois, M.P. and Sairam, M.R. (1994) Inhibition of FSH but not LH affects spermatogenesis in the mature ram. *Animal Reproduction Science* 34, 253–264.
Killeen, I.D. and Caffrey, G.J. (1982) Uterine insemination of ewes with the aid of a laparoscope. *Australian Veterinary Journal* 59, 95.
Knight, T.W. (1972) A study of factors which affect the potential fertility of the ram. PhD Thesis, University of Western Australia.
Knight, T.W. (1977) Methods for the indirect estimation of testes weight and sperm numbers in Merino and Romney rams. *New Zealand Journal of Agricultural Research* 20, 291–295.
Kupferschmied, H.U. (1992) A look over the borders: France. *KB-Mitteilungen* 30(3), 55.
Lees, J.L. (1966) Variations in the time of onset of the breeding season in Clun ewes. *Journal of Agricultural Science, Cambridge* 67, 173–179.
Lees, J.L. (1978) Functional infertility in sheep. *Veterinary Record* 102, 232–236.
L'Haridon, R.M., Huynh, L., Assal, N.E. and Martal, J. (1995) A single intrauterine infusion of sustained recombinant ovine interferon-τ extends corpus luteum lifespan in cyclic ewes. *Theriogenology* 43, 1031–1045.
Lillo, A. (1989) Results of artificial insemination of sheep in the 1988–89 breeding season. *Farskotsel* 69(9), 12.
Lincoln, G.A. (1976) Secretion of LH in rams exposed to different photoperiods. *Journal of Reproduction and Fertility* 47, 351–356.
Lubbers, L.S. and Jackson, G.L. (1993) Neuroendocrine mechanisms that control seasonal changes of luteinizing hormone secretion in sheep are sexually differentiated. *Biology of Reproduction* 49, 1369–1376.
McDonald, M.F., Anwar, N. and Keogh, R.G. (1994) Reproductive performance of ewes after grazing on G27 red clover, a low formononetin selection in cultivar Pawera. *Proceedings of the New Zealand Society of Animal Production* 54, 231–234.
McKelvey, W.A.C., Robinson, J.J., Aitken, R.P. and Henderson, G. (1985) The evaluation of laparoscopic insemination technique in ewes. *Theriogenology* 24, 519–535.
McKelvey, W.A.C., Robinson, J.J. and Aitken, R.P. (1988) The use of reciprocal embryo transfer to separate the effects of pre- and post-mating nutrition on embryo survival and growth of the ovine conceptus. *Proceedings of the 11th International*

Congress Animal Reproduction and AI (Dublin) 2, Paper 176.
McKenzie, F.F. and Terrill, C.E. (1937) Estrus, ovulation and related phenomena in the ewe. *Research Bulletin Missouri Agricultural Experiment Station* No. 254.
MacLaren, A.P.C. (1988) Ram fertility in south-west Scotland. *British Veterinary Journal* 144, 45–54.
Magyar, K. (1994) New insemination pipette for the intrauterine laparoscopic insemination of sheep. *Magyar Allatorvosok Lapja* 49, 478–479.
Marshall, F.H.A. (1905) Fertility in Scottish sheep. *Proceedings Royal Society* B 77, 58.
Martin, G.B., Tjondronegoro, S. and Blackberry, M.A. (1994) Effects of nutrition on testicular size and the concentrations of gonadotrophins, testosterone and inhibin in plasma of mature male sheep. *Journal of Reproduction and Fertility* 101, 121–128.
Martinod, S., Maurer, R.R., Siegenthaler, B., Gerber, C. and Hansen, P.J. (1991) The effect of recombinant bovine interferon-alpha on fertility in ewes. *Theriogenology* 36, 231–239.
Martyniuk, E. and Rzepecki, R. (1995) Sheep husbandry in Poland – an outline. *Cahiers Options Mediterraneennes* 11, 121–131.
Matthews, L.R., Uljee, A.E., Bremner, K.J., Painting, A.M., Cate, L.R. and Smith, J.F. (1991) Development of a self-drafting system for oestrus ewes. *Proceedings of the New Zealand Society of Animal Production* 51, 315–318.
Mattner, P.E., Braden, A.W.H. and George, J.M. (1971) Studies in flock mating of sheep. 4. The relation of libido tests to subsequent service activity of young rams. *Australian Journal Experimental Agricultural and Animal Husbandry* 11, 473–477.
Maxwell, W.M.C. (1984) Current problems and future potential of artificial insemination programmes. In: Lindsay, D.R. and Pearce, D.T. (eds) *Reproduction in Sheep*. Cambridge University Press, Cambridge, pp. 291–297.
Maxwell, W.M.C., Curnock, R.M., Logue, D.N. and Reed, H.C.B. (1980) Fertility of ewes following artificial insemination with semen frozen in pellets or straws: a preliminary report. *Theriogenology* 14, 83–89.
Maxwell, W.M.C., Butler, L.G. and Wilson, H.R. (1984) Intrauterine insemination of ewes with frozen semen. *Journal of Agricultural Science, Cambridge* 102, 233–235.
Maxwell, W.M.C. and Salamon, S. (1993) Liquid storage of ram semen: a review. In: Symposium on Sperm Preservation and Encapsulation. *Reproduction, Fertility and Development* 5, 613–638.
Mekonnen, G., Boland, M. and Gordon, I. (1989) Effect of prostaglandin on semen production and libido in the ram. *Irish Veterinary Journal* 42, 56–59.
Menger, H., Bruckner, G. and Wenig, H. (1988) Investigations on seasonal effects on semen quality of AI rams and on variability and manipulability of semen. *Wissenschaftliche Zeitschrift-Karl-Marx Universitat Leipzig, Mathematisch-Naturwissenschaftliche Reihe* 37, 266–275.
Mesnil du Buisson, F. du (1994) Artificial insemination of domestic animals (except cattle) in France and its development. *Comptes Rendu de l'Academie d'Agriculture de France* 80(3), 98–106.
Molinia, F.C., Evans, G., Quintna Caseres, P.I. and Maxwell, W.M.C. (1994a) Effect of monosaccharides and disaccharides in Tris-based diluents on motility, acrosome integrity and fertility of pellet frozen ram spermatozoa. *Animal Reproduction Science* 36, 113–122.
Molinia, F.C., Evans, G. and Maxwell, W.M.C. (1994b) In vitro evaluation of zwitterion buffers in diluents for freezing ram spermatozoa. *Reproduction, Nutrition and Development* 34, 491–500.

Morcombe, P.W., Peet, R.L. and Bell, S.J. (1990) The sterility and productivity of young, short scrotum, Merino rams. *Australian Veterinary Journal* 67, 235–236.

Morley, F.H.W., White, D.H., Kenney, P.A. and Davis, I.F. (1978) Predicting ovulation rate from liveweight in ewes. *Agricultural Systems* 3, 27–45.

Mukasa-Mugerwa, E., Anindo, D., Lahlou-Kassi, A., Mutiga, E.R. and Sovani, S. (1993) Seasonal variation in ovarian and oestrous activity of tropical Menz sheep as affected by plane of nutrition. *Reproduction, Nutrition and Development* 33, 585–595.

Murray, P.J., Rowe, J.B. and Pethick, D.W. (1991) Effect of season and nutrition on scrotal circumference of Merino rams. *Australian Journal of Experimental Agriculture* 31, 753–756.

Nephew, K.P., McClure, K.E., Day, M.L., Xie, S., Roberts, R.M. and Pope, W.F. (1990) Effects of intramuscular administration of recombinant bovine interferon-α 1 during the period of maternal recognition. *Journal of Animal Science* 68, 2766–2770.

Nephew, K.P., Cardenas, H., McClure, K.E., Ott, T.L., Bazer, F.W. and Pope, W.F. (1994) Effects of administration of human chorionic gonadotropin or progesterone before maternal recognition of pregnancy on blastocyst development and pregnancy in sheep. *Journal of Animal Science* 72, 453–458.

Neville, W.E. and Neathery, M.W. (1970) Effect of natural differences in atmospheric temperature on the incidence of first estrus following anestrus in sheep. *Journal of Animal Science* 30, 242–249.

Nilsson, B. (1989) Artificial insemination of sheep in Sweden – organization and its importance for breeding. *Farskotsel* 69(9), 5–7.

Olesen, I. (1993) Effects of cervical insemination with frozen semen on fertility and litter size of Norwegian sheep. *Livestock Production Science* 37, 169–184.

Ortavant, R., Pelletier, J., Ravault, J.P., Thimonier, J. and Volland-Nail, P. (1985) Photoperiod: main proximal and distal factor of the circannual cycle of reproduction in farm mammals. *Oxford Reviews of Reproductive Biology* 7, 305–345.

Ortavant, R., Bocquier, F., Pelletier, J., Ravault, J.P., Thimonier, J. and Volland-Nail, P. (1988) Seasonality of reproduction in sheep and its control by photoperiod. *Australian Journal of Biological Science* 41, 69–85.

Payne, S.R., Oliver, J.E. and Upreti, G.C. (1994) Effect of antifreeze proteins on the motility of ram spermatozoa. *Cryobiology* 31, 180–184.

Pelletier, J. and Almeida, G. (1987) Short light cycles induce persistent reproductive activity in Ile-de-France rams. *Journal of Reproduction and Fertility* Suppl. 34, 215–226.

Pope, W.F., Cardenas, H. and McClure, K.E. (1994) Influence of short-term fasting on blastocyst morphology and survival in ewes. *Sheep Research Journal* 10, 16–19.

Pope, W.F., Cardenas, H., Wiley, T.M. and McClure, K.E. (1995) Dose-response relationships of exogenous progesterone shortly after ovulation on estrous cycle length, blastocyst development and fertility in sheep. *Animal Reproduction Science* 38, 109–117.

Price, E.O., Borgwardt, R. and Dally, M.R. (1995) Heterosexual experience differentially affects the expression of sexual behaviour in 6- and 8-month-old ram lambs. *Journal of Animal Science* 73 (Suppl. 1), p. 125.

Raadsma, H.W. and Edey, T.N. (1985) Mating performance of paddock-mated rams. I. Changes in mating performance, ejaculate characteristics and testicular size during the joining period. *Animal Reproduction Science* 8, 79–99.

Radford, H.M. (1961) Photoperiodism and sexual activity in Merino ewes. I. The effect of continuous light on the development of sexual activity. *Australian Journal of Agricultural Research* 12, 139–140.

Radford, H.M. (1966) Regulation of the breeding season in mammals. *Proceedings of the Australian Society of Animal Production* 6, pp. 19–31.

Ravault, J.P., Blanc, M., Ortavant, R., Pelletier, J. and de Reviers, M.M. (1980) Variations circaiennes et circannulles de la secretion de prolactine (PRL), LH et FSH chez les animaux domestiques males. In: Ortavant, R. and Reinberg, A. (eds) *Rythmes et Reproduction*. Masson, Paris, pp. 115–128.

Reichel, M.P., Baber, D.J., Armitage, P.W., Lampard, D., Whitley, R.S. and Hilbink, F. (1994) Eradication of *Brucella ovis* from the Falkland Islands 1977–1993. *Veterinary Record* 134, 595–597.

Reyna, A.G., Valencia Mendez, J., Foote, W.C. and Murphy, B.D. (1991) Hair sheep in Mexico: reproduction in the Pelibuey sheep. *Animal Breeding Abstracts* 59, 509–524.

Rhind, S.M. (1992) Nutrition: its effects on reproductive performance and its hormonal control in female sheep and goats. In: *Progress in Sheep and Goat Research*. CAB International, Wallingford.

Rhind, S.M., McMillen, S., McKelvey, W.A.C., Rodriguez-Herrejon, F.F. and McNeilly, A.S. (1989) Effect of the body condition of ewes on the secretion of LH and FSH and the pituitary response to gonadotrophin-releasing hormone. *Journal of Endocrinology* 120, 497–502.

Rival, M.D., Chenoweth, P.J. and McMicking, L.I. (1984) Semen deposition and fertility in ovine artificial breeding programmes. In: Lindsay, D.R. and Pearce, D.T. (eds) Reproduction in Sheep. Cambridge University Press, Cambridge, pp. 301–303.

Robertson, H.A. (1977) Reproduction in the ewe and the goat. In: Cole, H.H. and Cupps, P.T. (eds) *Reproduction in Domestic Animals*, 3rd edn. Academic Press, London, pp. 477–498.

Robertson, I.S., Fraser, H.M., Innes, G.M. and Jones, A.S. (1982) Effect of immunological castration on sexual and production characteristics in male cattle. *Veterinary Record* 111, 529–531.

Robinson, J.J. (1979) Intensive systems. In: *The Management and Diseases of Sheep*. Commonwealth Agricultural Bureau, Slough, pp. 431–446.

Robinson, J.J. (1990) The pastoral animal industries in the 21st century. *Proceedings of the New Zealand Society Animal Production* 50, pp. 345–359.

Robinson, T.J. (1952) Role of progesterone in the mating behaviour of the ewe. *Nature* 170, 373–374.

Robinson, T.J. (1965) Use of progestagen-impregnated sponges interserted intravaginally or subcutaneously for the control of the oestrous cycle in the sheep. *Nature* 206, 39–41.

Robinson, T.J. (1988) Controlled sheep breeding: update 1980–1985. *Australian Journal of Biological Science* 41, 1–13.

Robinson, T.J., Moore, N.W., Lindsay, D.R., Fletcher, I.C. and Salamon, S. (1970) Fertility following synchronization of oestrus in the sheep with intravaginal sponges. I. Effects of vaginal douche, supplementary steroids, time of insemination and numbers and dilution of sperm. *Australian Journal of Agricultural Research* 21, 767–770.

Russell, A.J.F. (1979) The nutrition of the pregnant ewe. In: *The Management and Diseases of Sheep*. Commonwealth Agricultural Bureau, Slough, pp. 221–241.

Ryder, M.L. (1981) Sheep in the Soviet Union. *Span* 24(1), 36–37.

Salamon, S., Maxwell, W.M.C. and Firth, J.H. (1979) Fertility of ram semen after storage at 5°C. *Animal Reproduction Science* 2, 373–385.

Scaramuzzi, R.J. and Martin, G.B. (1984) Pharmacological agents for manipulating oestrus and ovulation in the ewe. In: Lindsay, D.R. and Pearce, D.T. (eds) *Reproduction in Sheep*. Cambridge University Press, Cambridge, pp. 316–325.

Scaramuzzi, R.J., Downing, J.A., Campbell, B.K. and Cognie, Y. (1988) Control of fertility and fecundity of sheep by means of hormonal manipulation. *Australian Journal of Biological Science* 41, 37–45.

Scaramuzzi, R.J. and Murray, J.F. (1994) The nutrient requirements for the optimum production of gametes in assisted reproduction in ruminant animals. *Proceedings of the 10th Meeting European Embryo Transfer Association* (Lyon), pp. 85–103.

Schalue-Francis, T.K., Farin, P.W., Cross, J.C., Keisler, D. and Roberts, R.M. (1991) Effect of injected bovine interferon-alpha 1 on oestrous cycle length and pregnancy success in sheep. *Journal of Reproduction and Fertility* 91, 347–356.

Schanbacher, B.D. and Ford, J.J. (1979) Photoperiodic regulation of ovine spermatogenesis: relationship to serum hormones. *Biology of Reproduction* 20, 719–726.

Setchell, B.P. (1984) The functions of the testis and epididymis in rams. In: Lindsay, D.R. and Pearce, D.T. (eds) Reproduction in Sheep. Cambridge University Press, Cambridge, pp. 53–58.

Shelton, M. (1990) Influence of docking fat-tail (Karakul) sheep on lamb production. *Small Ruminant Research* 3, 73–76.

Shelton, M. (1995) Harnessing the biological potential of sheep in providing protein for growing world population. *Journal of Animal Science* 73 (Suppl. 1), p. 243.

Skinner, J.D. and Rowson, L.E.A. (1968) Puberty in Suffolk and cross-bred rams. *Journal of Reproduction and Fertility* 16, 479–488.

Smith, J.F., di Menna, M.E. and McGowan, L.T. (1986) Effect of fusarium culture and zearalenone on the reproductive performance of ewes. *Proceedings of the New Zealand Society for Animal Production* 46, pp. 255–258.

Smith, J.F. and Murray, G.R. (1995) Use of bovine oocytes for the evaluation of the fertility of ram semen. *Journal of Reproduction and Fertility* Abstract Series No. 15, p. 70.

Smyth, P. and Gordon, I. (1967) Seasonal and breed variations in the semen characteristics of rams in Ireland. *Irish Veterinary Journal* 21, 222–233.

Strittmatter, K. and Peter, W. (1991) Maintaining sheep AI. *Tierzuchter* 43(9), 379.

Suss, R. and Strittmatter, K. (1991) Sheep breeding in the Netherlands – impressions from an educational visit. *Tierzucht* 45, 216–219.

Synnot, A., Fulkerson, W.J. and Lindsay, D.R. (1981) Sperm output by rams and distribution amongst ewes under conditions of continual mating. *Journal of Reproduction and Fertility* 61, 355–361.

Tervit, H.R. (1984) Frozen semen and surgical insemination. In: Artificial Insemination of Sheep. *Proceedings of a Workshop at Ruakura Animal Research Station, New Zealand.*

Thimonier, J. and Mauleon, P. (1969) Variations saisonnieres du comportement d'oestrus et des activites ovariennes et hypophysaires chez les ovins. *Annales de Biologie Animale Biochimie Biophysique* 9, 233–238.

Thwaites, C.J. (1994) The effects of feeding supplements containing different amounts and sources of nitrogen on live weight and the testes of rams during and after mating. *Animal Feed Science and Technology* 48(3/4), 177–184.

Thwaites, C.J. (1995a) The comparative effects of undernutrition, exercise and frequency of ejaculation on the size and tone of the testes and on semen quality in the ram. *Animal Reproduction Science* 37, 299–309.

Thwaites, C.J. (1995b) Effect of undernutrition on the size and tone of the ram's testes. *Small Ruminant Research* 16, 283–286.

Thwaites, G.J. and Hannan, G.D. (1989) The effects of frequency of ejaculation and undernutrition on the size and tone of the ram's testes. *Animal Reproduction Science* 19, 29–35.

Timon, V.M. (1993) Strategies for sustainable animal agriculture in developing countries. In: Mack, S. (ed.) *FAO Animal Production and Health* Paper 107, pp. 7–22.

Treacher, T.T., Bahhady, F., Hreitani, H. and Termanini, A. (1994) A comparison of the performance of Turkish and Syrian strains of Awassi ewes at two levels of nutrition. *Proceedings of the British Society of Animal Production* (Winter Meeting), Paper No. 202.

Turner, H.N. (1969) Genetic improvement of reproduction rate in sheep. *Animal Breeding Abstracts* 37, 545–555.

Udala, J. (1993) The relationships between body weight, testis size and blood LH, testosterone and prolactin concentrations in the first year of life in Polish Longwool rams kept in natural or artificial light. *Advances in Agricultural Sciences* 2(1), 3–10.

Udala, J. (1993) Effect of GnRH on semen quality of Polish Longwool rams exposed to different light conditions. *Advances in Agricultural Sciences* 2(1), 11–18.

Udala, J., Seremak, B., Kozak, M., Kryzysztofik, M. and Lojko-Szydlowska, D. (1991) Effect of season on the quality of frozen semen of Polish Longwool sheep. *Medycyna Weterynaryjna* 47, 318–321.

Upreti, G.C., Oliver, J.E., Duganzich, D.M., Munday, R. and Smith, J.F. (1995) Development of a chemically defined ram semen diluent (RSD-1). *Animal Reproduction Science* 37, 143–157.

Villegas, N., Casco, S., Hernandez, E. and Fernandez Abella, D. (1992) Effects of vasectomy on semen traits in rams. *Boletin Technico de Ciencias Biologigicas Universidad de la Republica (Salto)* 2(1), 75–78.

Visser, D. and Salamon, S. (1974) Fertility following inseminations with frozen-thawed reconcentrated and unconcentrated ram semen. *Australian Journal of Biological Science* 27, 423.

Wallace, J.M. (1992) Artificial insemination and embryo transfer. In: Speedy, A.W.J. (ed.) *Progress in Sheep and Goat Research* CAB International, Wallingford, pp. 1–24.

Wallace, J.M., Aitken, R.P. and Cheyne, M.A. (1994) Effect of post-ovulation nutritional status in ewes on early conceptus survival and growth in vivo and luteotrophic protein secretion in vitro. *Reproduction, Fertility and Development* 6, 253–259.

Wassmuth, R. (1990) Methods of improving the competitiveness of sheep breeding. *Zuchtungskunde* 62, 431–440.

Watson, P.F. (1979) The preservation of semen in mammals. In: *Oxford Reviews of Reproductive Biology* vol. 1. Oxford University Press, Oxford, pp. 283–350.

Watson, R.H. (1962) Seasonal variation in occurrence of oestrus in Merino ewes in Southern Victoria. *Australian Veterinary Journal* 28, 310–315.

Watson, R.H. and Radford, H.M. (1955) A note on the hours of daylight associated with the seasonal increase in sexual activity in Merino ewes. *Australian Veterinary Journal* 31, 31–36.

Webster, J.R., Suttie, J.M., Veenvliet, B.A., Manley, T.R. and Littlejohn, R.P. (1991) Effect of melatonin implants on secretion of luteinizing hormone in intact and castrated rams. *Journal of Reproduction and Fertility* 92, 21–31.

Welch, R.A.S., Kilgour, R.J., Robson, G.A., Smith, M.E. and Williams, E.T. (1979) The effect of shearing ewes during the mating period on the subsequent lambing pattern. *Proceedings of the New Zealand Society of Animal Production* 39, pp. 100–102.

Wilkins, J.F. and Kilgour, R.J. (1978) Early reproductive performance of female progeny of rams selected on serving capacity. *Proceedings of the 10th Annual Conference Australian Society of Reproductive Biology* 22.

Windsor, D.P. (1995) Factors influencing the success of transcervical insemination in Merino ewes. *Theriogenology* 43, 1009–1018.

Windsor, D.P., Szell, A.Z., Buschbeck, C., Edward, A.Y., Milton, J.T.B. and Buckrell, R.C. (1994) Transcervical artificial insemination of Australian Merino ewes with frozen-thawed semen. *Theriogenology* 42, 147–157.

Yeates, N.T.M. (1949) The breeding season of the sheep, with particular reference to its modification by artificial means using light. *Journal of Agricultural Science, Cambridge* 39, 1–43.

Yeates, N.T.M. (1956) The effect of light on the breeding season, gestation and birth weight of Merino sheep. *Australian Journal of Agricultural Research* 7, 440–445.

Yellon, S.M., Foster, D.L., Longo, L.D. and Suttie, J.M. (1992) Ontogeny of the pineal melatonin rhythm and implications for reproductive development in domestic ruminants. *Animal Reproduction Science* 30, 91–112.

Zhao, Y.L., Zhao, X.N., Zhao, X.H., Hu, S.L. and Ming, J. (1992) The use of cryptorchid rams for detecting oestrus. *Chinese Journal of Animal Science* 28(4), 31.

2 The Ewe's Oestrous Cycle and Seasonal Breeding Activity

2.1. Introduction

The ewe is seasonally polyoestrous with oestrous cycles usually commencing in late summer and continuing through until the start of spring, unless pregnancy intervenes. During the breeding season, the ewe shows cycles of 16–17 days; in general, most cycles range from 14 to 18 days with an average length of between 16.5 and 17.5 days (McKenzie and Terrill, 1937; Asdell, 1964; Hafez, 1952). Heape (1990) described the phases of the oestrous cycle as pro-oestrum, oestrus, met-oestrum and di-oestrum. More commonly, the oestrous cycle is divided into two phases, the follicular phase and the luteal phase. The follicular phase has a duration of 2–3 days and is characterized by the exhibition of oestrous behaviour, the preovulatory luteinizing hormone (LH) surge and ovulation. The transition from the follicular to the luteal phase is marked by ovulation. However, the transition from luteal to follicular is more complex. Should the sheep become pregnant, the corpus luteum will persist throughout pregnancy.

Prostaglandin and progesterone levels

If the animal is non-pregnant, prostaglandin (PG) $F_{2\alpha}$, which is produced by the uterus, is secreted; the subsequent action of PG is to induce regression of the corpus luteum, reducing progesterone concentrations in the circulation and allowing a new follicular phase to start (Fig. 2.1). The luteal phase has a duration of 14–15 days and is characterized by the secretion of progesterone from the corpus luteum. The oestrous cycle length seems to be more variable in the second half of the breeding season as a result of the luteal phase of the cycle increasing its duration (Hammond, 1944).

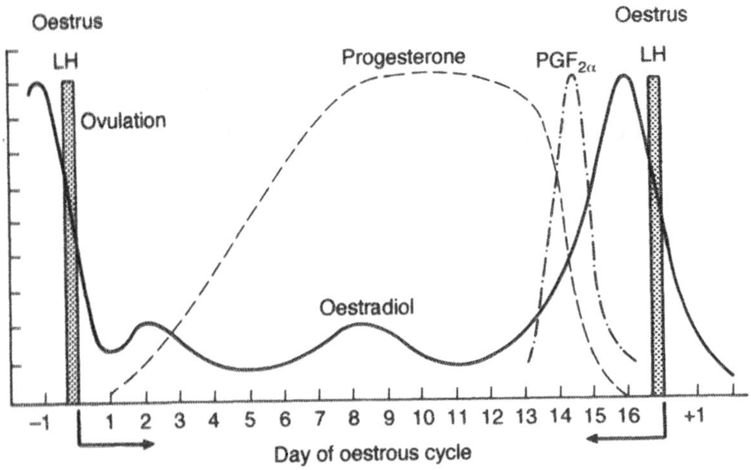

Fig. 2.1. Diagrammatic representation of changes in hormone levels during the ewe's oestrous cycle (from Caldwell *et al.*, 1972).

2.2. Oestrus and the Oestrous Cycle

The duration of oestrus in the ewe, looking at the reports in the literature, generally fall within the range of 1–1.5 days (Asdell, 1964); detailed observation in several British breeds in a 3-year study by Hafez (1952) showed the heat period to be, on average, about 35h in length. There have been those who have observed that long heat periods are more intense than short ones and that the first oestrus of the breeding season may be shorter and less intense than subsequent periods (Grant, 1934); the duration of oestrus is generally shortest in ewe-lambs and of intermediate duration in yearling sheep (McKenzie and Terrill, 1937; Hafez, 1952).

2.2.1. Oestrus and ovulation

The ewe is a spontaneous ovulator, although estimates of the time at which ovulation occurs, relative to the onset of oestrus, have varied. Reviewing the literature at the time, Robinson (1959) concluded that ovulation occurs at about the end of oestrus, regardless of the duration of the sexually receptive period. The phenomenon of oestrus in the ewe is generally recognized as being rather more complex than was at one time thought. It is now evident that the continuous presence of rams can reduce the duration of behavioural oestrus (Parsons and Hunter, 1967; Parsons *et al.*, 1967) and advance the time of ovulation, relative to the onset of oestrus (Lindsay *et al.*, 1975; Signoret, 1975) by advancing the preovulatory surge of LH. In the absence of the male, presumably the preovulatory surge of LH awaits the build-up of follicular oestradiol to an appropriate concentration before it is triggered off.

LH surge to ovulation interval

Duration of the heat period can be markedly influenced by breed. It should also be noted that the interval between the start of oestrus and the time of preovulatory LH discharge varies between and within breeds (Thimonier and Pelletier, 1971). The interval is reported by Land *et al.* (1973) to be greater in highly prolific sheep (18h) than in less prolific breeds (6–7h). The maximum concentration of preovulatory LH varies from animal to animal but duration of the surge is 8–12h (Cunningham *et al.*, 1975; Legan and Karsh, 1979); ovulation itself occurs about 24h after the onset of the LH surge (Cumming *et al.*, 1971).

Behavioural symptoms of oestrus

The behavioural symptoms displayed by the ewe during oestrus have been described in several reports (McKenzie and Terrill, 1937; Hafez, 1952); in effect, signs are very few and consist of the ewe remaining close to the ram and standing to be mounted (Fig. 2.2). It should be noted that ewes display a strong ram-seeking behavioural pattern (Inkster, 1957; Lindsay and Fletcher, 1972) so that contact between the sexes is not wholly dependent on the ram's activity; in fact, studies have demonstrated that about 75% of ewes in oestrus seek out and remain near rams (Matthews *et al.*, 1991). The absence of any clear symptoms make detection of oestrus, other than in the presence of a ram, extremely difficult. Some reports mention that the vulva may be oedematous and that a mucus discharge from the vagina may be evident (Kelley, 1937); occasionally, the oestrous ewe will move its tail vigorously as part of a display pattern when it is with the ram.

Differences in the sexual 'attractiveness' of oestrous ewes to rams has been studied by Australian workers (Tilbrook, 1987a,b; Tilbrook and Lindsay, 1987), who found that the components of ewe attractiveness are characteristic of the ewe and are relatively stable over at least two heat periods; the stage of oestrus had no significant influence and the soliciting behaviour of the ewe was not a major factor determining her sexual attractiveness.

Ewe response to the ram

According to studies reported by Blissitt *et al.* (1990) from Edinburgh, rams are able to differentiate between sexually receptive and non-receptive ewes by the difference in urine odour. It appears that this odour subsequently wanes to become undetectable 4 days after oestrus. Ewes that are not in oestrus usually urinate in response to the approach of a ram, whereas oestrous ewes do not (Bland and Jubilan, 1987). According to Blissitt *et al.* (1994), urination by ewes may be a non-contact communication that allows rams to determine efficiently the oestrous status of the ewe under field conditions; ewes avoid disturbance by signalling that they are not sexually receptive. The ram concentrates attention on the voided urine and exhibits the 'flehmen' response,

Fig. 2.2. Oestrus symptoms are not easy to detect in the ewe. In the absence of rams or sterile teasers, it is usually impossible to distinguish a ewe that is in oestrus from one that is not – simply using powers of observation.

which is believed to facilitate the entry of urine into the vomeronasal organ for chemosensory analysis.

Ram sexual behaviour

Various authors have reviewed the sequence of events leading up to mounting in the ram (Pelletier *et al.*, 1977). Some of these movements are illustrated in Fig. 2.3. In more recent times in Japan, Odagiri *et al.* (1995) attempted to analyse sexual behaviour in rams using a time-lapse video recorder. Sexual behaviour fell into one or other of eight movement categories; these were: (i) following or approaching the ewe; (ii) chin-resting; (iii) flehmen; (iv) mounting; (v) nosing; (vi) nudging; (vii) pushing; and (viii) twisting. The results of the study showed that the most common sequence of movements was: following/approaching – twisting – nudging – chin-resting – mounting.

It may be noted that in several mammalian species, including sheep, sexual interactions have been reported to increase LH and testosterone levels in the male (Schanbacher *et al.*, 1987). Studies on the effect of sexual receptivity of ewes and the sexual experience of rams has shown that rams react to the presence of ewes by changes in hormone levels regardless of their experience or the sexual state of the ewe (Gonzalez *et al.*, 1991). However, experienced rams respond more strongly to receptive than to non-receptive ewes, suggesting that they have previously learnt to determine the physiological and/or behavioural state of the ewe's receptivity.

Fig. 2.3. Some of the movements associated with ram sexual behaviour (from Pelletier *et al.*, 1977).

Follicular oestrogen and sexual activity

Oestrus is the behavioural response of the ewe to the action of ovarian (follicular) oestrogen on specific receptors in the hypothalamus; the plasma concentration of oestradiol reaches its peak at about the start of oestrus (Scaramuzzi and Land, 1978), after which it rapidly disappears. This means that the ewe spends much of oestrus in the absence of ovarian oestrogen and progesterone in the circulation. Progesterone concentration in the circulation is lowest during oestrus but begins to rise immediately after formation of the corpus luteum and reaches a maximum later in the cycle.

Oestrus in pregnancy

Oestrus may occur in the pregnant cow, although this would certainly be uncommon. Although no evidence has been recorded in Irish studies of oestrus in the pregnant ewe (Gordon, 1983), the phenomenon has been recorded in several reports (Williams *et al.*, 1956; Bichard *et al.*, 1974; Younis and Afifi, 1979); some authors have gone so far as suggesting that oestrus in the pregnant ewe occurs so frequently as to make the practice of oestrus detection in identifying non-pregnant ewes questionable. This would seem to be very much against the evidence of the literature as a whole. Certainly, Irish experience lends little credence to the suggestion.

2.2.2. Physiology and endocrinology of the oestrous cycle

The nature of the hormonal mechanisms involved in the regulation of the ewe's oestrous cycle has been the subject of many investigations, particularly over the past 25 years when extremely sensitive radioimmunoassays (RIAs) have been available.

Luteal phase

After ovulation, the ruptured follicle is transformed into a corpus luteum (CL). According to Jablonka-Shariff *et al.* (1993), growth of sheep corpora lutea is extremely rapid, linear from day 2 to day 12. Such growth is primarily due to hyperplasia; there is a high rate of cell proliferation, but this is associated primarily with non-steroidogenic cells, a large proportion of which appear to be endothelial cells. It is believed that there are at least two morphologically and biochemically distinct steroidogenic luteal cell types in the ewe (Fitz *et al.*, 1982; Rodgers and O'Shea, 1982). The two cell types differ markedly in size and are referred to as small and large luteal cells. The small luteal cells are spindle-shaped and are 12–22 μm in diameter; the large cells are 22–50 μm in diameter and are spherical in shape.

Corpora lutea and progesterone levels

The sheep's CL attains full secretory activity by about the sixth to eighth day of the oestrous cycle and continues secreting progesterone at a fairly constant level until about day 15; progesterone concentrations in the circulation during the cycle follow the growth and development of the corpus luteum, the maximum levels being reached at about day 8 and then beginning to fall a day or two before the next oestrus (Cunningham *et al.*, 1975). It is known that season and nutrition, as well as breed and ovulation rate, can have an influence on the maximum concentration of steroid secreted. There is only a marginal increase in the progesterone level with two corpora lutea rather than one, so accurate detection of double ovulations in sheep is not possible by measuring the steroid.

Maintenance of the corpus luteum

Data in the 1960s suggested that there was some decline in progesterone levels between the 10th and 15th days of the cycle (Thorburn *et al.*, 1969); electron microscopy studies at that time also provided some evidence of incipient regression of the corpus luteum commencing as early as day 12/13 (Deane *et al.*, 1966). Such data were in agreement with earlier observations on the morphology of the corpus luteum in the late luteal phase of the oestrous cycle

(Warbritton, 1934; Hutchinson and Robertson, 1966). The secretory activity of the CL is believed to be maintained by luteotrophic hormones released by the anterior pituitary. There is evidence that both LH and prolactin (PRL) contribute to the maintenance of the functional activity of the sheep's CL during the oestrous cycle (Denamur *et al.*, 1973). It is also recognized that the normal growth and development of the ovine corpus luteum can be markedly affected by the administration of exogenous progesterone in the early days of the oestrous cycle (Ginther, 1968; Thwaites, 1971); this can result in a much reduced span of activity. Presumably, progesterone treatment at this time adversely affects the hormonal mechanisms responsible for the normal establishment and maintenance of luteal tissue.

The progesterone secreted by the CL has many functions both before and after ovulation and fertilization. It has been ably demonstrated by Australian workers that the behavioural and vaginal responses of the ewe to oestradiol are dependent upon a progesterone priming regime (Moore, 1988); pretreatment with progesterone is known to prevent a refractory state of responsiveness and maintains the ewe in a state in which it is highly sensitive to minute quantities of oestradiol.

Follicular phase

The follicular phase of the cycle is characterized by the growth and development of ovarian follicles and by the decline in progesterone concentrations, associated with the regression of the corpus luteum. There is unanimity among reports showing that there is a precipitous decline in the circulatory levels of progesterone at about day 15; this decline is associated with the abrupt termination of the functional activity of the corpus luteum that occurs at the end of the luteal phase of the cycle.

It is believed that $PGF_{2\alpha}$ is the luteolytic agent in the ewe, and that its pulsatile secretion during early luteolysis is dependent on the binding of ovarian oxytocin to endometrial receptors (Flint and Sheldrick, 1983). $PGF_{2\alpha}$ secretion starts as a series of small pulses around day 12/13 of the cycle and the frequency of these pulses increases to reach a peak by the 14th day of the cycle. It is believed that the sheep corpus luteum contains both high and low affinity $PGF_{2\alpha}$ receptors; activation of the high affinity receptors selectively releases oxytocin without any effect on progesterone secretion while activation of the low affinity receptors increases luteal oxytocin secretion and decreases luteal progesterone secretion. It is believed that the initial PG release occurs in response to elevated progesterone levels (for 7–10 days) and the subsequent release is associated with the decline in progesterone and rise in oestradiol concentrations.

Oxytocin receptor concentrations

The increase in the level of uterine oxytocin receptors that occurs at the end of the luteal phase is of primary importance in determining whether the CL regresses. Oestradiol and progesterone are known to have a controlling influence on the oxytocin receptor concentration and the uterine $PGF_{2\alpha}$

response to oxytocin (Vallett *et al.*, 1990). During a normal oestrous cycle, following a period of progesterone inhibition, oestradiol stimulates the luteolytic mechanism by ensuring that the uterus responds to oxytocin with increased PG secretion. A positive feedback loop is thought to exist between luteal oxytocin and uterine $PGF_{2\alpha}$ with each hormone stimulating the secretion of the other. Studies by Mann and Lamming (1995) indicate that although luteal oxytocin does not provide the trigger for PG episodes, it does appear to be involved in regulating the pattern of PG release.

Role of oestrogens in luteolysis

The effect of exogenous and endogenous oestrogen on the luteolytic process in the ewe has been demonstrated in various ways. Treatment of ewes with high doses of oestradiol during the mid-luteal phase induced premature luteal regression (Hixon and Flint, 1987). It has also been evident for some time, that exogenous oestradiol given in appropriate doses will induce luteolysis when given near the end of the cycle; the secretion of endogenous oestradiol starts about 48h before the onset of oestrus. This agrees with evidence showing that the destruction of ovarian follicles by X-irradiation prevented any increase in oestrogen secretion and prevented luteal regression (Hansel, 1975). Results reported by Beard and Lamming (1994) for ovariectomized ewes treated with progesterone and oestradiol have indicated that a high concentration of oestradiol may be associated with a stronger and earlier luteolytic response. This may have implications for the mated ewe in terms of the efficiency of mechanisms for the maternal recognition of pregnancy.

First corpora lutea

It is now firmly established that the first corpora lutea of the sheep's breeding season may regress prematurely. Studies at Nottingham has shown that progesterone pretreatment of the ovariectomized ewe alters the subsequent steroid hormone control of oxytocin receptor concentrations and have identified delayed decline in oxytocin receptor concentration as the potential cause of premature luteolysis in ewes not pretreated with progesterone (Beard and Hunter, 1994b). Work elsewhere by Morgan *et al.* (1993) has shown that adequate progesterone exposure during the early to mid-luteal phase of the sheep's oestrous cycle is essential for initiation of ovarian-uterine mechanisms that lead to luteolysis in ewes.

Apoptosis and luteal regression

Whatever the mechanisms involved in corpus luteum regression at the end of the cycle, it is known that the plasma concentration of progesterone sinks rapidly to a negligible value and that this remains true during oestrus and until the fresh corpus luteum forms after ovulation; the extremely low basal concentration of progesterone is believed to be of adrenal origin (Robertson, 1977). As observed by Auletta *et al.* (1994), the regressing corpus luteum of the ewe displays an orderly series of morphological events bearing many characteristics of apoptosis; these workers showed that apoptotic cells are clearly identifiable in the sheep CL and such evidence supports the view that apoptotic events are an integral component of the luteolytic process.

2.2.3. Folliculogenesis and follicular dynamics

The ovaries of the ewe have two important functions, namely, the cyclic production of fertilizable oocytes and the production of steroid and peptide hormones that maintain the reproductive tract.

Folliculogenesis

Follicle development in the ewe is a dynamic process and commences with the growth of some of the follicles present in the primordial follicle reserve. These are inactive primordial follicles that are formed during prenatal life. The factors responsible for the stimulation of follicle growth are still not fully understood (Cahill, 1984). However, it is clear that the follicle reserve continually supplies follicles throughout life or until the reserve has become exhausted. The development from inactive to active primordial follicle occurs at regular intervals and the follicles either mature and ovulate or undergo atresia at some point along the way.

Some reports have estimated that follicular development, from resting stage to preovulatory status, requires about six months, with 2–3 follicles leaving the pool of resting follicles to begin growth each day (Cahill and Mauleon, 1980). The ovary of the ewe contains several hundred growing follicles, of which 10–40 are visible on its surface (McNatty *et al.*, 1982). An average of 44 visible vesicular follicles at different stages was recorded by Kaulfuss *et al.* (1994) during the oestrous cycle; in that work, the number of follicles did not differ significantly between the left and right ovaries. The control mechanisms that allow only one or two of the developing follicles to complete growth and ovulate at the end of each oestrous cycle have yet to be elucidated (Fig. 2.4). Studies in the USA by Murdoch (1995) indicate that there is a progressive increase in the number of apoptotic cells in ovarian tissues as ovulation approaches, which suggests that apoptosis may be a potential determinant of ovulation in the ewe.

Role of IGF-I and IGF-binding proteins

Changes in insulin-like growth factor-I (IGF-I) and IGF-binding proteins during follicular development in the ewe have been reported by various authors. Changes in the profile of IGF-binding proteins have suggested that these proteins may be an important regulator of IGF-I action on cell proliferation and steroidogenesis within the ovary (Khalid and Haresign, 1995).

Role of EGF and FGF

In Australia, Scaramuzzi and Downing (1995) have reported evidence of paracrine and autocrine effects of epidermal growth factor (EGF) and fibroblast growth factor (FGF) on follicle and luteal function in the ewe.

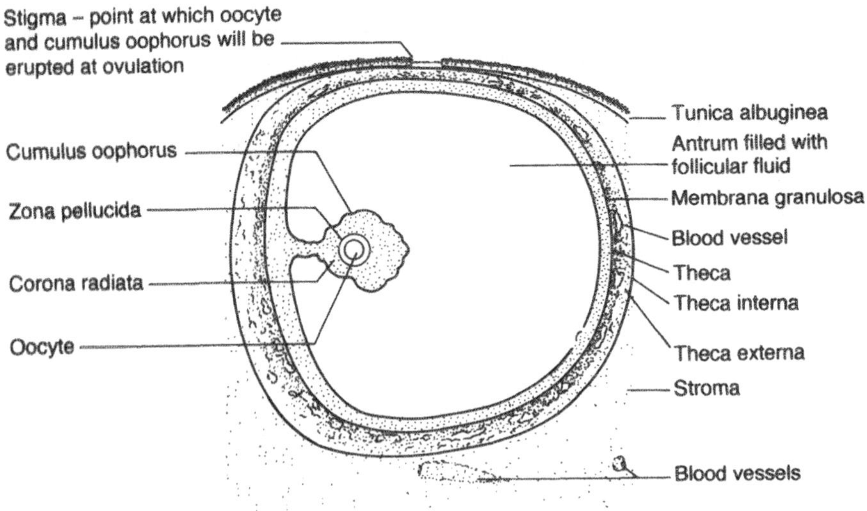

Fig. 2.4. Diagrammatic representation of an ovine preovulatory follicle (adapted from Freeman and Bracegirdle, 1968).

Follicular dynamics

It is now well accepted in cattle that growth of large vesicular follicles occurs in two or three distinct waves during the oestrous cycle; each wave is characterized by the simultaneous emergence of a cohort of follicles and the establishment of a dominant follicle (DF) that continues to develop, while apparently suppressing the growth of other follicles in the cohort. Studies in sheep in the 1960s and early 1970s led to the conclusion that vesicular follicles grow early in the cycle and then remain in the ovary to enlarge prior to ovulation, or that there may be several phases of vesicular follicle growth during the cycle (Smeaton and Robertson, 1971).

Studies reported by Driancourt *et al.* (1990) confirmed that follicular waves also occur in sheep by analysing the ovulatory response of ewes to human chorionic gonadotrophin (hCG); whatever the physiological stage (prepubertal, anoestrous, luteal phase), the proportion of ewes with LH-sensitive follicles was high (> 80%), suggesting that waves of follicular growth occur continuously in the ewe. As observed by Driancourt (1991), this is in contrast to what is known to occur in pigs and primates, where growth of ovulatory follicles is apparently limited to the follicular phase of the cycle. The implication is that mechanisms controlling follicular growth are different in sheep and other farm ruminants (cattle and goats) from those operating in pigs. In more recent times, Noel *et al.* (1993), by way of daily laparoscopic examinations, studied ovarian follicular dynamics in Suffolk ewes in Belgium at different times of the breeding season and during anoestrus; they recorded three waves of follicle growth and atresia during the oestrous cycle (two waves during the luteal phase; one wave during the follicular phase) and observed a similar turnover of follicles during anoestrus.

Follicular growth waves

Other studies in Belgium also recorded three waves of follicular growth and atresia during the cycle, this time in work that described the effects of an exogenous progestagen (fluorogestone acetate; FGA) and gonadotrophin (PMSG) on follicular dynamics in sheep (Noel *et al.*, 1994). In Canada, Leyva *et al.* (1995) used transrectal ultrasound to monitor ovarian activity and ovulation in cyclic ewes and reported that most ewes (80%) showed three follicular waves in a normal 17-day cycle; the last wave resulted in ovulation.

In data reported by Ravindra *et al.* (1994), who employed ultrasonic scanning of the ovaries, significantly more vesicular follicles were observed emerging from a pool (> 2mm diameter) on days 2 and 11 than at other times of the cycle; the authors suggested that the two phases of follicle emergence may be similar to the waves of growth found in cattle. However, it did appear that follicular dominance was a weak phenomenon in sheep compared with the cow; support for this view also came from previous observations made by Schrick *et al.* (1993). In contrast to cattle, some authors have found no obvious connection between follicle emergence and FSH secretion in the ewe (Ravindra *et al.*, 1994); on the other hand, others, such as Ginther *et al.* (1995) have reported evidence supporting a cause-and-effect relationship between transient increases in FSH and follicle stimulation.

Emergence of dominant follicles

Studies in Edinburgh by Souza *et al.* (1995) led them to conclude that the size of vesicular follicles alone was not an adequate marker of follicle health and that the secretory status of the follicle must be taken into account in assessing the dynamics of follicular growth in sheep. These workers found clear evidence of a wave-like pattern of oestradiol secretion over a period of 4–5 days. It is clear that the development of large vesicular follicles in the ewe is continuous throughout its adult reproductive life; this includes the ewe anoestrus as well as the normal breeding season. However, the fact that one of the key determinants of follicular health is ability to secrete oestradiol, means that ovary scanning alone may not necessarily reveal the full story of follicle dominance. In the USA, Ginther *et al.* (1995) have recorded that FSH fluctuations (an increase followed by a decrease) and the emergence of follicular waves both occur at approximately 4-day intervals in sheep.

In France, Driancourt (1994) attempted to identify the mechanisms used by follicles that are destined to establish dominance over other follicles in the ovary. This worker concluded from his study that no between-follicle interaction was detectable and no negative role of the dominant follicle in differentiation of other follicles (measured by aromatase activity) could be found.

Follicular activity during anoestrus

In Canada, Bartlewski *et al.* (1995) used ultrasonography to study follicular dynamics in ewes during the mid-anoestrus period; they found that ovarian follicles continued to develop, reaching diameters similar to those observed during normal oestrous cycles in the ewe. Results reported by Rubianes *et al.* (1995) indirectly support the view that waves of follicle development occur in ewes during the sheep anoestrus by showing that, as in cattle, the presence of a large follicle (DF) at the time of gonadotrophin administration affects the superovulatory response.

2.2.4. Action of gonadotrophic hormones

It is the joint action of FSH and LH on the ovaries of the ewe that determines the number of follicles that begin preovulatory development and which follicles ovulate at the end of the oestrous cycle. Much more information has accumulated on the actions of LH than FSH over the past 25 years.

LH levels during the cycle
The concentration of LH in the blood of the ewe is determined by the operation of two distinct systems; preovulatory LH in the female is controlled by one group of neurons whereas a basal (tonic) secretion of LH is regulated in both sexes by neurons in another part of the hypothalamus. In the sheep, the tonic level of LH is apparently maintained in the face of luteal activity, whereas the preovulatory surge is blocked by progesterone but released in response to oestrogen at the time of oestrus (Goding *et al.*, 1970). It has been suggested by Whisnant and Goodman (1994) that the inhibitory influence of progesterone on LH secretion is mediated largely via opioid neurons within the medial basal hypothalamus.

As progesterone concentration decreases at the end of the cycle, tonic LH levels rise to reach values at least fivefold greater than the base-line by the time of the onset of the preovulatory LH surge. The rise in LH reflects an increase in the frequency of pulsatile LH discharges and constitutes an increase in tonic LH secretion separate from the LH surge. The sustained increase in tonic LH, covering a period of about 48h, is accompanied by a fivefold increase in oestradiol secreted by the rapidly growing preovulatory follicles in the ovaries. It is this sudden increase in oestradiol that triggers the massive preovulatory surge of LH and which, some hours earlier, resulted in the onset of oestrus (see Fig. 2.5).

Oestradiol and the GnRH surge
Results of studies by Moenter *et al.* (1990) have clearly demonstrated that regardless of season, a rise in oestradiol concentration to late follicular phase levels acts centrally upon the GnRH neurosecretory system to initiate a large and abrupt release of GnRH coincident with the onset of the preovulatory LH surge; these workers note that this increase in secretion is so pronounced that it may be termed a GnRH surge. This surge is apparently in excess of what is required to produce the LH surge (Bowen *et al.*, 1995).

The preovulatory LH surge, which occurs in the early hours of the heat period, triggers events in those follicles destined for ovulation at that oestrus

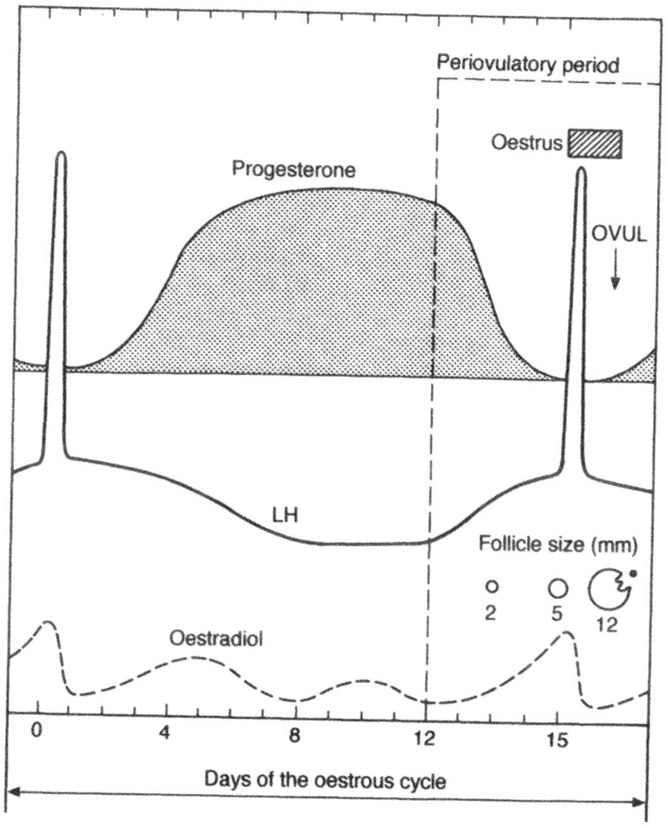

Fig. 2.5. Diagrammatic representation of changes in hormone levels during the ewe's oestrous cycle (after Hauger *et al.*, 1977).

(see Table 2.1); among these events is the nuclear and cytoplasmic maturation of the oocyte and rupture of the follicle. Such changes result in the conversion of the follicle from a predominantly oestrogen-secreting to a progesterone-secreting structure and the reprogramming of the oocyte to support both fertilization and the early stages of embryonic development.

LH surge and ovulation

Ovulation in the ewe occurs about 24h after LH release (Cumming *et al.*, 1971), which would agree with observations that this event generally coincides with the end of heat (Parsons *et al.*, 1967; Holst and Braden, 1972; Cumming *et al.*, 1973). This timing also agrees with data on the occurrence of ovulation after administering hCG in early oestrus (Dziuk, 1965) and with the times recorded for the completion of nuclear maturation in the sheep oocyte cultured *in vitro* (Crosby and Gordon, 1971).

Table 2.1 Compartmental changes in preovulatory follicles induced by the LH surge (from Osborn and Moor, 1985).

Somatic compartment	Germinal compartment
Increase in follicular fluid volume	Resumption of meiosis
Temporary increase and then inhibition of androgen and oestrogen secretion	Changes in:
Stimulation of progesterone secretion	membrane transport
Reduction in LH receptors on theca and granulosa cells	protein phosphorylation
Increased synthesis of prostaglandins PGF_2 and PGE_2	pattern of protein synthesis
Disaggregation of granulosa and cumulus cells	carbohydrate metabolism
	Migration of cortical granules
	Redistribution of mitochondria
	Acquisition of developmental competence

Control of tonic LH levels
It appears that the oestrous cycle of the ewe is characterized by genuine day-to-day variations in the concentration of LH; as indicated in Fig. 2.5, levels are relatively high during the first few days after the preovulatory surge, decrease to a low point during the mid-luteal phase and then rise progressively to a high concentration during the last day or two of the cycle (Hauger *et al.*, 1977).

Oestrogen effects
According to evidence provided by Goodman *et al.* (1980), it is believed that oestradiol contributes to the determination of LH levels in two ways; by acting alone in the late cycle (follicular phase) in partially suppressing tonic LH and in the luteal phase by acting in concert with progesterone in controlling LH. These two effects of oestradiol, together with the known inhibitory effects of progesterone on LH levels, would seem to account for the way in which tonic LH is secreted during the sheep's oestrous cycle (Karsh *et al.*, 1984); it is believed that ovarian secretions other than oestradiol and progesterone play little, if any, physiologically important role in controlling tonic LH in the ewe.

There is evidence to support the view that during the pre-LH-surge period in the ewe, oestradiol, in a dose-dependent manner, induces a qualitative change in the pattern of GnRH release in addition to stimulating GnRH pulse frequency and reducing pulse amplitude (Evans *et al.*, 1994, 1995). As observed by Karsh (1995), the most striking qualitative alteration is evident during the preovulatory LH surge, when GnRH output increases massively but not in a manner that is overtly episodic. It appears that around the onset of the LH surge, significant secretion of GnRH begins to emerge between episodic discharges. As the surge develops further, interpulse GnRH is greatly augmented and sustained, with values in the pituitary portal blood remaining continuously elevated for some hours after completion of the preovulatory LH discharge. Although oestradiol is known to have powerful feedback actions at the level of the hypothalamus, GnRH neurons themselves do not contain oestradiol receptors. The regulatory actions of oestradiol on GnRH secretion appear to be indirect, being relayed to GnRH neurons via oestrogen sensitive interneurons.

FSH levels during the cycle

As to the role of FSH during the cycle, it remains ill-defined at this time; the factors controlling release of this hormone have yet to be fully understood. It is known that the synthesis and secretion of FSH is stimulated by GnRH and inhibited by oestradiol and inhibin from the ovary. FSH assay studies have provided evidence of two different control mechanisms for FSH and LH operating during the oestrous cycle; one control centre, presumably acting via hypothalamic releasing hormone, appears to regulate large sudden changes that affect LH secretion more than FSH; the second control, presumably acting by way of ovarian oestrogen and inhibin, regulates the gradual changes that occur in FSH and LH levels in a similar way.

Secretion of FSH by the pituitary is under the dual inhibitory control of oestradiol and inhibin, both of these hormones being synthesized by the granulosa cells of large dominant follicles. The secretion of oestradiol is under the acute stimulatory control of LH but it is evident from work in cyclic ewes (McNeilly *et al.*, 1989) and sheep in anoestrus (Campbell *et al.*, 1989) that this is not the mechanism controlling inhibin secretion.

Action of inhibin

Results of work by Mann *et al.* (1989) led them to conclude that, in the ewe, large vesicular follicles are responsible for most, if not all, the secretion of inhibin by the ovary at all stages of the oestrous cycle, and that the corpus luteum secretes little or no inhibin; they also concluded that inhibin was probably not responsible for the decline in FSH observed during the follicular phase of the ewe's cycle, oestradiol being more important in this aspect of FSH regulation. The relative importance of inhibin and oestradiol in controlling FSH secretion during the follicular phase of the sheep's oestrous cycle was examined by Campbell *et al.* (1990a,b); they also concluded that oestradiol, and not inhibin, is the major component of the inhibitory feedback loop controlling the pattern of FSH secretion at this stage of the cycle.

A further report by Campbell *et al.* (1991) led them to conclude that inhibin is secreted by large oestrogenic follicles and also in appreciable quantities by large non-oestrogenic and small follicles; they suggest that this range of follicular inhibin sources may explain the lack of variation in inhibin secretion at different stages of the oestrous cycle. There is strong evidence of both autocrine and paracrine actions of inhibin in enhancing gonadotrophin stimulated steroid secretion by granulosa and thecal cells (Campbell and Webb, 1995).

Negative feedback actions of oestradiol and inhibin

Baird *et al.* (1991) have suggested that inhibin with its long half-life sets the overall level of negative feedback, while oestradiol controls the diurnal fluctuations in the concentration of FSH, which determines the number of ovulatory follicles. The number of large vesicular follicles in the ovaries is controlled by the negative feedback of inhibin on FSH; the number of follicles selected for ovulation is similarly controlled by oestradiol.

FSH isoforms

It is now clear that FSH, like many other protein hormones, consists of a family of related isoforms characterized by differences in metabolic clearance rates and biological activities. Studies reported by Padmanabhan (1995) have shown major shifts in the distribution pattern of circulating FSH isoforms and an associated increase in FSH bioactivity during experimentally-induced puberty in sheep; studies characterizing FSH secretory patterns at the pituitary site have shown that FSH is secreted in discrete pulses. In sheep, the changes in plasma concentrations of FSH throughout the oestrous cycle relate to changes in follicle activity, reflecting the output of inhibin and oestradiol.

FSH and LH release patterns

In New Zealand, Phillips *et al.* (1994) determined bioactive FSH concentrations during the sheep's cycle and showed these to be elevated during and after the preovulatory LH surge and to be significantly lower during the late-luteal to mid-follicular phases of the cycle compared with the mid-luteal phase. It is evident that there are fundamental differences in release patterns between FSH and LH. During the oestrous cycle of the ewe, up to 50% of the pituitary content of FSH may be released each day, whereas only 1–5% of LH is released, in the form of pulses (McNeilly *et al.*, 1995). There is much to suggest that factors other than GnRH play an important role in controlling the secretion of FSH. LH is known to be stored in secretory granules which are released in response to pulses of GnRH; FSH, in contrast, is released episodically during the luteal phase, but such release is only partially dependent on GnRH stimulus (Van Cleef *et al.*, 1995); other factors, which may be pituitary or hypothalamic in origin, control FSH release, at a rate which is closely related to the rate of synthesis. Much remains to be learnt about the nature of the mechanisms controlling the different pathways of gonadotrophin release.

2.2.5. Oestrous activity in tropical sheep breeds

Various authors have reported on oestrous activity in tropical breeds of sheep. In Egypt, Aboul-Naga *et al.* (1987) have demonstrated that the subtropical Rahmani breed of ewe does not have a distinct period of acyclicity as in the seasonal temperate breeds; about 33% of the ewes showed continuous ovarian activity throughout the year, whilst almost all ewes had periods of behavioural anoestrus. The average duration of the oestrous cycle in the Rahmani sheep was 18 days and about 93% of animals showed cycles of normal length (16.5–19.5 days). The characteristics of oestrus and ovulation in Awassi sheep in Syria have been described by Kassem *et al.* (1990); they recorded average cycle lengths of 16.4 and 17.4 days for ewe-lambs and ewes, respectively. In Ethiopia, Mukasa-Mugerwa and Lahlou-Kassi (1995) have reported on the reproductive performance of Menz sheep; they note that 65% of the ewes lambed three times in two years in a study of data covering a 7-year period.

2.2.6. Stress and oestrous activity

The disruption of oestrous behaviour in ewes by management-related stress was examined by Ehnert and Moberg (1991) in California. Isolation of a ewe from other flock members induced significant increases in corticosteroids and blocked expression of oestrus in 33% of animals; 8h of transportation similarly increased corticosteroid concentrations and blocked oestrus in most animals. Such results indicated that management-related stress can block oestrous behaviour. The effect of transport on LH surge release in ewes during the follicular phase of the cycle was examined in a study reported by Phogat *et al.* (1995) in the UK; it was found that transport reduced the self-priming effect of GnRH on the pituitary and delayed the onset of the preovulatory LH surge.

2.3. The Breeding Season of the Ewe and Ram

Knowledge of the important bearing that seasonal changes in daylength has on the behaviour and function of farm livestock has been steadily accumulating during the past 50 years. In the 1920s, attention was first drawn to the phenomenon of photoperiodism in plants, showing how their reproductive phase may be related to length of day. In this same period, it was shown that the annual rhythm of breeding in birds was not due to seasonal changes in temperature, as had been commonly accepted, but to changes in light. In the 1930s, investigations were extended to mammals; in the farming world, attention was focused by the observation of Marshall (1937) that when British sheep were transported across the equator from one hemisphere to another, they finally altered by six months the time of the calendar year at which they breed, so conforming with the seasons in their new environment. This was strong evidence that the seasonal nature of breeding in the ewe must be the result of some environmental factor.

Later work, much of it in Cambridge, showed the factor to be daylength. Ewes start to show breeding activity when days are shortening and are sometimes termed 'short-day breeders' in contrast to farm species such as horses and chickens, which would be termed 'long-day breeders'. Other work at Cambridge clearly demonstrated that the length of the breeding season differs among breeds (Hafez, 1952). In general, mountain and hill breeds (e.g.

Blackface) have short seasons, whereas lowland breeds (e.g. Suffolks) have longer seasons.

Practical implications of sheep anoestrus

For the sheep farmer in Ireland and the UK, as well as in many other countries, a major restriction to intensifying lamb production or otherwise improving flock profitability may be imposed by the seasonal anoestrus in the ewe. For that reason, in the various attempts made to increase the productivity of sheep, studies have been conducted to determine how environmental and other factors modify ovarian activity as assessed by different criteria. The general view is that daylength is the predominant environmental factor regulating breeding activity. There have been experiments reported by several workers showing that decreasing daylength promotes the onset of the sexual season whereas increasing daylength is associated with the start of anoestrus. The hormonal events which are responsible for the start and cessation of breeding activity in the ewe have still to be fully elucidated, but much information has accumulated in the past 25 years, much of it due to the activities of Michigan workers and their associates in other institutions (Karsh, 1984).

In Ireland, with the normal range of breeds used in commercial sheep farming, the breeding season has a duration of some six months, spanning the period from September to the following March (Gordon, 1983). The general rule is that breeding seasons grow shorter if one goes north from Ireland and grow longer as one approaches the equator. In Iceland, for example, Eldon (1994) records the average onset of the breeding season in Icelandic sheep as being 15 November and its cessation on 20 February, a period of some three months.

Factors regulating the season

Dealing with factors regulating the onset of the breeding season in sheep, Malpaux *et al.* (1989) put forward the following three conclusions, drawn from their studies with Suffolk ewes in Michigan:

1. the lengthening photoperiod between the winter and summer solstices is required for the occurrence of the breeding season in the autumn;
2. the time of initial exposure to this lengthening photoperiod provides an important cue for determining when the reproductive period occurs;
3. the time of onset of the breeding season does not depend upon the decreasing photoperiod after the summer solstice, nor does it require the photoperiod to stop increasing as the summer solstice approaches.

According to the concepts advanced by these workers, there is a critical role for increasing photoperiod to initiate a process in the late winter–spring which ultimately leads to the start of the autumn breeding season. As observed by Woodfill *et al.* (1994), it is of interest that sheep, which are consistently exposed to photoperiodic information, appear to be dependent on the longer photoperiods perceived during the spring/summer period to synchronize their endogenous circannual rhythm of reproductive neuroendocrine activity; these workers speculate that such synchronization in the spring and summer in sheep ensures that they will be ready to mate in the autumn, allowing the ewe to bear her young at a beneficial time in the following spring.

2.3.1. Hormonal events involved in the ewe anoestrus

The ewe in normal commercial sheep breeds in Ireland shows a clear non-breeding season, whereas the ram remains capable of reproduction throughout the year. Much more needs to be examined both in ewes and rams in order to explain why reproduction periodically halts in the ewe but continues without interruption in the ram.

LH levels in anoestrus

During the ewe anoestrus, it is known that ovarian follicles develop, produce steroids and are capable of ovulating; many of the positive and negative feedback effects of ovarian steroids on secretion of LH continue as in the autumn breeding season.

Evidence reviewed by Legan and Karsh (1979) suggested that anoestrus was not primarily a result of a deficiency in the LH surge mechanism; neither was it a result of the behavioural patterns exhibited at oestrus nor due to an absence of oestradiol production. The essential difference appeared to be in the lack of sustained increase in tonic LH secretion which normally followed the decrease of progesterone in the cyclic ewe.

Sensitivity to oestrogen

It has been suggested that seasonal anoestrus is a result of light-induced changes in the sensitivity of the hypothalamic-pituitary axis to the negative feedback action of ovarian steroids. Workers in Michigan suggested that the main change was an increased response to the negative feedback action of oestradiol on tonic LH (Legan and Karsh, 1979). In agreement with this, results of a study by Sakurai *et al.* (1995) have shown that sensitivity to the negative feedback effects of oestradiol on secretion of gonadotrophins is most marked in ewes during the non-breeding season; conversely, the stimulatory effect of oestradiol on gonadotrophin secretion was most marked during the breeding season.

According to the increased sensitivity to oestrogen explanation, once the last corpus luteum of the breeding season starts to regress, tonic LH secretion rises, leading to an increase in follicular oestradiol. However, unlike events in the cyclic ewe, this elevated oestradiol inhibits secretion by its negative feedback action, thereby preventing the normal occurrence of the sustained, 48h rises in tonic LH secretion. The Michigan workers have presented evidence showing that changes in the response of the hypothalamic–pituitary axis to this negative feedback action of oestradiol is governed with considerable precision by the environmental photoperiod; short days were associated with the breeding season and high tonic LH levels, long days with anoestrus and low LH levels, even when artificial light regimes were completely out-of-phase with the natural environmental photoperiod.

Changes in LH pulse frequency
The tonic secretion of LH in the sheep is pulsatile and it is known that such pulses occur more frequently in the breeding season than they do during anoestrus. These LH pulses are followed by corresponding pulses of ovarian oestrogen and seasonal changes in the sensitivity of the hypothalamus to the negative feedback effects presumably limit the frequency of the LH pulses. According to Brinkley (1981), for the ewe to pass from the anoestrous to the cyclic condition, the longer intervals between the LH pulses have to shorten to increase the growth of follicles and to aid in the production of oestradiol; the oestrogen eventually triggers the release of the preovulatory LH surge and ovulation occurs. Subsequently, the sequence of events necessary for a succession of oestrous cycles is continued until events once more alter the frequency of the LH pulses and bring the breeding season to an end.

Although the seasonal changes in the feedback effects of oestradiol observed by the Michigan workers may well occur, the precise relationship of such changes to variations in the photoperiodic environment may vary according to other factors. Studies with ewes born and reared under a 24h continuous light regimen have shown, in the presence of a ram, fairly regular periods of sexual activity interspersed with periods of anoestrus (Robertson, 1977). It may be that there is an inherent physiological mechanism in the ewe which creates a rhythmical pattern and that no one external factor, such as photoperiodism, can completely control breeding activity without there being some alternation between a sexual and non-sexual season.

FSH and PRL levels
Although it is true that follicular development to the preovulatory stage is arrested during the ewe anoestrus, studies have indicated that FSH levels are not significantly different from those in the cyclic animal.

There have been reports showing that anoestrus in the sheep is associated with elevated levels of PRL. McNeilly and Land (1979) concluded that the problem of anoestrus was basically a failure in the normal preovulatory development of ovarian follicles due to inadequacies in the LH pulse frequency rather than difficulties arising from prolactin.

First cycle of the breeding season
It should be mentioned that at the start of the ewe's breeding season, it is only after the end of the first ovarian cycle that behavioural oestrus is exhibited by the sheep. The initial ovulations of the season, and the establishment of corpora lutea occur at a 'silent' heat (Grant, 1933). It is now well accepted that progesterone produced by these initial corpora lutea, play a crucial role in sensitizing hypothalamic receptors so that the ewe becomes capable of responding subsequently to the minute quantities of follicular oestrogen by

exhibiting the behavioural symptoms of oestrus. As previously noted, corpora lutea formed after the initial ovulations of the breeding season often regress prematurely 5–6 days after their formation (Oldham and Martin, 1978); such regression is generally followed by further ovulations, usually without oestrus, but the resulting corpora lutea persist for the normal period.

Seasonally anoestrous ewes can be induced to ovulate with multiple injections of GnRH; however, a low proportion of ewes develop short lifespan corpora lutea, which produce only a transient increase in progesterone concentration before regressing prematurely (Southee *et al.*, 1988). The induction of these short lifespan corpora lutea has provided a useful model for studying the causes of premature luteolysis. Studies by Beard and Hunter (1994a) have demonstrated that oestradiol plays a key regulatory role in the control of functional oxytocin receptors and the occurrence of premature luteolysis in GnRH-treated anoestrous ewes. These authors believe that increased oestrogenic stimulation of the luteolytic mechanism may be involved in the initiation of premature luteolysis and that an increase in oestradiol receptors in ewes not pretreated with progesterone may increase the effectiveness of any oestradiol available.

2.3.2. Endocrine events in the ram during the ewe anoestrus

Although there is ample evidence to show that Suffolk rams in Ireland and the UK do not apparently suffer from any obvious loss of libido nor any serious reduction in fertility during the ewe anoestrus (Fig. 2.6), distinct seasonal changes in reproductive ability and behaviour have been recorded in the primitive Soay breed; such changes in the Soay can be related to specific hormonal events which can be controlled by way of the photoperiodic environment (Lincoln, 1976; Lincoln and Davidson, 1977; Lincoln *et al.*, 1977).

Pulsatile release of LH in the ram

In the mature ram, LH is released from the pituitary in pulses that occur every few hours at random throughout the day; in the peripheral circulation, this is reflected in the form of transitory surges in LH concentration. This episodic secretory pattern of LH develops at puberty and in Soay rams has been shown to change in relation to the breeding season and to be capable of alteration by manipulation of the photoperiod. Each pulse of LH is known to be related to a corresponding episodic release of GnRH, which is presumed to be the result of the agent being released from the GnRH secretory neurons.

As mentioned above, Soay rams show much more pronounced seasonal changes in reproductive ability compared with the normal commercial sheep breeds, such as the Suffolk; this is reflected in the fact that the Soay's testes regress under normal daylengths to about 20% of their maximum size (Lincoln *et al.*, 1977). Even in the Soay ram, as already noted for the female sheep, there is evidence that an inherent reproductive cycle operates. Lincoln and Davidson

Fig. 2.6. Studies conducted in Ireland over a period of many years have failed to provide evidence of seriously reduced breeding performance in Suffolk rams during the summer months – although it is evident that some semen characteristics may be lower than those recorded in the full breeding season.

(1977) found it necessary to envisage the role of the photoperiod in sheep as being to control the time of the sexual season rather than to cause it to occur. Under normal circumstances, the stimulatory (autumn) and inhibitory (spring) effects of daylength would serve to align the breeding season to these changes in the environment. Studies have shown that testicular growth and regression in the ram may occur under conditions of a constant photoperiod (Howles *et al.*, 1980).

PRL concentrations

In the ram, as in the ewe, conspicuous changes in the blood levels of prolactin are known to occur during the year, the highest concentrations being evident during the summer months and the lowest levels in the winter. The timings of these seasonal changes can be readily modified by alteration of the photoperiod. There is some evidence that daylength regimes that favour prolactin secretion (long days) have the reverse effect on gonadotrophin secretion (Lincoln *et al.*, 1977). Under natural daylight conditions, therefore, the implication is that there is an inverse relationship between prolactin levels and the release of gonadotrophins. In all of this, it is necessary to keep in mind that the findings obtained in primitive sheep breeds such as the Soay may not operate to the same extent in the improved mutton breeds which have been selected over many generations. However, to those perusing the scientific

Table 2.2 Seasonal variation in the total number of sperm per ejaculate (from Smyth and Gordon, 1967).

Breed	Total sperm ejaculated (× 10^9 per collection)			
	Autumn	Winter	Spring	Summer
Galway	3.78	3.65	3.01	2.49
Suffolk	3.84	3.47	2.51	1.69
Wicklow	3.58	3.32	1.62	1.99
Dorset Horn	4.10	3.91	3.22	3.01

literature, the distinction between improved and unimproved sheep breeds may not appear to be as obvious as they are to those who practise commercial sheep farming.

Seasonal changes in semen production

McKenzie and Terrill (1937) in the USA were among the first to report seasonal variations both in the production and quality of ram semen. The number of sperm ejaculated and the percentage of abnormal cells recorded varied according to a definite seasonal pattern. The incidence of abnormal sperm was highest and sperm density lowest during the warm summer months in Missouri; similar results were to be found in the reports of other American authors, but it may be as well to remember that in such instances the temperature component of the environment may have been an important factor determining semen quality. Smyth and Gordon (1967), in an examination of semen production in various breeds of rams throughout the year, certainly did find evidence of seasonal fluctuations (see Table 2.2), but these were not such as to render the ram incapable of an acceptable breeding performance in the spring and summer period.

In France, using tracer techniques, workers in the 1950s reported that in the ram testes, the rate of division of primary spermatocytes and time of subsequent maturation were relatively unaffected by the photoperiod but that the number of spermatids which survive the complete maturation was affected; under conditions of increasing daylength, the failure rate in spermatids was found to be high. Later studies by the same workers reported that a 12h photoperiod provided the optimal short-term light stimulus for spermatogenesis in the ram; it was evident, however, that the maintenance of constant daylengths for periods up to 40 days resulted in a decline in sperm production (Ortavant *et al.*, 1988).

2.3.3. Management history and the breeding season

There may be commercial sheep farming circumstances in which an attempt is made to breed ewes towards the end of their natural breeding season rather than at the beginning. In the UK, for example, there are farmers who have adopted May lambing (rather than February/March) in an effort to reduce labour and feed costs. Although much research has been devoted to methods of advancing the breeding season of the ewe, much less is known about the duration of the natural breeding season and how this may be influenced by previous breeding management. Work with prolific sheep breeds has shown that the lambing percentage declines after the winter solstice (21 December) but ability to conceive is not affected. Studies at Aberdeen reported by King *et al.* (1995) showed that oestrous cycles have a tendency to become naturally synchronized towards the end of the breeding season (March) without the use of teaser rams or hormones; the observed seasonal decline in ovulation rate may result in a lower lambing percentage, which is one factor to be taken into account in commercial practice. Other work at Aberdeen with Bluefaced Leicester × Scottish Blackface ewes reported by Mitchell *et al.* (1996) also demonstrated a decline in ovulation rate during the latter part of the breeding season (February); this decrease was apparently independent of body condition.

2.3.4. Role of the ram in stimulating breeding activity in ewes

It has been known for the past 50 years that the unaccustomed presence of rams can induce ovulation and hasten the start of breeding activity in anoestrous ewes. Shortly after their contact with rams, it is known that the frequency of LH pulses in ewes increases (Martin *et al.*, 1980) which results in a preovulatory surge of LH occurring within 48h of the introduction of rams (Pearce and Oldham, 1984). The induced ovulation is not accompanied by oestrus and it is evident that corpora lutea formed after ovulation often regress prematurely (Oldham and Martin, 1978).

2.3.5. Role of the pineal and thyroid glands

Pineal gland

In the past 20 years, it has become clear that, at least in some mammals, the pineal gland is heavily involved in the regulation of photoperiodic responses and that the indoleamine, melatonin, is the hormone responsible for transfer of daylength information to the hypothalamic-pituitary axis (Arendt, 1986). Numerous studies have reported evidence that melatonin is secreted during the dark period of the day; its level is very low during the light period. The duration of the nocturnal increase in melatonin level may constitute the message whereby the pineal mediates photoperiodic regulation of hormonal secretion.

Whatever the mode of action of melatonin, its administration, either by infusion, ingestion, injection or by constant-release devices allows short days to be mimicked. In the UK, many studies have been conducted on the effect of treating lowland ewes with melatonin on the time of mating and their reproductive performance (Arendt *et al.*, 1983; Haresign, 1992a,b). Further afield, in Greece, Laliotis and Vosniakou (1993) have carried out studies showing that oral administration or implantation of melatonin at the end of the sheep anoestrus, with or without oestrus synchronization treatment, brought an earlier onset to the sheep breeding season.

Thyroid gland

Studies in the UK have shown that hypothyroidism affects reproductive performance in sheep. It is now known that anoestrus in sheep is abolished by removal of the thyroid gland (thyroidectomy). Follett and Potts (1990) found that the breeding season lasted an average of 122 days in hypothyroid ewes compared with controls. It was concluded that hypothyroidism (induced by feeding methylthiouracil) affected the development of the refractoriness that terminates the reproductive season in the sheep; these effects were similar to those found after thyroidectomy, although less pronounced. In the USA, results published by Moenter *et al.* (1991) supported the view that the thyroid is necessary for the seasonal neuroendocrine pulsatile LH release and that removal of the gland blocked seasonal suppression of reproductive activity in ewes.

Other work reported in that same year from Michigan tested the hypothesis that the thyroid is required for the breeding season to end and that this effect was due to the secretion of thyroxine (Webster *et al.*, 1991); results lent further support to the concept that the thyroid gland plays a fundamental role in seasonal reproduction in the ewe. In later studies at Michigan, Dahl *et al.* (1994) reported results suggesting that, in the absence of the thyroid gland, the reproductive neuroendocrine axis is uncoupled from the photoperiodic influence between the pineal and the GnRH neurosecretory system. Further work by Dahl *et al.* (1995) led them to conclude that the seasonal increase in circulating thyroid hormone in the ewe does not drive the transition from the breeding season to anoestrus. It is believed that thyroid hormones act permissively to allow the neuroendocrine changes that lead to the seasonal cessation of reproductive activity.

2.4. References

Aboul-Naga, A.M., Aboul-Ela, M.B., El-Nakhla, S.M. and Mehrez, A.Z. (1987) Oestrous and ovarian activity of subtropical fat-tailed Rahmani sheep and their response to light treatment. *Journal of Agricultural Science, Cambridge* 108, 617–621.

Arendt, J. (1986) Role of the pineal gland and melatonin in seasonal reproductive function in mammals. *Oxford Reviews of Reproductive Biology* 8, 266–320.

Arendt, J., Symons, A.M., Laud, C.A. and Pryde, S.J. (1983) Melatonin can induce early onset of the breeding season in ewes. *Journal of Endocrinology* 97, 395–399.

Asdell, S.A. (1964) *Patterns of Mammalian Reproduction*. Cornell University Press, Ithaca, New York.

Auletta, F.J., Williams, R.E., Kelm, L.B. and Kenny, N. (1994) In situ detection of apoptotic cells in the regressing sheep corpus luteum. *Journal of Reproduction and Fertility* Abstract Series No. 13, p. 30.

Baird, D.T., Campbell, B.K., Mann, G.E. and McNeilly, A.S. (1991) Inhibin and oestradiol in the control of FSH secretion in the sheep. *Journal of Reproduction and Fertility* Suppl. 43, 125–138.

Bartlewski, P.M., Beard, A.P. and Rawlings, N.C. (1995) Ovarian follicular dynamics during anoestrus in ewes. *Journal of Reproduction and Fertility* Abstract Series No. 15, p. 14.

Beard, A.P. and Hunter, M.G. (1994a) Effects of bovine follicular fluid and exogenous oestradiol on the GnRH-induced short luteal phase in anoestrous ewes. *Journal of Reproduction and Fertility* 100, 211–217.

Beard, A.P. and Hunter, M.G. (1994b) Effects of progesterone pretreatment on the oxytocin receptor concentration and the response to oxytocin during the simulated early luteal phase in the ovariectomized ewe. *Journal of Reproduction and Fertility* 102, 57–63.

Beard, A.P. and Lamming, G.E. (1994) Oestradiol concentration and the development of the uterine oxytocin receptor and oxytocin-induced $PGF_{2\alpha}$ release in ewes. *Journal of Reproduction and Fertility* 100, 469–475.

Bichard, N., Younis, A.A., Forrest, P.A. and Cumberland, P.H. (1974) Analysis of production research from a lowland flock. *Animal Production* 19, 177–191.

Bland, K.P. and Jubilan, B.M. (1987) Correlation of flehmen by male sheep with female behaviour and oestrus. *Animal Behaviour* 35, 735–738.

Blissitt, M.J., Bland, K.P. and Cottrill, D.F. (1990) Discrimination between the odours of fresh oestrous and non-oestrous ewe urine by rams. *Applied Animal Behaviour Science* 25, 51–59.

Blissitt, M.J., Bland, K.P. and Cottrell, D.F. (1994) Detection of oestrous-related odour in ewe urine by rams. *Journal of Reproduction and Fertility* 101, 189–191.

Bowen, J., Dahl, G.E., Evans, N.P., Thurn, L.A. and Karsh, F.J. (1995) Does the GnRH surge amplitude exceed that required for the production of the LH surge? *Journal of Reproduction and Fertility* Abstract Series No. 15, p. 39.

Brinkley, H.J. (1981) Endocrine signalling and female reproduction. *Biology of Reproduction* 24, 22–43.

Cahill, L.P. (1984) Folliculogenesis and ovulation rate in sheep. In: Lindsay, D.R. and Pearce, D.T. (eds) *Reproduction in Sheep*. Cambridge University Press, Cambridge, pp. 92–98.

Cahill, L.P. and Mauleon, P. (1980) Influence of season, cycle and breed on follicular growth rates in sheep. *Journal of Reproduction and Fertility* 58, 321–328.

Caldwell, B.V., Tillson, S.A., Brock, W.A. and Speroff, L. (1972) The effects of exogenous progesterone and oestradiol on prostaglandin F levels in ovariectomized ewes. *Prostaglandins* 1, 217–228.

Campbell, B. and Webb, R. (1995) Evidence that inhibin has autocrine and paracrine actions in controlling ovarian function in sheep. *Journal of Reproduction and Fertility* Abstract Series No. 15, pp. 48–49.

Campbell, B.K., McNeilly, A.S. and Baird, D.T. (1989) Episodic ovarian inhibin secretion is not due to LH pulses in anoestrous ewes. *Journal of Endocrinology* 123, 173–179.
Campbell, B.K., McNeilly, A.S., Picton, H.M. and Baird, D.T. (1990a) The effect of a potent gonadotropin-releasing hormone antagonist on ovarian secretion of oestradiol, inhibin and androstendendione and the concentration of LH and FSH during the follicular phase of the sheep oestrous cycle. *Journal of Endocrinology* 126, 377–384.
Campbell, B.K., Mann, G.E., McNeilly, A.S. and Baird, D.T. (1990b) The pattern of ovarian inhibin, estradiol and androstenedione secretion during the estrous cycle of the ewe. *Endocrinology* 127, 227–235.
Campbell, B.K., McNeilly, A.A., Mann, G.E. and Baird, D.T. (1991) The effect of stage of estrous cycle and follicular maturation on ovarian inhibin production in sheep. *Biology of Reproduction* 44, 483–490.
Crosby, T.F. and Gordon, I. (1971) Timing of nuclear maturation in oocytes cultured in growth medium. *Journal of Agricultural Science, Cambridge* 76, 373–374.
Cumming, I.A., Brown, J.M., Blockey, M.A., Winfield, C.G., Baxter, R. and Goding, J.R. (1971) Consistency of interval between luteinizing hormone release and ovulation in the ewe. *Journal of Reproduction and Fertility* 24, 134–135.
Cumming, I.A., Brown, J.M., Cerini, J.G., Cerini, M.E., Chamley, W.A., Findlay, J.K. and Goding, J.R. (1973) Ovine luteinizing hormone release induced by a synthetic gonadotrophin-releasing factor. *Journal of Reproduction and Fertility* 32, 340.
Cunningham, N.F., Symons, A.M. and Saba, N. (1975) Levels of progesterone, LH and FSH in the plasma of sheep during the oestrous cycle. *Journal of Reproduction and Fertility* 45, 177–180.
Dahl, G.E., Evans, N.P., Moenter, S.M. and Karsh, F.J. (1994) The thyroid gland is required for reproductive neuroendocrine responses to photoperiod in the ewe. *Endocrinology* (Philadelphia) 135, 10–15.
Dahl, G.E., Evans, N.P., Thrun, L.A. and Karsh, F.J. (1995) Thyroxine is permissive to seasonal transitions in reproductive neuroendocrine activity in the ewe. *Biology of Reproduction* 52, 690–696.
Deane, H.W., Hay, M.F., Moor, R.M., Rowson, L.E.A. and Short, R.V. (1966) The corpus luteum of sheep: the relationships between morphology and function during the oestrous cycle. *Acta Endocrine Copenhagen* 51, 245.
Denamur, R., Martinent, J. and Short, R.V. (1973) Pituitary control of the ovine corpus luteum. *Journal of Reproduction and Fertility* 32, 207–220.
Driancourt, M.A. (1991) Follicular dynamics in sheep and cattle. *Theriogenology* 35, 55–79.
Driancourt, M.A. (1994) Lack of between-follicle interactions in the sheep ovary. *Reproduction, Nutrition and Development* 34, 249–260.
Driancourt, M.A., Bodin, L., Boomarov, O., Thimonier, J. and Elsen, J.M. (1990) Number of mature follicles ovulating after a challenge of human chorionic gonadotropin in different breeds of sheep at different physiological stages. *Journal of Animal Science* 68, 719–724.
Dziuk, P.J. (1965) Response of sheep and swine to treatments for control of ovulation. *Proceedings of a Conference on Oestrous Cycle Control in Domestic Animals* USDA Publication, 1005, pp. 28–38.
Ehnert, K. and Moberg, G.P. (1991) Disruption of estrous behavior in ewes by dexamethasone or management-related stress. *Journal of Animal Science* 69, 2988–2994.

Eldon, J. (1994) Time of onset and potential length of the breeding season of Icelandic sheep: luteal activity. *Animal Reproduction Science* 34, 101–109.

Evans, N.P., Dahl, G.E., Glover, B.H. and Karsh, F.J. (1994) Central regulation of pulsatile gonadotropin-releasing hormone (GnRH) secretion by estradiol during the period leading up to the prevoluatory GnRH surge in the ewe. *Endocrinology* (Philadelphia) 134, 1806–1811.

Evans, N.P., Dahl, G.E., Mauger, D. and Karsh, F.J. (1995) Estradiol induces both qualitative and quantitative changes in the pattern of gonadotropin-releasing hormone secretion during the presurge period in the ewe. *Endocrinology* (Philadelphia) 136, 1603–1609.

Fitz, T.A., Mayan, M.H., Sawyer, H.R. and Niswender, G.D. (1982) Characterization of two steroidogenic cell types in the ovine corpus luteum. *Biology of Reproduction* 27, 703–711.

Flint, A.P.F. and Sheldrick, E.L. (1983) Evidence for a systemic role for ovarian oxytocin receptor in the choice between cyclicity and gestation in ruminants. *Journal of Reproduction and Fertility* 67, 215–225.

Follett, B.K. and Potts, C. (1990) Hypothyroidism affects reproductive refractoriness and the seasonal oestrous period in Welsh Mountain ewes. *Journal of Endocrinology* 127, 103–109.

Freeman, W.H. and Bracegirdle, B. (1968) *An Atlas of Histology*. Heinemann, London.

Ginther, O.J. (1968) Influence of exogenous progesterone and the uterus on ovarian activity in sheep. *Endocrinology* 83, 613–615.

Ginther, O.J., Kot, K. and Wiltbank, M.C. (1995) Associations between emergence of follicular waves and fluctuations in FSH concentrations during the estrous cycle in ewes. *Theriogenology* 43, 689–703.

Goding, J.R., Blockey, M.A., Brown, J.M., Catt, K.J. and Cumming, I.A. (1970) The role of oestrogen in the control of the oestrous cycle in the ewe. *Journal of Reproduction and Fertility* 21, 368–369.

Gonzalez, R., Orgeur, P., Poindron, P. and Signoret, J.P. (1991) Female effect in sheep. I. The effects of sexual receptivity of females and the sexual experience of rams. *Reproduction, Nutrition and Development* 31, 97–102.

Goodman, R.L., Legan, S.J., Ryan, K.D., Foster, D.L. and Karsh, F.J. (1980) Two effects of estradiol that normally contribute to the control of tonic LH secretion in the ewe. *Biology of Reproduction* 23, 415–422.

Gordon, I. (1983) *Controlled Breeding in Farm Animals*. Pergamon Press, Oxford, 436 pp.

Grant, R. (1933) Occurrence of ovulation without heat in the ewe. *Nature* (London) 131, 802.

Grant, R. (1934) Studies on the physiology of reproduction in the ewe. *Transactions of the Royal Society of Edinburgh* 58, 1–47.

Hafez, E.S.E. (1952) Studies on the breeding season and reproduction of the ewe. *Journal of Agricultural Science, Cambridge* 42, 189–265.

Hammond, J. Jr (1944) On the breeding season in the sheep. *Journal of Agricultural Science, Cambridge* 34, 97–105.

Hansel, W. (1975) Luteal regression in domestic animals. *Annales de Biologie animale Biochimie Biophysique* 15, 147–160.

Haresign, W. (1992a) The effect of implantation of lowland ewes with melatonin on the time of mating and reproductive performance. *Animal Production* 54, 31–39.

Haresign, W. (1992b) Responses of ewes to melatonin implants: importance of the interval between treatment and ram introduction on the synchrony of mating, and

effects on ovulation rate. *Animal Production* 54, 41–45.
Hauger, R.L., Karsh, F.J. and Foster, D.L. (1977) New concept for control of the estrous cycle of the ewe based on the temporal relationships between luteinizing hormone, estradiol and progesterone in peripheral serum and evidence that progesterone inhibits tonic LH secretion. *Endocrinology* 101, 807–817.
Heape, W. (1990) The sexual season of mammals. *Quarterly Journal of Microscopical Science* 44, 1-44.
Hixon, J.E. and Flint, A.P.F. (1987) Effects of a luteolytic dose of oestradiol benzoate on uterine oxytocin receptor concentrations, phosphoinositide turnover and PG $F_{2\alpha}$ secretion in sheep. *Journal of Reproduction and Fertility* 79, 457–467.
Holst, P.J. and Braden, A.W.H. (1972) Ovum transport in the ewe. *Australian Journal of Biological Science* 25, 167–173.
Howles, C.M., Webster, G.M. and Haynes, N.B. (1980) The effect of rearing under a long or short photoperiod on testis growth, plasma testosterone and prolactin concentrations, and the development of sexual behaviour in rams. *Journal of Reproduction and Fertility* 60, 437–447.
Hutchinson, J.S.M. and Robertson, H.A. (1966) The corpus luteum in sheep. *Research in Veterinary Science* 7, 17.
Inkster, I.J. (1957) The mating behaviour of sheep. *Massey College Farming Annual* 163.
Jablonka-Shariff, A., Grazul-Bilska, A.T., Redmer, D.A. and Reynolds, L.P. (1993) Growth and cellular proliferation of ovine corpora lutea throughout the estrous cycle. *Endocrinology* (Philadelphia) 133, 1871–1879.
Karsh, F.J. (1984) Endocrine and environmental control of oestrous cyclicity in sheep. In: Lindsay, D.R. and Pearce, D.T. (eds) *Reproduction in Sheep.* Cambridge University Press, Cambridge, pp. 10–15.
Karsh, F.J. (1995) Neuroendocrine signals for ovulation – fitting pieces to an unsolved puzzle. *Journal of Reproduction and Fertility* Abstract Series No. 15, p. 1.
Karsh, F.J., Bittman, E.L., Foster, O.L., Goodman, R.L., Legan, S.J. and Robinson, J.E. (1984) Neuroendocrine basis of seasonal reproduction. *Recent Progress in Hormone Research* 40, 185–232.
Kassem, R., Owen, J.B. and Fadel, I. (1990) A note on the characteristics of oestrus and ovulation in Awassi ewes. *Animal Production* 50, 198–201.
Kaulfuss, K.H., Richter, A. and Schulz, J. (1994) Patterns of ovarian follicle growth during the oestrous cycle in Mutton Merino sheep monitored by laparoscopy. *Reproduction in Domestic Animals* 29, 22–37.
Kelley, R.B. (1937) Studies of fertility in sheep. *Bulletin of the Council of Scientific and Industrial Research of Australia*, pp. 112–114.
Khalid, M. and Haresign, W. (1995) Relationships between concentrations of steroids, IGF-1 and IGF binding proteins during follicular development in the ewe. *Proceedings of the British Society of Animal Science* (Winter Meeting), Paper 56.
King, M.E., Mitchell, L.M., Aitken, R.P. and Walace, J.M. (1995) The effect of management history on oestrous cyclicity in Mule ewes. *Proceedings British Society Animal Science* (Winter Meeting), Paper 94.
Laliotis, V.N. and Vosniakou, A.G. (1993) Role of melatonin in reproduction of ewes. *Epitheorese Zootehnikes Epistemes* 22, 117–133.
Land, R.B., Pelletier, J., Thimonier, J. and Mauleon, P. (1973) A quantitative study of genetic differences in the incidence of oestrus, ovulation and plasma LH concentration in the sheep. *Journal of Endocrinology* 58, 305–317.
Legan, S.J. and Karsh, F.J. (1979) Neuroendocrine regulation of the oestrous cycle and

seasonal breeding of the ewe. *Biology of Reproduction* 20, 74–85.

Leyva, V., Walton, J.S., Buckrell, B.C., Buhr, M.M., King, W.A. and Gartley, C. (1995) Ultrasound examination of ovarian activity and ovulation regulated by exogenous progesterone in cycling ewes. *Journal of Animal Science* 73 (Suppl. 1), p. 226.

Lincoln, G.A. (1976) Photoperiodic control of reproduction in the ram: time-lags from stimulus to response. *Annales de Biologie Animale Biochimie Biophysique* 16, 170–180.

Lincoln, G.A. and Davidson, W. (1977) The relationship between sexual and aggressive behaviour, and pituitary and testicular activity during the seasonal sexual cycle of rams, and the influence of photoperiod. *Journal of Reproduction and Fertility* 49, 267–276.

Lincoln, G.A., Peet, M.J. and Cunningham, R.A. (1977) Seasonal and circadian changes in the episodic release of follicle-stimulating hormone, luteinizing hormone and testosterone in rams exposed to artificial photoperiods. *Journal of Endocrinology* 72, 337–349.

Lindsay, D.R. and Fletcher, I.C. (1972) Ram-seeking activity associated with oestrous behaviour in ewes. *Animal Behaviour* 20, 452–456.

Lindsay, D.R., Cognie, Y., Pelletier, J. and Signoret, J.P. (1975) Influence of the presence of rams on the timing of ovulation and discharge of LH in ewes. *Physiology and Behaviour* 15, 423–426.

McKenzie, F.F. and Terrill, C.E. (1937) Estrus, ovulation and related phenomena in the ewe. *Research Bulletin of the Missouri Agricultural Experiment Station* No. 264.

McNatty, K.P., Gibb, M., Dobson, C., Ball, K., Coster, D., Heath, D. and Thurley, D.C. (1982) Preovulatory follicular development in sheep treated with PMSG and/or prostaglandin. *Journal of Reproduction and Fertility* 65, 111–123.

McNeilly, A.S. and Land, R.B. (1979) Effect of suppression of plasma prolactin on ovulation, plasma gonadotrophins and corpus luteum function in LH–RH-treated anoestrous ewes. *Journal of Reproduction and Fertility* 56, 601–609.

McNeilly, A.S., Swanston, I.A., Crow, W., Tsonis, C.G. and Baird, D.T. (1989) Changes in plasma concentrations of inhibin throughout the normal sheep oestrous cycle and after infusion of FSH. *Journal of Endocrinology* 120, 295–305.

McNeilly, A.S., Brooks, J., McNeilly, J.R. and Brown, P. (1995) Synthesis and release of FSH. *Journal of Reproduction and Fertility* Abstract Series No. 15, p. 2.

Malpaux, B., Robinson, J.E., Wayne, N.L. and Karsch, F.J. (1989) Regulation of the onset of the breeding season of the ewe: importance of long days and of an endogenous reproductive rhythm. *Journal of Endocrinology* 122, 269–278.

Mann, G.E. and Lamming, G.E. (1995) The role of luteal oxytocin in episodic secretion of prostaglandin $F_{2\alpha}$ at luteolysis in the ewe. *Biology of Reproduction* 52 (Suppl. 1), p. 197.

Mann, G.E., McNeilly, A.S. and Baird, D.T. (1989) Source of ovarian inhibin secretion during the oestrous cycle of the sheep. *Journal of Endocrinology* 123, 181–188.

Marshall, F.H.A. (1937) On the change-over in the oestrous cycle in animals after transference across the equator, with further observations on the incidence of the breeding seasons and the factors controlling sexual periodicity. *Proceedings of the Royal Society* B 122, 413.

Martin, G.B., Oldham, C.M. and Lindsay, D.R. (1980) Increased plasma LH levels in seasonally anovular Merino ewes following the introduction of rams. *Animal Reproduction Science* 3, 125–132.

Matthews, L.R., Uljee, A.E., Bremner, K.J., Painting, A.M., Cate, L.R. and Smith, J.F. (1991) Development of a self-drafting system for oestrus ewes. *Proceedings of the*

New Zealand Society of Animal Production 51, 315–318.

Mitchell, L.M., King, M.E., Aitken, R.P. and Wallace, J.M. (1996) Effect of mating season and body condition on ovulation, fertilization and pregnancy rates in crossbred ewes. *Theriogenology* 45, 293.

Moenter, S.M., Caraty, A. and Karsh, F.J. (1990) The estradiol-induced surge of gonadotrophin-releasing hormone in the ewe. *Endocrinology* 127, 1375–1384.

Moenter, S.M., Woodfill, C.J.I. and Karsh, F.J. (1991) Role of the thyroid gland in seasonal reproduction: thyroidectomy blocks seasonal suppression of reproductive neuroendocrine activity in ewes. *Endocrinology* (Philadelphia) 128, 1337–1344.

Moore, N.W. (1988) The ovariectomized ewe: its contribution to controlled breeding. *Australian Journal of Biological Science* 41, 15–22.

Morgan, G.L., Geisert, R.D., McCann, J.P., Bazer, F.W., Ott, T.L., Mirando, M.A. and Stewart, M. (1993) Failure of luteolysis and extension of the interoestrous interval in sheep treated with the progesterone antagonist mifepristone (RU 486). *Journal of Reproduction and Fertility* 98, 451–457.

Mukasa-Mugerwa, E. and Lahlou-Kassi, A. (1995) Reproductive performance and productivity of Menz sheep in the Ethiopian highlands. *Small Ruminant Research* 17, 167–177.

Murdoch, W.J. (1995) Programmed cell death in preovulatory ovine follicles. *Biology of Reproduction* 53, 8–12.

Noel, B., Bister, J.L. and Paquay, R. (1993) Ovarian follicular dynamics in Suffolk ewes at different periods of the year. *Journal of Reproduction and Fertility* 99, 695–700.

Noel, B., Bister, J.L., Pierquin, B. and Paquay, R. (1994) Effects of FGA and PMSG on follicular growth and LH secretion in Suffolk ewes. *Theriogenology* 41, 719–727.

Odagiri, K., Matsuzawa, Y. and Yoshikawa, Y. (1995) Analysis of sexual behaviour in rams (*Ovis aries*). *Experimental Animals* 44, 187–192.

Oldham, C.M., and Martin, G.B. (1978) Stimulation of seasonally anovular Merino ewes by rams. II. Premature regression of ram-induced corpora lutea. *Animal Reproduction Science* 1, 291–295.

Ortavant, R., Bocquier, F., Pelletier, I., Ravault, J.P., Thimonier, J. and Volland-Nail, P. (1988) Seasonality of reproduction in sheep and its control by photoperiod. *Australian Journal of Biological Sciences* 41, 79–85.

Osborn, J.C. and Moor, R.M. (1985) Oocyte maturation and developmental competence. In: In Vitro Fertilization and Donor Insemination. *Proceedings of the 12th Study Group of the Royal College of Obstetricians and Gynaecologists* pp. 101–114.

Padmanabhan, V. (1995) Neuroendocrine control and physiologic relevance of FSH heterogeneity. *Journal of Reproduction and Fertility* Abstract Series No. 15, p. 2.

Parsons, S.D. and Hunter, G.L. (1967) Effect of the ram on duration of oestrus in the ewe. *Journal of Reproduction and Fertility* 14, 61–66.

Parsons, S.D., Hunter, G.L. and Ravner, A.A. (1967) Use of probit analysis in a study of the effect of the ram on time of ovulation in the ewe. *Journal of Reproduction and Fertility* 14, 71–80.

Pearce, D.R. and Oldham, C.M. (1984) The ram effect, its mechanism and application to the management of sheep. In: Lindsay, D.R. and Pearce, D.T. (eds) *Reproduction in Sheep.* Cambridge University Press, Cambridge, pp. 24–26.

Pelletier, J., Signoret, J.P., Cahill, L., Cognie, Y., Thimonier, J. and Ortavant, R. (1977) Physiological processes in oestrus, ovulation and fertility in sheep. In: *Symposium on Management of Reproduction in Sheep and Goats* (Madison), Sheep Industry Development Program, pp. 1–14.

Phillips, D.J., Hudson, N.L., Gentle, L.R. and McNatty, K.P. (1994) Bioactive follicle stimulating hormone concentrations in the plasma during the estrous cycle of the ewe. *Biology of Reproduction* 51, 1292–1298.

Phogat, J.B., Smith, R.F. and Dobson, H. (1995) Effect of transport or adrenocorticotrophin hormone (ACTH) on LH released by exogenous GnRH and/or oestradiol during the follicular phase in the ewe. *Journal of Reproduction and Fertility* Abstract Series No. 15, p. 38.

Ravindra, J.P., Rawlings, N.C., Evans, A.C.O. and Adams, G.P. (1994) Ultrasonographic study of ovarian follicular dynamics in ewes during the oestrous cycle. *Journal of Reproduction and Fertility* 101, 501–509.

Robertson, H.A. (1977) Reproduction in the ewe and doe. In: Cole, H.H. and Cupps, P.T. (eds) *Reproduction in Domestic Animals*. Academic Press, New York, pp. 475–498.

Robinson, T.J. (1959) Estrous cycle of the ewe and doe. In: Cole, H.H. and Cupps, P.T. (eds) *Reproduction in Domestic Animals*. Academic Press, New York.

Rodgers, R.J. and O'Shea, J.D. (1982) Purification, morphology, and progesterone production and content of three cell types isolated from the corpus luteum of the sheep. *Australian Journal of Biological Science* 35, 441–445.

Rubianes, E., Ibarra, D., Ungerfeld, R., Carbajal, B. and de Castro, T. (1995) Superovulatory response in anestrous ewes is affected by the presence of a large follicle. *Theriogenology* 43, 465–472.

Sakurai, H., Adams, B.M. and Adams, T.E. (1995) Gonadotroph responsiveness in orchidectomized sheep. IV. Effect of estradiol infusion during the breeding and anoestrous seasons. *Biology of Reproduction* 52, 382–389.

Scaramuzzi, R.J. and Land, R.B. (1978) Oestradiol levels in sheep plasma during the oestrous cycle. *Journal of Reproduction and Fertility* 53, 167–171.

Scaramuzzi, R.J. and Downing, J.A. (1995) The *in vivo* effects of fibroblast growth factor and epidermal growth factor on the secretion of oestradiol, androstenedione and progesterone by the autotransplanted ovary in the ewe. *Journal of Endocrinology* 146, 301–311.

Schanbacher, B.D., Orgeur, P., Pelletier, J. and Signoret, J.P. (1987) Behavioural and hormonal responses of sexually experienced Ile de France rams to estrus females. *Animal Reproduction Science* 14, 293–300.

Schrick, F.N., Surface, R.A., Pritchard, J.Y., Dailey, R.A., Townsend, E.C. and Inskeep, E.K. (1993) Ovarian structures during the oestrous cycle and early pregnancy in ewes. *Biology of Reproduction* 49, 1133–1140.

Signoret, J.P. (1975) Influence of the presence of rams on the luteinizing hormone surge after oestradiol benzoate injection in ovariectomized ewes. *Journal of Endocrinology* 64, 589–590.

Smeaton, T.C. and Robertson, H.A. (1971) Studies on the growth and atresia of graafian follicles in the ovary of the sheep. *Journal of Reproduction and Fertility* 25, 243–252.

Smyth, P. and Gordon, I. (1967) Seasonal and breed variations in the semen characteristics of rams in Ireland. *Irish Veterinary Journal* 21, 222–233.

Southee, J.A., Hunter, M.G., Law, A.S. and Haresign, W. (1988) Function of abnormal corpora lutea in vivo after GnRH-induced ovulation in the anoestrous ewe. *Journal of Reproduction and Fertility* 84, 131–137.

Souza, C.J.H., Campbell, B.K. and Baird, D.T. (1995) Dynamics of sheep follicular growth during anoestrus. *Journal of Reproduction and Fertility* Abstract Series No. 15, p. 13.

Thimonier, J. and Pelletier, J. (1971) Differences genetique dons la decharge ovulante (LH) chez les brebis de race Ile-de-France: relations avec le nombre of ovulations. *Annales Biologie Animale Biochimie Biophysique* 11, 559–567.
Thorburn, G.D., Bassett, J.M. and Smith, I.D. (1969) Progesterone concentration in the peripheral plasma of the sheep during the oestrus cycle. *Journal of Endocrinology* 45, 459–469.
Thwaites, C.J. (1971) Exogenous progesterone and oestrous cycle length in the ewe. *Journal of Agricultural Science, Cambridge* 77, 147–149.
Tilbrook, A.J. (1987a) Physical and behavioural factors affecting the sexual 'attractiveness' of the ewe. *Applied Animal Behaviour Science* 17, 109–115.
Tilbrook, A.J. (1987b) The influence of factors associated with oestrus on the sexual 'attractiveness' of ewes to rams. *Applied Animal Behaviour Science* 17, 117–128.
Tilbrook, A.J. and Lindsay, D.R. (1987) Differences in the sexual 'attractiveness' of oestrous ewes to rams. *Applied Animal Behaviour Science* 17, 129–138.
Vallett, J.L., Lamming, G.E. and Batten, M. (1990) Control of endometrial oxytocin receptor and uterine response to oxytocin by progesterone and oestradiol in the ewe. *Journal of Reproduction and Fertility* 90, 625–634.
Van Cleeff, J., Dahl, G.E., Evans, N.P., Mauger, D.T., Karsh, F.J. and Padmanabhan, V. (1995) Existence of a GnRH-independent component of episodic FSH release in the natural luteal phase ewe. *Journal of Reproduction and Fertility* Abstract Series No. 15, pp. 41–42.
Warbritton, V. (1934) The cytology of the corpora lutea of the ewe. *Journal of Morphology* 56, 181–202.
Webster, J.R., Moenter, S.M., Woodfill, C.J.I. and Karsh, F.J. (1991) Role of the thyroid gland in seasonal reproduction. II. Thyroxine allows a season-specific suppression of gonadotropin secretion in sheep. *Endocrinology* (Philadelphia) 129, 176–183.
Whisnant, C.S. and Goodman, R.L. (1994) Effect of anterior hypothalamic deafferentation on the negative feedback of gonadal steroids on luteinizing hormone pulse frequency in the ewe. *Domestic Animal Endocrinology* 11, 151–159.
Williams, S.M., Garrigus, U.S., Norton, H.W. and Nalbandov, A.V. (1956) The occurrence of oestrus in pregnant ewes. *Journal of Animal Science* 15, 978–983.
Woodfill, C.J.I., Wayne, N.L., Moenter, S.M. and Karsh, F.J. (1994) Photoperiodic synchronization of a circannual reproductive rhythm in sheep: identification of season-specific time cues. *Biology of Reproduction 50* 965–976.
Younis, A.A. and Afifi, E.A. (1979) On the occurrence of oestrus in pregnant Barki and Merino ewes. *Journal of Agricultural Science, Cambridge* 92, 505–506.

3 Artificial Control of Oestrus and Ovulation

3.1. Introduction

Attempts to control the occurrence of oestrus and ovulation in sheep, whether in the full breeding season or in the ewe's anoestrus, are usually based on trying to simulate the activity of the cyclic sheep's corpus luteum, especially its action in producing progesterone in quantity for about two weeks and in shutting off production sharply and completely at the end of the oestrous cycle. In most out-of-season applications, it is also considered essential to augment the supply of endogenous gonadotrophin by administering a follicle-stimulating preparation on completion of the progestagen treatment. The cheapest, most readily available and consistently effective gonadotrophin is pregnant mare serum gonadotrophin (PMSG).

Up until 1964, most attempts at controlled breeding in sheep involved repeated doses of progesterone or oral administrations of potent progestagens; the time and labour involved in giving these agents constituted a serious obstacle to any general acceptance of the techniques into commercial sheep farming. Some discussion of the limitations of injection and oral methods has been attempted elsewhere (Gordon, 1971a, 1983); in brief, it is difficult to achieve a smooth, steady input of progestagen or to obtain a sharp predictable end-point by such procedures.

3.1.1. Development of intravaginal devices

Progestagen administration was eventually made commercially feasible by the efforts of Robinson and his associates at Sydney (Robinson, 1964) using the progestagen impregnated sponge inserted intravaginally. The subsequent monograph by Robinson and co-workers contained an impressive analysis of factors involved in oestrus control by progestagens (Robinson, 1967). Although initial laboratory trials established that a high level of fertility was possible in cyclic Merino ewes after intravaginal treatment, the subsequent outcome of widespread testing in Australia was far from satisfactory. In retrospect, it now appears that much of the low fertility observed was the result of the testing being carried out in severe drought years, which markedly influenced the nutritional environment. It is also clear that sponges with less than adequate doses of progestagen were used in many instances and semen for the AI programme was often diluted to the limit (Robinson, 1974).

This Australian experience serves to illustrate the important principle that controlled reproduction techniques should only be applied in the appropriate feeding, breeding and management setting. Certainly, in Ireland, and probably elsewhere, because of these unforeseen problems associated with the intravaginal sponge technique, there was some tendency to dismiss the method as being of little commercial value. As subsequent events proved, this assessment was premature, and the progestagen impregnated sponge became the cornerstone in almost all controlled reproduction applications in France, Ireland, the UK and many other countries. In the early 1980s, a silastic device (controlled internal drug release, CIDR; AHI Plastic Mouldings) containing progesterone was shown to be effective by New Zealand workers when used intravaginally for the control of oestrus and ovulation.

3.2. Prostaglandin $F_{2\alpha}$ and Analogues

As already noted in Chapter 2, there is ample evidence to show that during the ovulatory cycle of the ewe, prostaglandin $F_{2\alpha}$ synthesized in and released from the endometrium of the uterus, causes regression of the corpus luteum. The literature available on the use of prostaglandins in cyclic sheep is much less than that available for cattle. Part of this is due to the fact that PGs are not relevant to controlled sheep reproduction during the anoestrous period.

Natural vs. induced regression of corpora lutea

From some of the reports which appeared on the ultrastructure and function of sheep corpora lutea in the normal cycle and after prostaglandin treatment (Corteel, 1975; Stacey *et al.*, 1976) there were indications that PG treatment could have a very rapid and dramatic effect on steroid synthesis in the lutein cell whereas normal luteolysis would seem to involve more gradual regression of the gland. This may have some relevance in explaining the oestrous response and fertility of sheep after this form of treatment; there is evidence that fertility may be quite variable when sheep are bred after PG-induced oestrus synchronization (Boland *et al.*, 1978a,b).

3.2.1. Treatment protocols and PG dose levels

Early reports included those of Douglas and Ginther (1973) and Hawk (1973) who showed that doses of 10–15mg $PGF_{2\alpha}$ in a single intramuscular injection could induce regression of the corpus luteum and result in the oestrus control in cyclic sheep. Workers elsewhere have also shown that a dose of 100μg (Trounson *et al.*, 1976) or 125μg (Fairnie *et al.*, 1976) of the analogue cloprostenol is similarly effective in inducing regression.

Responsive period

The corpus luteum of the ewe is only responsive to PG between about day 4 and 14 of the oestrous cycle (Chamley *et al.*, 1972); to ensure that all sheep in the flock are at an appropriate stage of the cycle to respond, the usual recommendation has been that two PG doses should be given 9–14 days apart. Reports in the 1970s showed that oestrus is exhibited about 40h after PG with ovulation occurring about 70h from the time of administering the agent. It need hardly be mentioned that PGs can only be effective in ewes that are already cyclic; it is possible that in flocks to be bred in the early weeks of the breeding season, this may not always be the case.

Interval between PG doses

According to some reports in the 1970s, the time interval between the two PG doses may influence fertility. In studies reported by Fairnie *et al.* (1978) in Australia, fertility in sheep treated with two doses of 125μg cloprostenol at 12-day intervals was much lower than in other groups treated at 14–15-day intervals. There was also an earlier report showing that very poor fertility was shown by ewes treated with two PG doses spaced 8 days apart compared with a 14-day interval (Fairnie *et al.*, 1977). The Australian workers concluded that the time interval between the two PG doses was critical and should not be reduced to less than 13–14 days, otherwise acceptable fertility to breeding by AI may not be achieved. However, if a 14-day interval is used, then the problem may be that the two-dose PG application may not find all ewes responsive to the second injection.

Dose levels of PG

Studies in Ireland in the 1970s suggested that, in comparison with the oestrous response after progestagen treatment in ewes, that which follows PG may be markedly lower (Boland *et al.*, 1978a,b), regardless of whether PG was administered at 9- or 14-day intervals. There has also been some evidence that the oestrous response may be influenced by PG dose level. In Canada, studies reported by Hackett and Robertson (1980) showed that a dose of 20mg $PGF_{2\alpha}$ induced oestrus in all sheep treated between day 4 and 15, whereas only 70% responded when 15mg was employed; the same workers reported that 67% of ewes lambed to natural service at the synchronized oestrus, which was comparable with the normal level of fertility expressed by these ewes. When the natural $PGF_{2\alpha}$ agent is employed, the accepted luteolytic dose of 15mg is about 60% of that required in the cow; using the cloprostenol analogue, 100μg has been employed as the luteolytic dose, which is only 20% of that employed in the cow.

In South Africa, Greyling and Van der Westhuysen (1979) reported that oestrus could be effectively synchronized by two doses of 250μg cloprostenol given at a 10-day interval; lower doses of the analogue (125μg) were often insufficient to induce complete luteolysis, as indicated by an initial decline in progesterone level followed by a gradual rise in the steroid, suggesting some recovery of luteal function. This ability of the corpus luteum to 'recover' after PG had previously been recorded by Thorburn and Nichol (1971). The South Africans found that with 125μg doses of cloprostenol, only about 80% of their ewes were in oestrus, as compared with 100% at the 250μg dose level.

3.2.2. Fertility and ovulation rate

As already mentioned, variable fertility has been reported after the use of PG or its analogues for the synchronization of oestrus; however, the treatment does not appear to influence ovulation rate. Although some reports in the 1970s showed little evidence of any adverse effect on fertility after the use of PG in oestrus synchronization, other results were not always reassuring. Australian data did indicate that fertilization and lambing rates may be depressed in ewes bred by AI at the PG controlled oestrus (Fairnie *et al.*, 1976; Lightfoot *et al.*, 1976).

In Ireland, comparisons between PG- and progestagen-treated ewes showed a marked decrease in oestrus response after PG and poor fertility both when natural service or AI was the method of breeding (Jennings, 1975; Boland *et al.*, 1978a,b). Unlike Australian experiences (Fukui and Roberts, 1977; Fairnie *et al.*, 1978), the work in Ireland showed little advantage in employing a long rather than a short time interval (14 vs. 9 days) between PG doses. There has been some indication that PG treatment may influence the efficiency of sperm transport in the ewe; studies in the USA by Hawk and Conley (1975) did show some evidence of partial inhibition of sperm transport in the cervix and of disturbed sperm movement in the oviducts. On the other hand, there have been those whose experiences with PG in sheep led them to maintain that fertility should not be adversely affected (Haresign, 1980).

The Irish experience, on the other hand, would be in accord with the observations of Scaramuzzi and Martin (1984), who noted that as an alternative to progestagen synchronization, PGs appear to have no real advantages. It seems clear that in the comparative studies that have been made, PGs have been shown to be less effective in synchronizing oestrus than progestagen pessaries (Hackett *et al.*, 1981; Acritopoulou-Fourcroy *et al.*, 1982; Henderson *et al.*, 1984). When PMSG was used in conjunction with synchronizing agents, an effect was evident in ewes treated with fluorogestone acetate (FGA) sponges but not in sheep treated with $PGF_{2\alpha}$ (Hackett and Hidiroglou, 1983).

3.2.3. Progestagen–PG combinations

One way of overcoming the need for two PG doses in oestrus synchronization would be in using short-term progestagen treatment prior to administering

PG. In this context, the use of 7–9-day intravaginal progestagen treatments with 15mg $PGF_{2\alpha}$ at their withdrawal (Fukui and Roberts, 1979) or 31µg cloprostenol (Greyling *et al.*, 1979) has been reported.

In Ireland, progestagen treatments as short as 4 days have been tested (Gordon, 1983). In Wales, Beck *et al.* (1993) studied the effect of combining a prostaglandin analogue with a 5-day progestagen treatment in synchronizing oestrus in cyclic sheep. The studies were conducted over two breeding seasons and 100% of ewes were mated by rams within a 3-day period. The authors concluded that PG combined with progestagen in this way would produce levels of oestrus synchronization and fertility comparable with other methods. Obviously, with synchronization treatments spanning periods as short as 5 days, there is a saving in time compared with some of the other options available.

3.3. Progestagens and a Standard Breeding Procedure

Progesterone and progestagens can be administered in several ways to sheep, including injections, oral treatments, the use of implants or by various intravaginal devices. Some of the common progestagens and their synonyms are given in Table 3.1.

3.3.1. The oral route

The highly potent progestagen, medroxyprogesterone acetate (MAP) was used in sheep in the USA by several workers in the 1960s, the compound being administered in daily oral doses (Evans *et al.*, 1962; Hinds *et al.*, 1964). In Australia, Lindsay *et al.* (1967) used either 40 or 80mg doses of MAP per sheep daily over a 16-day period and recorded oestrus in only 58% of cyclic ewes after the treatment. In the 1970s, Norwegian workers used MAP in daily doses of 50mg over a 10-day period and recorded 89% coming in oestrus within six days and 74% of these becoming pregnant at the controlled heat

Table 3.1 Common progestagens and their synonyms.

Name	Synonyms	Full name
Medroxyprogesterone acetate	MAP, methyl acetoxyprogesterone, Veramix, Repromix, Provera	17α-Acetoxy-6α-methyl pregn-4-ene-3, 20-dione
Flurogestone acetate	FGA, Cronolone, SC9880 Chronogest, Synchro-mate	17α-Acetoxy-9-fluoro-113-hydroxy pregn-4-ene-3, 20-dione
Norgestomet	SC 21009	19α-Acetoxy-113-methyl 19-nor pregn-4-ene-3, 20-dione

period (Velle and Helle, 1979); although the authors maintained that such results were better than most after controlled breeding, including those after using intravaginal sponges, the various considerations involved in oral dosing (time, labour, cost) make such methods commercially impracticable. In Ireland, for example, using the intravaginal sponge, it would not be uncommon to obtain an oestrous response in excess of 95% and conception rates in excess of 75% in cyclic ewes in the breeding season (Gordon, 1983).

Use of melegestrol acetate in sheep

According to Umberger *et al.* (1994), the decreasing size of the sheep industry and the limited use of artificial insemination with sheep in the USA has resulted in a reluctance on the part of manufacturers of synthetic progestagens and suppliers of exogenous sources of gonadotrophins (PMSG) for oestrus synchronization to make the substantial investment required for product approval in that country. For that reason, products such as melengestrol acetate (MGA), administered orally for suppressing oestrus in feedlot heifers, have been used for the control of oestrus and ovulation in seasonally anoestrous ewes (Safranski *et al.*, 1992; Umberger and Lewis, 1992; Jabbar *et al.*, 1993). The introduction of a preparation licenced for use in pigs (PG-600) has provided a commercially available supply of PMSG (5ml PG-600 contains 400IU PMSG + 200IU human chorionic gondatrophin (hCG)). A paper by Umberger *et al.* (1994) reports on the use of MGA fed daily at the rate of 0.3mg over 10 days with PG-600 given at treatment withdrawal. Although the reasons for such 'make-shift' treatment arrangements might be understandable in the commercial context of sheep farming in the USA, they must surely have little appeal to a wider audience.

3.3.2. Implant treatments

An alternative approach to the intravaginal route for sustained progestagen administration in sheep is the use of the subcutaneous implant, the earliest form of which was the silicone rubber progesterone-impregnated device (Dziuk and Cook, 1966). Reports on the use of this implant (Sil-Estrus, Abbott; impregnated with 375mg progesterone) appeared in the 1970s from workers in the USA and elsewhere, especially in Greece (Xenoulis *et al.*, 1972). The Greek workers observed that the use of these implants required greater skill and experience than sponges. Working with the Sil-Estrus implants in Ireland, it was not possible to match the speed and simplicity of the intravaginal sponge technique (O'Reilly, 1972; Keane, 1974; Gordon, 1975a,b).

A subsequent approach took the form of a much smaller implant designed for use in the ear and impregnated with the potent progestagen, norgestomet. However, results for the norgestomet ear-implant in Irish studies did not make the device a realistic alternative to the intravaginal sponge (Boland *et al.*, 1979). Elsewhere, the 3mg norgestomet ear implant has been employed over a 10-day period, combined with an initial oestrogen/progestagen injection (0.5mg oestradiol valerate + 1.5mg norgestomet), administered at implant insertion (Spitzer and Carpenter, 1979); 95% of sheep were in oestrus within five days and 62% of all ewes treated became pregnant at the controlled heat.

Implant vs. sponge comparisons

In Canada, Ainsworth and Wolynetz (1982) concluded that 3mg norgestomet implants presented a useful and acceptable alternative to the 40mg FGA-impregnated sponge for the induction of a fertile synchronized oestrus in ewes. However, they did stress that further detailed and systematic studies were required before the potential and applicability of the implant treatment could be realized. In more recent times, a comparison of implants (2mg norgestomet) and 60mg MAP sponges for out-of-season breeding in sheep was reported by Tritschler *et al.* (1991) in Massachusetts; ewes were treated with 500IU PMSG at time of implant/pessary withdrawal. The authors found no significant difference in oestrous response, pregnancy rate and litter size between the two treatments.

In the USA, because the Synchromate B (norgestomet ear implants) treatment is licenced for commercial use in oestrus synchronization in beef cattle, it has been available for commercial use in out-of-season breeding in sheep (Carpenter and Spitzer, 1981). As mentioned earlier, agents that are routinely employed in oestrus control in sheep around the world (FGA; MAP; CIDRs) are not available in the USA due to the costs involved in getting them licensed. However, the 3mg norgestomet implant can be used in conjunction with PG-600 at treatment withdrawal as one treatment option (Umberger *et al.*, 1994).

3.3.3. Standard intravaginal sponge treatment

In considering the most appropriate hormonal technique for sheep fertility control, it should be emphasized that this is but one of several elements necessary for a successful mating response and lambing outcome. The basic sponge technique, as adopted in Ireland in the early 1970s, involved intravaginal treatment for 12 days with a suitable progestagen, dusting the sponges with an appropriate antibiotic powder at insertion and administering 500–750IU PMSG by intramuscular injection at sponge withdrawal. This is followed by introduction of sexually proven rams (1:10 ratio) 48h after terminating treatment and running the sheep at pasture in groups not exceeding 50 ewes and their five accompanying rams.

Such a technique has been employed in Ireland in all months of the year, and among ewes in all the different physiological states in which they are found in normal farm practice (prepubertal, anoestrous and cyclic). This is not the same as saying that the controlled breeding technique as described is equally effective in all conditions, but simply that considerable data are available on the level of response that may be expected (see Table 3.2). In looking at these

Table 3.2 Lambing outcome in response to standard progestagen–PMSG treatment in Ireland (from Gordon, 1983).

	Season		
	Spring	Summer	Autumn
Groups	83	594	41
Ewes treated	2508	21,545	1600
In oestrus (%)	93.0	97.0	97.2
Ewes lambing	871	13,795	1206
Conceptions (%)	37.0	66.0	77.6
Lambs born	1375	22,396	2088
Lambs/conception	1.58	1.62	1.73
% Ewes pregnant (first oestrus)	34.7	64.0	75.4
% Ewes pregnant (first and second oestrus)	35.0	79.6	90.5

figures, it should be borne in mind that spring-treated ewes would have lambs at foot after a February/March lambing, summer-treated ewes would be dry and anoestrous and autumn-treated sheep would be dry and cyclic.

3.3.4. Progestagens employed

Studies in the 1960s and 1970s in Australia (Robinson, 1967), France (Colas, 1975) and Ireland (Gordon, 1975a,b) led workers to agree that a high level of progestagen, followed by its rapid withdrawal and adequate ovarian stimulation, is a necessary prerequisite for acceptable fertility in sheep. It is now well accepted that only compounds with characteristics identical to progesterone, especially in having a short duration of activity, are suitable (Robinson, 1982, 1988). The Sydney group concentrated their efforts on one such compound (FGA) and large-scale French applications in controlled breeding have been based on this progestagen.

In Ireland, as a result of many comparative studies, it became clear that MAP at the 60mg dose level (Veramix sponge; Upjohn) could give equally good results when natural service was the method of breeding (Gordon, 1974). When FGA (30mg) and MAP (60mg) sponges were compared in ewes bred by AI rather than natural service, a small but significant advantage in favour of FGA was found; such data, based on the reduced sperm doses employed in AI, may be taken as a more sensitive indicator of the progestagen's effect on fertility.

3.3.5. Progestagen dose level and impregnation procedure

As well as the question of the particular progestagens that are to be regarded as acceptable for intravaginal applications (FGA; MAP), there are two other important considerations: dose level of compound and method of impregnation employed in the preparation of sponges. These questions are important because of the need to maintain an appropriate concentration of progestagen in the sheep's circulation to mimic the action of the corpus luteum.

Robinson (1968) suggested that many of the progestagen dosages employed to control oestrus prior to that time had probably been too low to duplicate fully the action of the natural corpus luteum. He drew attention to evidence that a progestagen dose which will inhibit ovulation in the cyclic ewe is lower than that required to condition for oestrus and that the dose required for such conditioning is lower than that required for full fertility. This was a view different to that previously expressed by Lamond (1964), who tended to the view that optimal fertility in synchronized sheep was likely to be associated with minimal doses of progesterone.

Robinson *et al.* (1968) reported that the rate of FGA absorption from the intravaginal sponge could be significantly affected by the impregnation procedure and by the initial dose of compound; absorption rate significantly affected the percentage of ewes in oestrus and the number of sheep lambing to service at the controlled heats. In Ireland, Gordon (1971b) reported that a significantly higher mating response and lambing outcome resulted from thorough dispersion of a 30mg dose of FGA in the sponge matrix; such dispersion of the progestagen in fine crystal form presumably ensured a higher uptake of the agent. The amount of FGA released from a sponge may be determined by impregnation procedure rather than by the actual dose employed. It has been shown, for example, that sponges impregnated with a 15mg dose of FGA in a well-dispersed form can result in a significantly higher conception rate than sponges carrying double that progestagen dose (30mg) but with the compound poorly dispersed (Robinson *et al.*, 1968).

FGA dose levels in sponges

As to the dose of FGA which can be regarded as optimal, the reports available would suggest that it lies in the range of 20–40mg. Early work by Robinson *et al.* (1967) showed a significant effect of FGA dose level on fertility over a dose range of 5–20mg but little effect between a 20mg and 40mg dose. French workers using AI as the breeding method in 'dry' ewes recommended FGA doses of 30mg in the seasonal anoestrus and 40mg in the breeding season (Thimonier and Cognie, 1971; Colas *et al.*, 1973) although the experimental grounds for such recommendations were not apparent. In Ireland, Smith *et al.* (1981) were unable to establish a difference in lambing outcome between 30mg and 45mg of FGA when employed in cyclic ewes prior to breeding by AI. Work with FGA in synchronizing oestrus in ewes in the UK has usually been on the basis of 30mg doses (J.J. Robinson, 1974; Vipond and King, 1979); in Canada, 40mg doses were employed in various studies reported

(Heaney *et al.*, 1980). In the USA, apparently as a requirement of the drug regulatory agencies, the permissible dose of FGA was held at 20mg.

MAP-impregnated sponges

With reference to the alternative MAP compound, work with sponges in the 1960s generally employed 40mg or 60mg doses; the 60mg dose appears to have been that adopted for commercial applications. A paper by Greyling *et al.* (1994) deals with the use of different doses of MAP (30–60 mg) in sponges applied to Merino ewes outside the normal breeding season. In the induction of oestrus in Tuj sheep in Turkey during the non-breeding-season, the MAP compound was employed in the 14-day intravaginal treatment period (Bekyurek, 1994); the oestrous response was enhanced in this work by 500IU of PMSG at sponge withdrawal.

3.3.6. Progesterone sponges

Much work was conducted in Ireland at one time using the natural steroid in sponges, usually at dose levels of 500mg or 1000mg (see Table 3.3). It was found, in extensive field trials, that the 500mg sponge could be employed in the induction of oestrus in the late stages of the non-breeding season (for early lamb production).

Progestagen without PMSG

Results achieved in the UK (Vipond and King, 1979) and in Ireland (Quirke, 1979) showed that a very high degree of control could be achieved over both mating and lambing times merely by the use of progestagen sponges alone

Table 3.3 Relative efficiency of progestagen and progesterone sponge treatments used in early breeding trials (from Gordon, 1983).

	60mg Medroxyprogesterone acetate	30mg Cronolone	500mg Progesterone	1000mg Progesterone
Sheep	181	176	175	184
Pessary loss				
(no.)	–	1	–	11
(%)	nil	0.56	nil	5.9
Sheep bred				
(no.)	171	172	161	158
(%)	94.5	98.3	92.0	91.3
Ewes lambing (no.)	129	125	109	113
Conception rate	75.4	72.3	67.7	71.5
Lambs born	210	197	175	161
Lambs/conception	1.63	1.58	1.60	1.42

during the breeding season. However, it is essential to ensure that progestagen treatments are not applied until it is certain that the sheep in question are showing spontaneous reproductive activity; this will vary with the breed or cross of ewe and other factors, such as age. In Ireland, this means for Suffolk-crossbreeds, not starting intravaginal treatment until at least the end of the first week of August, for Galways, no earlier than mid-September and for Cheviots, no earlier than October.

Other considerations

The placement of the intravaginal sponge may affect the incidence of sponge loss, which in normal circumstances should not exceed 0.5%; it is important that the device is placed up against the cervix as deep in the vagina as possible. Dusting of sponges with an antibiotic powder is recommended on the basis of studies in Ireland (Joyce and Gillespie, 1973). At withdrawal, there will be the release of a small volume of fluid; this is an accumulation of vaginal fluid which does not interfere with the ewe's welfare, nor does it affect fertility. Applicators should always be washed off after each sponge insertion. Sponges should never be lubricated, as this can markedly influence their loss; ensure that the sponge draw-strings are clearly visible.

The outcome of controlled breeding using the intravaginal sponge and PMSG can be markedly influenced by the body condition and age of sheep, and by the degree of stress (social or physical) to which the animals are exposed around the time of mating. These same factors are well enough recognized in sheep fertility under any circumstances; they apply to rams as well as ewes and are embraced by the term 'stockmanship'. Ewes in the early months after parturition and those nursing lambs are clearly in a category of their own and require separate consideration.

Use of progestagen sponges in milking sheep

A paper by Karagiannidis (1995) in Greece notes that in that country ewes are often synchronized during lactation and that the milk is used for human consumption. This author examined ewes treated with Veramix (60 mg MAP) and recorded that 0.08% of the MAP was excreted into the milk from the moment of sponge insertion until 5 days after its removal; no attempt is made to assess the implications of this in terms of human health. Large numbers of dairy ewes are treated annually with progestagen-impregnated sponges in countries such as France.

3.3.7. Controlled internal drug releasing device

In the 1980s, trials were conducted in New Zealand and several other countries, including Ireland, on the use of a controlled internal drug releasing (CIDR) dispenser, an intravaginal device constructed of a progesterone-impregnated silicone elastomere moulded over a nylon core. The CIDR was developed in New Zealand as an intravaginal device for oestrus and ovulation

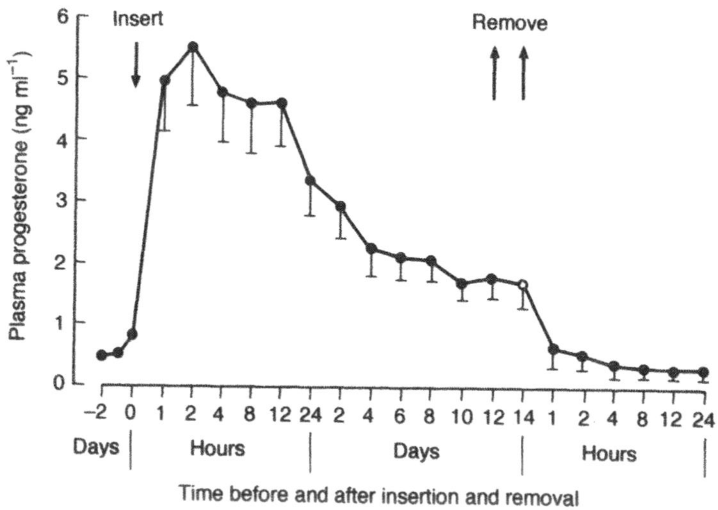

Fig. 3.1. Plasma progesterone levels (±SEM) in spayed ewes before, during and after treatment with CIDRs (from Ainsworth and Downey, 1986).

control in sheep (Welch *et al.*, 1984). The CIDR is suitable for parous ewes and ewe-lambs. Plasma progesterone level increases rapidly after insertion of the device and reaches the highest concentration 3 days after insertion and then decreases gradually (see Fig. 3.1).

Potential advantages of CIDRs

One incentive for work with CIDRs lies in the belief that it may prove simpler to gain clearance from drug regulatory authorities for progesterone-based intravaginal devices, rather than for those based on the synthetic progestagens (FGA; MAP). A further point which is often emphasized is that withdrawal of CIDRs is not accompanied by the fluid discharge seen at sponge withdrawal, and for such reasons the CIDR is aesthetically more pleasant to handle than a sponge (Carlson *et al.*, 1989).

In New Zealand, Knight *et al.* (1988) compared the effects of using CIDRs with progesterone sponges for the synchronization of oestrus in Romney ewes. They record that ewes treated with CIDRs had an earlier onset of oestrus than ewes treated with sponges. During the first 10 days after treatment, fewer ewes treated with CIDRs mated (87% vs. 94% in sponge-treated ewes); this apparently was due to a higher loss rate of CIDRs (6.3%) compared with sponges (0.8%). A report by Scott and Montgomery (1990) dealt with ovulation rates of CIDR-synchronized ewes over the peak of the breeding season in New Zealand; their results suggested that fluctuating ovulations did not occur in CIDR-treated ewes in contrast to previous observations in unsynchronized sheep.

CIDR trials in Ireland

In Ireland, Crosby *et al.* (1988) evaluated the use of four intravaginal progestagen/progesterone treatments combined with varying PMSG dose levels in anoestrous and cyclic ewes and substantiated preliminary findings using large-scale field trials on commercial farms.

In one trial, these authors found that in anoestrous sheep, oestrous response, lambing rate and litter size did not differ significantly among the four intravaginal treatments (see Table 3.4); however, the percentage of ewes mated and the lambing rate were significantly lower in the progesterone-treated cyclic sheep than in those receiving FGA or MAP.

In a second trial with ewes during late anoestrus/early breeding season, the same authors recorded a significant effect of intravaginal treatment on lambing rate to first service (see Table 3.5); the difference was evident on all the ten farms participating in the study. It was concluded that the CIDR device was effective in dealing with anoestrous ewes but could not match the synthetic progestagens in cyclic ewes. Further information on progestagen/progesterone treatments, PMSG dose levels and mating management systems in anoestrous sheep is given in a report by Crosby *et al.* (1991).

Table 3.4 Oestrous response and lambing outcome in relation in intravaginal treatment employed (from Crosby *et al.* 1988).

	Anoestrus				Cyclic			
	FGA	MAP	PROG	CIDR	FGA	MAP	PROG	CIDR
Ewes treated	61	62	62	62	25	25	25	25
% mated	98	97	92	95	100	92	84	88
% lambing	70	68	70	80	88	80	68	56
Litter size	1.63	1.91	1.71	1.71	2.05	1.70	1.59	1.71

Devices compared: FGA sponge; MAP sponge; 500mg progesterone sponge; CIDR.
Seasons compared: late anoestrus; normal breeding season.

Table 3.5 Oestrous response and lambing outcome in ewes treated with FGA sponges or CIDRs in late anoestrus (from Crosby *et al.* 1988).

Treatment	FGA	CIDR
No. present at lambing	403	403
% lambing (1st service)	77.2	66.7
Litter size	1.92	1.90

CIDR studies in North America

Studies reported by Ainsworth and Downey (1986) from Canada led them to the conclusion that a 12- or 14-day treatment with CIDR dispensers presented a useful and acceptable alternative to the FGA-impregnated sponge for the control and synchronization of the oestrous cycle of adult cyclic ewes. However, they note that data for the device were not available over the range of environmental and physiological conditions covered in FGA sponge applications; they suggested that further more detailed and systematic studies were required before the overall effectiveness of the CIDR could be assessed.

In Minnesota, Hamra *et al.* (1986) used 12% and 9% CIDR dispensers in ovariectomized sheep and recorded that they maintained higher progesterone levels over a longer period than did progesterone-impregnated sponges; they reported the CIDRs to be easy to insert and that 87% were retained. Out-of-season fertility trials, in which progesterone sponges and CIDRs (type S) were compared with 30mg FGA sponges, were reported in a further paper by Hamra *et al.* (1989); the authors record in these trials, that the percentages of ewes that were bred and lambed tended to favour the FGA sponges over the progesterone devices. In the breeding season, Carlson *et al.* (1989) evaluated CIDR-S dispensers in terms of their ability to produce a fertile, synchronized oestrus; following a 12-day treatment period, they record that 91% of ewes were bred and 67% lambed in response to matings within five days. Such data would be in agreement with CIDR experiences in Ireland (see Table 3.5).

In more recent studies in Minnesota, Wheaton *et al.* (1993), again with intact ewes, reported using CIDRs to synchronize oestrus in the breeding season; 74% of treated ewes lambed within a 6-day period and 20% lambed 16 days later. The same authors compared CIDRs with progesterone sponges, both in combination with PMSG, for out-of-season breeding and reported both treatments to be effective.

Elsewhere in the USA, Hunnicutt *et al.* (1995) note that oestrous synchronization procedures available to them in that country have not been effective or cost-efficient enough for use by farmers who have large numbers of sheep under range conditions. These authors tested the effectiveness of CIDRs applied over a 12-day period and record that they outperformed $PGF_{2\alpha}$ treatment (two PG doses at an interval of 7 days) as a method of synchronizing oestrus in mature range ewes. The CIDR at that time had not been licensed for commercial use in the USA.

CIDRs in South Africa

A report by Greyling and Brink (1987) in South Africa compared MAP sponges with CIDRs in mature Karakul ewes in the breeding season. They record a higher incidence of loss with the CIDR (13.5% vs. 6.7%) and a somewhat lower pregnancy rate (72.2% vs. 79.3%) but concluded that the device warranted further research.

Onset of oestrus after CIDR treatment

Most of the reports for CIDRs have shown an earlier onset of oestrus, relative to the time of withdrawing the device, than that found with progestagen sponges. In New Zealand, for example, Smith *et al.* (1991a,b) dealing with several thousand ewes, recorded that sheep treated with sponges had a significantly longer (+8h) interval to heat onset than those treated with CIDRs; they concluded that this time variation was large enough to affect conception rates when AI was carried out on a fixed-time basis. Other work in New Zealand, reported by Shackell (1991), recorded that the time from intravaginal device withdrawal to oestrus was significantly shorter for CIDRs (30h) than for MAP (42h) or FGA (40h); this earlier onset was associated with an earlier preovulatory LH surge and time of ovulation. Studies reported by Knight *et al.* (1992) noted that the interval from CIDR removal to ovulation was about 8h earlier in cyclic ewes injected with 400IU PMSG.

Re-use of CIDRs

The efficacy of used CIDR devices for synchronization of oestrus has been examined by some workers. In one such study in New Zealand, groups of sheep in each month of the year were treated for 14 days with CIDRs or a CIDR that had been used for 14 days and washed (Smith *et al.*, 1991a,b); an injection of 400IU PMSG was given at time of CIDR withdrawal. The authors found that the incidence of oestrus was significantly higher for new CIDRs (88%) than for used devices (68.5%); pregnancy rate was also 7% higher with new CIDRs. Various authors have reported on the re-insertion of CIDRs in an attempt to enhance pregnancy rates by providing additional progesterone in sheep that are already bred. In Japan, for example, Fukui *et al.* (1994) re-inserted CIDRs seven days after breeding and left them in place for 14 days; they recorded no effect on pregnancy or lambing rates.

3.3.8. *Oestrus synchronization of sheep in the tropics*

The characteristics of oestrus and ovulation in Awassi sheep in Syria were recorded after intravaginal progestagen treatment by Kassem *et al.* (1990). Elsewhere, trials were conducted with ewes of several breeds by Godfrey *et al.* (1995) on using oestrus synchronization procedures (CIDRs or PG injections) in the tropics; the authors concluded that both these methods could be used in sheep in the tropics without adversely affecting fertility.

3.4. The Use of PMSG in Conjunction with Intravaginal Progestagen Treatment

As already mentioned, the intravaginal progestagen treatment alone (FGA; MAP) is adequate in controlling oestrus among cyclic ewes in the full breeding season (Robinson, 1988). After progestagen withdrawal in such sheep, there presumably is a surge of gonadotrophin from the anterior pituitary sufficient to initiate the sequence of hormonal events that result in oestrus and ovulation. However, for progestagen treatment to be effective in the induction of oestrus in the non-breeding season, there does need to be sufficient gonadotrophin available to initiate these preovulatory events and this necessitates the augmentation of endogenous gonadotrophin with some amount of exogenous FSH; the usual agent used for this purpose is PMSG.

3.4.1. PMSG doses

Several considerations may influence the decision as to whether an FSH-type preparation should be employed and, if employed, the dose level and timing of its administration. Although, as already mentioned, cyclic ewes can be expected to come in oestrus shortly after progestagen withdrawal in the absence of exogenous gonadotrophin, a low dose of PMSG (375IU) can result in a more predictable and precise synchronization of oestrus/ovulation, which can have a favourable effect on the outcome of set-time AI applications (Colas *et al.*, 1973; Jennings and Quirke, 1976).

In certain breeds of sheep, PMSG can also have the additional merit of inducing a mild superovulatory response; there is evidence that PMSG can bring the twinning percentage of breeds characterized by low litter size up to a more acceptable level (Gordon, 1975b). As to the choice of dose level of gonadotrophin, this would appear to lie within the range 375–750IU PMSG; it is possible, however, to depress rather than enhance conception rate after progestagen–PMSG treatment in sheep by increasing the dose beyond a certain point (Larson *et al.*, 1970; Gordon, 1971b; Botha *et al.*, 1975). It is believed that overstimulating the ewe can mean 5–6 ova are shed which may result in subsequent embryo mortality and the birth of a single lamb. In Ireland, for example, Crosby *et al.* (1991) found that increasing PMSG dose from 500 to 1000IU depressed litter size (1.90 vs. 1.52) in progestagen-treated cyclic sheep.

Although there is evidence in cyclic sheep that superovulatory response may be substantially increased when PMSG is given several days prior to withdrawal of intravaginally administered progestagen (Gordon, 1969a), this does not necessarily apply to anoestrous ewes; Gordon (1969b) reported a reduced ovulatory and oestrous response in sheep receiving 500IU PMSG several days prior to progestagen withdrawal in the ewe anoestrus. On the question of administering the gonadotrophin, South African evidence has shown that an intramuscular injection of PMSG resulted in a superovulatory response more than double that observed after a subcutaneous dose (Boshoff and Burger, 1973).

PMSG source

There have been occasions when the source of PMSG has been shown to have an effect on the results of oestrus control measures. In Ireland, oestrus was synchronized in 2112 ewes with progestagen sponges (FGA/MAP) and doses of 500IU PMSG using two different commercial preparations (Al-Kamali *et al.*, 1990); the authors recorded that PMSG source significantly affected conception rate and litter size when the sheep were subsequently bred by AI. It is clearly necessary to have confidence in a PMSG preparation if optimal results are to be obtained.

3.4.2. Refractoriness to PMSG

Over the years, several workers have examined the question of possible refractoriness in sheep to repeated doses of PMSG. Clearly, in some commercial applications (e.g. early lamb production), it may be expected that individual ewes would be treated with PMSG on several occasions in the course of their reproductive life. It has been shown in some studies, however, that even after 17 injections of PMSG given 17 days apart, ewes did not show evidence of refractoriness or the development of antibodies (Gherardi and Lindsay, 1980). In the USA, Diekman *et al.* (1995) injected PMSG after norgestomet implant treatment of ewes in three successive years and noted no antibody production in the sheep.

3.4.3. Onset of oestrus after progestagen–PMSG

Several factors may influence the extent of the interval between the end of progestagen treatment and the start of the controlled heat period; generally, it is an interval of about 36h, although some ewes may be in oestrus as early as 24h or as late as 48h (Gordon, 1975a). The use of PMSG at progestagen withdrawal will certainly shorten the interval to oestrus, as shown in South African studies (Botha *et al.*, 1975).

There has even been some suggestion that the time of day (morning or evening) at which intravaginal sponges are removed may influence the interval to oestrus: Robinson (1980) noted that a diurnal pattern of mating was evident in normal cyclic sheep and suggested that oestrus may be affected in some similar way by the time of day when progestagen treatment ceases. Elsewhere, and this time with progesterone implants, it was reported that the time of day of removing the implant in anoestrous ewes had an effect on the interval from removal to the preovulatory LH surge (Cunningham *et al.*, 1977). In further studies, Cunningham *et al.* (1980) found that although there was a 12h difference in the time of removing implants from ewes (17.30 vs. 05.30), the times of matings differed by no more than 2h; the authors suggested that there may be diurnal variations in the responsiveness of neural centres responsible for oestrus. For those who may be considering the use of sheep AI on a fixed-time basis, clearly it is important to have information about all factors that may influence the time interval between progestagen–PMSG and oestrus/ovulation.

3.5. Rams and Mating Management

In commercial sheep farming in Ireland, ewes are normally bred by natural service; the same is true in most other sheep-producing countries. As already mentioned, the success or otherwise of controlled breeding in sheep is not alone a question of influencing the reproductive processes in the ewe. The outcome is also likely to depend on the capability of the ram (activity and semen quality). Part of the difficulty, in fact, is created in keeping treatment procedure in ewes as simple as possible; to ensure this, the standard practice involves the simultaneous insertion and removal of sponges in all the sheep selected for treatment. This results in oestrus occurring at the same time in ewes, about 36h after progestagen withdrawal. Such a synchronization of heat periods can result in the ram being confronted with a situation radically different from any that faces him in the normal course of events (Fig. 3.2).

3.5.1. Ram management practices

Several factors are known to be important for successful controlled breeding by natural service in sheep. First and foremost, the ram should be in good health, sexually experienced and possess a clear record of achievement in

Fig. 3.2. Ewes in oestrus after progestagen–PMSG pose novel problems for the ram. Rams that are to be used in controlled matings should always be males that have previously worked well among sheep. Provided the testes and genitalia are normal there is nothing to be gained by semen examinations unless these are carried out by someone well experienced in this area.

producing pregnancies in the natural breeding season. Ram fertility depends directly on semen production rate, reserves and quality of sperm and upon libido and mating competence. Fortunately for those concerned with oestrus control in sheep, the ram belongs to a species which is noted for its large testicular size and high sperm-production rate. As noted by Harvey and May (1989), the ram holds sperm reserves equivalent to about 95 ejaculates in contrast to species such as the rabbit with reserves of about 30 and humans with less than two.

Ram management procedures previously described in the 1960s (Gordon, 1963) suggested that progestagen–PMSG-synchronized ewes should be joined with rams at the end of treatment, that sheep should be confined to small fields or paddocks in the first few days after treatment, that a ram to ewe ratio of 1:10 should be employed and that the size of the sheep groups should not exceed 50 ewes and their accompanying five rams.

Timing of ram introduction

Towards the end of the 1960s, comparisons were made between the standard practice of introducing rams at progestagen withdrawal to an alternative system in which ram introduction was delayed until 48h later; it appeared that conception rate could be improved at the controlled oestrus by this relatively simple adjustment (Gordon, 1971a). Subsequent reports in Ireland by Joyce (1972) and Boland and Gordon (1979) provided further evidence in support of the finding (see Table 3.6). Elsewhere, Bryant and Tomkins (1976) in the UK recorded observations that were not at variance with the Irish data.

Distribution of sperm doses

It should be mentioned that a substantial body of data had been collected in the 1960s and 1970s on the timing of oestrus (onset and cessation) in progestagen–PMSG treated sheep. In introducing rams at 48h it was known that they were joining a group of ewes, most of which would be several hours into oestrus (Gordon, 1971a,b). On the basis of well-established sheep behavioural patterns, as reported by authors such as Hulet (1966), in studies

Table 3.6 Effect of timing of ram introduction on the conception rate of FGA–PMSG-treated ewes in late anoestrus (from Boland and Gordon, 1979).

	Ram in at pessary withdrawal	Ram in at 48h after pessary withdrawal
Sheep	79	80
Bred at induced oestrus	76 (96.2%)	77 (96.3%)
Conceptions at induced oestrus	30 (39.5%)	52 (67.5%)
Lambs born	43 (1.43)	80 (1.54)

Trials with 16 pens of sheep – 10 ewes and Suffolk ram per paddock ($P < 0.01$).

in which rams were found to prefer breeding ewes fresh in oestrus and not recently mated, it is to be expected that the initial ejaculates (with the highest sperm doses) might be distributed more uniformly than in the usual ram joining system, in which males may use up their sperm reserves unduly dealing with the first few ewes in oestrus in the group.

Changing conditions during oestrus

When ewes are perhaps 12 and more hours into the controlled oestrus before mating takes place, conditions in the reproductive tract of the ewe may also be more favourable to sperm transport/survival after the delayed introduction of the ram. It is known, for instance, that the flow of cervical mucus can be markedly influenced by progestagen treatment, and that this is probably a consequence of changes in the way in which oestrogen is being produced (Smith and Allison, 1971). It may be that a heavy flow of cervical mucus in the early hours of oestrus may serve to dilute ram semen to the point at which it is more difficult for an adequate cervical population of sperm to become established. It is known that the cervix is one of the sperm reservoirs in the ewe (Mattner, 1966) and it seems possible that factors that affect the retention and survival of sperm in this site may influence fertility.

Mention should be made that mating in the late stages of oestrus may have its own share of problems; Mattner and Braden (1969) and Killeen and Moore (1970) showed that the efficiency of sperm transport declines in the late stages of the natural oestrus. This may be one of the sheep's protective mechanisms aimed at preventing the fertilization of aged oocytes.

3.5.2. Ewe : ram ratio

In regard to the ram : ewe ratio, this was held at 1:10 in most of the controlled breeding applications in Ireland in the 1960s and 1970s. For most flock conditions, it would not be considered commercially feasible to employ a ratio less than this, although French workers recommended one ram for every five ewes under some flock conditions (Colas *et al.*, 1974) and Bryant and Tomkins (1975) in the UK suggested that an individual ram should not be allowed to serve more than six progestagen-treated ewes if high levels of fertility are sought. There is much anecdotal evidence suggesting that results improve with greater ram numbers, but convincing evidence is more difficult to find.

Some workers have demonstrated the possibility of employing a normal ram : ewe ratio of 1 : 50 among synchronized sheep (Galindez *et al.*, 1977); however, this involves 'staggering' both the start and termination of intravaginal treatments and it would not be favoured by farmers anxious to have a very compact lambing period.

One practical problem which the 1:10 ram:ewe ratio raises is that of the farmer finding enough rams to meet his needs. In Ireland, ram sharing arrangements have been set up as the means of overcoming this difficulty in some instances. In this, rams were moved to a farm a day or two before the ewes were due to be bred and worked away in the flock on the day of mating; with such systems, the males can be employed once weekly for a period of several weeks.

3.5.3. Hand-mating techniques

Workers in the State research organization (Agricultural Institute) in Ireland concentrated time and attention in ram management routines to be used in association with controlled breeding (Jennings and Crowley, 1972; Jennings, 1976, 1977); the suggestion was made in several reports that near-normal conception rates in progestagen–PMSG-treated ewes may be achieved by 'hand-mating' each ewe on one or two occasions during the controlled oestrus. The Irish workers produced evidence indicating that substantial discrepancies may exist between the number of synchronized ewes 'marked' by rams and the number actually inseminated; it was suggested that much of the subfertility recorded previously among progestagen-treated sheep was a reflection of inadequacies among the rams, either in libido or fertility, in coping with a situation in which a group of ewes are in oestrus simultaneously.

Undoubtedly, there is some truth in this view, but it may not always be clear what 'hand-mating' may mean in terms of ram supply and in the time and labour involved in its application. Australian stud farmers accustomed to breeding Merinos by hand-service regard the practice as expensive, time-consuming and generally less effective in getting ewes pregnant than paddock mating. Used with synchronized sheep which are housed and accustomed to handling and with plenty of rams to choose from, the hand-mating system can be the means of inseminating larger than usual sperm doses into the ewe's tract at stated intervals during oestrus. In Scotland, J.J. Robinson (1974), in his intensive lamb production studies at the Rowett Research Institute, achieved excellent conception rates in Finn × Dorset crossbreds after two 'hand-matings' during the synchronized oestrus with an interval of 12h separating the two services.

With ewes and rams accustomed to pastoral conditions, it becomes a matter of establishing clearly just how far the lambing outcome can be improved beyond that found in unsupervised mating systems and whether this justifies the time and labour involved in its application. In fact, evidence reported by Joyce (1972) made it clear that when rams are kept away from ewes until 48h after terminating oestrus control treatments, paddock matings resulted in conception rates equal to those achieved by a 'hand-mating' routine (one hand-service at 48h and ewes thereafter exposed to rams in the field). In more recent times, Crosby *et al.* (1991) examined the effect of mating management systems (hand or paddock mating) in breeding progestagen/progesterone–PMSG-treated ewes in late anoestrus; they record that mating method did not affect the percentage of ewes mating or lambing, although there were indications that paddock mating resulted in a higher litter size.

3.5.4. Other considerations

Of the several possible methods that are commercially available for the control of oestrus and ovulation in the sheep, the intravaginal sponge (impregnated with FGA or MAP) or CIDR are the simplest to apply; indeed, these treatments can be handled by the farmer after a minimum of training. Alternatives such as the implantation method, using the norgestomet ear-implant, may not have this same advantage. Other agents, such as prostaglandin $F_{2\alpha}$ and its analogues, have probably even less appeal, on the basis of the variable fertility data which is associated with them.

Rams and fertilization rates

There is ample evidence to show that any subfertility associated with progestagen treatment in sheep is probably the result of a failure in the fertilization process rather than anything else; this failure stems from some impairment of normal transport and survival of sperm in the cranial part of the cervix (Robinson, 1973). For such reasons, fertilization failure can possibly be avoided in certain ewe categories ('dry' anoestrous and 'dry' cyclic sheep) by employing a much larger than usual number of sperm, whether by delaying ram introduction until 48h, using a 'hand-mating' routine or an appropriate AI procedure. In considering ram management routines, it may be relevant to question whether 'hand-mating' under certain conditions may constitute an additional stress factor which may reduce rather than enhance conception at the synchronized oestrus; it is possible that with certain flock conditions or perhaps with some ewes and rams, the application of the 'hand-mating' technique may be counter-productive. Similar arguments could be used with reference to AI carried out under conditions in which ewes are subjected to unusual stresses and strains.

3.5.5. Irish rams in controlled breeding

It should be noted that the Irish results for ram performance in synchronized sheep flocks relate mainly to males of the Suffolk breed, the predominant fat-lamb sires in the country. They also relate to those rams that are known to have previously 'worked' satisfactorily in flocks. As mentioned at several points previously, it would appear that Suffolks and rams of other breeds in Ireland produce semen apparently of good quality (see Table 3.7), and maintain their sex drive at a high level throughout the year. This is not to say that the most intense sexual activity and highest quality semen is not to be found in the autumn season, but there is no evidence that serious ram inadequacies occur in the ewe anoestrus comparable to those recorded elsewhere (Colas *et al.*, 1974; Colas, 1979).

Table 3.7 Effect of season on the incidence of sperm abnormalities in Texel rams (from Gordon, 1963).

Season	No. of collections	Morphologically abnormal sperm			
		Primary (%)	Secondary (%)		
			Free heads	Bent tails	Total
Winter	43	0.18	2.24	0.30	2.54
Spring	41	0.21	2.96	0.46	3.42
Summer	42	Nil	2.29	0.20	2.49
Autumn	44	Nil	1.57	0.14	1.71
Totals	170	0.10	2.26	0.29	2.55

French workers have reported a significantly higher proportion of secondary sperm abnormalities (especially sperm showing the cytoplasmic droplet in the distal position) among Ile-de-France rams when daylight increased than when it decreased (22% vs. 10%).

3.6. Compactness of Lambings

The application of controlled reproduction, when ewes in the flock are bred on the one day, can mean that up to 80% of lambs born in the flock are produced within a period of little more than a week, thus giving an extremely compact lambing. As mentioned previously, the sheep which can be expected to give the most compact lambings, on a flock basis, are likely to be those belonging to highly prolific breeds that usually have a high conception rate to first service.

Compact lambings can have several advantages, especially in flocks of less than 100 ewes where the hours spent attending ewes can be literally halved and the potential for saving lambs around the time of birth can be increased. At an earlier stage, of course, scanning by ultrasound for identifying ewes with singles and multiples is that much easier in flocks where ewes are at similar and accurately known stages of pregnancy. A compact rather than a protracted lambing also ensures that supplementary meal feeding can be more accurately timed according to the stage of pregnancy, producing at time of birth a more even-sized batch of lambs. The cross-fostering of lambs at birth is made much easier than usual and management of lambs for clostridial disease control and castration/tail-docking can be carried out at optimum times. Advantages carry on through until marketing, when the lambs for sale are likely to be ready at much the same time.

Other advantages of compactness

Sheep farmers in Ireland traditionally join rams with the ewe flock for a 6–8-week period during the autumn breeding season, a practice which may often result in a very protracted lambing season. For that reason, the concentration of both mating and lambing into a short and predictable period has several advantages, over and above those already mentioned. A foreknowledge of the mating and lambing dates can permit economies in feed requirements, both at 'flushing' in the autumn mating period and during the supplementary period prior to lambing; parturition in ewes can be supervised more closely and optimum use made of overnight housing and shelter, thereby assisting in reducing lamb losses.

3.7. References

Acritopoulou-Fourcroy, S., Pappas, V., Peclaris, G. and Zerras, M. (1982) Synchronization of oestrus in ewes with Provera sponges/PMSG, prostaglandin $F_{2\alpha}$ or the prostaglandin analogue, ICI 80996 and fertility following natural mating or artificial insemination. *Reproduction, Nutrition and Development* 22, 345–354.

Ainsworth, L. and Downey, B.R. (1986) A controlled internal drug-release dispenser containing progesterone for control of the estrous cycle of ewes. *Theriogenology* 26, 847–856.

Ainsworth, L. and Wolynetz, M.S. (1982) Synchronization of estrus and reproductive performance of ewes treated with synthetic progestagens administered by subcutaneous ear implant or by intravaginal sponge pessary. *Journal of Animal Science* 54, 1120–1127.

Al-Kamali, A.A., Crosby, T.F., Boland, M.P., Kelleher, D.L. and Gordon, I. (1990) Effects of progestagen type and PMSG source on lambing outcome in ewes following artificial insemination. *Irish Veterinary Journal* 43, 99–103.

Beck, N.F.G., Davies, B. and Williams, S.P. (1993) Oestrous synchronization in ewes: the effect of combining a prostaglandin analogue with a 5-day progestagen treatment. *Animal Production* 56, 207–210.

Bekyurek, T. (1994) Induction of oestrus in Tuj sheep during anoestrus. *Doga, Tuk Veterinerlik ve Hayyancilik Dergisis* 18(1), 11–15.

Boland, M.P. and Gordon, I. (1979) Effect of timing of ram introduction on fertility in progestagen-PMSG treated anoestrous ewes. *Journal of Agricultural Science, Cambridge* 92, 247–249.

Boland, M.P., Gordon, I. and Kelleher, D.L. (1978a) The effect of treatment by prostaglandin analogue (ICI-80996) or progestagen (SC-9880) on ovulation and fertilization in cyclic ewes. *Journal of Agricultural Science, Cambridge* 91, 727–730.

Boland, M.P., Lemainque, F. and Gordon, I. (1978b) Comparison of lambing outcome in ewes after synchronization of oestrus by progestagen or prostaglandin treatment. *Journal of Agricultural Science, Cambridge* 91, 765–766.

Boland, M.P., Kelleher, D. and Gordon, I. (1979) Comparison of control of oestrus and ovulation in sheep by an ear implant (SC-21009) or by intravaginal sponge. *Animal Reproduction Science* 1, 275–283.

Boshof, D.A. and Burger, F.J.L. (1973) Limitation of multiple-ovulations in Karakul ewes after the use of PMSG. *South African Journal of Animal Science* 3, 79–81.

Botha, H.K., Van Niekerk, C.H. and Pagel, R.F.E. (1975) Influence of synchronization of the oestrous period, PMSG administration and flushing on oestrus and conception of S. African mutton Merionos. *South African Journal of Animal Science* 5, 231–233.

Bryant, M.J and Tomkins, T. (1975) The flock-mating of progestagen-synchronized ewes. 1. The influence of ram-to-ewe ratio upon mating behaviour and lambing performance. *Animal Production* 20, 381–390.

Bryant, M.J. and Tomkins, T. (1976) The flock mating of progestagen-synchronized ewes. 2. The influence of time of ram introduction upon mating behaviour and lambing performance. *Animal Production* 22, 379–384.

Carlson, K.M., Pohl, H.A., Marcek, J.M., Muser, R.K. and Wheaton, J.E. (1989) Evaluation of progesterone controlled internal drug release dispensers for synchronization of estrus in sheep. *Animal Reproduction Science* 18, 205–218.

Carpenter, R.H. and Spitzer, J.C. (1981) Response of anestrous ewes to norgestomet and PMSG. *Theriogenology* 15, 389–393.

Chamley, W.A., Buckmaster, J.M., Cain, M.D., Cerini, J., Cerini, M.E., Cunningham, I.A. and Goding, J.R. (1972) The effect of $PGF_{2\alpha}$ on progesterone, oestradiol and LH secretion in sheep with ovarian transplants. *Journal of Endocrinology* 55, 253–263.

Colas, G.(1975) The use of the progestagen SC-9880 as an aid for AI in ewes. *Annales de Biologie animale Biochimie Biophysique* 15, 353–363.

Colas, G. (1979) Fertility in the ewe after AI with fresh and frozen semen at the induced oestrus, and influence of the photoperiod on the semen quality of the ram. *Livestock Production Science* 6, 153–166.

Colas, G., Thimonier, J., Courot, M. and Ortavant, R. (1973) Fertility, prolificacy and fecundity during the breeding season of ewes artificially inseminated after treatment with fluorogestone acetate. *Annales Zootechnica* 22, 441–451.

Colas, G., Brice, G. and Guerin, Y. (1974) Acquisitions recentes en matier d'insemination artificielle ovine. *Bulletin Technical Information Ministry Agriculture* 294, 795–800.

Corteel, M. (1975) Luteolysis induced by $PGF_{2\alpha}$ compared with natural luteolysis in the ewe. *Annales de Biologie animale Biochimie Biophysique* 15, 175–180.

Crosby, T.F., Boland, M.P., Murray, B.M. and Gordon, I. (1988) Effect of progestagen/progesterone treatment on the induction of pregnancy in ewes. *Proceedings of the 11th International Congress Animal Reproduction and AI* (Dublin) 4, p. 428.

Crosby, T.F., Boland, M.P. and Gordon, I. (1991) Effect of progestagen treatments on the incidence of oestrus and pregnancy rates in ewes. *Animal Reproduction Science* 24, 109–118.

Cunningham, N.F., Saba, N. and Millar, P.G. (1977) The effects of progesterone and oestradiol-17B treatment on plasma hormone levels and on the reproductive behaviour of ewes in late anoestrus and early in the breeding season. *Research in Veterinary Science* 22, 324–329.

Cunningham, N.F., Saba, N., Boarer, C.D.H. and Hattersley, J.J.P. (1980) Plasma hormone levels and reproudctive behaviour in anoestrous ewes after treatment with progesterone and PMSG. *Journal of Reproduction and Fertility* 60, 177–185.

Diekman, M.A., Neary, M.K. and Kelly, G.R. (1995) Repeated injections of pregnant mare serum gonadotrophin (PMSG) failed to induce antibody production in fall-lambing ewes. *Journal of Animal Science* 73 (Suppl. 1), 51.

Douglas, R H. and Ginther, O.J. (1973) Luteolysis following a single injection of $PGF_{2\alpha}$ in sheep. *Journal of Animal Science* 37, 990–993.

Dziuk, P.J. and Cook, B. (1966) Passage of steroids through silicone rubber. *Endocrinology* 78, 208–211.

Evans, J.S., Cutt, R.H. and Simpson, E.C. (1962) Breeding performance in ewes after synchronizing estrus by feeding MAP. *Journal of Animal Science* 21, 804–807.

analogue ICI 80996 to synchronize ovulation in sheep in an AI programme. *Proceedings of the Australian Society of Animal Production* 11, pp. 133–136.
Fairnie, I.J., Wales, R.G. and Gherardi, P.B. (1977) Time of ovulation, fertilization rate and blastocyst formation in ewes following treatment with prostaglandin analogue (ICI 80996). *Theriogenology* 8, 183.
Fairnie, I.J., Martin, E.R. and Rogers, S.C. (1978) The lambing performance of Merino ewes following synchronization of ovulation with cloprostenol, a prostaglandin analogue (80996). *Proceedings of the Australian Society of Animal Production* 12, p. 256.
Fukui, Y. and Roberts, E.M. (1977) Fertility of ewes treated with $PGF_{2\alpha}$ and artificially inseminated at predetermined intervals thereafter. *Australian Journal of Agricultural Research* 28, 891–887.
Fukui, Y. and Roberts, E.M. (1979) Comparison of methods for estrous synchronization in sheep. *Japanese Journal of Animal Reproduction* 25, 131–135.
Fukui, Y., Tabuchi, K., Yamada, A., Hayashi, N. and Tanaka, K. (1994) Effect of insertion periods of controlled internal drug release device (CIDR) on conception rate by fixed-time intrauterine insemination with frozen semen in seasonally anoestrous ewes. *Journal of Reproduction and Development* 40, 221–226.
Galindez, F.J., Prud'Hon, M. and Reboul, G. (1977) Reproductive performance of group-synchronized Merino and Romanov crossbred ewes. *Animal Production* 24, 113–116.
Gherardi, P.B. and Lindsay, D.R. (1980) The effect of season on the ovulatory response of Merino ewes to serum from pregnant mares. *Journal of Reproduction and Fertility* 60, 425–429.
Godfrey, R.W., Gray, M.L. and Collins, J.R. (1995) Estrus synchronization of sheep in the tropics using either controlled internal drug release (CIDR) dispensers or prostaglandin $F_{2\alpha}$ (PGF). *Journal of Animal Science* 73 (Suppl. 1), p. 232.
Gordon, I. (1963) The induction of pregnancy in the anoestrous ewe by hormonal therapy. *Journal of Agricultural Science, Cambridge* 60, 31–79.
Gordon, I. (1969a) Controlled reproduction in sheep and cattle. *Journal of the Irish Department of Agriculture and Fisheries* (Dublin) 66, 184–211.
Gordon, I. (1969b) Factors affecting response of anoestrous sheep to progestagen treatment. *Journal of the Irish Department of Agriculture and Fisheries* (Dublin) 66, 232–267.
Gordon, I. (1971a) Control of reproduction in sheep: towards programmed lamb production. *Journal of the Irish Department of Agriculture and Fisheries* (Dublin) 68, 3–51.
Gordon, I. (1971b) Induction of early breeding in sheep by standard and modified progestagen-PMS treatments. *Journal of Agricultural Science, Cambridge* 76, 337–341.
Gordon, I. (1974) Controlled breeding in sheep. *Irish Veterinary Journal* 28, 118–126.
Gordon, I. (1975a) The use of progestagens in sheep bred by natural and artificial insemination. *Annales de Biologie animale Bichimie Biophysique* 15, 303–315.
Gordon, I. (1975b) Hormonal control of reproduction in sheep. *Proceedings of the British Society of Animal Production* 4, 79–93.
Gordon, I. (1983) *Controlled Breeding in Farm Animals*. Pergamon Press, Oxford, pp. 181–195.
Greyling, J.P.C. and Brink, W.C.J. (1987) Synchronization of oestrus in sheep: the use of controlled internal drug release (CIDR) dispensers. *South African Journal of Animal Science* 17, 128–132.

Greyling, J.P.C. and Van der Westhuysen, J.M. (1979) The synchronization of oestrus in sheep. II. Dose effect of prostaglandin in the double injection regime. *South African Journal of Animal Science* 9, 193–195.

Greyling, J.P.C., Van der Westhuysen, J.M. and Van Niekerk, C.H. (1979) The synchronization of oestrus in sheep. I. Dosage and time of prostaglandin administration following progestagen pretreatment. *South African Journal of Animal Science* 9, 185–187.

Greyling, J.P.C., Kotze, W.F., Taylor, G.J., Hagenduk, W.J. and Cloete, F. (1994) Synchronization of oestrus in sheep: use of different doses of progestagen outside the normal breeding season. *South African Journal of Animal Science* 24, 33–37.

Hackett, A.J. and Hidiroglou, M. (1983) Effects of PMSG on progesterone levels in ewes treated with fluorogestone acetate or prostaglandin $F_{2\alpha}$. *Animal Reproduction Science* 6, 191–197.

Hackett, A.J. and Robertson, H.A. (1980) Effect of dose and time of injection of prostaglandin $F_{2\alpha}$ in cycling ewes. *Theriogenology* 13, 347–351.

Hackett, A.J., Robertson, H.A., Penner, P. and McLaughlin, G.R. (1981) Comparison of two methods of synchronizing oestrus and subsequent lambing in a commercial sheep flock. *Canadian Journal of Animal Science* 61, 67–72.

Hamra, A.H., Massri, Y.G., Marcek, J.M. and Wheaton, J.E. (1986) Plasma progesterone levels in ewes treated with progesterone-controlled internal drug-release dispensers, implants and sponges. *Animal Reproduction Science* 11, 187–194.

Hamra, A.H., McNally, J.W., Marcek, J.M., Carlson, K.M. and Wheaton, J.E. (1989) Comparison of progesterone sponges, cronolone sponges and controlled internal drug release dispensers on fertility in anestrous ewes. *Animal Reproduction Science* 18, 219–226.

Haresign, W. (1980) Controllng reproduction in sheep. *Span* 23, 88–91.

Harvey, P.H. and May, R.M. (1989) Copulation dynamics: out for the sperm count. *Nature* 337, 508–509.

Hawk, H.W. (1973) Uterine motility and sperm transport in the estrous ewe after prostaglandin induced regression of corpora lutea. *Journal of Animal Science* 37, 1380–1385.

Hawk, H.W. and Conley, H.H. (1975) Involvement of the cervix in sperm transport failures in the reproductive tract of the ewe. *Biology of Reproduction* 13, 322–328.

Heaney, D.P., Ainsworth, L., Batra, T.R., Fiser, P.S., Langford, G.A., Kee, A.J. and Hackett, A.J. (1980) Research for intensive total confinement sheep production systems. *Animal Research Institute Technical Bulletin* No. 2, Agriculture Canada Publication.

Henderson, K.M., Downing, J.M., Beck, N.F.G. and Lees, J.L. (1984) Oestrus synchronization in ewes. A comparison of prostaglandin $F_{2\alpha}$ than salt with a progestagen pessary. *Animal Production* 39, 229–233.

Hinds, F.C., Dziuk, P.J. and Lewis, J.M. (1964) Control of estrus and lambing performance in cyclic ewes fed MAP. *Journal of Animal Science* 23, 782.

Hulet, C.F. (1966) Behavioural, social and psychological factors affecting mating time and breeding efficiency in sheep. *Journal of Animal Science* 25 (Suppl.), 5–10.

Hunnicutt, L.K., Stobart, R.H., Townsend, R.S. and Aimone-Lupher, C.E. (1995) EAZI-breed CIDR G devices versus prostaglandin $F_{2\alpha}$ for synchronization of estrus in mature range ewes. *Journal of Animal Science* 73 (Suppl. 1), 241.

Jabbar, G., Umberger, S.H. and Lewis, G.S. (1993) A comparison of melengestrol acetate and norgestomet used alone or in combination with zeranol or PMSG and

hCG for improved spring-breeding performance in anestrous ewes. *Journal of Animal Science* 71 (Suppl. 1), 242.
Jennings, J.J. (1975) Effect of ICI 80996 on conception rate in ewes. *Annual Animal Production Research Report* (An Foras Taluntais), 34–35.
Jennings, J.J. (1976) Mating behaviour of rams in anoestrus. *Irish Journal of Agricultural Research* 15, 301–307.
Jennings, J.J. (1977) Influence of mating behaviour of rams on fertility in progestagen-PMS treated anoestrous ewes. *Irish Journal of Agricultural Research* 16, 155–162.
Jennings, J.J. and Crowley, J.P. (1972) The influence of mating management on fertility in ewes following progesterone-PMS treatment. *Veterinary Record* 90, 495–498.
Jennings, J.J. and Quirke, J.F. (1976) Artificial insemination in sheep. *Annual Report Animal Production Division* (An Foras Taluntais), 35–36.
Joyce, M.J.B. (1972) A comparison of three different mating systems. *Proceedings of the 7th International Congress Animal Reproduction and AI* (Munich), 2, pp. 935–938.
Joyce, M.J.B. and Gillespie, J.B. (1973) Antibiotic treatment of intravaginal pessaries. *Irish Veterinary Journal* 26, 248–250.
Karagiannidis, A.K. (1995) Excretion of MPA in the milk of lactating ewes treated for synchronization of estrus. *Theriogenology* 43, 605–613.
Kassem, R., Owen, J.B. and Fadel, I. (1990) A note on the characteristics of oestrus and ovulation in Awassi ewes. *Animal Production* 50, 198–201.
Keane, M.G. (1974) Effect of progestagen-PMS hormone treatment on reproduction in ewe lambs. *Irish Journal of Agricultural Research* 13, 39–48.
Killeen, I.D. and Moore, N.W. (1970) Transport of spermatozoa and fertilization in the ewe following cervical and uterine insemination early and late in oestrus. *Australian Journal of Biological Science* 23, 1271–1277.
Knight, T.W., Hall, D.R.H. and Smith, J.F. (1988) Effects of immunisation with polyandroalbumin (Fecundin), pasture allowance, post-mating shearing, and method of synchronization on reproductive performance of Romney and Marshall Romney ewes. *New Zealand Journal of Agricultural Research* 31, 243–247.
Knight, T.W., O'Neill, K., Ridland, M., Hamilton, G., Death, A. and Wyeth, T. (1992) Effects of month and PMSG on the interval from CIDR removal to ovulation in Romney and Merino ewes. *Proceedings New Zealand Society Animal Production* 52, 261–263.
Lamond, D.R. (1964) Synchronization of ovarian cycles in sheep and cattle. *Animal Breeding Abstracts* 32, 269–285.
Larson, W.M., Banbury, E.D. and Spaeth, C.W. (1970) Effect of previous lambing rate on response to PMSG. *Journal of Animal Science* 31, 225–238.
Lightfoot, R.J., Croker, K.P. and Marshall, R. (1976) Use of a prostaglandin analogue (ICI 80996) for the synchronization of estrus and lambing in Merino ewes. *Proceedings of the International Sheep Breeding Congress* (Muresk), pp. 449–454.
Lindsay, D.R., Moore, N.W., Robinson, T.J., Salamon, S. and Shelton, J.N. (1967) The evaluation of an oral progestagen (Provera:MAP) for the synchronization of oestrus in the entire cyclic Merino ewe. In: Robinson, T.J. (ed.) *Control of the Ovarian Cycle in the Sheep.* Sydney University Press, Sydney, pp. 155–165.
Mattner, P.E. (1966) Formation and retention of the spermatozoan reservoir in the cervix of the ruminant. *Nature* 212, 1479.
Mattner, P.E. and Braden, A.W.H. (1969) Comparison of the distribution of the motile and immotile spermatozoa in the ovine cervix. *Australian Journal of Biological Science* 22, 1069–1070.

O'Reilly, P.J. (1972) Fertility resulting from AI or natural mating using pessaries or implants to synchronize oestrus in ewes. *Proceedings of the 7th International Congress Animal Reproduction and AI* (Munich), 2, pp. 941–944.

Quirke, J.F. (1979) Control of reproduction in adult ewes and ewe lambs and estimation of reproductive wastage in ewe lambs following treatment with progestagen impregnated sponges and PMSG. *Livestock Production Science* 6, 295–305.

Robinson, J.J. (1974) Intensifying ewe productivity. *Proceedings of the British Society of Animal Production* 3, pp. 31–40.

Robinson, T.J. (1964) Synchronization of oestrus in sheep by intravaginal and subcutaneous application of progestin impregnated sponges. *Proceedings of the Australian Society of Animal Production* 8, 47–49.

Robinson, T.J. (1967) Conclusions. In: Robinson, T.J. (ed.) *Control of the Ovarian Cycle in the Sheep.* Sydney University Press, Sydney, pp. 237–244.

Robinson, T.J. (1968) The synchronization of the oestrous cycle and fertility. *Proceedings of the 6th International Congress of Animal Reproduction and AI* (Paris), 2, pp. 1347–1383.

Robinson, T.J. (1973) Contraception and sperm transport in domestic animals. *INSERM* 26, 453–478.

Robinson, T.J. (1974) The present status of applied reproductive physiology in animal production. *Proceedings of the New Zealand Society Animal Production* 34, 37–44.

Robinson, T.J. (1980) Programmed year-round sheep breeding. *Australian Journal of Experimental Agriculture and Animal Husbandry* 20, 667–673.

Robinson, T.J. (1982) Hammond Memorial lecture. The magic of Hammond. *Journal of Reproduction and Fertility* 66, 397–410.

Robinson, T.J. (1988) Controlled sheep breeding: update 1980–1985. *Australian Journal of Biological Science* 41, 1–13.

Robinson, T.J., Salamon, S., Moore, N.W. and Smith, J.F. (1967) The evaluation of SC-9880-impregnated intravaginal sponges for the synchronization of oestrus for large-scale AI of Merino ewes in summer and autumn. In: Robinson, T.J. (ed.) *The Control of the Ovarian Cycle of the Sheep.* Sydney University Press, Sydney, pp. 208–236.

Robinson, T.J., Quinlivan, T.D. and Baxter, C. (1968) The relationship between dose of progestagen and method of preparation of intravaginal sponges on their effectiveness for the control of ovulation in the ewe. *Journal of Reproduction and Fertility* 17, 471–476.

Safranski, T.J., Lamberson, W.R. and Keisler, D.H. (1992) Use of melengestrol acetate and gonadotrophins to induce fertile estrus in seasonally anestrous ewes. *Journal of Animal Science* 70, 2935–2939.

Scaramuzzi, R.J. and Martin, G.B. (1984) Pharmacological agents for manipulating oestrus and ovulation in the ewe. In: Lindsay, D.R. and Pearce, D.T. (eds) *Reproduction in Sheep.* Cambridge University Press, Cambridge, pp. 316–325.

Scott, I.C. and Montgomery, G.W. (1990) Ovulation rates of synchronized Coopworth ewes over the peak of the breeding season. *New Zealand Journal of Agricultural Research* 33, 443–447.

Shackell, G.H. (1991) The timing of oestrus, LH surge and ovulation in ewes following synchronization with MAP sponge, FGA sponges or CIDRs. *Proceedings of the New Zealand Society Animal Production* 51, pp. 73–77.

Smith, J.F. and Allison, A.J. (1971) The effect of exogenous progestagen on the production of cervical mucus in the ewe. *Journal of Reproduction and Fertility* 24, 279–282.

Smith, J.F., Konlechner, J.A. and Parr, J. (1991a) The efficacy of used CIDR devices for synchronization of oestrus and post-mating treatment. *Proceedings of the New Zealand Society Animal Production* 51, 111–115.

Smith, J.F., Konlechner, J.A. and Parr, J. (1991b) Factors influencing the time to onset of oestrus after synchronization treatment in ewes. *Proceedings of the New Zealand Society Animal Production* 51, 117–121.

Smith, P.A., Boland, M.P. and Gordon, I. (1981) Effect of dose of Cronolone in intravaginal sponges on lambing outcome to fixed-time A.I. *Journal of Agricultural Science, Cambridge* 96, 253–254.

Spitzer, J.C. and Carpenter, R.H. (1979) Synchronized breeding of cyclin ewes to produce fetuses of known gestational age. *Laboratory Animal Science* 29, 755–758.

Stacy, B.D., Gemmell, R.T. and Thorburn, G.D. (1976) Morphology of the corpus luteum in the sheep during regression induced by prostaglandin $F_{2\alpha}$. *Biology of Reproduction* 14, 280–291.

Thimonier, J. and Cognie, Y. (1971) Acceleration des mises bas et conduite d'elebvage chez les ovins. *Bulletin Technical Information Ministry Agriculture* 257, 1–10.

Thorburn, G.D. and Nicol, D.H. (1971) Regression of ovine corpus luteum after infusion of $PGF_{2\alpha}$ into ovarian artery and vein. *Journal of Endocrinology* 51, 751–752.

Tritschler, J.P., Duby, R.T., Parsons, E.M., Parsons, M.J. and Giordano, D.J. (1991) Comparison of two protestagens during out-of-season breeding in a commercial ewe flock. *Theriogenology* 35, 943–952.

Trounson, A.O., Willadsen, S.M. and Moor, R.M. (1976) Effect of prostaglandin analogue cloprostenol on oestrus, ovulation and embryonic viability in sheep. *Journal of Agricultural Science, Cambridge* 86, 609–611.

Umberger, S.H. and Lewis, G.S. (1992) Melengestrol acetate (MGA) for estrous synchronization and induction of estrus in spring-breeding ewes. *Sheep Research Journal* 8, 59–62.

Umberger, S.H., Jabbar, G. and Lewis, G.S. (1994) Seasonally anovulatory ewes fail to respond to progestogen treatment in the absence of gonadotropin stimulation. *Theriogenology* 42, 1329–1336.

Velle, W. and Helle, O. (1979) Experience with estrus synchronization in sheep over a twelve-year period. *Journal of Animal Science* 48, 1015–1019.

Vipond, J.E. and King, M.E. (1979) Synchronization of oestrus as an aid to management in small flocks. *Animal Production* 28, 447–451.

Welch, R.A.S., Andrewes, W.D., Barnes, D.R., Bremner, K. and Harvey, T.G. (1984) CIDR dispensers for oestrus and ovulation control in sheep. *Proceedings of the 10th International Congress Animal Reproduction and AI* (Illinois) 3, No. 354.

Wheaton, J.E., Carlson, K.M., Windels, H.F. and Johnston, L.J. (1993) CIDR: a new progesterone-releasing intravaginal device for induction of estrus and cycle conttrol in sheep and goats. *Animal Reproduction Science* 33, 127–141.

Xenoulis, P.C., Minotakis, C.S. and Tsamis, C. (1972) The evaluation of progesterone implants and MAP-impregnated sponges for the advancement of the breeding season in ewes. *Proceedings of the 7th International Congress of Animal Reproduction and AI* (Munich) 2, 990–994.

4 Fixed-Time Sheep Artificial Insemination

4.1. Introduction

Artificial insemination (AI) of sheep has been applied most extensively in the former Soviet Union and associated countries of Eastern Europe where the technique has been used on a vast scale (Ozin, 1968; Jheltobruch, 1979). However, the sheep AI procedure as employed in the former USSR involved the use of teaser rams to detect those sheep in oestrus and inseminating each ewe as she exhibited her natural heat period; most ewes have been bred using 0.05 ml of undiluted semen within 20 min of taking the collection from the ram. Such a system would not have been practicable but for the large-flock conditions that existed in the former USSR which permitted an acceptable number of ewes to be available for insemination on a daily basis.

For such reasons, the application of sheep AI in Ireland and many other countries was, until the 1970s, confined to experimental work. For any serious thought of applying AI under the same flock conditions of Ireland, it is essential that any need for oestrus detection be eliminated and that all members of the flock or selected group be inseminated at a pre-determined hour.

Much valuable information on AI in sheep is to be found in the excellent book of Evans and Maxwell (1987).

4.1.1. *Advantages of sheep AI*

Artificial insemination could offer the sheep farmer three possible advantages over natural service in controlled breeding applications: it could be used to overcome the need to gather a bunch of rams to meet the 1:10 ram : ewe ratio, to avoid instances of poor conception rates in sheep arising from ram subfertility and it could have merit in making rams of superior genetic merit freely available to farmers. There is also the important possibility of employing sheep AI in breeding improvement programmes in order to identify the genetically superior sires (see Fig. 4.1).

In Ireland, as in many other countries, breed improvement is effected by what is done in the pedigree flocks where the pure-bred sires are produced for sale to the commercial flock-owners. Any improvement that is achieved in the pedigree flocks will eventually be reflected in the quality of the commercial sheep population. However, if there is little improvement in the quality of pedigree stock, there can be little movement in the commercial flocks using rams from such sources. There are many ways in which AI can play a valuable role in the progeny testing of rams.

Faster progress in improving genetic quality required

The desirability of achieving faster progress in improving the genetic quality in sheep in New Zealand has been discussed by Gretton (1992). This author studied production parameters of the New Zealand sheep flock between 1950 and 1980 showing that during this period lambing percentages changed from 95.6% to 98%, average wool yield from 5.1 to 5.2kg and lamb carcass weight from 16.0 to 13.7kg; failure to fully utilize the available genetic techniques is identified as the cause for this slow rate of change. The application of progeny testing of rams, using sire referencing, and other means of improving genetic quality are examined by the author. It should be noted that sire referencing

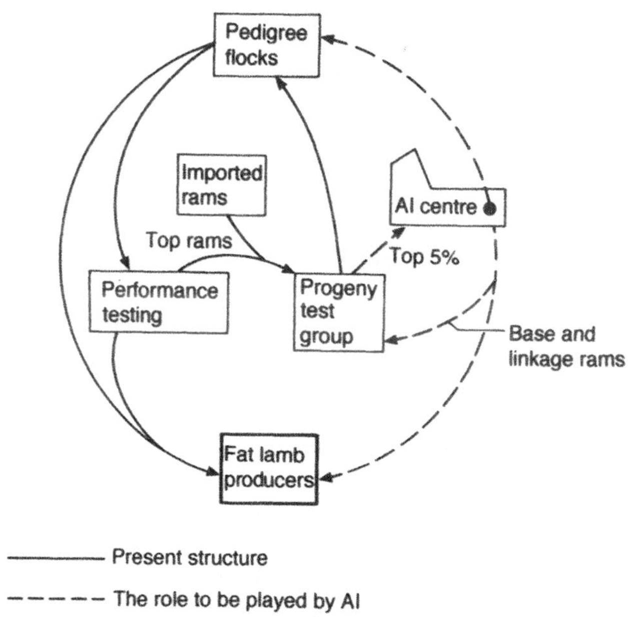

Fig. 4.1. Diagrammatic representation of a proposed breeding and testing system for the use of AI in Suffolk ram breeding (from Smith, 1977).

techniques are used to establish an overall breeding value ranking of the rams being tested in a number of separate flocks.

A report by Hanrahan (1995) in the Irish Republic presented results of an evaluation of 'commercial' (i.e. non-pedigree) Suffolk rams and 'superior' (i.e. those chosen by top pedigree breeders) used as terminal sires in lamb production; the author found essentially no difference between the two ram categories, in terms of progeny growth rate. The author suggests that pedigree breeders need to examine critically their present breeding policies and implement more effective improvement programmes. Artificial insemination could be the means of implementing such programmes. In France, for example, where it is estimated that well in excess of 700,000 ewes are bred annually by fixed-time AI (Mesnil du Buisson, 1994), the technique is employed to the greatest extent in the milking Lacaune breed, very much with a view to making the greatest use of the superior sires that have been identified by progeny testing through the use of AI.

4.1.2. AI and conception rates

As well as making genetically desirable rams available, sheep AI should also be capable of establishing pregnancies in sheep as readily as when rams are employed in breeding by natural service. Certainly, if the technique is to be widely used in commercial practice, the cost and labour involved in its application can only be justified if there is a consistent and acceptably high conception rate to the AI service. However, AI is not a magic wand that can be waved over the flock; it does have technical problems of its own (especially storage and dilution of semen) and there is the important question of cost. Until the 1970s, conception rates with AI in sheep after controlled breeding procedures were quite unacceptable in the great majority of reports.

4.2. Progestagen–PMSG Techniques

French investigators were the first to develop satisfactory semen handling procedures and to show that it was possible to achieve acceptable conception rates, using the intravaginal sponge and pregnant mare serum gonadotrophin (PMSG) to control ovulation, and a two-dose insemination technique (Colas, 1975, 1979). In their work, there was no attempt to detect oestrus and the timing of the two inseminations was based on progestagen withdrawal rather than on the time of oestrus onset; the times chosen were 50 and 60h after terminating treatment.

In using AI after the progestagen–PMSG technique, it was known that the oestrus control measures could be applied without reducing the ewe's fertility to any great extent. Although the studies of the 1960s had shown that pre-treatment of ewes with progestagens could result in an impairment of sperm transport with a consequent reduction in fertility (Quinlivan and Robinson,

1969), later studies showed that this need not always be so (Allison and Robinson, 1970): it was a matter of the particular physiological state of the ewe at time of treatment. Applying an optimal controlled breeding technique to cyclic ewes in the autumn breeding season could result in ewes exhibiting a degree of fertility much the same as that shown by untreated sheep.

4.2.1. *Irish efforts in sheep AI*

In Ireland, farm applications of AI were first attempted in the early 1970s, using the semen processing techniques described by Colas *et al.* (1968) as the basis of the work. The sheep AI application involved two inseminations, each of 200 million sperm in 0.2ml volume, with an interval of 10–14h separating the two sperm doses. Smith *et al.* (1977a) used fluorogestone acetate (FGA)–PMSG to control oestrus in 853 cyclic sheep, the ewes being either naturally mated or artificially inseminated (see Table 4.1). There was no significant difference between the two breeding methods in the number of sheep conceiving to first service (AI, 70%; natural service, 75.5%); these results were comparable with those reported elsewhere for a double insemination procedure. Thus, French studies reported average conception rates varying from 60.5% to 75% (Colas *et al.*, 1968; Colas, 1975); in the UK, the Meat and Livestock Commission in their field trials, involving 1000 sheep treated with medroxyprogesterone acetate (MAP)–PMSG, reported conception rates ranging from 52% to 78%, with an overall average of 59% (Anon, 1978).

4.2.2. *Progestagen and PMSG*

Oral progestagens

Some studies still report the use of AI in conjunction with the oral dosing of sheep with progestagens. In Mexico, Quispe *et al.* (1994) synchronized 540 cyclic ewes with melengestrol acetate (MGA) given at the level of 0.22mg daily

Table 4.1. Effect of PMSG dose level on lambing outcome in ewes conceiving at the controlled oestrus (from Smith *et al.* 1977a).

Breeding method	PMSG (IU)	No. ewes	Litter size 1	2	3	4+	Ewes lambing	(%)	Lambs born	Lambs ewe
Artificial insemination	375	214	70	65	10	1	146	(68.2)	234	1.60
	750	210	48	77	20	6	151	(71.9)	287	1.89
Natural service	375	214	15	80	17	2	156	(72.9)	277	1.77
	750	215	45	83	30	10	168	(78.1)	343	2.04

for two weeks, with the ewes being inseminated with fresh or frozen semen at the first or second oestrus. Artificial insemination after oestrus detection during the controlled oestrus resulted in a substantial decrease in fertility using fresh (28% vs. 80% in untreated controls) and frozen semen (12% vs. 28%). The authors note that the detrimental effect of MGA on fertility could be partially offset by programming the breeding period to start on the second oestrus post-treatment. For good fertility after breeding at the progestagen controlled oestrus, regard to the type of progestagen used and the dose of PMSG associated with it are of paramount importance.

Progestagen sponges

In Ireland, several trials have been conducted in examining the effect of PMSG dose level and the particular progestagen employed in the intravaginal sponge. Such studies have shown a small but significant difference in conception rate to AI in favour of FGA (30mg dose level) when compared with MAP (60mg dose level) and no material improvement in conception rate when 750IU rather than 375IU was the gonadotrophin dose level employed (Smith *et al.*, 1981). Further work reported by Al-Kamali *et al.* (1990) found FGA and MAP sponges to be equally effective (Table 4.2). These data are of some interest, in view of the fact that most French efforts in sheep AI are apparently based on the use of sponge pessaries impregnated with 30–40mg FGA (Colas, 1975, 1979) whereas efforts in the UK by the Meat and Livestock Commission and in sheep AI work in Iceland were usually based on the 60mg MAP intravaginal sponge (Barlow *et al.*, 1974; Dyrmundsson, 1977).

Need for PMSG

The importance of using PMSG in conjunction with intravaginal FGA in improving the precision of ovulation was mentioned by Canadian workers

Table 4.2. Lambing outcome in relation to progestagen employed (all sheep received 500IU PMSG at sponge withdrawal) (from Al-Kamali *et al.*, 1990).

Progestagen	FGA	MAP
No. ewes treated	878	1176
No. ewes lambed	628	846
No. lambs born	1157	1150
% ewes lambing	71.5	71.9
Lambs born per ewe lambing	1.80	1.80
% ewes with singles	35.2	34.7
% ewes with twins	48.8	50.3
% ewes with triplets or more	16.0	15.0
Lambs per 100 ewes treated	131	131

FGA: fluorogestone acetate.
MAP: medroxyprogesterone acetate.

(Langford and Hackett, 1980; Langford, 1982); they also reported that very low doses of serum gonadotrophin (150–300IU) were effective in their particular sheep breeds. Those who have compared PMSG with human chorionic gonadotrophin (hCG), have clearly shown serum gonadotrophin to be the appropriate preparation. In Brazil, for example, Perrenoud *et al.* (1991) used 12–14-day treatments with 50mg MAP sponges followed by either 300IU of PMSG or hCG at sponge withdrawal. Sheep were inseminated at 51–60h after sponge withdrawal; of ewes injected with PMSG, 73% did not return to oestrus versus 34% of those injected with hCG.

4.3. Collection and Preparation of Semen for Set-Time AI

Remarks in this section are based on experiences in Ireland in operating a small-scale sheep AI service for sheep farmers over a period of some years. This involved the collection and preparation of semen during most months of the year, but especially in the autumn season, when most inseminations on farms were conducted (Fig. 4.2).

According to workers in the Meat and Livestock Commission in the UK, although it is possible to collect acceptable semen ejaculates from rams of British breeds throughout the year (Anon, 1982), semen production was highest between September and January (ejaculate volume for Suffolk rams = 1.1ml) and lowest between January and June (volume = 0.8ml).

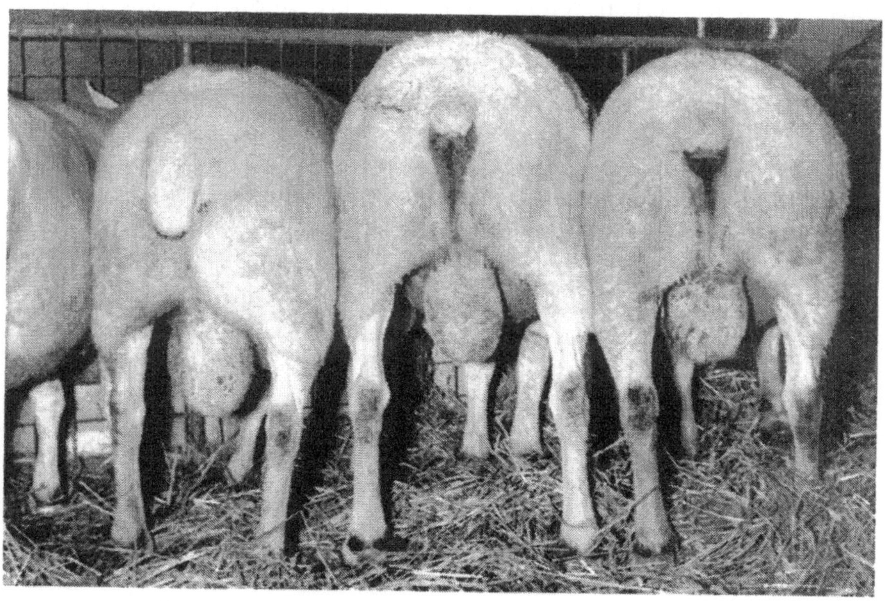

Fig. 4.2. Sheep AI service for making available Texel rams in Ireland.

4.3.1. Semen collection

Rams should be carefully trained for semen collection by the artificial vagina. Most rams can be readily trained, using 'teaser' ewes that have been artificially induced to show oestrus (spayed ewes treated with intravaginal progestagen and a dose of 25–50 μg oestradiol benzoate at sponge withdrawal). A method for markedly prolonging oestrus in intact ewes has been described by Lewis and Goebel (1993); this involved injections of zeranol which doubled the length of oestrus, both in ewes showing spontaneous or progestagen-regulated oestrus. Ram training takes between one and two weeks during the normal breeding season. It is important that rams are not excited and become used to the semen collector's presence. There have been some reports emphasizing the importance of preparing rams for the breeding season. In Russia, German (1988) recorded a pregnancy percentage of 66.4 in Karakul ewes inseminated with semen from prepared rams in contrast to 20.9 for those not prepared. The preparation procedure took the form of collecting from rams 2–3 times daily for three weeks before the AI season.

Teaser ewe effects

Various methods may be employed to increase sperm output per collection in rams. In Ireland, for example, McGrath *et al.* (1979) showed that the presence of an oestrous rather than a non-oestrous teaser ewe significantly increased total sperm numbers collected by 17% (see Table 4.3); a beneficial effect was also gained by exposing rams to the sight of collections being taken from their companions. Work elsewhere has shown that testicular function may be enhanced under certain conditions if rams are constantly kept in close proximity to oestrous ewes (Sanford and Yarney, 1980). There is ample evidence to show that the artificial vagina, rather than electro-ejaculation, is the preferred method of collection for semen destined for use in AI (Memon and Ott, 1981).

In passing, it might be mentioned that studies in the USA have shown that removal of the filiform appendage increases semen collection efficiency in rams

Table 4.3. Effect of oestrous teaser on total sperm output in rams (from McGrath *et al.*, 1979).

Rams		Oestrous teaser	Non-oestrous teaser
Breed	No.	Total sperm output (sperm/ejaculate × 10^9)	Total sperm output (sperm/ejaculate × 10^9)
Suffolk	(5)	2.10	1.85
Texel	(5)	1.63	1.23
Dorset Horn	(5)	2.29	2.02
Total	(15)	2.01	1.70

without affecting semen quality (Thomas *et al.*, 1991). Back in Bakewell's day in the eighteenth century, master breeders amputated the filiform appendage ('worming the ram') before putting spare rams in the market in the belief that this rendered them sterile.

Semen doses

In work in Ireland, the usual procedure is to take one or two collections per day from each ram during the working week (Monday–Friday) and to provide the males with a 2-day rest period over the weekend; at this rate of semen collection, it is possible to plan on each ram providing 50 semen doses per week (400–500 million sperm per dose) for just as long as the AI season lasts. Elsewhere, semen collection studies in the UK carried out with Suffolk rams by the Meat and Livestock Commission showed them to be capable of producing some 220 semen doses weekly during the breeding season as compared with only about 50 doses in the ewe anoestrus (Anon, 1978); this is some measure of semen production as between autumn and spring.

As mentioned elsewhere, French workers have reported that ram semen collected in the spring contains a significantly higher incidence of morphologically abnormal sperm than that collected in the autumn and have suggested that this may markedly influence fertility of the males (Colas, 1979, 1983); these studies were with Ile-de-France rams and it is not always clear how far these findings apply to other breeds. Certainly, in Ireland, there has been no evident increase in the morphological abnormalities in the semen produced by rams in the spring.

4.3.2. Preparation and processing of semen

In operating an AI service in Ireland, semen was collected from rams and used to inseminate ewes within 6h; only those rams yielding ejaculates of high quality were allowed to contribute to the day's AI programme. The quality of semen was assessed on the basis of motility, concentration and morphology of the sperm. Motility of sperm, or wave motion as occurs in undiluted semen under low-power microscopy, was described by a grade (0 = no wave motion; 5 = very fast moving waves). Smith *et al.* (1979) used semen samples with a wave-motion grading of 3.0 or above in their AI work; this involved rejection of about 20% of samples collected from Suffolk, Texel and Dorset Horns in their programme.

Pooled semen and milk diluent

Pooled semen (at least three rams contributing) was extended in a skim-milk diluent, the formulation being based on that described in the report of Colas *et al.* (1968). Most diluents for ram semen have either egg yolk or milk or a combination of the two as a basic ingredient. The reconstituted skim-milk diluent was heated (92°C for 10 min) to inactivate a spermicidal factor (lactenin) present in the protein fraction of the milk before adding trace amounts of antibiotics and other agents (penicillin, streptomycin, sulphanilamide, catalase). Semen was then diluted to provide a standard sperm density of 2000 million per ml and the diluted semen cooled from 30 to 15°C over a 30-min period before being loaded into 0.25ml capacity Cassou straws and packed into vacuum containers for transportation to the farm. A storage temperature of 15°C was maintained up to the time of carrying out the inseminations on the farm.

Elsewhere, Canadian workers reported using skim-milk and adjusting the sperm density to a value of 900 million sperm ml^{-1}; a 0.5ml volume of diluted semen was employed to provide a sperm dose of 450 million for insemination (Langford and Hackett, 1980). The influence of storage time and temperature on the fertility of rams was reported by Langford and Fisher (1980); at sperm doses of 450 million in 0.5ml, storage at 4°C allowed semen to be kept for 24h without reducing fertility, whereas at 15°C, fertility decreased markedly when storage time exceeded 6h.

When ram semen is preserved in the non-frozen state, the protective properties of diluents depend very much on temperature and duration of storage. At 15°C, and for a short storage period, reconstituted cow's skimmed milk is regarded as more efficient than egg-yolk-based media (Colas, 1983); at 4°C, the reverse is true. For that reason, skim-milk diluents should not be used in cooling ram sperm down to 4°C.

4.3.3. *One vs. two inseminations*

An important advance in the development of a commercially acceptable sheep AI technique came by way of reducing from two inseminations to a single one in suitable categories of sheep (Colas, 1979; Smith *et al.*, 1978). In adopting this simplification, however, it was apparently necessary for ewes to be inseminated with the same total sperm numbers (400–500 million) as employed with the two-insemination procedure (2 × 200 million or more). Data from some of the Irish attempts to simplify procedures are given in Table 4.4.

In regard to insemination timing, it proved to be a question of using the

Table 4.4. Effect of sperm doses on outcome of sheep AI (from Gordon, 1983).

Season	Ewes to AI	First services only: Lambed	Lambs born	Conception rate (%)	Litter size	Sperm doses (million)
Autumn breeding season	5072	3461	6171	68.2	1.78	2 × 200
	396	250	456	63.2	1.82	1 × 200–300
	478	354	601	74.0	1.70	1 × 400

Table 4.5. Lambing outcome to sheep AI with Texel ram semen in counties Roscommon and Wexford (single inseminations) (from Gordon, 1983).

County	Year	Sheep to AI	Ewes lambing to AI			
			Lambed	Lambs born	Conceptions to AI	Litter size
Roscommon	1980	759	561	1023	73.9%	1.82
	1981	639	465	809	72.8%	1.74
Wexford	1980	428	329	650	70.3%	1.91
	1981	265	183	356	69.1%	1.94
Totals	–	2145	1538	2838	71.7%	1.85

figure mid-way between the two times previously employed (making it 55–57h after progestagen–PMSG for most single-dose AI techniques). In Ireland, experimental results were borne out in subsequent farm applications in which conception rates of 70% have been achieved after breeding cyclic ewes at 56h after progestagen withdrawal (see Table 4.5).

Simplification of the AI technique to a single insemination greatly eased the cost and labour involved in applying this breeding procedure on the farm and opened the way to greater exploitation of AI in sheep breeding improvement programmes. In the Irish experience, the denser semen (2000 million per ml) employed in single inseminations has a shorter storage life in the 0.25 straw than the semen previously employed in double inseminations (1000 million ml^{-1}); to meet this difficulty, a 6h rather than a 12h storage period was adopted.

The timing of AI, when two inseminations have been employed after progestagen–PMSG treatment, differed in a minor way from country to country. French workers inseminated ewes at 50 and 60h after sponge withdrawal whereas workers in the UK inseminated at 48 and 58 (McClelland and Quirke, 1971) or 48 and 64h (Barlow *et al.*, 1974). In Ireland, two variants were employed in timing double inseminations. In one system, the first insemination was timed for 09.00h and the second for 19.00h; this was on the second day after sponge withdrawal (i.e. at 48 and 58h). The alternative to this was a system where the first insemination was at 19.00 on the second day and the final one at 09.00h on the third day (48 and 62h). The percentage of ewes conceiving after AI was much the same in the two systems (Smith *et al.*, 1977b).

Insemination and ovulation

As already mentioned, much is known about the timing of oestrus and ovulation after progestagen–PMSG treatment in sheep in Ireland. It is usual for oestrus to commence about 36h after sponge withdrawal, for the heat period to be about 36h in length and for ovulation to occur about 70h or so after progestagen withdrawal (Boland *et al.*, 1978). Thus, a single insemination at 56h should permit sperm to be available in the reproductive tract for some hours before ovulation.

For ease of applying the AI technique on the farm, there would be much in favour of carrying out insemination at 48 rather than 56h. In the AI service operated in Ireland in the 1970s and 1980s, progestagen sponges were withdrawn at 08.00h to permit a 16.00 insemination two days later. In view of evidence showing that ovulation in FGA/MAP-treated ewes occurs about 70h after sponge removal (Boland *et al.*, 1978), carrying out the insemination at 48h would not seem appropriate, although such timing would greatly facilitate the planning of on-farm operations.

Elsewhere, with different breeds and conditions, it should be noted that there may be considerable variation from flock to flock in fertility results after AI. In France, Maurel *et al.* (1992) used ELISA kits to detect LH peaks in the milk of two sheep breeds receiving FGA–PMSG treatment; they detected an earlier occurrence of the LH peak in Manech ewes than in Lacaune sheep which could explain fertility differences. These authors suggested that the LH kit could be a useful and simple tool in dealing with some infertility problems, especially in investigating the timing of ovulation in different sheep breeds after oestrus control. In Australia, using 30mg FGA sponges and 400IU PMSG at their removal, Maxwell (1986a) recorded the median time of ovulation as being 59.7h after pessary withdrawal.

Sheep AI by way of travelling rams

In the UK, albeit on a very limited scale, some farmers have provided a fresh semen AI service from rams in the top 5% of rams in sire reference schemes in that country. This involves transporting rams and equipment to commercial flocks where ram isolation pens, semen collection facilities and insemination stalls are set up. Semen is collected from rams about six times a day and diluted with an equal volume of skim-milk for immediate use in AI. For insemination of progestagen–PMSG-treated ewes, a 0.1ml volume of semen is employed. Under such an arrangement, it is possible for a ram to produce enough semen for 100 ewes and more. Although excellent conception rates to AI have been claimed (75%), in view of remarks made elsewhere about possible adverse effects of stress in the handling and insemination of sheep, it would not be surprising if the travelling ram AI system ran into difficulties on that score under some farm conditions.

4.3.4. Use of controlled internal drug releasing dispensers

As mentioned elsewhere, the CIDR is a nylon core over which a progesterone containing medical elastomere is moulded. Initial work in New Zealand employed CIDRs containing 12% (0.5g) of progesterone, but trials showed that the steroid level could be reduced to 9% (0.38g) and this is the progesterone content of the devices that are available commercially. A report by Harvey *et al.* (1984) used CIDRs without PMSG in synchronizing oestrus in cyclic ewes for AI; results indicated that a higher conception rate might be achieved by inseminating after oestrus detection rather than by way of a fixed-time schedule. As mentioned earlier, PMSG has been regarded as essential for fixed-time AI systems.

In Japan, Fukui *et al.* (1993a) reported on conception rates by fixed-time AI with frozen semen in seasonally anoestrous ewes treated with MAP-impregnated pessaries or CIDRs. An injection of PMSG was given the day prior to removal of the intravaginal devices and AI was intrauterine with the aid of laparoscopy. The time of onset of oestrus was significantly earlier for sheep treated with the CIDRs than with the sponges (24.9 vs. 30.0h). Results indicated that the optimum time for intrauterine insemination is different for the two devices.

Other work by this author and co-workers examined the effects of self-made progesterone (500mg) sponges and CIDRs when used in anoestrous ewes that were inseminated 44–52h after treatment using a laparoscope (Fukui *et al.*, 1993b). There was no significant difference in lambing outcome between the two intravaginal treatments. A semen dose (0.05ml) containing 36 million sperm resulted in more than 60% of CIDR-treated ewes lambing by intra-uterine insemination with frozen-thawed semen.

A study by McMillan (1994) in New Zealand attempted to identify some of the factors (PMSG, time of year, ewe body weight) influencing the interval to oestrus following CIDR treatment. It was concluded that the actual timing of AI to achieve near maximum fertility is dependent on the average interval to oestrus in a given flock. If this is unknown, the 'universal' timing for a single fixed-time AI during the breeding season is 42–46h after CIDR removal.

4.4. Insemination Procedures

In cervical sheep AI, because of the type of cervical canal found in this species, semen can only be deposited inside the cervix or into its first fold. As well as that limitation, the retention capacity of the ovine cervix is very low, only 0.1–0.2ml. One major difficulty with fixed-time AI procedures is the fact that a very large sperm dose is required (400–500 million total sperm in one or two inseminations), and this has to be given in as low a volume as possible. In normal AI, as employed, for example, in the former USSR, the sperm dose was of the order of 120–150 million. In Australia, various workers have shown in AI studies that something of the order of 120 million sperm could be regarded as the minimal sperm dose for acceptable conception rates in spontaneously ovulating cyclic sheep.

Fixed-time AI applications in France and Ireland has involved using the Cassou sheep inseminating gun to deposit semen in the first fold of the cervix (Colas, 1975; Gordon, 1975). Attempts were made by Smith *et al.* (1978) in Ireland, using a specially designed insemination instrument, to deposit semen deeper than usual in the cervical folds but this significantly depressed rather than enhanced the conception rate. However, it should be noted that work elsewhere with frozen-thawed semen recorded evidence of an improvement in conception rates with increased depth of deposition in the cervix (Eppleston *et al.*, 1994).

4.4.1. Inseminator effects

As in cattle, the ability of the inseminator can have a marked effect on the outcome of sheep AI; this has been recorded in France (Aguer and Le Provost, 1976) and elsewhere (Gordon and Crosby, 1980; Schakell *et al.*, 1990). In the UK, trials were conducted on the possibility of suitably trained farm staff carrying out inseminations with semen supplied from a central ram stud (Anon, 1978). Working with 444 MAP–PMSG-treated ewes and using semen stored for up to 8h, the average conception rate achieved by farm staff as inseminators was 53% compared with 66% for professional inseminators. The work did show, however, that in certain instances, farm staff were as capable as the professional inseminators.

4.4.2. Restraining ewes

Work in Ireland showed that sheep-handling facilities on the farm need not be elaborate; the usual method employed for restraining was to hold the ewe over a straw bale (Fig. 4.3); an alternative method is shown in Fig. 4.4. The actual insemination, although requiring some degree of skill and experience on the part of the operator, generally occupied less than one min per animal. Efficiency in locating the mouth of the cervix may vary with the ewe's reproductive history (parous/non-parous) and the equipment used. In the Irish experience, it was regarded as essential that ewes for AI were handled as gently as possible at the time of insemination and the flock as a whole was never subjected to any unnecessary agitation (barking dogs and such). French experience appeared to agree with this and workers in that country suggested that the group awaiting insemination should not exceed 40–50, no matter how many sheep are to be dealt with (Aguer and Le Provost, 1976). This group size phenomenon is presumably a measure of the agitation which may arise among individual sheep if they are constantly being disturbed as members of a larger group.

4.4.3. Stress and sheep AI

There is evidence that ewes subjected to relatively short-term nutritional or handling stress around the time of mating can show markedly reduced fertility, whether this stems from problems that interfere with fertilization or from a higher than usual loss of embryos in the early weeks of pregnancy (Gunn and

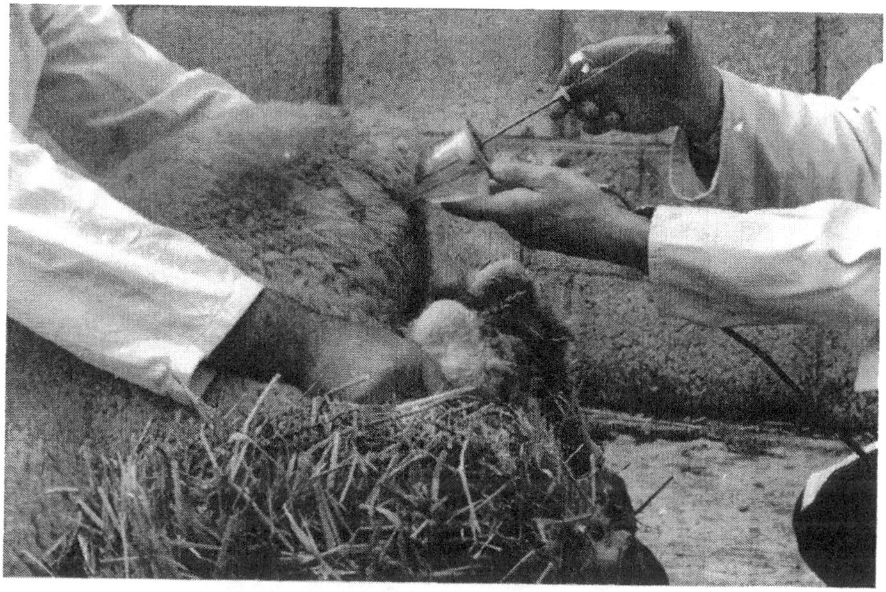

Fig. 4.3. Standard method adopted for restraining the ewe for AI using the straw bale. Among the various considerations in sheep AI is the amount of time and labour involved in carrying out the actual insemination. What is needed is a simple straightforward method of restraint, which involves a minimum of discomfort to the ewe – and which is capable of permitting a rapid throughput of sheep. Every effort should be taken to prevent undue agitation of the ewes around the time of AI – on the understanding that such care will assist the fertilization process.

Doney, 1975; Doney *et al.*, 1976). It has always been a rule, with controlled breeding techniques in Ireland, that farmers should handle the ewes (and rams) with the minimum of disturbance around the time of oestrus; the observance of this principle probably becomes all the more important when breeding sheep by AI rather than natural service.

Robinson (1973) was one worker who noted that the passage of sperm through the ewe's cervix may be impaired in stressed sheep; such stress could arise from having ewes in unfamiliar surroundings, the presence of dogs and even the act of insemination itself. Manipulations of the cervix to collect mucus in progestagen–PMSG-treated sheep bred by AI apparently decreased conception rate in a study reported by Le Roux (1976). It should also be mentioned that, other things being equal, it would seem preferable to operate sheep AI on the basis of a single insemination technique rather than one requiring multiple inseminations, for the reason of minimizing stress effects; such considerations might be especially relevant in dealing with sheep unaccustomed to regular handling.

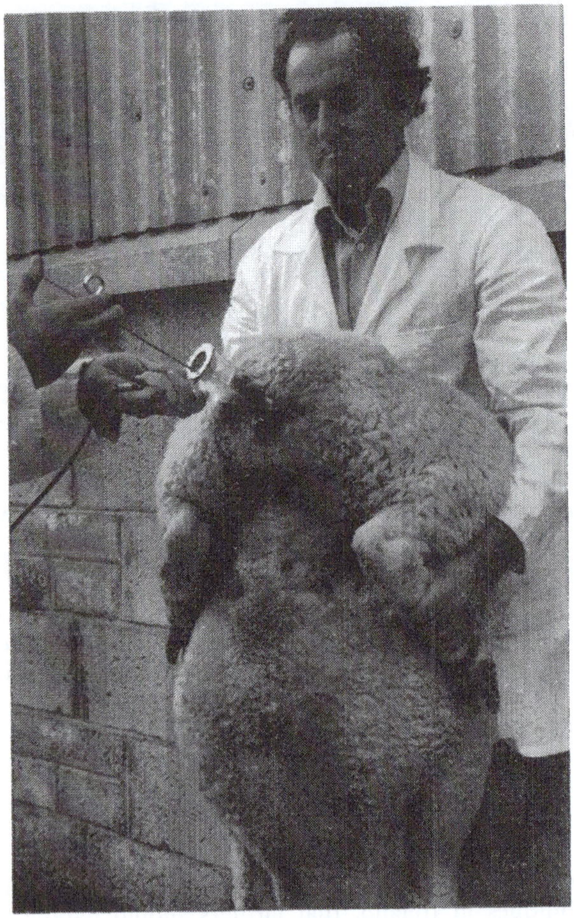

Fig. 4.4. Alternative method of restraint employed for ewes bred by AI. This method is not designed for use with large breeds of ewe – but it can be useful with light breeds that are accustomed to handling and which do not respond to such treatment by showing undue agitation.

4.4.4. Sheep AI and prostaglandins

The results from some studies in Ireland did indicate that a single-set-time insemination at 56h (after the end of treatment) was much less effective in achieving fertilization in prostaglandin (PG)-treated ewes than in those synchronized by the FGA-intravaginal sponge (Boland *et al.*, 1978). It was felt that there was need for further investigations before PGs could be recommended as an alternative to intravaginal progestagen treatment when breeding was by AI.

Elsewhere, in the UK, one study reported by the Meat and Livestock Commission involved a comparison between MAP–progestagen-treated ewes and sheep receiving two-dose prostaglandin analogue treatment (9-day interval); the work involved 300 ewes, and conception rates were 26% for prostaglandin and 61% for MAP-treated ewes, respectively. In Canada, Hackett *et al.* (1980) also presented evidence of variable fertility when prostaglandin treatment was followed by timed AI in different breeds of sheep; it appeared that breed type influenced the timing of ovulation in the ewe. In Australia, a study by Wales and Fairnie (1984) evaluated the potential for the use of the PG analogue, cloprostenol, in Merino sheep AI programmes in that country. The PG was administered in two doses at intervals of 8–14 days; because of the wide variability in the response of sheep to the use of the analogue, the authors concluded that there was little merit in the use of PG in sheep AI programmes. Conclusions drawn from such PG studies indicated that this approach did not offer a viable alternative to intravaginal sponge treatment when fixed-time AI was to be the method of breeding.

4.4.5. Breeding ewe-lambs by AI

It is well accepted that reproductive performance of ewe-lambs is generally poor and varies among breeds. Although much of the subfertility shown by ewe-lambs has been attributed to a high incidence of embryo mortality, there have been reports suggesting that poor oestrous behavioural responses of ewe-lambs to rams may also reduce or inhibit mounting behaviour in rams. In Canada, Langford (1986) attempted to eliminate such ram behavioural problems by inseminating ewe-lambs with fresh semen; results indicated that even when ewe-lambs were bred by AI, sheep with heavier body weights showed the higher pregnancy rates. Such results supported the recommendation of Thimonier and Cognie (1971) that ewe-lambs selected for breeding should weigh two-thirds of the mature body weight.

4.4.6. Other considerations

Considerable effort has been made over the past 30 years in bringing the reproductive processes of the ewe under close and effective control. Having gone to the trouble and cost of doing this, it may seem a retrograde step to leave the question of conception to the behavioural whims and variable fertility of rams under a system of natural mating. Fixed-time AI, as developed in France and Ireland during the 1970s, remains the simplest form of AI possible for any of the farm species. There was the problem that insemination at 56h after progestagen–PMSG made careful planning on the farm necessary, both in terms of terminating treatment and conducting the AI. However, the insemination procedure itself is technically less demanding than either cattle or pig AI, so that farmers or shepherds could be expected to carry out the procedure themselves, where necessary. There is no reason why further improvements in the AI technique cannot be expected, either as adjustments to the oestrus control procedure or to certain aspects of semen quality.

Of course, there is much more to the introduction of sheep AI than the technical problems associated with it. For example, there needs to be an organization within the country to offer a widespread and continuing service. Sheep farmers in France were fortunate to have a national agency such as INRA to implement their AI programmes.

4.5. Intrauterine Insemination and Frozen Semen

4.5.1. *Frozen semen limitations*

Frozen semen has a number of potential advantages for use in a sheep AI programme:

1. it allows an increased use of superior sires;
2. with intrauterine insemination it permits lower sperm doses to be employed;
3. it gives an opportunity to accumulate stocks of semen from rams for use in the breeding season;
4. it enables countries to import and export ram semen.

Studies in Australia have shown that ram semen extended in Tris-based diluents, cooled in 'one-step' and frozen in PVC mini-straws is capable of maintaining adequate motility on thawing in comparison with pellet-frozen semen (Hunton *et al.*, 1987); such semen, used in intrauterine inseminations, was also capable of giving satisfactory pregnancy rates. A study by Molinia *et al.* (1994) examined the cryoprotective effects of DMSO, ethylene glycol, glycerol and propanediol alone and in combinations with each other in Tris–citrate–glucose diluents on sperm characteristics after pellet-freezing; they concluded that glycerol was the single most effective cryprotectant and there was nothing to be gained by the addition of the other compounds.

Limited usage of frozen ram semen

It is clear from the literature that frozen semen has been used on a limited scale in sheep breeding programmes. In France, for example, although the sheep inseminations in 1986 numbered 452,293, only 8000 of these were with frozen semen. While there is ample evidence in the literature showing that fresh semen AI permits a reproductive performance in the ewe comparable with that obtained by natural mating, results also clearly show that the performance achievable with fresh semen was seldom obtainable with frozen semen (Colas, 1972, 1979; Smith *et al.*, 1975; Langford *et al.*, 1979; Dyrmundsson and Olafsson, 1989).

Decline in semen fertility with storage

In South Africa, Grobbelaar and Joubert (1986) reported acceptable conception rates after cervical inseminations of MAP–PMSG-synchronized ewes (54–63%) using frozen Ile-de-France semen imported from France; however, the same workers recorded a significant decline in conception rates (25–28%) after the semen had been stored in liquid nitrogen for a further 15–20 months. It should be noted, however, that such experiences have not been recorded for bull semen stored in liquid nitrogen.

Depth of semen deposition in cervix

A paper by Salamon and Maxwell (1995a) reviewed the literature on the freezing of ram semen and the fertility of ewes after such semen is used in cervical inseminations. The authors note that conception rates after cervical insemination with frozen semen have been low in comparison with those obtained with fresh diluted semen. The authors concluded that cervical insemination with frozen-thawed semen has limited application in sheep. A further review article by the same authors (Salamon and Maxwell, 1995b) dealt with the causes of low fertility after cervical insemination and the attempts that have been made to try and improve conception rates. They note that the most effective of such attempts is to increase the depth of deposition of the frozen-thawed semen in the cervical canal. However, despite all the attempts made to date, only intrauterine insemination by laparoscopy can be relied upon to give acceptable and reliable lambing results.

Reduced sperm viability after freeze-thawing

It is believed that frozen ram semen fails primarily because of the reduced viability of the cells and the impairment of sperm transport through the cervix; this results in a marked reduction in the number reaching the site of fertilization (Mattner *et al.*, 1969; Lightfoot and Salamon, 1970). Avoiding the cervix by way of uterine insemination, whether via laparotomy (Lightfoot and Salamon, 1970; Boland *et al.*, 1978) or by laparoscopic intrauterine insemination (Killeen *et al.*, 1982) initially appeared to give similar fertilization rates for both fresh and frozen semen.

Incidence of embryonic mortality

Langford *et al.* (1979), however, did report some evidence of increased embryonic mortality when using frozen rather than fresh semen (33% wastage vs. 6% in the period from day-18 and term); differences between frozen and fresh semen were also noted in some previous studies (Lightfoot and Salamon, 1970). In more recent times, Haresign (1992) records experiences at Nottingham showing that conception rates after intrauterine AI with frozen semen were considerably below those achieved with fresh semen (55% vs. 80–90%), even when the same number of motile sperm were employed. On that basis, the author suggests that the fertilizing capacity of the sperm may be reduced, even though they show good motility.

4.5.2. Pregnancy rates after intrauterine AI

Since the early 1980s there have been reports from several countries by workers evaluating the intrauterine insemination of sheep with frozen semen.

Intrauterine insemination with fresh and frozen semen was reported from New Zealand by Tervit *et al.* (1984); these workers used a laparoscopic procedure to introduce volumes of 0.03–0.04 ml semen into the lumen of each uterine horn. A lambing rate of 83% for fresh and 38% for frozen semen was recorded. Elsewhere, at that time, lambing rates of about 40–60% were being achieved with frozen semen (Killeen *et al.*, 1982; Maxwell *et al.*, 1983; Maxwell, 1984, 1986b; Maxwell and Hewitt, 1986). Investigations on the possible application of intrauterine AI using frozen semen in the UK were reported by Haresign *et al.* (1986), the authors concluding that the technique justified further research to identify optimum times of insemination and sperm doses to achieve optimal conception rates in sheep.

Progestagen effects

Studies by Eppleston and Roberts (1986) examined the effect of insemination timing on fertility as well as the effect of type of progestagen treatment, following intrauterine insemination with frozen semen. They recorded a significantly greater proportion of ewes lambing after treatment with FGA sponges compared with ewes treated with MAP sponges (55.4% vs. 43.4%); no significant effect of time of AI on the pregnancy rate was recorded (41.5% vs. 57% vs. 49.7% for 48, 60 and 72h, respectively).

Timing of AI

Earlier work by Maxwell *et al.* (1983) had shown a linear relationship between fertility and the time of intrauterine insemination. Fertilization rates with fresh semen increased from 71% to 97% as the time of intrauterine insemination increased from 24 to 48h after sponge withdrawal; with frozen semen, high fertilization rates were achieved by AI at 48 or 55h after pessary withdrawal (Maxwell *et al.*, 1984a,b). A study of Hunton *et al.* (1987) indicated that intrauterine insemination with semen diluted at a high rate prefreezing (up to 24-fold) and frozen in straws, may be as effective as insemination with semen diluted at a low rate prefreezing and frozen in pellets. The authors note that the use of high prefreezing dilution rates and packaging in straws permits more accurate measurement of the inseminate dose and more convenient storage and transport of semen.

Pellets, straws and minitubes

A report by Maxwell *et al.* (1995) has compared post-thaw survival and fertility of ram sperm frozen in pellets, PVC straws and German minitubes (Minitub GmBH, Tiefenbach, Germany); they found that the 0.25ml minitubes provided a useful alternative to pellets as a storage package for ram sperm, and they allowed for individual dose identification and easier storage while maintaining fertility indistinguishable from that obtained with pellet-frozen semen. Ewes were synchronized with 60mg MAP pessaries and received 400IU PMSG at sponge removal; intrauterine inseminations were performed by laparoscopy 60h after sponge removal, using 0.08ml per uterine horn (10 million sperm per horn). In contrast to the findings of Hunton *et al.* (1987), they did find pellet freezing to be superior to PVC straws in terms of fertility testing (71% vs. 55–57%) but not superior to minitubes (65%).

Semen quality considerations

In Brazil, Luz and Neves (1991) studied the effect of quality of frozen semen on the conception rate of ewes bred by an intrauterine insemination 60h after MAP-sponge withdrawal and PMSG injection; doses of 50 million sperm were used and conception rate was significantly higher in ewes bred with sperm showing the higher post-thawing motility percentages.

4.5.3. Timing of AI and sperm doses

In Scotland, McKelvey *et al.* (1985) evaluated a laparoscopic insemination technique in sheep and in a later report, McKelvey (1988) described the development of novel laparoscopic techniques for intrauterine insemination, embryo recovery and embryo transfer. According to this author, intrauterine AI gave fertilization rates of > 90% (fresh semen) and 55% (frozen-thawed). Elsewhere in the UK, experiments were reported by Findlater *et al.* (1991) in which ewes in commercial flocks were synchronized, using intravaginal sponges and an injection of 500IU PMSG at sponge withdrawal, and bred by intrauterine insemination with frozen semen. Maximum conception rates were achieved when inseminations were carried out at 54–60h after sponge removal; the overall conception rate to AI for two trials, each involving 900 ewes, was 56% and 58%. The authors found no significant difference in conception rate when motile sperm numbers were reduced from 52.2 million to 13 million spermatozoa per uterine horn. Investigations reported by Taljaard *et al.* (1991) in South Africa showed that the technique of intrauterine AI did not affect the length of subsequent oestrous cycles.

In Brazil, Souza *et al.* (1991a,b) inseminated synchronized Texel ewes with varying doses of frozen sperm by means of laparoscopy; they used intrauterine AI doses containing 30, 50, 100 and 150 million sperm and reported finding no significant difference in non-return rates. In France, Vallet *et al.* (1992) reported results showing a significantly higher pregnancy rate after uterine insemination (10 or 20 million sperm) than after cervical insemination with 200 million sperm (61.5 and 63.7% vs. 43.9%).

In Spain, Lopez Sebastian (1992) reported intrauterine AI with frozen semen of ewes synchronized by treatment with FGA sponges and doses of 300–500IU PMSG at sponge withdrawal; conception rates (55–58%) and litter sizes after laparoscopic AI were similar for sheep inseminated in October, April and August. Other studies in Spain were reported by Garde *et al.* (1993) who inseminated 675 synchronized ewes; for ewes inseminated into the uterus with frozen semen, the conception rate was significantly higher (67.8% vs. 51.1%) than for sheep cervically inseminated with fresh semen.

Australian AI research
In Australia, Eppleston *et al.* (1994) examined the depth of cervical insemination and site of intrauterine insemination and their relationship to the fertility of frozen-thawed ram semen. It was possible to penetrate the cervix deeper in adult than in maiden ewes and the mean depth of insemination increased with age of ewe (4–7 years). A significant linear increase in fertility was evident as the depth of insemination increased but there was no advantage in dividing the inseminate and inseminating 12 and 24h after oestrus detection. The pregnancy rates on day 35 after AI (16.4–27.7%) were not significantly different among ewes inseminated with varying sperm doses (80–320 million sperm). In terms of semen deposition, fertilization rates were not significantly different for AI into the common body compared with the left uterine horn. There was, however, a linear increase in fertilization rate as motile sperm doses increased; a dose of 65 million gave a fertilization rate of 72.8%.

4.5.4. Ram effects

In a further study, Eppleston and Maxwell (1995) examined the sources of variation in the reproductive performance of ewes inseminated with frozen-thawed ram semen by laparoscopy; only the total and motile sperm numbers inseminated per ewe were found to be correlated with fertility. The same authors regarded the most important finding of their study to be the large variation in the pregnancy rate of ewes inseminated with the semen from different rams; fertility ranged from 17.3% to 86.1% (69 percentage points) but such differences between males were not detectable using conventional microscopic methods of quality assessment.

Frozen semen in breeding improvement programmes
Considerable progress has been made in the use of frozen sheep semen in breeding improvement programmes. This has been possible by using intrauterine rather than cervical inseminations. A paper by Moroz *et al.* (1993) recorded results of the first stage of an Australian–Russian project in which laparoscopic insemination of synchronized ewes with frozen semen from Australian Merino rams was carried out on farms in Russia; results indicated that high conception rates can be achieved under optimal conditions.

In 1992, staff from the Scottish Agricultural College carried out the insemination of 16,000 ewes, with most of the inseminations being carried out in pedigree flocks to support sire-referencing schemes. It was reported that intrauterine insemination with frozen semen by the laparoscopic route achieved a success rate of about 64%; this compared with 80% for fresh semen and 65–70% for cervical AI using fresh semen.

CIDR treatment
In Japan, Fukui *et al.* (1994) examined the effect of varying insertion periods with CIDRs on conception rate by fixed-time intrauterine insemination with

frozen semen in seasonally anoestrous Suffolk ewes. In this, ewes were treated with CIDRs for 6, 9 or 12 days, with PMSG injected the day prior to CIDR withdrawal. Ewes were inseminated 42–50h after PMSG with frozen semen; lambing rates were not affected by the duration of the insertion period.

Unilateral AI

In Chile, Correa *et al.* (1994) have reported on fertilization rates in sheep unilaterally inseminated with frozen semen at 60h after the cessation of 60mg MAP treatment, with 400IU PMSG given at sponge withdrawal; fertilization rate in such sheep was 84% as compared with 19% for ewes inseminated cervically.

Welfare aspects of intrauterine insemination

In the UK, there are government advisory committees that deliberate on the ethical issues raised by emerging farm animal breeding technologies. Intrauterine insemination in sheep has been one of the issues considered and there have been suggestions that routine use of this technique should not be permitted. Current codes of practice in the UK require the use of sedation and local anaesthesia before carrying out intrauterine AI. An attempt to measure objectively the stress responses of sheep to laparascopic procedures under local anaesthesia was reported by Haresign *et al.* (1995); results of their study showed that sheep express a temporary stress response to laparoscopy but that this was little above that attributable to restraint alone. The magnitude of responses (heart rate and cortisol levels) was similar to those reported in the literature for such husbandry procedures as handling, penning, visual isolation and introduction to a new flock. The authors suggest that their data show that the stress imposed by laparoscopy may be much less than the public perception of the technique might indicate.

4.5.5. Transcervical AI in sheep

There were some reports during the 1970s of attempts to deposit ram semen in the lumen of the uterus non-surgically by forcing a catheter through the cervical canal. The success rates claimed for such attempts at cervical penetration were low (Anderson *et al.*, 1973; Fukui and Roberts, 1977). In more recent times, work in Canada has resulted in what is known as the Guelph System for Transcervical Artificial Insemination (GST-AI) technique or sheep AI. Contrary to earlier reports (Dun, 1955; More, 1984). Canadian workers found that transcervical passage is possible in the ewe using instruments for insemination (Halbert *et al.*, 1990a). However, the same authors showed that there are a number of reasons why such passage may be difficult and in some instances impossible. These include the presence of blind spaces from folds of vaginal tissue which make identification of the cervical opening difficult and various other features.

A technique was developed by Halbert *et al.* (1990b) in which ewes were positioned in dorsal recumbency with their hindquarters elevated; the vagina was dilated using a duck-billed speculum, the cervix was grasped and retracted using forceps, and an inseminating instrument introduced into the cervical opening and manipulated through the cervical canal. The authors described the difficulty in locating the cervical opening, the force required to retract the cervix and the time required to insert the instrument into the uterus were recorded. The authors reported that uterine penetration was achieved in 82% of 89 multiparous ewes.

The field evaluation of the Guelph AI technique is reported in a paper by Halbert *et al.* (1990c) who report lambing rates of 50%, 55% and 40% in three flocks that were inseminated transcervically with fresh semen and 65% for ewes inseminated laparoscopically. The authors note that transcervical insemination is not likely to be possible with all ewes and that a time limit of perhaps 5 min should be established for the manipulation, to avoid unnecessary trauma to the cervical canal. Despite such limitations, the authors suggested that such results demonstrated that the technique was suitable for further evaluation using frozen semen.

Field trials with GST-AI technique

Such evaluation, using previously frozen semen, was reported by Buckrell *et al.* (1994). In this, 2060 parous ewes on 65 farms were inseminated over a two-year period. As in the previous work, oestrus was synchronized using progestagen-impregnated vaginal sponges and PMSG, with AI at 54h from pessary withdrawal. Success of penetration increased from 76.3% in the first 500 ewes to 97.9% in the last 500 ewes and was higher in ewes from an accelerated lambing programme (92%) than in those from an annual lambing programme (82.4%). The lambing rate for ewes bred during the autumn breeding season was 50.7% compared with 24.4% for sheep inseminated at other periods. The average time required for handling and insemination decreased from 8.62 min in the first 500 ewes to 3.62 min in the last 500 ewes. The authors note that the average time required per sheep (5.8 min) compared favourably with the laparoscopic technique under small-farm conditions.

Oxytocin and relaxin to dilate the cervix

Elsewhere, a report by Khalifa *et al.* (1992) in Virginia indicated that exogenous oxytocin was effective in dilating the ovine cervix to the extent that a stainless steel rod could be passed into the uterus in 33 of 43 (77%) ewes. In a more recent report, Khalifa (1993) has reported non-surgical intrauterine AI in Rahmani sheep in Egypt using exogenous oxytocin. According to this report up to 100% of ewes could be penetrated after administering 200 USP units of oxytocin. The author does, however, refer to depositing semen in the os cervix if the inseminating pipette could not be passed through the cervix within 10 min; this hardly seems a feasible proposition for a farmer with several hundred ewes awaiting insemination. In the USA, Akinbami *et al.* (1990) has reported unsuccessful attempts at cervical dilation, using injections of porcine relaxin at 24 and 36h after the cessation of MAP-sponge–PMSG synchronization treatment. In Argentina, Veksler Hess and Cisale (1992) examined the possibility of selecting the most suitable ewes for AI on the basis of the conformation of their cervix; it was concluded that this was not possible on the basis of external cervical orifice measurements.

4.6. References

Aguer, D. and Le Provost, L. (1976) *L'Insemination artificielle ovine*. Searle Laboratory Publication, 123 pp.

Akinbami, M.A., Meredith, S., Warren, J.E. Jr., Anthony, R.V. and Day, B.N. (1990) Cervical dilation, conception rate, and concentrations of progesterone and estradiol-17β in postpartum ewes treated with porcine relaxin. *Theriogenology* 34, 927–940.

Al-Kamali, A.A., Crosby, F., Boland, M., Kelleher, D.L. and Gordon, I. (1990) Effects of progestagen type and PMSG source on lambing outcome in ewes following artificial insemination. *Irish Veterinary Journal* 43, 99–103.

Allison, A.J. and Robinson, T.J. (1970) The effect of dose level of intravaginal progestagen on sperm transport, fertilization and lambing in the cyclic Merino ewes. *Journal of Reproduction and Fertility* 22, 515–521.

Anderson, V.K., Aamdal, J. and Fougner, J.A. (1973) Intrauterine and deep cervical insemination with frozen semen in sheep. *Zuchthygiene* 8, 113–118.

Anon (1978) *Artificial Insemination and Reproduction*. Meat and Livestock Commission, 11th Annual Report, p. 25.

Anon (1982) Sheep artificial insemination. *MLC Veterinary Services, Technical Bulletin* 15 pp.

Barlow, M., Pryce-Jones, D. and Reed, H.C.B. (1974) MLC sheep AI field trials: a comparison of milk and egg-yolk diluents. *Veterinary Record* 94, 159–160.

Boland, M.P., Gordon, I. and Kelleher, D.L. (1978) The effect of treatment by prostaglandin analogue (ICI-80996) or progestagen (SC-9880) on ovulation and fertilization in cyclic ewes. *Journal of Agricultural Science, Cambridge* 91, 765–766.

Buckrell, B.C., Buschbeck, C., Gartley, C.J, Kroetsch, T., McCutcheon, W., Martin, J., Penner, W.K. and Walton, J.S. (1994) Further development of a transcervical technique for artificial insemination in sheep using previously frozen semen. *Theriogenology* 42, 601–611.

Colas, G. (1972) Ewe fertility following insemination with liquid or deep frozen semen in the course of oestruses induced by progestins during the breeding season. *Proceedings of the 7th International Congress Animal Reproduction and AI* (Munich) 2, 925–930.

Colas, G. (1975) The use of progestagen SC 9880 as an aid for AI in ewes. *Annales de Biologie animale Biochimie Biophysique* 15, 317–327.

Colas, G. (1979) Fertility in the ewe after AI with fresh and frozen semen at the induced oestrus, and influence of the photoperiod on the semen quality of the ram. *Livestock Production Science* 6, 153–166.

Colas, G. (1983) Semen technology in the ram. In: Courot, M. (ed.) *The Male in Farm Animal Reproduction*. CEC Seminar, pp. 219–236.

Colas, G., Dauzier, L., Courot, M., Ortavant, R. and Signoret, J.P. (1968) Results obtained while investigating some important factors in AI in sheep. *Annales Zootechnica* 17, 47–57.

Correa, J.E., Bergmann, B. and Gatica, R. (1994) Fertilization rate in sheep unilaterally inseminated with frozen semen. *Small Ruminant Research* 13, 99–101.

Doney, J.M., Smith, W.F. and Gunn, R.G. (1976) Effects of post-mating environmental stress or administration of ACTH on early embryonic loss in sheep. *Journal of Agricultural Science, Cambridge* 87, 133–136.

Dun, R.B. (1955) The cervix of the ewe – its importance in artificial insemination of sheep. *Australian Veterinary Journal* 31, 101–103.

Dyrmundsson, O.R. (1977) Synchronization of oestrus in Iceland ewes with special reference to fixed-time artificial insemination. *Acta Agriculturae Scandinavica* 27, 250–252.

Dyrmundsson, O.R. and Olafsson, T. (1989) Sexual development, reproductive performance, artificial insemination and controlled breeding. *Reproduction, Growth and Nutrition in Sheep, Reykjavik* pp. 97–104.

Eppleston, J. and Maxwell, W.M.C. (1995) Sources of variation in the reproductive performance of ewes inseminated with frozen-thawed ram semen by laparoscopy. *Theriogenology* 43, 777–788.

Eppleston, J. and Roberts, E.M. (1986) The effect of progestagen, PMSG and time of insemination on fertility in ewes following intra-uterine insemination with frozen semen. *Australian Veterinary Journal* 63, 124–125.

Eppleston, J., Salamon, S., Moore, N.W. and Evans, G. (1994) The depth of cervical insemination and site of intrauterine insemination and their relationship to the fertility of frozen-thawed ram semen. *Animal Reproduction Science* 36, 211–225.

Evans, G. and Maxwell, W.M.C. (1987) *Salamon's Artificial Insemination of Sheep and Goat.* Butterworths, London.

Findlater, R.C.F., Haresign, W., Curnock, R.M. and Beck, N.F.G. (1991) Evaluation of intrauterine insemination of sheep with frozen semen: effects of time of insemination and semen dose on conception rates. *Animal Production* 53, 89–96.

Fukui, Y. and Roberts, E.M. (1977) Sperm transport after non-surgical intrauterine insemination with frozen semen in ewes treated with prostaglandin $F_{2\alpha}$. *Journal of Reproduction and Fertility* 51, 141–143.

Fukui, Y., Hirai, H., Honda, K. and Hayashi, K. (1993a) Lambing rates by fixed-time intrauterine insemination with frozen semen in seasonally anoestrous ewes treated with a progestogen-impregnated sponge or CIDR device. *Journal of Reproduction and Development* 39, 1–5.

Fukui, Y., Fujii, M. and Tashiro, Y. (1993b) Insemination doses of frozen-thawed semen in seasonally anestrous ewes treated with two different progesterone-impregnated intravaginal devices. *Journal of Reproduction and Development* 39, 269–273.

Fukui, Y., Tabuchi, K., Yamada, A., Hayashi, N. and Tanaka, K. (1994) Effect of insertion periods of controlled internal drug release device (CIDR) on conception rate by fixed-time intrauterine insemination with frozen semen in seasonally anoestrous ewes. *Journal of Reproduction and Development* 40, 221–226.

Garde, J., Perez, S., Garrido, D., Aguado, M.J., Angulo, C., Perez Guzman, M.D., Jimenez, J. and Montoro, V. (1993) Use of frozen semen for the intrauterine insemination of selected Manchega ewes in the breed selection scheme. Preliminary results. *Proceedings of the 5th International Symposium Animal Reproduction* (Portugal) 2, 293–298.

German, V.F. (1988) The preparation of rams for the mating season. *Ovsevodstvo* No. 4, 17–18.

Gordon, I. (1975) The use of progestagens in sheep bred by natural service and AI.

Annales de Biologie animale Biochimie Biophysique 15, 303–316.
Gordon, I. (1983) *Controlled Breeding in Farm Animals*. Pergamon Press, Oxford, pp. 197–208.
Gordon, I. and Crosby, T.F. (1980) AI in sheep promises well. *Irish Farmers' Journal* 32(25), 12–13.
Gretton, S. (1992) Accelerated progress through sheep breeding. *Proceedings of the 44th Ruakura Farmers' Conference* (Ruakura), 47–50.
Grobbelaar, J.A.N. and Joubert, J.J. (1986) Lambing results obtained with imported Ile de France ram semen. *South African Journal of Animal Science* 16, 219–220.
Gunn, R.G. and Doney, J.M. (1975) The interaction of nutrition and body condition at mating on ovulation rate and early embryo mortality in Scottish Blackface ewes. *Journal of Agricultural Science, Cambridge* 85, 465–470.
Hackett, A.J., Langford, G.A. and Robertson, H.A. (1980) Fertility and prolificacy of confined ewes treated with PG $F_{2\alpha}$ and bred by AI. *Journal of Animal Science* 51 (Suppl. 1), Abstract 282.
Halbert, G.W., Dobson, H., Walton, J.S. and Buckrell, B.C. (1990a) The structure of the cervical canal of the ewe. *Theriogenology* 33, 977–992.
Halbert, G.W., Dobson, H., Walton, J.S. and Buckrell, B.C. (1990b) A technique for transcervical intrauterine insemination of ewes. *Theriogenology* 33, 993–1010.
Halbert, G.W., Dobson, H., Walton, J.S., Sharpe, P. and Buckrell, B.C. (1990c) Field evaluation of a technique for transcervical intrauterine insemination of ewes. *Theriogenology* 33, 1231–1243.
Hanrahan, J.P. (1995) Evaluation of Suffolk and Texel rams from different sources. *Proceedings of the Irish Grassland and Animal Production Association* (21st Meeting), pp. 105–106.
Haresign, W. (1992) Manipulation of reproduction in sheep. *Journal of Reproduction and Fertility* Suppl. 45, 127–139.
Haresign, W., Read, S.R., Curnock, R.M. and Reed, H.C.B. (1986) A note on the use of laparoscopy for intrauterine insemination of frozen-thawed semen in the ewe. *Animal Production* 43, 553–556.
Haresign, W., Williams, R.J., Khalid, M. and Rodway, R. (1995) Heart rate responses and plasma cortisol and β-endorphin concentrations in ewes subjected to laparoscopy and its associated handling procedures. *Proceedings of the British Society Animal Science* (Winter Meeting), Paper 58.
Harvey, T.G., Johnson, D.L., Tervit, H.R. and Welch, R.A.S. (1984) Synchronization and artificial insemination of ewes – techniques which have possible commercial application. *Proceedings of the New Zealand Society Animal Production* 44, 7–9.
Hunton, J.R., Flecker, S.E. and Maxwell, W.M.C. (1987) Pregnancy rates following intra-uterine insemination with pellet or straw-frozen ram semen. *Journal of Agricultural Science, Cambridge* 109, 189–191.
Jheltobruch, N.A. (1979) AI of sheep in the Soviet Union. In: Haresign, W. (ed.) *Sheep Breeding* 2nd edn. Butterworths, London, pp. 565–570.
Khalifa, R.M. (1993) Non-surgical intrauterine artificial insemination in sheep using exogenous oxytocin. *Egyptian Journal of Animal Production* 30, 55–61.
Khalifa, R.M.E., Sayre, B.L. and Lewis, G.S. (1992) Exogenous oxytocin dilates the cervix in ewes. *Journal of Animal Science* 70, 38–42.
Killeen, I.D., Caffery, G. and Holt, N. (1982) Fertility of ewes following intra-uterine insemination with the aid of a laparoscope. *Proceedings 14th Annual Conference Australian Society for Reproductive Biology* p. 104.
Langford, G.A. (1982) Influence of PMSG and time of artificial insemination on

fertility of progestagen-treated sheep in confinement. *Journal of Animal Science* 54, 6–10.

Langford, G.A. (1986) Influence of body weight and number of inseminations on fertility of progestogen-treated ewe lambs raised in controlled environments. *Journal of Animal Science* 62, 1058–1062.

Langford, G.A. and Hackett, A.J. (1980) Dose-related effects of PMSG in breeding confined sheep by artificial insemination. *Canadian Journal of Animal Science* 60, 562–563.

Langford, G.A. and Fisher, P.S. (1980) Influence of storage temperature and duration of storage on the fertilizing capacity of extended ram semen. *Journal of Animal Science* 51 (Suppl. 1), Abstract 295.

Langford, G.A., Marcus, G.J., Hackett, A.J., Ainsworth, L., Wolynetz, M.S. and Peters, H.F. (1979) A comparison of fresh and frozen semen in the insemination of confined sheep. *Canadian Journal of Animal Science* 59, 685–691.

Le Roux, P.J. (1976) The conception rate of MAP and MAP–PMSG-treated Karakul ewes inseminated with diluted semen. *South African Journal of Animal Science* 6, 1–5.

Lewis, G.S. and Goebel, K.A. (1993) Development of a method for prolonging estrus in ewes. *Sheep Research Journal* 9, 59–61.

Lightfoot, R.H. and Salamon, S. (1970) Fertility of ram spermatozoa frozen by the pellet method. 1. Transport and viability of spermatozoa within the genital tract of the ewe. *Journal of Reproduction and Fertility* 22, 385–398.

Lopez Sebastian, A. (1992) Intrauterine insemination of ewes with frozen semen. *ITEA, Produccion Animal* 88A(1), 70–75.

Luz, S.L.N. and Neves, J.P. (1991) Effect of the quality of frozen semen on the conception rate of ewes inseminated into the uterus by means of laparoscopy. In: *Anais, IX Congresso Brasileiro de Reproducao Animal* 2, 440.

McClelland, T.H. and Quirke, J.F. (1971) Artificial insemination and natural service at a pre-determined time in cyclic sheep treated with SC-9880-progesterone sponges. *Animal Production* 13, 323–328.

McGrath, P.E., Boland, M.P. and Gordon, I. (1979) Effect of sexual preparation procedures on semen characteristics in the ram. *Journal of Agricultural Science, Cambridge* 93, 761–763.

McKelvey, W.A.C. (1988) Studies on the establishment of pregnancy in the ewe. *Index to Theses Accepted for Higher Degrees in the Universities of Great Britain and Ireland* 37(3), 1241.

McKelvey, W.A.C., Robinson, J.J., Aitken, R.P. and Henderson, R.G. (1985) The evaluation of a laparoscopic insemination technique in ewes. *Theriogenology* 24, 519–535.

McMillan, W.H. (1994) Timing single fixed-time inseminations in ewes: some new concepts. *Proceedings of the New Zealand Society of Animal Production* 54, pp. 45–49.

Mattner, P.E., Entwhistle, K.W. and Martin, I.C.A. (1969) Passage, survival and fertility of deep frozen ram semen in the genital tract of the ewe. *Australian Journal of Biological Science* 22, 181–187.

Maurel, M.C., Arranz, J.M., Belloc, M., Brice, G., Briois, M. and Herve, D. (1992) Detection of the preovulatory LH peak in milk: application of an ELISA kit in lactating ewes. *Proceedings of the 8th Meeting European Embryo Transfer Association* (Lyon), 184.

Maxwell, W.M.C. (1984) Current problems and future potential of artificial insemination programmes. In: Lindsay, D.R. and Pearce, D.T. (eds) *Reproduction in Sheep*. Cambridge University Press, Cambridge, pp. 291–298.

Maxwell, W.M.C. (1986a) Artificial insemination of ewes with frozen-thawed semen at a synchronised oestrus. 1. Effect of time of onset of oestrus, ovulation and insemination on fertility. *Animal Reproduction Science* 10, 301–308.

Maxwell, W.M.C. (1986b) Artificial insemination of ewes with frozen-thawed semen at a synchronized oestrus. 2. Effect of dose of spermatozoa and site of intrauterine insemination on fertility. *Animal Reproduction Science* 10, 309–316.

Maxwell, W.M.C. and Hewitt, L.J. (1986) A comparison of vaginal, cervical and intrauterine insemination of sheep. *Journal of Agricultural Science, Cambridge* 106, 191–193.

Maxwell, W.M.C., Butler, L.G. and Wilson, H.R. (1983) Fertility after intra-uterine insemination with frozen ram semen. *Proceedings of the 15th Annual Conference of the Australian Society for Reproductive Biology* p. 92.

Maxwell, W.M.C., Butler, L.G. and Wilson, H.R. (1984a) Intra-uterine insemination of ewes with frozen semen. *Journal of Agricultural Science, Cambridge* 102, 233–235.

Maxwell, W.M.C., Wilson, H.R. and Butler, L.G. (1984b) Fertility of ewes after intrauterine insemination with frozen semen. *Proceedings of the Australian Society of Animal Production* 15, 448–451.

Maxwell, W.M.C., Landers, A.J. and Evans, G. (1995) Survival and fertility of ram spermatozoa frozen in pellets, straws and minitubes. *Theriogenology* 43, 1201–1210.

Memon, M.A. and Ott, R.S. (1981) Methods of semen preservation and artificial insemination in sheep and goats. *World Review Animal Production* 17, 19–24.

Mesnil du Buisson, F. du (1994) Artificial insemination of domestic animals (except cattle) in France and its development. *Comptes Rendus de l'Academie d'Agriculture de France* 80(3), 89–106.

Molinia, F.C., Evans, G. and Maxwell, W.M.C. (1994) Incorporation of penetrating cryoprotectants in diluents for pellet-freezing ram spermatozoa. *Theriogenology* 42, 849–858.

More, J. (1984) Anatomy and histology of the cervix uteri of the ewe: new insights. *Acta Anatomica* 120, 156–159.

Moroz, V.A., Burdukoskaya, T.K., Rabochev, V.K., Aibazov, M.M., Mamytov, G.A., Purvis, I., Maksvell, K.H., Osborn, D., Vil'son, G., Moor, P., Maxwell, C., Osborne, D., Wilson, G. and Moore, P. (1993) Results of the first stage of an Australian-Russian experiment. *Ovtsevodstvo* No. 3, 10–14.

Ozin, F.V. (1968) The role of artificial insemination in the re-organization of sheep breeding in the U.S.S.R. *Zhivotnovodstvo* 30(7), 64–69.

Perrenoud, V.M., Salvador, S.M. and Rodrigues, J.L. (1991) Efficiency of hCG treatment in sheep laparoscopy AI programmes. In: *Anais IX Congresso Brasileiro de Reproducao Animal* 2, p. 361.

Quinlivan, T.D. and Robinson, T.J. (1969) Numbers of spermatozoa in the genital tract after artificial insemination of progestagen-treated ewes. *Journal of Reproduction and Fertility* 19, 73–96.

Quispe, T., Zarco, L., Valencia, J. and Ortiz, A. (1994) Estrus synchronization with Melengestrol acetate in cyclic ewes. Insemination with fresh or frozen semen during the first or second estrus post treatment. *Theriogenology* 41, 1385–1392.

Robinson, T.J. (1973) Contraception and sperm transport in domestic animals. *INSERM* 26, 453–478.

Salamon, S. and Maxwell, W.M.C. (1995a) Frozen storage of ram semen. I. Processing, freezing, thawing and fertility after cervical insemination. *Animal Reproduction Science* 37, 185–249.

Salamon, S. and Maxwell, W.M.C. (1995b) Frozen storage of ram semen. II. Causes of low fertility after cervical insemination and methods of improvement. *Animal Reproduction Science* 38, 1–36.

Sanford, L.M. and Yarney, T.A. (1980) Social environment and seasonality in reproductive processes of rams. *Canadian Journal of Animal Science* 60, 1050–1055.

Schakell, G.H., Kyle, B. and Littlejohn, R.P. (1990) Factors influencing the success of a large-scale artificial insemination programme in sheep. *Proceedings of the New Zealand Society Animal Production* 50, pp. 427–430.

Smith, J.F., Boys, P.T.S., Drost, H. and Wilson, S.G. (1975) AI of sheep with frozen semen. *Proceedings of the New Zealand Society Animal Production* 35, p. 71.

Smith, P.A. (1977) Studies in the artificial insemination of sheep. PhD thesis, National University of Ireland, Dublin.

Smith, P.A., Boland, M.P. and Gordon, I. (1977a) Lambing outcome in farm flocks following set-time AI and natural service. *Journal of the Department of Agriculture and Fisheries* (Dublin) 74, 44–49.

Smith, P.A., Boland, M.P. and Gordon, I. (1977b) Effect of timing of inseminations on lambing to a double AI. *Journal of the Department of Agriculture and Fisheries* (Dublin) 74, 50–55.

Smith, P.A., Boland, M.P. and Gordon, I. (1978) Conception rate in ewes: effect of method of breeding and number of inseminations. *Journal of Agricultural Science, Cambridge* 91, 511–512.

Smith, P.A., Boland, M.P. and Gordon, I. (1979) Studies in ram semen collection for use in AI during the breeding season. *Journal of the Irish Department of Agriculture* (Dublin) 74, 56–65.

Smith, P.A., Boland, M.P. and Gordon, I. (1981) Effect of type of intravaginal progestagen on the outcome of fixed-time AI. *Journal of Agricultural Science, Cambridge* 96, 243–245.

Souza, C.J.H., Chagas, L.M., Bartolini, M., Aguiar, P.R.L. and Rodrigues, J.L. (1991a) Effect of time of onset of synchronized oestrus in ewes on the NR rate to insemination with frozen semen. In: *Anais IX Congresso Brasileiro de Reproducao Animal* 2, p. 359.

Souza, C.J.H., Chagas, L.M., Bertolini, M., Aguiar, P.R.L. and Rodrigues, J.L. (1991b) Effect of sperm concentration on the non-return rate of ewes inseminated by means of laparoscopy. In: *Anais, IX Congresso Brasileiro de Reproducao Animal* 2, p. 360.

Taljaard, T.L., Terblanche, S.J., Bertschinger, H.J. and Vuuren, L.J. Van (1991) The effect of the laparoscopic insemination technique on the oestrous cycle of the ewe. *Journal of South African Veterinary Association* 62, 60–61.

Tervit, H.R., Goold, P.G. and James, R.W. (1984) The insemination of sheep with fresh or frozen semen. *Proceedings of the New Zealand Society of Animal Production* 44, 11–13.

Thimonier, J. and Cognie, Y. (1971) Acceleration des mises-bas et conduite d'elevage chez les ovines. *Bulletin Technical Information Ministry Agriculture* 257, 187.

Thomas, M.G., Merilan, C.P. and Keisler, D.H. (1991) The effects of filiform appendage removal on semen collection in the ram. *Theriogenology* 35, 309–316.

Vallet, J.C., Baril, G., Leboeuf, B. and Perrin, J. (1992) Intrauterine insemination by

laparoscopy in domestic small ruminants. *Annales de Zootechnie* 41, 305–309.
Veksler Hess, J.D. and Cisale, H. (1992) Selection of ewes for artificial insemination on the basis of the conformation of the cervix. *Veterinaria Argentina* 9(90), 680–685.
Wales, R.G. and Fairnie, I.J. (1984) The use of a PGF analog in Australian sheep AI programs. *Proceedings of the 10th International Congress Animal Reproduction and AI* (Illinois) 3, Communication No. 318.

5 Advancing the Sheep Breeding Season

5.1. Introduction

Although it is generally held that the beginning and end of the sheep's breeding season is controlled by natural daylength changes, the precise onset of the season in any one year is probably the result of many modifying factors. The breed, age and previous reproductive history of ewes, changes in the environmental temperature and the sudden introduction of the ram to ewes can all have a modifying effect on the occurrence of oestrus. It is generally believed that the onset of breeding is probably not materially affected by the nutritional status of the ewe, unless it is exceptionally low (Hafez, 1952; Ducker and Boyd, 1974).

As mentioned elsewhere (Section 2.3), the relationship between photoperiodicity and ovarian activity in the ewe is probably more complex than once was assumed. Almost 20 years ago, Robertson (1977) noted that no continuously successful system of breeding ewes on a production basis at 6–8-month intervals by the use of controlled lighting alone had been perfected; the same remains true today. In the various studies which have been made with a view to advancing the breeding season, it has been a matter of looking at possible light manipulations, at hormonal treatments and at developing breeds and crossbreeds which are capable of showing oestrous cycles earlier than usual.

Reviewing technology for out-of-season breeding with New Zealand sheep, Smith *et al.* (1989) noted that stimulation of earlier ovulation by the sudden introduction of rams has been much studied, but with highly variable results. They observe that progestagen–pregnant mare serum gonadotrophin (PMSG) treatment is effective but that ewes failing to conceive may show anoestrus. In terms of using melatonin, it is clear that such treatment can induce earlier oestrus, but its efficacy varies with genotype and depth of anoestrus.

5.2. Manipulation of the Light Environment

According to Robinson (1990), the role of photoperiod in timing the breeding season of ewes is in synchronizing a rhythm in reproductive neuroendocrine function, including luteinizing hormone (LH) and prolactin secretion, rather than to generate this rhythm. Only part of the annual photoperiodic cycle is necessary for this entrainment, and it appears to be long daylengths which time the onset of breeding. Short days stimulate ovarian activity in sheep from temperate latitudes, but if used for a long time, cause refractoriness. This refractoriness may be overcome by exposing ewes to long days; it is evident that an alternation between long days and short days is essential for the photoperiodic control of seasonal reproduction in the sheep. Malpaux *et al.* (1989) reported findings which led them to conclude that the lengthening photoperiod between the winter and summer solstices is required for the occurrence of the autumn breeding season. French workers have shown that it is possible to induce consistent reproductive activity in rams by monthly alternation between long days and short days (Chemineau *et al.*, 1992); such treatment greatly improved sperm production and quality. In the ewe, however, such treatment did not abolish seasonality of ovulatory activity.

Gradual or abrupt light changes

There are two main types of artificial daylength control that are capable of influencing reproductive activity in ewes: it may be a matter of providing a gradual decrease or increase in artificial daylength, similar to that occurring under a natural daylength environment or it may be done by subjecting the ewes to an abrupt decrease on one day and thereafter maintaining them at that daylength until a response is shown (Hafez, 1952; Fraser and Laing, 1969).

Inevitably, the response of ewes to any light manipulation is not immediate and it may often be a matter of months rather than weeks before a response is evident. However, the time of year at which light treatment is applied is known to have a marked effect on the speed of response (Ducker and Bowman, 1970b); the greater the decrease or increase in daylength applied, the faster the ewes responded (Ducker and Bowman, 1970a; Ducker *et al.*, 1970). Inducing ewes in the UK to breed by light control soon after the longest day in June was reported by Ducker and Bowman (1970b) to be much more effective than if attempts were made when natural daylength is increasing and ewes have just entered anoestrus (in the March to May period). One practical disadvantage of using light control is the fact that individual ewes show oestrus after varying intervals; several weeks may elapse between the time the first and last sheep in the flock comes into heat.

5.2.1. Lighting systems employed

As to lighting regimes employed by different investigators, Duckler and Bowman (1972) used one involving the abrupt extension of daylength to 22h either in late pregnancy or at parturition followed by a reduction to that of natural daylength; such a system had the merit of dispensing with the need for a light-proof building, which would be difficult to justify on the basis of the high costs involved (requirement for forced mechanical ventilation, among other things).

Newton and Betts (1972), on the other hand, used a regimen involving an abrupt increase in daylength to 18h for one month during late pregnancy followed by an abrupt decrease to a constant level of 8h; this proved effective in inducing a fertile oestrus within about 3 months after parturition in March-lambing ewes. Despite experimental work in the UK in developing such light control measures, it appeared that only one large-scale commercial unit in England employed the light control approach to advancing the breeding season (Murdock, 1975); the unit's practical objective was to achieve matings in the month of June for early lamb production.

5.3. Employing the Ram Effect

Seasonally anoestrous ewes of many breeds, if preconditioned by a period of isolation from rams, respond to the re-introduction of the male by exhibiting a relatively well-synchronized heat period about one oestrous cycle interval later, as shown in studies with Merinos in Australia (Underwood *et al.*, 1944; Fulkerson *et al.*, 1981), with Romneys in New Zealand (Edgar and Bilkey, 1963; Knight *et al.*, 1981) and with many flocks elsewhere. This phenomenon, known as the 'ram effect' has been employed in commercial sheep flocks with varying degrees of success (Fig. 5.1).

Effect on ovulation rate

There have been suggestions that the 'ram effect' may elicit two different responses: advancing the breeding season and a higher ovulation rate. In France, Cognie *et al.* (1980) observed a higher lambing rate in Ile-de-France ewes that were exposed to vasectomized rams as compared with non-'teased' sheep; Oldham (1980) also recorded a higher ovulation rate in ewes at first oestrus than at later oestrous cycles. Elsewhere, however, studies (Knight *et al.*, 1980; Hudgens *et al.*, 1987; Thompson *et al.*, 1990) have revealed little evidence of an effect on the lambing rate.

Rams in New Zealand sheep farming

Coop and Clark (1968) suggested that an acceptable result for their South island conditions in New Zealand was to have 80% of Romney ewes mated in six days, 55–66% of ewes lambing within a week, 12% of ewes lambing in one day at the peak of lambing and this date predictable to two days. Results from ram introductions conducted elsewhere in New Zealand, however, particularly in the North island, showed that there was no response among ewes to the ram effect in some seasons yet a satisfactory response in others (McDonald, 1971). There was evidence that some ram breeds (Dorset) could induce a greater

Fig. 5.1. Use of the ram to stimulate an earlier start to breeding in the ewe. The introduction of the ram among ewes during late anoestrus has been shown by many workers to be effective in inducing ovulation and subsequently oestrus. The response of the ewe to the ram will certainly vary according to a number of factors and it is unwise to rely on such an approach under Irish conditions.

response than others (Romney) among Romney ewes (Tervit *et al.*, 1977); an even greater effect, in which Dorset 'teaser' rams initiated cyclic breeding in ewes about two weeks earlier than Romney teasers was recorded in similar work conducted by Meyer (1979).

Working with ewes in New Zealand, Knight *et al.* (1983) observed that although stimulation of the ewe with teaser rams caused an earlier onset of ovulatory activity it did not increase the number of heat periods shown by the ewe during their breeding season; earlier onset of breeding was accompanied by an earlier return to anoestrus. In more recent times, Scott and Johnstone (1994) have recorded variations between years in the ram effect when Coopworth or Dorset rams were introduced to seasonally anoestrous Coopworth ewes; the authors concluded that the variation in expression of the ram effect is probably governed by the depth of anoestrus of the ewes at the time of their exposure to rams.

5.3.1. Ram pheromonal effects

It became apparent some years ago that rams do not need to be in physical or visual contact with the ewe to produce an effect (Watson and Radford, 1960) and that ewes with their sense of smell impaired would not exhibit oestrus after being stimulated by rams (Morgan *et al.*, 1972). Such evidence led to the suggestion that pheromones may be produced by the rams which led to a stimulation of breeding activity. There was the possibility that differences between Dorset and Romney rams in their ability to stimulate ewes might arise from differences in the production of pheromones (Tervit and Peterson, 1978).

Knight and Lynch (1980) also recorded that pheromones present in the wool and/or wax of entire rams were able to produce a response in ewes, further strong evidence that olfactory cues constituted the main sensory input from the ram at teasing. As a means of alleviating the problem of a slow onset of oestrous activity in New Zealand Romneys early in the breeding season, Knight and Lynch (1980) suggested the isolation and subsequent commercial use of this pheromone in concentrated form to stimulate ewes to exhibit oestrus earlier and as the means of achieving partial synchronization of heat periods.

5.3.2. *Timing of oestrus and ovulation*

As already noted, the introduction of rams in the latter weeks of the ewe's non-breeding season will often stimulate some proportion of anoestrous ewes to ovulate within 2–3 days (Knight *et al.*, 1978; Oldham *et al.*, 1979) although behavioural oestrus is not exhibited at this time but usually after about three weeks. Subsequently, it became evident that the response of ewes, in ovarian terms, was rather more complex than initially thought. It became apparent that there were often two peaks of oestrous activity in response to the ram effect, the first at about 18 days after contact with the male and the second at about 22–24 days (see Fig. 5.2). If ovulation did occur within three days of exposure to the ram and the normal lifespan of the ovine corpus luteum is taken as 14 days, then clearly some other factor was operating in ewes that showed oestrus after three weeks or later.

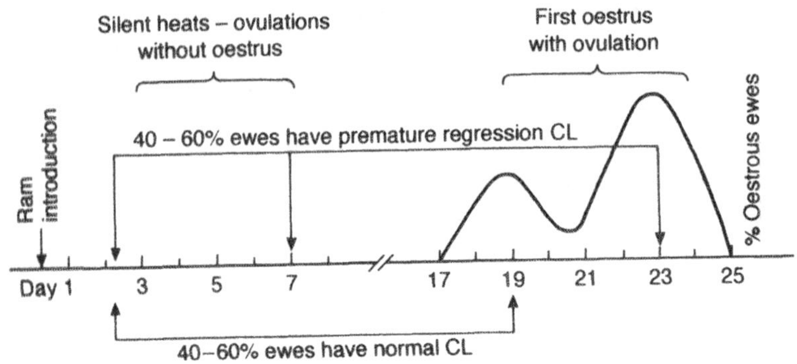

Fig. 5.2. Schematic diagram of the time after ram introduction when ovulation and regression of the corpus luteum (CL) occurs (after Knight, 1983).

An explanation of this delayed oestrus was advanced in the report of Tervit *et al.* (1977), who drew attention to the fact that many of the Romneys in their study ovulated twice before exhibiting oestrus; it appeared that the first corpus luteum had regressed after 6–8 days. This was in agreement with earlier work elsewhere in which premature regression of corpora lutea by day 10 had been recorded in Merinos stimulated to ovulate by rams (Oldham and Martin, 1979) and was supported in later studies reported by Knight *et al.* (1981). In the latter report, a small peak of progesterone indicated that such corpora lutea did secrete some amount of the steroid. It was well accepted by this time that adequate progesterone priming is essential if behavioural oestrus is to accompany oestrus (Robinson, 1968).

5.3.3. Ram introduction and progesterone treatment

In fact, when progesterone was administered to anoestrous ewes prior to teasing, there was no evidence of premature regression of corpora lutea (Hunter *et al.*, 1971); this suggested to Oldham and Martin (1979) that a progestational phase not only facilitates behavioural oestrus but also prevents premature regression of the ram-induced corpora lutea. A subsequent report by Oldham and Pearce (1984) dealt with various strategies for the induction of oestrus using the 'ram effect' in combination with progesterone or progestagens. They noted the work of Cognie *et al.* (1982) who found that a single dose of 20 mg progesterone, injected immediately prior to ram introduction, was the simplest and cheapest method of preventing the premature regression of corpora lutea and of inducing oestrus in anovulatory ewes within a short time span.

Variable results with progestagen/progesterone treatments

In more recent times, Hanrahan and O'Riordan (1990) in Ireland exploited the ram effect in inducing early breeding, treating ewes with medroxyprogesterone acetate (MAP) for 11 days and introducing rams at sponge removal. Rodriguez Iglesias *et al.* (1992) in Argentina reported on the daily distribution of teaser-induced heat periods in Corriedale ewes injected with progesterone or MAP. Elsewhere, in Turkey, Aksoy *et al.* (1994) concluded from studies with anoestrous Merino sheep that a combination of ram introduction and progesterone treatment is an effective method for the induction of oestrus; in France, Lassoued *et al.* (1995) reported studies from which they concluded that 20 mg progesterone administered on the day of ram introduction was effective in preventing the occurrence of short ovarian cycles. In the USA, Umberger *et al.* (1994) reported on the ram effect in seasonally anovulatory ewes treated with melengestrol acetate (MGA) or norgestomet; they found that ewes treated with progestagens without exposure to rams showed little luteal activity. The workers concluded that progestagens alone were not adequate for inducing oestrus in anovular ewes. Other studies in the USA, reported in the same year by Wheaton and Windels (1994), indicated that a single progesterone injection prior to ram introduction in August did not effectively substitute for a 12-day controlled internal drug release (CIDR) treatment.

Ram effect and LH levels

According to Michigan workers (Legan and Karsh, 1979), lack of ovulation in ewes during the anoestrous period is the result of an increased negative feedback action of oestrogen upon tonic secretion of LH. Ram introduction is believed to result in a sustained increase in gonadotrophin by way of increased LH pulse frequency. Chesworth and Tait (1974) recorded increases in LH concentrations within one hour of exposing Greyface ewes to the ram just prior to the breeding season. The LH pulse frequency was recorded to increase markedly within minutes of ram introduction in studies reported by Martin *et al.* (1980).

In Australia, Atkinson and Williamson (1985) showed that the ram was capable of stimulating ovarian follicular development in anoestrous ewes when the concentrations of circulating gonadotrophins were low. Further studies by Atkinson *et al.* (1986) led them to suggest that a lack of hCG binding sites in ram-induced follicles may be the cause of poor luteinization and suboptimal development of luteal tissue after induced ovulation in ewes during anoestrus. In terms of breed differences in response of ewes to rams, there is evidence that LH pulse frequency may be greater in Merinos than in some other breeds during anoestrus; this may explain why breeds such as the Merino are capable of responding to the ram even in mid-anoestrus (Schinkel, 1954).

Ewes stimulated by the ram effect apparently experience a preovulatory LH surge similar to that of spontaneously ovulating sheep and have been observed to ovulate about 40h after exposure to the male (Oldham *et al.*, 1979). It would seem possible that ram-induced ovulations occur because there is a lowering of the sensitivity of the hypothalamus to the negative feedback action of oestrogen (Martin *et al.*, 1980); this permits an increased LH pulse frequency, which may be easier to activate in Merinos than in some other breeds.

5.3.4. Ram sexual activity

As well as pheromonal effects, the sexual behaviour of the ram may be important in initiating ovarian activity in anoestrous ewes. In California, Perkins and Fitzgerald (1994) recorded that a significantly higher percentage of ewes exposed to rams exhibiting a high level of sexual activity (on the basis of serving capacity tests) ovulated than of sheep exposed to low-activity rams (95% vs. 78%). It was also found that on the first day of exposure, high-activity rams spent significantly more time near ewes than did the males of low activity. Ram sexual drive may be a further factor to add to those variables that influence the outcome of studies on the 'ram effect'.

A review by Haynes and Haresign (1987) suggested possible practical ways in which the ram effect may be strengthened so as to obtain a good response in ewes. These included:

1. the use of older rather than younger rams;
2. the use of high libido rams (identified by mating tests);
3. running the rams with ewes prior to their introduction to the main flock.

Enhancing ram sexual activity

Mindful that fluctuations in the concentrations of LH and testosterone have been associated with variations in ram sexual activity (Schanbacher and Lunstra, 1976), workers at Aberdeen used injections of human chorionic gonadotrophin (hCG) to enhance the secretion of testosterone and to improve the mating performance of vasectomized 'teaser' rams (Ibraheem *et al.*, 1990).

Use of oestrus-induced ewes

Work in New Zealand by Muir *et al.* (1989) led them to conclude that social facilitation by the introduction of 10% oestrous ewes (progesterone + PMSG) could improve the ram effect, particularly when rams were introduced prior to the commencement of the breeding season. Work in Ireland reported by O'Callaghan *et al.* (1994) found no effect from introducing oestrous ewes to other anoestrous ewes in late summer on subsequent LH pulse frequency or time of ovulation. Presumably, when rams are present, the increased sexual activity is what influences the response of anoestrous sheep.

5.3.5. Overexposure to rams

Schinckel (1954) did observe that ewes which had been continually exposed to sterile rams during the non-breeding-season resumed their seasonal reproductive activity at a later date than sheep isolated from males during anoestrus. It is known that individual ewes differ in their sensitivity to oestrogen (Robinson and Moore, 1956; Moore, 1988) and it is possible that the continuous presence of the ram during the anoestrous season might modify the sensitivity of the neural mechanisms mediating oestrous behaviour. There have been reports which indicate that when ewes are subjected to the ram stimulus for several months, adaptation to this stimulus occurs and such sheep apparently become less sensitive to stimuli which induce breeding activity (Underwood *et al.*, 1944; Lishman, 1975). As noted by Pearce and Oldham (1984), the precise requirements of isolation from rams have not been quantified, but isolation from the sight and smell of rams for one month has been the general recommendation. The practical point here is that farmers should ensure that rams are well and securely separated from the ewes during the summer months if they are to seriously consider using the ram to induce an early start to breeding. This has been the general principle observed in work in Ireland.

Not all authors may agree with this. In California, for example, Cushwa *et al.* (1992) concluded that the induction of oestrus in ewes by the presence of rams could be achieved without prior isolation of ewes from rams. The same authors recorded ovulation in anoestrous ewes occurring 8 days after joining with rams, which would appear to be markedly different from most reports.

5.3.6. Possible ram substitutes

The introduction of vasectomized 'teaser' rams was employed at one time in New Zealand flocks to induce early breeding (Edgar and Bilkey, 1963). However, vasectomy is a relatively expensive procedure to employ and more care is necessary in managing rams than wethers or ewes; for such reasons, an alternative system in which ewes or wethers are employed as the teasers could be of advantage. Lishman *et al.* (1969) and D'Occhio and Brooks (1976) were among authors who suggested that hormone-treated castrate male sheep (wethers) may be reasonable substitutes for vasectomized rams.

Following on from Australian studies showing the oestrus-inducing activity of testosterone in the ewe (Lindsay and Robinson, 1961), Marit *et al.* (1979) demonstrated that treatment with doses of 50 mg testosterone propionate at 2-day intervals for 20 days (induction regimen) and at 10-day intervals thereafter (maintenance regimen) would produce male-type behaviour in ewes; not only did testosterone-treated ewes exhibit male behaviour but other ewes regarded them as such. It is believed that testosterone is aromatized to oestrogen as a result of the particular capacity of neural receptor sites. Other studies reported by Fulkerson *et al.* (1981) also showed that wethers treated with oestrogen or testosterone, but not untreated wethers, could be employed effectively in the initiation of breeding activity in ewes in late anoestrus.

5.3.7. Ram exposure and prostaglandin treatment

Studies in the USA examined the possibility of sharpening the timing of oestrus onset in anoestrous Finn-cross ewes exposed to rams in mid-July by applying prostaglandin treatment 16 days after ram introduction (Smith *et al.*, 1986). As noted earlier, there are different patterns of luteal function in ewes exposed to the ram in late anoestrus. About half the sheep responding to the ram may be expected to have a normal luteal lifespan whereas the other half have a short luteal lifespan followed by a second silent ovulation and a normal luteal lifespan. By administering prostaglandin $F_{2\alpha}$ after 16 days, Smith *et al.* (1986) succeeded in synchronizing oestrus in the ewes much more closely than ram exposure alone.

5.4. Use of Progestagen–PMSG Treatments

In the summer of 1973, lowland sheep farmers in most counties in the Irish Republic could telephone their local Cattle AI Centre and request the services of a technician to treat sheep for 'early lamb production' with intravaginal sponges (fluorogestone acetate (FGA)/MAP) and PMSG. The 60p per sheep charge covered two farm visits by the technician: the first to insert sponges, the second, 12 days later, to remove them and administer the dose of 500–750IU PMSG. On the basis of considerable experimental evidence, the farmer could expect about 60–70% of ewes to conceive to first services and to lamb within the space of about one week in late December or early January; between the first and second services, some 80% of the flock would produce early lambs. The sheep 'sponging' service, based on natural mating, was used by several hundred farmers in that first year of operation (Fig. 5.3); sponges and PMSG have been used routinely by Irish sheep farmers ever since.

The introduction of a cheap, effective and simple controlled breeding technique for advancing the sheep breeding season was the result of considerable research and development work during the late 1960s and early 1970s by staff in the state agricultural research organization (An Foras Taluntais) and in University College, Dublin (see Table 5.1). Due credit is also deserved by the State Department of Agriculture, the body that initially sponsored extensive field-testing of the progestagen–PMSG technique and arranged with cattle AI centres to provide countrywide coverage in 1973.

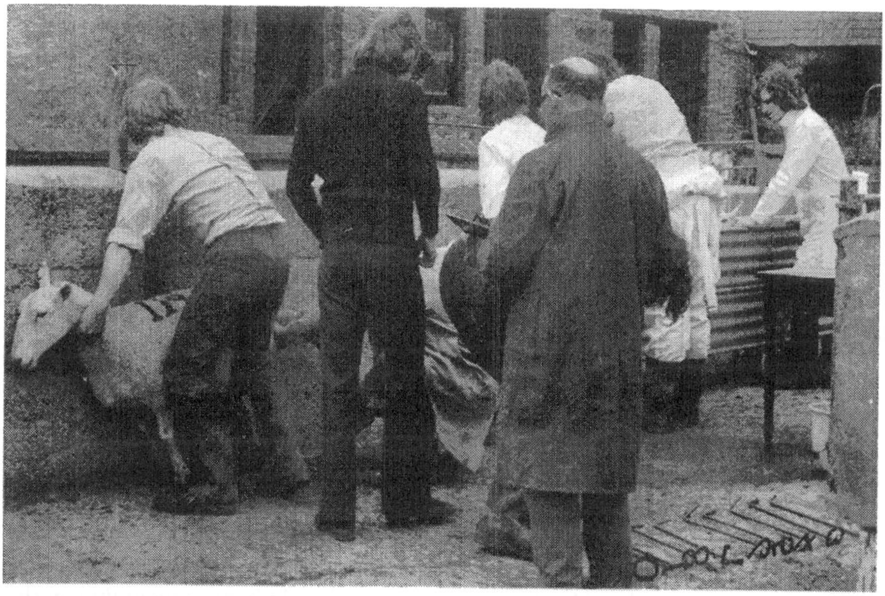

Fig. 5.3. Progestagen–PMSG treatment for early breeding made available to sheep farmers in the Irish Republic in 1973.

Table 5.1. Lambing outcome in trials sponsored by State Department of Agriculture on early breeding in sheep.

			Birth of 'early-lambs'					
			Lambed to first service			Lambed to first and second services		
Year	No. flocks	Sheep	% ewes	Lambs per conception	Lambs per 100 ewes treated	% ewes	Lambs per conception	Lambs per 100 ewes treated
1972	207	8314	61.6	1.6	100	80.2	1.6	126
1973	307	11,159	65.6	1.6	105	79.3	1.6	124

5.4.1. Late anoestrus in Irish sheep

As already mentioned, daylength plays an important role in maintaining seasonal reproductive activity in Irish and British sheep breeds, although other factors, such as temperature, ram effect and possibly nutrition, may exert an effect on the initiation of the mating season. Irish sheep which lamb at their usual time in February/March and nurse their lambs for 12–16 weeks certainly do not show evidence of spontaneous heats in the month of July; these are the sheep which are generally treated for early-lamb production.

The controlled breeding treatment offers the sheep farmer three possible advantages:

1. the birth of most lambs at an accurately predicted time in December/January;
2. a litter size which is in keeping with the mid-season performance of the sheep breed; and
3. that most lambs will be born in a compact period of about one week.

It should be mentioned here that ewes breeding spontaneously show a significantly lower litter size at the start of their season than they do a month or two later (see Fig. 5.4); the use of an appropriate dose of gonadotrophin as part of the controlled breeding technique can eliminate that problem. The Irish experience is that the hormonal approach gives an acceptable early-lambing result when applied to adult sheep. Whether the sheep farmers gain financially is very much a question of their adopting low-cost lamb rearing systems.

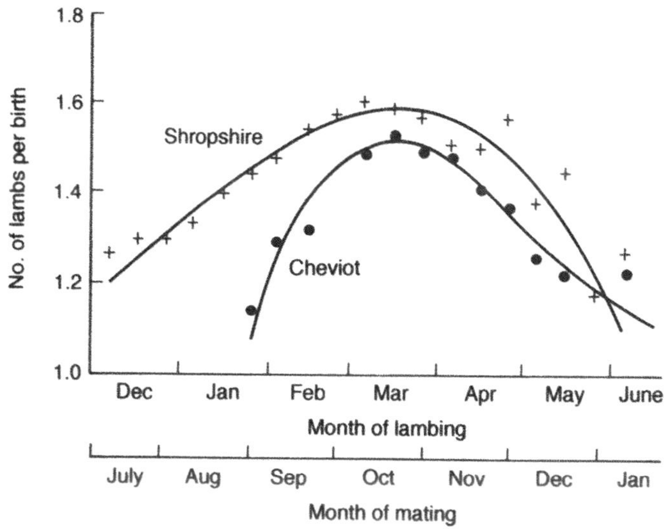

Fig. 5.4. Seasonal variation in fertility in Swedish Shropshire and Cheviot sheep (from Johansson and Hansson, 1943).

5.4.2. Time of treatment in anoestrus

The usual period of treatment for early lambs in the Irish Republic extends for some six weeks, sponges being inserted from about the end of June until mid-August. Within this period, the percentage of ewes becoming pregnant (first and second services) should rise as the summer progresses; to the farmer, the lambing outcome can almost be the same in sheep bred in early July as in those bred a month later in August (see Table 5.2). The fact that the response of ewes to the progestagen–PMSG treatment is little affected during the last several weeks of anoestrus is of general interest because it agreed with observations made in the 1960s (Gordon *et al.*, 1969) in which dry anoestrous ewes, treated in May and June (mid-anoestrus), were often found to exhibit a second (spontaneous) heat period when conception did not occur at the controlled oestrus. Clearly, cyclical release of pituitary gonadotrophins may be initiated some months prior to the normal autumn season, and not only during the last few weeks of the ewe anoestrus.

As already noted, the general view is that daylength is the over-riding environmental factor controlling reproductive activity in the ewe; the results from the early-lamb studies suggest that hypothalamic centres involved in gonadotrophin release can be activated to maintain cyclical activity, even when a factor such as daylength is not operating favourably. This may suggest that much of the ewe's early anoestrus is really a refractory period in which neural centres become unduly sensitive to the negative feedback effects of ovarian oestrogen; however, such refractoriness presumably lessens to the extent that certain stimuli become capable of initiating ovarian activity some time prior to

Table 5.2. Early-lambing outcome in relation to time of treatment during late anoestrus (from Gordon, 1983).

			Lambing to 1st service		Lambing to 1st + 2nd services	
	Time of treatment	Sheep	No.	%	No.	%
Early	30 June to 14 July	2632	1524	58	2065	79
Medium	15 July to 28 July	4278	2719	63	3487	81
Late	29 July to 11 August	1085	726	67	916	84

that usual with decreasing daylength as the main controlling exteroceptive factor.

5.4.3. *Sheep breed considerations*

Of the various breed-types of sheep entered by Irish farmers for early-lamb production, using the intravaginal progestagen–PMSG treatment, the best results were achieved with the more prolific ewes (e.g. Half-Breds or Greyfaces); they have shown the highest conception rates and litter sizes in response to a standard FGA–PMSG regimen (see Table 5.3). It was also observed that the prolific Half-Bred/Greyface type of sheep often showed a greater readiness than some other breeds to maintain cyclical activity, when conception did not occur at the induced oestrus; this resulted in a high percentage (86%) of the treated ewes becoming pregnant to first and second services. It may be that the gonadotrophin levels associated with prolific sheep permitted them to maintain cycles in the summer in a way less likely with other breeds.

Treatment with CIDRs

The development of early lamb production systems in New Zealand has been reported by some authors. Lowe *et al.* (1988) used CIDRs, followed by PMSG with 1402 Romney ewes and recorded 71% of the sheep conceiving at the controlled oestrus. The authors suggested that early lambing systems may offer economic benefit in areas of that country characterized by good winter and spring growth rates.

Enhancing fertility at the progestagen–PMSG controlled oestrus

In Japan, attempts were made by Fukui *et al.* (1991) to improve the fertility of Suffolk ewes induced to show oestrus using 60mg MAP sponges and PMSG treatments during anoestrus; it was concluded that an injection of gonadotrophin releasing hormone (GnRH) or hCG at the controlled oestrus had no

Table 5.3. Lambing outcome after early breeding in relation to breed or cross of ewe (from Jennings, 1973; Gordon, 1974).

	Data from Jennings (1973)				Data from Gordon (1974)			
Breed or cross of ewe	PMS dose (IU)	No. of ewes	% lambed	Litter size	No. of ewes	% lambed	Litter size	% lambing to 1st and 2nd services
Galway	500	991	52.9	1.55	7974	63.8	1.54	77.4
	750	1027	59.2	1.65				
Suffolk-Cross	500	827	58.6	1.67	5964	64.4	1.71	82.5
	750	789	60.7	1.77				
Cheviot	500	316	64.9	1.51	1743	61.7	1.58	76.5
	750	332	65.2	1.75				
Border Leicester × Blackface/ Border Leicester × Cheviot	500	293	65.0	1.72	1234	68.1	1.71	85.8
	750	290	69.0	1.85				

beneficial effect; the same authors used successive injections of hCG and MAP sponge treatment after mating (attempted progesterone supplementation), also without effect.

5.4.4. Progestagen–PMSG in Fecundin immunized anoestrous ewes

A paper by Robinson and Scaramuzzi (1994) in Australia has dealt with the induction of breeding in anoestrous ewes previously immunized against androstenedione-7-human serum albumin (Fecundin) which were treated with FGA pessaries and PMSG at sponge withdrawal. The response to immunization was affected by the interval from the booster injection to the induced oestrus; an interval of 21 days gave the best results. The Fecundin treatment markedly increased ovulation rate, did not reduce the percentage of ewes lambing, and resulted in a 47% increase in lambs born to ewes from which lambs had recently been weaned.

Use of GnRH

The work of McLeod and Haresign (1984) demonstrated that the administration of low doses of GnRH, either as pulsed injections or as a continuous intravenous infusion over 48h was effective in inducing a fertile oestrus in progestagen-primed seasonally anoestrous ewes. According to Haresign

(1992a), such treatment gave more normal ovulation rates than PMSG but it has not yet proved possible to develop a form of GnRH treatment suitable for field application. On a cost basis, the application of GnRH in the form of a biodegradable slow release implant may also be a consideration.

5.5. Treatments with Melatonin

In the 1980s, the importance of the pineal gland and hormone melatonin in the control of the breeding season in the ewe became increasingly apparent. In the 1970s, it had been shown that the secretion of this indole amine had a distinctive diurnal rhythm, with increased levels coinciding with darkness. It was evident that the pattern of melatonin secretion could afford the ewe a mechanism by which changing photoperiod could be monitored. In the 1980s, it became clear that exposure to short daylength or administration of melatonin by daily treatment or by implants significantly advanced the sheep breeding season and could also increase ovulation rate and embryo survival.

5.5.1. Prenatal effects in sheep

Studies have shown that results from using exogenous melatonin vary according to factors such as recent photoperiodic history, season of birth and age. For example, it is not only in postnatal life that female sheep are influenced by melatonin; photoperiod, and therefore the melatonin signal, is even more critical for the timing of the first oestrous cycle at puberty. There is now evidence that the photoperiodic history of a ewe commences *in utero*, the fetus being receptive to the maternal melatonin signal from early in pregnancy; the photoperiod experienced by the dam is apparently transmitted to the fetus and modifies the subsequent response of the lamb to the photoperiod.

5.5.2. Research and reviews of research

Various workers have demonstrated that melatonin may be given by daily injection, oral administration, by a soluble glass bolus containing melatonin (designed for intraruminal administration) and by different forms of slow-release subcutaneous implants. Treatment of ewes during late anoestrus with melatonin has clearly been demonstrated to be effective in advancing the start of the breeding season (Kennaway *et al.*, 1982; Arendt *et al.*, 1983; Nowak and Rodway, 1985; English *et al.*, 1986; Luhman and Slyter, 1986; Wigzell *et al.*, 1986; Amir *et al.*, 1987; Poulton *et al.*, 1986, 1987, 1988; Poulton and Kelly, 1988; Wallace *et al.*, 1988; Guerin *et al.*, 1989, 1994; Ronayne *et al.*, 1989; Yang *et al.*, 1989; Haresign *et al.*, 1990; Robinson *et al.*, 1991; Croker *et al.*, 1992; Staples *et al.*, 1992; Daveau *et al.*, 1994; Donovan *et al.*, 1994; Skinner *et al.*, 1994).

The availability of melatonin also permitted a reappraisal of photostimulation as a treatment for breeding sheep in the early period of anoestrus for autumn lambings (Williams and Ward, 1988); in this, the hormone could be employed as an alternative to the costly long night phase of treatment. There have also been some number of papers in which the use of melatonin in controlled sheep breeding has been reviewed (Poulton, 1988; Haresign, 1990, 1992a,b,c; Lincoln, 1992; O'Callaghan *et al.*, 1992; Chemineau *et al.*, 1993; Laliotis and Vosniakou, 1993; Williams and Helliwell, 1993; Williams, 1994).

5.5.3. Melatonin in advancing the breeding season

In New Zealand, McMillan and Sealey (1989) studied the effect of time of joining rams on the reproductive performance of Coopworth ewes implanted with melatonin one month before ram introduction; they concluded that the response to melatonin was very sensitive to the time of joining. However, it was found that joining treated ewes in late-December to mid-January could result in 20–40 more lambs per 100 ewes joined, compared with untreated animals. Various designs of slow-release melatonin implants (daily release rates of 0.25–1.7mg) were tested by Durotoye *et al.* (1991) at Leeds University with Swaledale × Blueface Leicester ewes in May and July. There was little difference between the various melatonin dose levels but implants given in May were less effective than those given in June or July.

A paper by Williams *et al.* (1992) reviewed results of five research trials and 108 clinical trials in three countries in which subcutaneous melatonin implants were employed to enhance the reproductive performance of sheep. They report that melatonin treatment of ewes resulted in more ewes conceiving early in the breeding season and that the duration of the lambing period was reduced. Work reported by Haresign (1992b) showed that increasing the interval from melatonin implantation to ram introduction from 4 to 6 weeks was associated with a progressive and significant reduction in the time from ram introduction to mating as well as a reduction in the spread of mating. It was concluded that melatonin influenced the pattern of mating and increased litter size in ewes. Further work by the same author (Haresign, 1992c) suggested that the optimum implantation date was mid-May to mid-June for Suffolk-crossbreds and mid-June to mid-July for Mule ewes.

Lack of response under some conditions

It is of interest to note that melatonin treatment has not always been effective in stimulating an earlier onset of the sheep breeding season. In Iceland, Eldon (1993) found that melatonin treatment for 75 days had no significant effect on the onset or duration of the breeding season of ewes. The author draws attention to the fact that Icelandic sheep came from Norway in the ninth and tenth centuries and constitute an isolated, genetically pure, highly domesticated breed living for many centuries under great seasonal change in photoperiod (summer photoperiod of almost 24h). It has been suggested that the lack of response to melatonin may be the result of entrainment of the reproductive system by the large seasonal changes in daylength that the ewes had experienced both before and during treatment.

Greek and South African studies

In Greece, Laliotis and Vosniakou (1993) have presented results of trials carried out in different areas of that country; these showed that oral administration or implantation of melatonin at the end of anoestrus stimulated oestrous activity and shortened the average interval to first oestrus. In South Africa, Nowers *et al.* (1994) conducted work to determine whether melatonin implants in Dohne Merino ewes during spring was a more successful alternative to increase lambing percentages than traditional management practices such as flushing and the use of teaser rams; they concluded that melatonin treatment enhanced seasonal oestrus and increased conception and ovulation rates during spring matings.

Melatonin × nutrition interactions

In Spain, Forcada *et al.* (1995) examined the effect of exogenous melatonin (18mg ear-implant) and plane of nutrition after weaning on oestrous activity in Salz ewes lambing in the seasonal anoestrus. The Salz breed is a genotype obtained by crossing the prolific Romanov breed with a local breed and usually displays a short anoestrous period between May and June. Melatonin treatment was found to reduce significantly the interval from weaning to first oestrus as compared with that found in untreated ewes (51 vs. 88 days); at the second oestrus, ovulation rate was significantly higher in the melatonin-treated animals and there was a significant melatonin × nutrition interaction. Although it is evident that melatonin treatment can increase ovulation rate early in the breeding season, sheep exposed to a constant melatonin treatment signal eventually become refractory and become anoestrous.

Attempts to extend the breeding season

A study was undertaken by Jordan *et al.* (1990) in Ireland with Scottish Blackface ewes to determine whether melatonin treatment in the mid-breeding season would extend the duration of the current breeding season or affect the onset and duration of the subsequent breeding season; results showed that such treatment did not extend the season but it did advance the onset of ovarian activity in the following season. As noted elsewhere, an unexpected finding in this study was a significant decrease in ovulation rate after prolonged administration of melatonin.

5.5.4. Effect of administration method

As noted by Robinson *et al.* (1992a), unless melatonin is administered by a method that stimulates a marked advance of the breeding season, it may offer little benefit over alternative treatments such as the progestagen–PMSG technique. There has, however, been evidence suggesting that the magnitude of the advance in the breeding season that can be achieved by a slow-release implant may be considerably less than that for daily administration. The insertion of melatonin implants in ewes in mid-May in the UK was found to be ineffective in advancing the breeding season (Nowak and Rodway, 1985), whereas daily dosing with 3 mg melatonin at 15.00 h from early March (Wigzell *et al.*, 1986) or late March (Wallace *et al.*, 1988) resulted in behavioural oestrus in late May–early June, four months ahead of untreated controls. The only method of administering melatonin that ensures a marked advance of the breeding season in sheep is by way of a daily oral dose in mid-afternoon (Robinson *et al.*, 1992b), which is not practically feasible under commercial conditions.

Prolonged melatonin administration in ewes

An unexpected adverse effect of melatonin treatment in ewes during the breeding season, as a consequence of prolonged exposure to high circulating levels of the hormone, was reported by workers in Ireland (Jordan *et al.*, 1990); they recorded a significant decrease in ovulation rate. The explanation for such an effect was not apparent.

Melatonin treatment and the ram effect

Various authors have studied the effect of ram introduction on the response of melatonin-treated ewes. In Spain, Brunet *et al.* (1995) found that treated and untreated anoestrous ewes responded to the ram effect with a significant increase in LH pulse frequency but there was no effect on pulse amplitude or mean concentration of the gonadotrophin. A preovulatory surge of LH was detected 8–26 h after ram introduction, but melatonin treatment affected neither the mean interval from ram introduction to the LH peak nor the magnitude of the LH response.

5.5.5 Melatonin vs. progestagen–PMSG comparisons

Some number of authors have reported comparisons between melatonin treatments and intravaginal progestagen pessaries with PMSG in advancing the breeding season in sheep. In Ireland, Crosby and O'Callaghan (1988) conducted a comparison between melatonin administered by way of a soluble glass rumen bolus and as standard progestagen–PMSG regime. The lambing data for these sheep are presented in Table 5.4. An obvious feature of the results is the wide spread in lambing dates in the melatonin-treated animals and the concentration of lambings towards the end of the lambing period; the authors concluded that the progestagen–PMSG technique was the treatment of choice for early lamb production under the conditions of their trial.

In the UK, Rajkumar *et al.* (1989) inserted intravaginal implants containing melatonin in silastic tubing eight weeks ahead of the target mating date in August; MAP sponges were used for a 12-day period with 500 IU PMSG given at their removal. The authors concluded that the melatonin treatment was an effective method of advancing the breeding season.

Table 5.4. Lambing outcome in melatonin and progestagen–PMSG-treated sheep (from Crosby and O'Callaghan, 1988).

	Melatonin		Progestagen–PMSG	
	Lambing no. (%)	Litter size	Lambing no. (%)	Litter size
26 December to 1 January	1 (2.5)	2.0	32 (80.0)	2.06
2 January to 8 January	4 (10.0)	1.75	3 (7.5)	1.33
9 January to 15 January	8 (20.0)	1.38	0 –	–
16 January to 22 January	11 (27.5)	1.55	1 (2.5)	2.0
23 January to 29 January	10 (25.0)	1.4	0 –	–
All periods	34 (85.0)	1.5	36 (90.0)	2.0

At Nottingham, Haresign (1992a) made various comparisons between the use of sponge-PMSG treatments and melatonin in the induction of early breeding in sheep (Table 5.5). In this, it was noted, for example, that implantation with melatonin must be undertaken 50–60 days before the proposed time of mating, whereas progestagen–PMSG treatment was a matter of starting two weeks before the mating date. Clearly, with oestrus-synchronized flocks, a marked increase in the number of rams is necessary, but that has not presented insurmountable problems to farmers who have routinely used progestagen sponges for the past two decades. Depending on flock size, there may be much in favour of having the compact lambings which follow from the progestagen–PMSG method. It is also mentioned that with sponges and PMSG, non-pregnant ewes revert to anoestrus and will not cycle again until the start of natural breeding. This apparently has not been the experience in Ireland, where a very high proportion of non-pregnant ewes have been found to recycle, even when treated several weeks ahead of the breeding season.

Melatonin and breeding ewe-lambs

Studies in New Zealand have shown that when melatonin was given at 3.5–4.5 months until eight months of age, it reduced the time of conception of spring-born Romney ewe-lambs (Moore *et al.*, 1984). Elsewhere, this time in the USA, Stellflug *et al.* (1989) drew attention to the fact that unlike spring-born lambs, which normally reach puberty at 6–8 months of age, lambs which are born in late summer or autumn reach the critical age for puberty in the spring under an inhibitory photoperiod. In Idaho, Stellflug *et al.* (1989) showed that exposure of autumn-born lambs to 12 weeks of long days 16h light : 8h darkness) before melatonin treatment improved reproductive efficiency at spring breeding over ambient controls or melatonin treatment alone.

Table 5.5. Comparison of progestagen–PMSG and melatonin treatments for the induction of early breeding in sheep (from Haresign, 1992a).

Attribute	Sponges plus PMSG	Regulin (melatonin)
Ability to induce early breeding	At any time during anoestrus but the efficacy is greatest just before the natural start of the season	Only during a limited treatment 'window' just before the natural start of the breeding season
Time of treatment before mating (days)	14	50–60
Ewe : ram ratio recommended	7 : 1 or 8 : 1	40 : 1
Synchrony of mating and lambing	Highly synchronized, but at the expense of fewer ewes pregnant	Spread evenly over 3–4 weeks, but with a very high proportion of ewes lambing
Litter size	Variation in both litter size and lamb birth weights is common owing to the superovulatory effects of PMSG in a proportion of the treated animals	The distribution of litter size and lamb birth weights is typical of that for the breed in the mid-breeding season with no induction of superovulation

5.5.6. Ram response to melatonin

There have been reports from the UK making the point that rams of most breeds in that country show marked seasonal fluctuations in reproductive performance, as revealed in decreased testicular size, libido and semen quality during the spring and summer months. It has been suggested that problems of subfertility frequently encountered in out-of-season breeding in ewes may, at least partly, be attributable to the ram. Although this is a view apparently at variance with the data of Gordon (1958, 1963) and with Irish experiences, it is of interest to record ways in which melatonin can be used to manipulate reproduction in the ram.

In London, Williams *et al.* (1990) showed that the continuous administration of melatonin to Suffolk rams from mid-March, after a six-week period of priming with artificial long days, resulted in improvements in ram reproductive performance; however, the authors recorded that the testes of the treated rams also regressed prematurely in early July. This obviously would not be a response favourable for the use of such rams in early lamb production. At Nottingham, Bateman and Haresign (1991) have reported results suggesting that implantation with melatonin in mid-May may well extend the period of peak fertility in Suffolk rams, making them more suitable for use in early-lambing flocks without compromising their fertility for autumn mating. They noted, however, that before practical recommendations could be made, more information was required for some of the other terminal sire breeds such as the Texel.

In Japan, Kusakari and O'Hara (1993) fed Suffolk rams melatonin pellets and presented results suggesting that such treatment increased their reproductive activity; scrotal circumference reached a maximum after 45 days of treatment and remained maximal for about 30 days after the cessation of treatment. The authors recorded, however, an abrupt decrease in scrotal size some 45 days after the cessation of melatonin feeding. As with the Nottingham work, the indications are that melatonin treatment of rams may not be capable of maintaining improved reproductive performance beyond a certain period.

5.6. Induced Hypothyroidism and Reproductive Activity in Sheep

It is now evident that thyroid hormones are necessary for the expression of seasonal changes in reproductive neuroendocrine activity in sheep. If the thyroid glands are removed from ewes before or during the breeding season, the sheep do not return to an anoestrous state due to a continued high rate of secretion of GnRH. It is believed that lack of the thyroid hormones affects the reproductive refractoriness that otherwise occurs in sheep. Refractoriness in this context may be defined as the physiological state developing after prolonged exposure to a fixed photoperiod and shows itself in the form of a spontaneous reversal in the prevailing neuroendocrine state, i.e. a spontaneous turning off after being turned on, or a spontaneous turning on after being off. Thus, sheep that are held on short days do not remain reproductively active continuously but after some months the activity ends spontaneously as refractoriness develops.

As noted by Follett and Potts (1990), experiments with birds have shown that both the acquisition of refractoriness under one daylength, and its dissipation under the opposite daylength, are dependent on thyroid hormones. Because of certain similarities between refractoriness in birds and mammals, thyroidectomy was extended to sheep and their reproductive activity was followed in thyroidectomized ewes exposed to various daylengths (Nicholls *et al.*, 1988); a high proportion of such ewes did not enter anoestrus at the same time as thyroid-intact sheep and some continued to show oestrous cycles for very long periods of time.

5.6.1. *Use of goitrogens*

Workers at Bristol extended such observations by using a goitrogen, thus providing a reversible means of altering thryoid function (Follett and Potts, 1990). Welsh Mountain ewes were rendered hypothyroid by daily treatment with methylthiouracil starting in early August and ending in late February. Treated sheep showed a longer breeding season than did the controls (122 vs. 91 days). Although the authors had hoped that such reversible thyroidectomy by treatment over the winter period with goitrogens might so delay the onset of anoestrus as to permit spring-lambing ewes to return in oestrus for out-of-season breeding, this did not prove to be so.

In Michigan, Dahl *et al.* (1992) showed that in the absence of thyroid hormones, the reproductive neuroendocrine axis appears to be uncoupled from photoperiodic influence between melatonin and the GnRH neurosecretory system. Further work reported by Dahl *et al.* (1994) noted that the decrease in the episodic release of GnRH, as ewes enter anoestrus, is independent of an increased secretion of thyrotrophin-releasing hormone (TRH).

Use of thyroxine

In Ireland, O'Callaghan *et al.* (1993) found evidence suggesting that the administration of thyroxine during the breeding season could advance the onset of anoestrus in the ewe but that the treatment did not affect the time of onset of the subsequent breeding season. The authors took such results as reinforcing the concept that the thyroid gland plays a role in seasonal reproduction in the ewe. Elsewhere, Porter *et al.* (1995) have recorded some evidence suggesting a seasonal variation in thyroxine metabolism; such variation may have a bearing on the timing of the non-breeding season.

5.6.2. Effects of thyroidectomy on the ram

Much less information is available on the effects of lowered thyroid activity in the male than in the female. However, there has been evidence in red deer that thyroidectomy almost abolished the non-breeding season, with males retaining full-grown testes throughout the year (Shi and Barrell, 1992). In Bristol, Parkinson and Follett (1994) have examined the effect of thyroidectomy upon testicular function in Welsh Mountain rams, maintained either on natural or controlled photoperiods; results supported the view that the thyroid gland exerts a crucial role in transitions between the breeding and non-beeding seasons of the ram. The authors reported the scrotal circumference of thyroidectomized rams to be significantly greater than that in intact males between April and August.

A further paper by the Bristol workers showed that thyroidectomy abolished seasonal testicular cycles of Soay rams (Parkinson and Follet, 1995). Other work in Bristol dealt with the responses of prepubertal and mature rams to thyroidectomy; in both categories, scrotal circumference increased significantly within five weeks of thyroidectomy (Parkinson *et al.*, 1995). However, although high FSH concentrations were maintained in mature rams, in prepubertal animals they decreased during the late spring; results were taken as indicating that the timing of puberty in seasonally breeding mammals is a thyroid-dependent phenomenon.

It is clear, from studies in both ewes and rams, that such information on the crucial role of thyroid hormones is likely to be valuable in understanding the means by which reproductive activity in sheep is inhibited in the non-breeding season. In terms of practical treatments for advancing the breeding season, however, such knowledge may have less impact.

5.7. References

Aksoy, M., Tekeli, T., Ozsar, S., Coyan, K., Guven, B., Semacan, A. and Ayar, A. (1994) Effect of ram introduction in combination with progesterone or cloprostenol on estrus induction rates of Konya Merino ewes in the anestrous season. *Reproduction in Domestic Animals* 29, 444–450.

Amir, D., Thimonier, J. and Gacitua, H. (1987) The effect of a light pulse and melatonin, alone or in combination, on the reproductive performance of Finn-cross ewes in spring in Israel. *Journal of Agricultural Science, Cambridge* 107, 273–279.

Arendt, J., Symons, A.M., Laud, C.A. and Pryde, S.J. (1983) Melatonin can induce the early onset of the breeding season in ewes. *Journal of Endocrinology* 97, 395–400.

Atkinson, S. and Williamson, P. (1985) Ram induced growth of ovarian follicles and gonadotrophin inhibition. *Journal of Reproduction and Fertility* 73, 185–189.

Atkinson, S., Williamson, P., Kang, C.L. and Carson, R.S. (1986) Steroid production and hCG binding by ram-induced ovarian follicles in seasonally anoestrous ewes. *Journal of Reproduction and Fertility* 78, 403–412.

Bateman, S.M. and Haresign, W. (1991) Response of Suffolk and Texel rams to implantation with melatonin. *Proceedings of the British Society Animal Production* (Winter Meeting), Paper No. 15.

Brunet, A.G., Sebastian, A.L., Picazo, R.A., Cabellos, B. and Goddard, S. (1995) Reproductive response and LH secretion in ewes treated with melatonin implants and induced to ovulate with the ram effect. *Animal Reproduction Science* 39, 23–34.

Chemineau, P., Malpaux, B., Guerin, Y., Maurice, F., Daveau, A. and Pelletier, J. (1992) Light and melatonin in the control of sheep and goat reproduction. *Annales de Zootechnie* 41, 247–261.

Chemineau, P., Berthelot, X., Daveau, A., Maurice, F., Viguie, C. and Malpaux, B. (1993) Does melatonin allow out-of-season reproduction in farm domestic mammals? *Contraception Fertilité Sexualité* 21, 733–738.

Chesworth, J.M. and Tait, A. (1974) A note on the effect of the presence of rams upon the amount of luteinizing hormone in the blood of ewes. *Animal Production* 19, 107–110.

Cognie, Y., Gayerie, F., Oldham, C.M. and Poindron, P. (1980) Increased ovulation rate at the ram-induced ovulation and its commercial application. *Proceedings of the Australian Society of Animal Production* 13, pp. 80–81.

Cognie, Y., Gray, S.J., Lindsay, D.R., Oldham, C.M., Pearce, D.T. and Signoret, J.P. (1982) A new approach to controlled breeding in sheep using the 'ram effect'. *Proceedings of the Australian Society of Animal Production* 14, pp. 519–522.

Coop, I.E. and Clark, V. (1968) Synchronization of oestrus in ewes. *Proceedings of the New Zealand Society of Animal Production* 28, p. 114.

Croker, K.P., Johns, M.A., Williams, A.H., McPhee, S.R. and Staples, L.D. (1992) Effect of treatment with melatonin implants in conjunction with teaser rams on the reproductive performances of Poll Dorset × Merino ewes joined in early summer in the south-west of Western Australia. *Australian Journal of Experimental Agriculture* 32, 1045–1049.

Crosby, T.F. and O'Callaghan, D. (1988) Effect of melatonin bolus or progestagen sponge plus pregnant mare serum gonadotrophin treatment on oestrus response and lambing outcome in ewes. *Proceedings of the 11th International Congress Animal Reproduction and AI* (Dublin), 4, p. 430.

Cushwa, W.T., Bradford, G.E., Stabenfeldt, G.H., Berger, Y.M. and Dally, M.R. (1992) Ram influence on ovarian and sexual activity in anestrous ewes: effects of isolation of ewes from rams before joining and date of ram introduction. *Journal of Animal Science* 70, 1195–1200.

Dahl, G.E., Evans, N.P. and Karsh, F.J. (1992) Thyroidectomy alters the reproductive response to photoperiod but not the photoperiodic effect on melatonin or prolactin secretion in ewes. *Journal of Animal Science* 70 (Suppl. 1), p. 269.

Dahl, G.E., Evans, N.P., Thrun, L.A. and Karsh, F.J. (1994) Does increased secretion of thyrotrophin-releasing hormone prolong the breeding season in thyroidectomized ewes? *Biology of Reproduction* 50 (Suppl. 1), 107.

Daveau, A., Malpaux, B., Tillet, Y., Roblot, G., Wylde, R. and Chemineau, P. (1994) Active immunization against melatonin in Ile-de-France ewes and photoperiodic control of prolactin secretion and ovulatory activity. *Journal of Reproduction and Fertility* 102, 285–292.

D'Occhio, M.J. and Brooks, D.E. (1976) The influence of androgens and oestrogens on mating behaviour in male sheep. *Theriogenology* 6, 614–620.

Donovan, A., Boland, M.P., Roche, J.F. and O'Callaghan, D. (1994) The effect of supplementary long days, a subcutaneous melatonin implant and exposure to a ram on the onset of the breeding season in sheep. *Animal Reproduction Science* 34, 231–240.

Ducker, M.J. and Bowman, J.C. (1970a) Photoperiodism in the ewe. 3. The effects of various patterns of increasing daylength on the onset of anoestrus in Clun Forest ewes. *Animal Production* 12, 465–471.

Ducker, M.J. and Bowman, J.C. (1970b) Photoperiodism in the ewe. 4. A note on the effect of onset of oestrus in Clun Forest ewes of applying the same decrease in daylength at two different times of the year. *Animal Production* 12, 513–516.

Ducker, M.J. and Bowman, J.C. (1972) Photoperiodism in the ewe. 5. An attempt to induce sheep of three breeds to lamb every eight months by artificial daylength changes in a non-light-proofed building. *Animal Production* 14, 323–334.

Ducker, M.J. and Boyd, J.S. (1974) The effect of daylength and nutrition on the oestrous and ovulatory activity of Greyface ewes. *Animal Production* 18, 159–167.

Ducker, M.J., Thwaites, C.J. and Bowman, J.C. (1970) Photoperiodism in the ewe. 1. The effects of decreasing daylength on the onset of oestrus in Clun Forest ewes. *Animal Production* 12, 115–123.

Durotoye, L.A., Rajkumar, R., Argo, C.M., Nowak, R., Webley, G.E., McNeil, M.E., Graham, N.B. and Rodway, R.G. (1991) Effect of constant-release melatonin implants on the onset of oestrous activity and on reproductive performance in the ewe. *Animal Production* 52, 489–497.

Edgar, D.G. and Bilkey, D.A. (1963) The influence of rams on the onset of the breeding season in ewes. *Proceedings of the New Zealand Society of Animal Production* 23, pp. 79–87.

Eldon, J. (1993) Effect of exogenous melatonin and exposure to a ram on the time of onset and duration of the breeding season in Icelandic sheep. *Journal of Reproduction and Fertility* 99, 1–6.

English, J., Poulton, A.L., Arendt, J. and Symons, A.M. (1986) A comparison of the efficiency of melatonin treatment in advancing oestrus in ewes. *Journal of Reproduction and Fertility* 77, 321–327.

Follett, B.K. and Potts, C. (1990) Hypothyroidism affects reproductive refractoriness and the seasonal oestrous period in Welsh Mountain ewes. *Journal of Endocrinology* 127, 103–109.

Forcada, F., Zarazaga, L. and Abecia, J.A. (1995) Effect of exogenous melatonin and plane of nutrition after weaning on estrous activity, endocrine status and ovulation rate in Salz ewes lambing in the seasonal anestrus. *Theriogenology* 43, 1179–1193.

Fraser, A.F. and Laing, A.H. (1969) Oestrus induction in ewes with standard treatments of reduced natural light. *Veterinary Record* 84, 427–430.

Fukui, Y., Kobayashi, K., Hirose, Y. and Ono, H. (1991) Effects of GnRH and hCG injections on lambing rate of estrus-induced ewes during the non-breeding season. *Japanese Journal of Animal Reproduction* 37, 243–250.

Fulkerson, W.J., Adams, N.R. and Gherardi, P.B. (1981) Ability of castrate male sheep treated with oestrogen or testosterone to induce and detect oestrus in ewes. *Applied Animal Ethology* 7, 57–66.

Gordon, I. (1958) Studies in the extra-seasonal production of lambs. *Journal of Agricultural Science, Cambridge* 50, 152–197.

Gordon, I. (1963) The induction of pregnancy in the anoestrous ewe by hormonal therapy. *Journal of Agricultural Science, Cambridge* 60, 43–66.

Gordon, I. (1974) Controlled breeding in sheep. *Irish Veterinary Journal* 28, 118–126.

Gordon, I. (1983) *Controlled Breeding in Farm Animals*. Pergamon Press, Oxford, pp. 181–195.

Gordon, I., Caffrey, W. and Morrin, P. (1969) Induction of early breeding in sheep following treatment with progestagen-impregnated pessaries and PMSG. *Journal of the Department of Agriculture and Fisheries* (Dublin) 66, 3–22.

Guerin, M.V., Watson, R., McLoughney, J., Earle, C., Seamark, R.F. and Matthews, C.D. (1989) The annual patterns of serum melatonin in Romney Marsh sheep held in natural photoperiodic conditions. *Advances in Pineal Research* 3, 137–141.

Guerin, M.V., Napier, A.J. and Matthews, C.D. (1994) Effect of exogenous melatonin and extending the dark period at dusk before the summer solstice on the onset of oestrus in Romney Marsh ewes. *Journal of Reproduction and Fertility* 101, 145–150.

Hafez, E.S.E. (1952) Studies on the breeding season and reproduction of the ewe. *Journal of Agricultural Science, Cambridge* 42, 189–265.

Hanrahan, J.P. and O'Riordan, E.G. (1990) Exploiting the ram effect for early breeding: effect of ewe age and breed. *Journal Irish Grassland and Animal Production Association* 24, 105–108.

Haresign, W. (1990) Controlling reproduction in sheep. In: *New Developments in Sheep Production*. Occasional Publication, British Society of Animal Production, No. 14, pp. 23–37.

Haresign, W. (1992a) Manipulation of reproduction in sheep. *Journal of Reproduction and Fertility* Suppl. 45, 127–139.

Haresign, W. (1992b) The effect of implantation of lowland ewes with melatonin on the time of mating and reproductive performance. *Animal Production* 54, 31–39.

Haresign, W. (1992c) Responses of ewes to melatonin implants: importance of the interval between treatment and ram introduction on the synchrony of mating, and effects on ovulation rate. *Animal Production* 54, 41–45.

Haresign, W., Peters, A.R. and Staples, L.D. (1990) The effect of melatonin implants on breeding activity and litter size in commercial sheep flocks in the UK. *Animal Production* 50, 111-121.

Haynes, N.B. and Haresign, W. (1987) Endocrine aspects of reproduction in the ram important to the male effect. *World Review of Animal Production* 23(1), 21–28.

Hudgens, R.E., Martin, T.G., Diekman, M.A. and Waller, S.L. (1987) Reproductive performance of Suffolk and Suffolk-cross ewes and ewe lambs exposed to vasectomized rams before breeding. *Journal of Animal Science* 65, 1173–1179.

Hunter, G.L., Belonje, P.D. and Van Niekerk, C.H. (1971) Synchronized mating and lambing in spring-bred Merino sheep flocks: the use of progestagen-impregnated intravaginal sponges and teaser rams. *Agroanimalia* 3, 133–140.

Ibraheem, M.F., Hutchinson, J.S.M., King, M.E., Mitchell, L.M. and Donald, M. (1990) Teasing to induce ovarian activity in anoestrous Mule ewes. *Proceedings of the British Society Animal Production* (Winter Meeting), Paper No. 166.

Jennings, J.J. (1973) Effect of progestagen treatment, number of matings and the time of mating on fertility in sheep. *Annual Report of the Animal Production Division*, An Foras Taluntais, Dublin.

Johansson, I. and Hansson, A. (1943) The sex ratio and multiple births in sheep. *Annals of the Agriculture College of Sweden* 11, 145–171.

Jordan, B.T., Hanrahan, J.P. and Roche, J.F. (1990) The effect of melatonin implantation in the middle of the breeding season on the subsequent reproductive activity of Scottish Blackface ewes. *Animal Reproduction Science* 23, 41–48.

Kennaway, D.J., Gilmore, T.A. and Seamark, R.F. (1982) Effect of melatonin feeding on serum prolactin and gonadotrophin levels and the onset of seasonal oestrous cyclicity in sheep. *Endocrinology* 110, 1766–1772.

Knight, T.W. (1983) Ram induced stimulation of ovarian and oestrous activity in anoestrous ewes – a review. *Proceedings of the New Zealand Society of Animal Production* 43, 7–11.

Knight, T.W. and Lynch, P.R. (1980) Source of ram pheromones that stimulate ovulation in the ewe. *Animal Reproduction Science* 3, 133–136.

Knight, T.W., Peterson, A.J. and Payne, E. (1978) The ovarian and hormonal response of the ewe to stimulation by the ram early in the breeding season. *Theriogenology* 10, 343.

Knight, T.W., Dalton, D.C. and Hight, G.K. (1980) Changes in the median lambing dates and lambing pattern with variation in time of joining and breed of teasers. *New Zealand Journal of Agricultural Research* 23, 281–285.

Knight, T.W., Tervit, H.R. and Fairclough, R.J. (1981) Corpus luteum function in ewes stimulated by rams. *Theriogenolgoy* 15, 183–190.

Knight, T.W., Hall, D.R. and Wilson, L.D. (1983) Effects of teasing and nutrition on the duration of the breeding season in Romney ewes. *Proceedings of the New Zealand Society Animal Production* 43, pp. 17–19.

Kusakari, N. and Ohara, M. (1993) Effect of melatonin feeding on testicular development, reproductive behaviour and sperm production in Suffolk rams. *Journal of Reproduction and Development* 39, 357–361.

Laliotis, V.N. and Vosniakou, A.G. (1993) Role of melatonin in reproduction of ewes. *Epitheorese Zootehnikes Epistemes* 22, 117–133.

Lassoued, N., Khaldi, G., Cognie, Y., Chemineau, P. and Thimonier, J. (1995) Effect of progesterone on ovulation rate and oestrus cycle length induced by the male effect in the Barbarine ewe and Tunisian local goat. *Reproduction, Nutrition and Development* 35, 415–426.

Legan, S.J. and Karsh, F.J. (1979) Neuroendocrine regulation of the estrous cycle and seasonal breeding in the ewe. *Biology of Reproduction* 20, 74–85.

Lincoln, G.A. (1992) Photoperiod–pineal–hypothalamic relay in sheep. *Animal Reproduction Science* 28, 203–217.

Lindsay, D.R. and Robinson, T.J. (1961) The oestrus inducing activity of testosterone in the ewe. *Nature* 192, 761–762.

Lishman, A.W. (1975) Reduced sensitivity to oestrogen in ewes continuously associated with rams. *South African Journal of Animal Science* 5, 235–238.

Lishman, A.W., de Lange, G.M. and Viljoen, J.T. (1969) Ability of masculized ewes to stimulate onset of the breeding season in maiden Merino ewes. *Proceedings of the South African Society Animal Production* 8, p. 141.

Lowe, K.I., Carter, M.L. and McCutcheon, S.N. (1988) Development of systems for out-of-season lambing at Limestone Downs. *Proceedings of the New Zealand Society Animal Production* 48, pp. 95–98.

Luhman, C.M. and Slyter, A.L. (1986) The effect of photoperiod and melatonin feeding on reproduction in the ewe. *Theriogenology* 26, 721–732.

McDonald, M.F. (1971) Factors associated with onset of the breeding season in sheep. *Sheepfarming Annual* Massey University, pp. 23–30.

McLeod, B.J. and Haresign, W. (1984) Induction of fertile oestrus in seasonally anoestrous ewes with low doses of GnRH. *Animal Reproduction Science* 7, 413–420.

McMillan, W.H. and Sealey, R.C. (1989) Do melatonin implants influence the breeding season in Coopworth ewes? *Proceedings of the New Zealand Society of Animal Production* 49, pp. 43–45.

Malpaux, B., Robinson, J.E., Wayne, N.L. and Karsh, F.J. (1989) Regulation of the onset of the breeding season of the ewe: importance of long days and of an endogenous reproductive rhythm. *Journal of Endocrinology* 122, 269–278.

Marit, G.B., Scheffrahn, N.S., Troxel, T.R. and Kesler, D.J. (1979) Sex behaviour and hormone responses in ewes administered testosterone propionate. *Theriogenology* 12, 375–381.

Martin, G.B., Oldham, C.M. and Lindsay, D.R. (1980) Increased plasma LH levels in seasonally anovular Merino ewes following the introduction of rams. *Animal Reproduction Science* 3, 125–132.

Meyer, H.H. (1979) Ewe and teaser breed effects reproductive behaviour and performance. *Proceedings of the New Zealand Society of Animal Production* 39, pp. 68–76.

Moore, N.W. (1988) The ovariectomized ewe: its contribution to controlled breeding. *Australian Journal of Biological Science* 41, 15–22.

Moore, R.W., Miller, C.M., Lynch, P.R., Welch, R.A.S., Barnes, D.R. and Hockey, H.-U.P., (1984) The effect of melatonin on the onset of the first oestrus in Romney ewe lambs. *Proceedings of the New Zealand Society of Animal Production* 44, pp. 21–23.

Morgan, P.D., Arnold, G.W. and Lindsay, D.R. (1972) A note on the mating behaviour of ewes with various senses impaired. *Journal of Reproduction and Fertility* 30, 151–152.

Muir, P.D., Smith, N.B. and Wallace, G.J. (1989) Early lambing in Hawkes Bay: use of the ram effect. *Proceedings of the New Zealand Society of Animal Production* 49, 271–275.

Murdock, D. (1975) Change of season for 1000 ewes. *Farmers' Weekly* 83, 62.

Newton, J.E. and Betts, J.E. (1972) A comparison between the effect of various photoperiods on the reproductive performance of Scotch half-bred ewes. *Journal of Agricultural Science, Cambridge* 78, 425–433.

Nicholls, T.J., Follett, B.K., Goldsmith, A.R. and Pearson, H. (1988) Possible homologies between photorefractoriness in sheep and birds: the effect of thyroi-

dectomy on the length of the ewe's breeding season. *Reproduction, Nutrition, Development* 28, 375–385.
Nowak, R. and Rodway, R.G. (1985) Effects of intravaginal implants of melatonin on the onset of ovarian activity in adult and prepubertal ewes. *Journal of Reproduction and Fertility* 74, 287–293.
Nowers, C.B., Coetzer, W.A. and Morgenthal, J.C. (1994) Effect of melatonin implants, flushing and teasing on the reproductive performance of spring-mated Dohne Merino ewes. *South African Journal of Animal Science* 24(1), 22–26.
O'Callaghan, D., Karsh, F.J. and Roche, J.F. (1992) Melatonin in ewes – a timekeeping hormone regulating seasonal reproductive transitions. *Agbiotech News and Information* 4, 101N–106N.
O'Callaghan, D., Wendling, A., Karsh, F.J. and Roche, J.F. (1993) Effect of exogenous thyroxine on timing of seasonal reproductive transitions in ewes. *Biology of Reproduction* 49, 311–315.
O'Callaghan, D., Donovan, A., Sunderland, S.J., Boland, M.P. and Roche, J.F. (1994) Effect of the presence of male and female flockmates on reproductive activity in ewes. *Journal of Reproduction and Fertility* 100, 497–503.
Oldham, C.M. (1980) Stimulation of ovulation in seasonally or lactationally anovular ewes by rams. *Proceedings of the Australian Society of Animal Production* 13, 93–99.
Oldham, C.M. and Martin, G.B. (1979) Stimulating seasonally anovular Merino ewes by rams. II. Premature regression of ram-induced corpora lutea. *Animal Reproduction Science* 1, 291–295.
Oldham, C.M. and Pearce, D.T. (1984) Alternative methods for synchronization of ewes in spring using the 'ram effect'. *Proceedings of the Australian Society Animal Production* 15, 158–170.
Oldham, C.M., Martin, G.B. and Knight, T.W. (1979) Stimulation of seasonally anovular Merino ewes by rams: time from introduction of the rams to the preovulatory LH surge and ovulation. *Animal Reproduction Science* 1, 283–290.
Parkinson, T.J. and Follett, B.K. (1994) Effect of thyroidectomy upon seasonality in rams. *Journal of Reproduction and Fertility* 101, 51–58.
Parkinson, T.J. and Follett, B.K. (1995) Thyroidectomy abolishes seasonal testicular cycles of Soay rams. *Proceedings of the Royal Society of London, Series B, Biological Sciences* 259(1354), 1–6.
Parkinson, T.J., Douthwaite, J.A. and Follett, B.K. (1995) Responses of prepubertal and mature rams to thyroidectomy. *Journal of Reproduction and Fertility* 104, 51–56.
Pearce, D.T. and Oldham, C.M. (1984) The ram effect, its mechanism and application to the management of sheep. In: Lindsay, D.R. and Pearce, D.T. (eds) *Reproduction in Sheep*. Cambridge University Press, Cambridge, pp. 26–34.
Perkins, A. and Fitzgerald, J.A. (1994) The behavioural component of the ram effect: the influence of ram sexual behaviour on the induction of estrus in anovulatory ewes. *Journal of Animal Science* 72, 51–55.
Porter, M.B., Cleaver, B.D., Robinson, G., Peltier, M., Shearer, L.C. and Sharp, D.C. (1995) Effect of thyroidectomy and thyroxine replacement on seasonal reproduction in ovarian-intact ewes: evidence for seasonal variation in T4 metabolism. *Journal of Animal Science* 73 (Suppl. 1), p. 226.
Poulton, A.L. (1988) The proposed use of melatonin in controlled sheep breeding. *Australian Journal of Biological Science* 41, 87–96.
Poulton, A.L. and Kelly, M.T. (1988) Plasma melatonin concentrations in ewes following administration of an intraruminal soluble glass/melatonin bolus. *Journal of Controlled Release* 7(2), 159–163.

Poulton, A.L., English, J., Symons, A.M. and Arendt, J. (1986) Effects of various melatonin treatments on plasma prolactin concentrations in the ewe. *Journal of Endocrinology* 108, 287–292.

Poulton, A.L., Symons, A.M., Kelly, M.I. and Arendt, J. (1987) Intraruminal soluble glass boluses containing melatonin can induce early onset of ovarian activity in ewes. *Journal of Reproduction and Fertility* 80, 235–239.

Poulton, A.L., Brown, D.C., Thomas, E.M., Kelly, M.I., Symons, A.M. and Arendt, J. (1988) Use of an intraruminal soluble glass bolus containing melatonin for early lamb production. *Veterinary Record* 122, 226–228.

Rajkamur, R.R., Argo, C.M. and Rodway, R.G. (1989) Fertility of ewes given either melatonin or progestogen sponges. *Veterinary Record* 215–217.

Robertson, H.A. (1977) Reproduction in the ewe and the goat. In: Cole, H.H. and Cupps, P.T. (eds) *Reproduction in Domestic Animals* 3rd edn. Academic Press, New York, pp. 477–498.

Robinson, J.E. (1990) Endogenous annual rhythms of luteinizing hormone secretion in the ewe and their entrainment by photoperiod. *Progress in Clinical and Biological Research* 342, 653–658.

Robinson, J.J., Wigzell, S., Aitken, R.P., Wallace, J.M., Ireland, S. and Robertson, I.S. (1991) The modifying effects of melatonin, ram exposure and plane of nutrition on the onset of ovarian activity, ovulation rate and the endocrine status of ewes. *Animal Reproduction Science* 26, 73–91.

Robinson, J.J., Wallace, J.M., Aitken, R.P. and Wigzell, S. (1992a) Effect of duration of melatonin treatment on the onset and duration of oestrous cyclicity in ewes. *Journal of Reproduction and Fertility* 95, 709–717.

Robinson, J.J., Wigzell, S., Aitken, R.P., Wallace, J.M., Ireland, S. and Robertson, I.S. (1992b) Daily oral administration of melatonin from March onwards advances by 4 months the breeding season of ewes maintained under the ambient photoperiod at 57°N. *Animal Reproduction Science* 27, 141–160.

Robinson, T.J. (1968) The synchronization of the oestrous cycle and fertility. *Proceedings of the 6th International Congress of Animal Reproduction and AI* (Paris) 2, 1347–1383.

Robinson, T.J. and Moore, M.W. (1956) The interaction of oestrogen and progesterone on the vaginal cycle of the ewe. *Journal of Endocrinology* 14, 97–109.

Robinson, T.J. and Scaramuzzi, R.J. (1994) Induction of breeding in anoestrous crossbred ewes with progestagen and PMSG with or without prior immunization against an androstenedione-protein conjugate. *Animal Reproduction Science* 35, 57–72.

Rodriguez Iglesias, R.M., Ciccioli, N.H. and Irazoqui, H. (1992) Daily distribution of teaser-induced oestrus in Corriedale ewes injected with progesterone or MAP. *Revista Argentina de Produccion Animal* 12(1), 65–70.

Ronayne, E., Jordan, B., Quirke, J.F. and Roche, J.F. (1989) The effect of frequency of administration of melatonin on the time of onset of the breeding season in anoestrous ewes. *Animal Reproduction Science* 18, 13–24.

Schanbacher, B.D. and Lunstra, D.D. (1976) Seasonal changes in sexual activity and serum levels of LH and testosterone in Finnish Landrace and Suffolk rams. *Journal of Animal Science* 43, 644–650.

Schinckel, P.G. (1954) The effect of the presence of the ram on ovarian activity of the ewe. *Australian Journal of Agricultural Research* 5, 465.

Scott, I.C. and Johnstone, P.D. (1994) Variations between years in the ram effect when Coopworth or Poll Dorset rams are introduced to seasonally anovular Coopworth

ewes. *New Zealand Journal of Agricultural Research* 37, 187–193.
Shi, Z.D. and Barrell, G.K. (1992) Requirement of thyroid function for the expression of seasonal reproductive and related changes in red deer (Cervus elaphus) stags. *Journal of Reproduction and Fertility* 95, 709–717.
Skinner, D.C., Maurice, F. and Malpaux, B. (1994) Is the pars tuberalis (PT) the site of action of melatonin in the ewe? *Journal of Reproduction and Fertility* Abstract Series No. 14, p. 7.
Smith, J.F., Andrewes, W.G.K., Knight, T.W., McMillan, W.H. and Quinlivan, T.D. (1989) A review of technology used for out-of-season breeding with New Zealand sheep breeds. In: *2nd International Congress for Sheep Veterinarians*. Massey University, pp. 1969–2003.
Smith, M.F., Swartz, H.A., Kiesling, D.O. and Warren, J.E. Jr. (1986) Effect of ram exposure and prostaglandin $F_{2\alpha}$ on the reproductive performance of anestrous ewes. *Theriogenolgoy* 26, 829–835.
Staples, L.D., McPhee, S., Kennaway, D.J. and Williams, A.H. (1992) The influence of exogenous melatonin on the seasonal patterns of ovulation and oestrus in sheep. *Animal Reproduction Science* 30, 185–223.
Stellflug, J.N., Fitzgerald, J.A. and Parker, C.F. (1989) Effect of melatonin and extended light on reproductive performance of fall-born Polypay ewe lambs and ewes during spring breeding. *Theriogenology* 32, 995–1005.
Tervit, H.R. and Peterson, A.J. (1978) Testosterone levels in Dorset and Romney rams and the effectiveness of these breeds in stimulating early onset of estrus in Romney ewes. *Theriogenology* 9, 279–294.
Tervit, H.R., Havik, P.G. and Smith, J.F. (1977) Effect of breed of ram on the onset of the breeding season in Romney ewes. *Proceedings of the New Zealand Society of Animal Production* 37, 142–148.
Thompson, L.H., Stookey, J.M., Giles, J.R. and Thomas, D.L. (1990) Reproductive response of mature ewes of different breeds to teasing prior to mating. *Small Ruminant Research* 3, 373–381.
Umberger, S.H., Jabbar, G. and Lewis, G.S. (1994) The ram effect and seasonally anovulatory ewes treated with melengestrol acetate or norgestomet. *Journal of Animal Science* 72 (Suppl. 1)/*Journal of Dairy Science* 77 (Suppl. 1), p. 340.
Underwood, E.J., Shier, F.L. and Davenport, N. (1944) Studies in sheep husbandry in W.A. V. The breeding season of Merino, Crossbred and British breed ewes in the agricultural districts. *Journal of Agriculture* (Western Australia) 2, 135–143.
Wallace, J.M., Robinson, J.J., Wigzell, S. and Aitken R.P. (1988) Effect of melatonin on the peripheral concentrations of LH and progesterone after oestrus, and on conception rate in ewes. *Journal of Endocrinology* 119, 523–530.
Watson, R.H. and Radford, H.M. (1960) The influence of rams on onset of oestrus in Merino ewes in the spring. *Australian Journal of Agricultural Research* 11, 65–71.
Wheaton, J.E. and Windels, H.F. (1994) Utilization of progesterone treatment for summer and spring breeding. *Sheep Research Journal* 10(1), 5–9.
Wigzell, S., Robinson, J.J., Aitken, R.P. and McKelvey, W.A.C. (1986) The effect of the oral administration of melatonin at two times of the year on ovarian activity in ewe. *Animal Production* 42, 448–449.
Williams, A.H., McPhee, S.R., Reeve, J.L. and Staples, L.D. (1992) Optimum use of subcutaneous melatonin implants to enhance the reproductive performance of seasonal and non-seasonal sheep joined in spring and early summer. *Animal Reproduction Science* 30, 225–258.
Williams, H. and Ward, S. (1988) Melatonin and light treatment of ewes for autumn

lambing. *Reproduction, Nutrition and Development* 28(2B), 423–429.
Williams, H.L. (1994) Sheep and melatonin. *Veterinary Annual* 34, 58–70.
Williams, H.L., Hanif, M. and Cairns, G. (1990) The use of light and melatonin treatments in the preparation of Suffolk rams for out-of-season breeding. In: *New Developments in Sheep Production*. Occasional Publications, British Society of Animal Production, No. 14, 142–145.
Williams, L.M. and Helliwell, R.J.A. (1993) Melatonin and seasonality in the sheep. *Animal Reproduction Science* 33, 159–182.
Yang, K-P., Lamming, G.E., Haynes, N.B. and Brooks, A.N. (1989) Failure of melatonin to influence endogenous opioid effects on LH secretion in the anoestrous ewe. *Journal of Reproduction and Fertility* 85, 397–403.

More Frequent Lambings in Sheep

6

6.1. Introduction

In economic terms, the maintenance costs of breeding females is much higher in cattle and sheep enterprises than in pigs or poultry production. This stems largely from the lower reproductive rates of the farm ruminants. Twice-yearly lambing or three lambings in two years is possible because sheep have a five-month gestation period. To lamb twice a year, ewes need to conceive approximately one month after lambing. Unfortunately, one postpartum period falls in close proximity to or during the seasonal anoestrus, making difficult those attempts to rebreed the ewe at that time.

6.1.1. Biological limits to reproduction in sheep

In sheep farming terms, production from sheep in most countries is very seasonal, meat and wool products coming available for marketing at regular dates once per year. Such seasonality of output places the sheep at some disadvantage compared with other farm animals such as cattle and pigs. At the same time, the ewe, with an estimated 'biological ceiling' of five lambs per pregnancy and a potential mean lambing interval of six months, has much further go to in achieving her full reproductive potential than other farm mammals.

From the farmer's point of view, the economic return from his or her sheep will depend primarily on their reproductive efficiency. Increasing the frequency of lambing under some sheep farming conditions may be the means of achieving greater reproductive efficiency, levelling out the flow of milk-fat lambs to the market and utilizing buildings, capital and labour more effectively (Hulet, 1977). However, because of the seasonal nature of breeding in sheep, any attempt to mate ewes at a greater frequency than once a year is likely to result in at least one mating during or near the ewe anoestrus in conventional seasonal breeding sheep. Despite a considerable amount of research dealing with the reproductive biology of the ewe, little practical use has been made of techniques designed to permit greater exploitation of the ewe's reproductive potential.

6.1.2. *Factors involved in increasing lambing frequency*

Although ewe productivity may be improved by increasing the frequency of lambing and modern technology now allows the farmer to manipulate breeding of the animal in ways not previously possible, there remain areas in which further research is required. It should also be remembered that the outcome of hormonal treatments (progestagen–pregnant mare serum gonadotrophin (PMSG)) may be markedly affected by factors such as postpartum interval, lactational status, nutrition, ram effect, daylength and the environment in general.

6.2. Physiology and Endocrinology of the Postpartum Ewe

If a ewe is to be bred twice yearly or three times in two years, the interval between lambing and rebreeding will be markedly shorter than usual. Thus, events in the postpartum ewe become of real importance in any consideration of more frequent lambings. There is also the question of when the sheep is giving birth, for this may be at times of the year other than the usual spring period. Ewes producing young in the spring give birth at a time when they would normally be about to enter anoestrus, even had no lambing occurred. The spring-lambing ewe, with lambs at foot, represents the most difficult category of ewe to deal with by way of controlled breeding procedures; she has both lactational and seasonal anoestrus to deal with in the early months after giving birth (Fig. 6.1).

6.2.1. *Endocrine events*

The resumption of ovulation and ovarian activity in the postpartum ewe is known to be influenced by factors such as season, lactation, suckling intensity, nutrition and breed (see Hunter, 1968; Nett, 1987; Peters and Lamming, 1990) but the endocrine basis of postpartum ovarian inactivity is not fully understood. However, it is evident that the postpartum interval is usually characterized by reduced pituitary luteinizing hormone (LH) secretion. During pregnancy in the sheep, it has been shown that the pituitary content of LH may be depleted to less than 20% of that found in non-pregnant animals (Chamley *et al.*, 1974a; Jenkin *et al.*, 1977); this was demonstrated in terms of the amount of LH released in response to a dose of gonadotrophin releasing hormone (GnRH) (Chamley *et al.*, 1974b).

In the USA, Newton and Edgerton (1989) found that Finn × Southdown

Fig. 6.1. Postpartum ewes with lambs at foot in the spring: a difficult category of sheep in which to induce pregnancy.

ewes lambing in January released significantly less LH on day 40 postpartum than ewes lambing in March and June; they concluded that seasonal modifications of the releasable pool of LH might mask or modify the effect of the postpartum interval upon this response. It has also been evident that pituitary responsiveness to GnRH increases with time postpartum and that maximum response may be achieved earlier in dry than in lactating ewes (Pelletier and Thimonier, 1975). The indications were that the lower response in lactating sheep was due to lower production of ovarian steroids, particularly oestrogens, which are known to play a role in sensitizing the pituitary to GnRH.

Breed differences in response to GnRH

It is believed that there may be breed differences in responsiveness to GnRH. In Clun Forest and Finn Landrace ewes lambing in anoestrus, normal responsiveness to GnRH had returned within 6–8 weeks of parturition (Jenkin *et al.*, 1977), whereas Romney ewes lambing in anoestrus only exhibited partial restoration of responsiveness to GnRH at a similar period (Chamley *et al.*, 1974a). It was not found, however, that GnRH responsiveness was of value in predicting the time at which breeding resumed after parturition in the ewe (Wright *et al.*, 1980).

At Nottingham, Fray *et al.* (1995) have recorded data demonstrating that a continuous infusion of GnRH can consistently induce ovulation in anovulatory postpartum lactating sheep treated during the breeding season, implying that inadequate GnRH release is the main factor preventing ovulation in such animals. The authors noted close similarities between the seasonally and the lactationally anoestrous ewe in their response to GnRH. However, it was evident that during the early postpartum period, behavioural oestrus and normal luteal function did not consistently follow ovulation in the progestagen-primed ewe.

LH levels in the postpartum ewe

The immediate postpartum period in the ewe is characterized by a gradual recovery of ovarian activity, high prolactin levels which gradually decrease after the first week (Lamming *et al.*, 1974) and a low tonic LH level which increases slowly (Restall and Starr, 1977). Lack of ovarian activity was thought to be due to an alteration in the response of the hypothalamic–pituitary axis to the negative feedback effect of oestrogen, similar to that shown by ovariectomized ewes during anoestrus (Legan *et al.*, 1977). In results presented by Wright *et al.* (1981), there were indications that in postpartum sheep there is an increased inhibitory (negative feedback) effect of oestradiol on LH release and a lower intrinsic frequency of pulsatile release of LH; the authors suggested that both seasonal anoestrus and postpartum anoestrus may involve suppression of tonic LH due to increased inhibition by oestradiol.

The oestradiol-induced LH surge

After lambing, the endocrine mechanisms controlling ovulation in the ewe fail to function in the normal way. It is known that the percentage of ewes responding to oestrogen injection with an LH surge increases with time postpartum. In Liverpool, Smart *et al.* (1994) investigated whether time postpartum or suckling or both influenced the occurrence of the oestradiol-induced LH surge and what part opioids may play in the response of ewes in the early postpartum period. In ewes, it may be noted, there has been some suggestion of a link between opioids and oestradiol positive feedback (Knight *et al.*, 1990). The Liverpool workers concluded that the onset of the oestradiol-induced LH surge was delayed in the early (day 7), but not in the late (day 14 or beyond) postpartum ewe. It was also recorded that suckling reduced the amplitude of the induced LH surge in both early and late postpartum ewes; lack of response to oestradiol immediately after lambing was apparently due to a non-opioidergic mechanism.

6.2.2. Suckling effects

Among the several factors known to influence the duration of the postpartum interval is suckling (Hunter, 1968), although there has been some debate in separating the effect of suckling from lactation. There have been those who have argued that the early weaning of lambs reduces the interval to the resumption of breeding (Mauleon and Dauzier, 1965) and delays uterine involution; others have maintained that lactation itself has little effect

(Fletcher, 1973) while others maintain that nutrition and season can modify the effects of suckling and lactation (Theriez and Molenat, 1975; Restall and Starr, 1977). Certainly, it would be generally accepted that lactational anoestrus is more pronounced at the end of the breeding season than at the beginning and where feeding is poor rather than generous (Fig. 6.2).

Fletcher (1971) showed that the frequency of suckling in the first two weeks after lambing was correlated with the duration of the postpartum interval and Cognie *et al.* (1975) reported that fertility at an induced oestrus in the postpartum ewe was reduced when two lambs rather than one was nursed. The stimulus of sucking in sheep can apparently result in elevated levels of prolactin in the circulation and this can influence the release of LH in the postpartum period (Kann *et al.*, 1977); it has also been shown by Kann and Martinet (1975) that the postpartum interval may be shorter (30–40 days) in milking ewes than in similar sheep nursing lambs (60–80 days). In France, a study of Prealpes de Sud ewes after an autumn lambing by Schirar *et al.* (1989a) recorded the interval to first oestrus to be shorter in non-suckling ewes (22 days) than in suckling ewes (35 days).

In Belgium, Mandiki *et al.* (1990) studied the effects of suckling mode on

Fig. 6.2. Production of a bonus lamb crop by progestagen–PMSG treatment. One category of sheep which can respond to the induction of pregnancy in the early part of anoestrus is the early-lambing ewe – especially when the birth of lambs is in December and the ewes are treated with progestagen–PMSG in February or so. Response is likely to vary with the body condition of the sheep and is improved if the lambs are taken off by the time of treatment. In Ireland, a 'bonus' crop of lambs can be taken from some proportion of early-lambing sheep; lambing in August or early-September means that they are ready to take the ram in October.

endocrine control of reproductive activity resumption in Texel ewes lambing in July or November. Resumption of ovarian and oestrous activity was much earlier after the November lambing than after lambing in July. In the early postpartum period, suckling inhibited the pulsatile secretion of LH, and consequently the first LH discharge. The authors suggested that inhibition of postpartum ovarian activity by suckling may be due to a temporary disturbance in hormonal balance (i.e. a rise in prolactin and cortisol secretion).

Prolactin levels

As previously noted, it is known that the concentration of prolactin in the ewe's circulation is elevated during the seasonal anoestrus, not only in ewes (Walton *et al.*, 1977) but also in rams (Ravault, 1976). It is evident that prolactin levels increase during lactation (Lamming *et al.*, 1974), particularly during and immediately after suckling (McNeilly *et al.*, 1972). Other authors have examined factors influencing prolactin concentrations in sheep and reported evidence suggesting that when plasma levels of prolactin are low, there is a greater likelihood of an earlier resumption of breeding activity in the ewe (Fitzgerald and Cunningham, 1981). Although such a relationship was not always evident, these findings lent support to data presented by Rhind *et al.* (1980) showing reduced fertility to be associated with high concentrations of prolactin.

Thyroid hormones and postpartum events

Studies reported from Slovakia by Bekeova *et al.* (1995) have suggested that decreased thyroxine and tri-iodothyronine secretion in certain phases of the postpartum period might be responsible for the decline in postpartum sexual activity in ewes.

Corpus luteum function

In the postpartum sheep, first ovulations may be characterized by a lower oestrogen peak in lactating than in non-lactating animals, a lower preovulatory LH peak and subsequent lower progesterone production by the corpus luteum (Cognie *et al.*, 1975) with a reduced lifespan (Restall, 1971). The occurrence of corpora lutea with a reduced lifespan has been recorded both in lactating and non-lactating postpartum ewes; such subnormal corpora lutea are known to occur in sheep in the absence of prior exposure to progesterone. A paper by Wallace *et al.* (1989b) confirmed and extended previous observations (Wallace *et al.*, 1989a) that induction of ovulation in the early postpartum ewe is associated with a high incidence of inadequate luteal function; it was also clear that this was not the result of inadequate progesterone and/or gonadotrophin priming.

In France, Schirar *et al.* (1989b) presented data suggesting that the frequent short duration of the first postpartum ovarian cycle in sheep was the result of:

1. the luteolytic influence of the involuting uterus due to an increased and

prolonged release of prostaglandin (PG) $F_{2\alpha}$ as found by other workers; and
2. an incomplete restoration of LH release leading to insufficient follicular growth and maturation.

Inadequacies in the uterine environment

However, although inadequate function is common in the early postpartum ewe, at least 60% of ewes with an induced oestrus 21 days postpartum were found to exhibit normal luteal function with progesterone concentrations identical to those found in ewes in which the interval from lambing was > 150 days (Wallace *et al.*, 1989a,b). This suggested that inadequate luteal function was not the only factor limiting the establishment of pregnancy in the postpartum ewe. These authors showed that the transfer of good quality embryos into postpartum recipient ewes induced to ovulate at 21 days after lambing and exhibiting normal luteal function failed to result in the establishment of pregnancy, indicating that the problem may lie with the environment provided by the involuting uterus. Results presented by Wallace *et al.* (1989c) have shown that sheep oocytes could be fertilized using laparoscopic intrauterine insemination as early as 24 days after parturition and that the resulting embryos were viable when transferred to a normal uterine environment.

In Belgium, Mandiki *et al.* (1995) have reported on the effects of progesterone treatment on ovarian and oestrous activity and on LH pulsatility in suckling and non-suckling Texel sheep; they recorded that progesterone supplementation induced earlier return to oestrus in suckling ewes and was not associated with subsequent inadequate luteal function. The same workers found progesterone treatment to be effective in reducing the interval between parturition and return to oestrus and to be more effective in dry than in suckling ewes.

6.2.3. Maternal recognition of pregnancy in the postpartum ewe

As noted previously, induction of ovulation in the early postpartum ewe is associated with a high incidence (30–40%) of premature luteolysis. It is believed that enhanced or premature release of $PGF_{2\alpha}$ from the uterus, mediated via the oxytocin receptor, is responsible for this. In the ewe, the maternal recognition of pregnancy depends on adequate secretion of interferon (IFN) by the conceptus into the uterine lumen. The IFN interacts with its receptor in the ewe's endometrium to locally suppress expression of the uterine oxytocin receptor. This is believed to be important in preventing the establishment of a positive feedback loop between episodic oxytocin secretion and uterine $PGF_{2\alpha}$ release which normally induces regression of the corpus luteum at the end of the oestrous cycle.

Studies at Aberdeen by Wallace *et al.* (1992a,b) indicated that an inability to completely suppress the endometrial oxytocin receptor and prevent luteolysis may be important in explaining pregnancy failure in ewes induced to ovulate in the early postpartum period. The authors suggest that factors underlying reduced growth rate of the conceptus and its suboptimal IFN secretion may involve defects in the actions of the many growth factors and cytokines that are known to play a role in pregnancy recognition in the sheep. Studies reported by Bettencourt *et al.* (1993) in the USA also led them to suggest that pregnancy failure in early postpartum ewes may result from an inappropriate uterine environment rather than from inadequate luteal support.

Response to exogenous hormones

In considering out-of-season breeding in sheep, whether in the early months after parturition or later, it is relevant to note the way in which season may influence the ewe's ability to respond to exogenous hormones. In progestagen–PMSG-treated ewes, oestrus and ovulation is the result of an interaction between the sensitivity of the appropriate neural centres and the quantity of ovarian oestrogen produced. Australian studies in the 1970s suggested that, compared with the autumn ewe, the quantity of oestrogen required to induce oestrus in the spring was increased and the amount of oestrogen produced per follicle reduced (Robinson, 1980); to correct for such seasonal differences, which might mean a late onset of oestrus and a shorter than usual period of sexual receptivity, Evans and Robinson (1980) suggested a 50% increase in the PMSG dose level. In Ireland, on the other hand, the regular dose of 500IU PMSG in conjunction with progestagen apparently induced oestrus as readily in the spring-lactating ewe as in the 'dry' sheep treated several months later (Gordon, 1975).

Failure to establish cyclical breeding activity

In the autumn season and in the normal breeding flock, ewes that do not become pregnant at a progestagen-controlled oestrus almost invariably repeat an ovarian cycle and are in a position to accept a second service. In the Irish sheep farming context, this results in about 90% of ewes conceiving at the first two heat periods. In the spring, the story is likely to be very different; in many flocks, none of the sheep may 'repeat' (return for a second service). Although these spring-mated ewes may have two-month-old lambs at foot, this is not the prime reason why they fail to repeat. This is evident in the fact that when autumn-lambing ewes are treated at a corresponding interval after parturition, they will repeat in much the same way as do dry cyclic ewes, and they conceive readily at the repeat oestrus, despite the fact that they usually find themselves in an increasingly difficult nutritional environment as the autumn months progress.

6.2.4. Involution of the uterus in postpartum sheep

When lambings occur at the usual time in the spring, the sheep's uterus has ample time (seven months) prior to the autumn breeding season in which to become prepared to sustain the next pregnancy. The question of restoration of uterine condition may be influenced by the season in which lambing occurs. In the spring, for example, Robinson (1959) reported finding in the early postpartum uterus what appeared to be blood undergoing autolysis; it appeared that material arising from postpartum haemorrhage had been retained through inability to pass the cervix. The author concluded that this might well be expected in sheep, a species that has a tightly interlocking cervix and which usually enters anoestrus on lambing in the spring with resultant uterine inactivity and cervical closure; this was a view supported in the evidence of McDonald and Rowson (1962) who reported the presence of detritus in the sheep's uterus in the early weeks after parturition in the spring. As noted later, it may be that the use of intravaginal progestagen sponges is not the most suitable form of treatment in postpartum ewes, in view of the fact that pessaries form a physical barrier to the passage of detritus originating from the involuting uterus and this may impair absorption of the progestagen by the vaginal mucosa.

Seasonal and suckling effects on uterine involution

Discussing some of the limitations to postpartum breeding in sheep, Van Niekerk (1976) noted that although ovulation may occur in some circumstances between 12 and 25 days after lambing, uterine involution and regrowth of epithelium may not be complete until the 26th day in the breeding season; during anoestrus, involution was delayed until the 30th day in 'dry' sheep and the 36th day in nursing ewes. Elsewhere, uterine involution has been estimated to be complete within 24 days (Call *et al.*, 1976) or from 35 days for non-lactating to 60 days for lactating ewes (Honmode, 1977). Clearly, for those striving to achieve a lambing interval of six months, such efforts may be counter productive. Although individual ewes may well conceive within six months of a previous conception, the above observations on involution would suggest that the time taken for the uterus to recover may be too long to permit a continuous six-month lambing interval on a flock basis; a more realistic minimum interval is probably about seven months.

Normal fertilization rates possible in postpartum ewes

Workers at Aberdeen have shown that the use of laparoscopic insemination to deposit semen at the tip of the uterine horn, thereby bypassing the involuting uterus, is effective in reducing the interval from lambing to fertilization in postpartum lactating ewes (McKelvey *et al.*, 1989); however, no pregnancy was established in ewes induced to ovulate and inseminated at 28 days after lambing. It was found that inadequate corpora lutea were prevalent, indicating that progesterone priming may itself have been inadequate. In a further report from Aberdeen, Wallace *et al.* (1989a) recorded the endocrine status of lactating ewes induced to ovulate 28, 35 or 42 days postpartum; they found the incidence of inadequate luteal function to be high in ewes induced to ovulate at 28 days postpartum, whereas luteal inadequacy was not a limiting factor to the establishment of pregnancy in the 35- and 42-day groups.

Hormonal attempts to influence rate of involution

A radiographic method, using radio-opaque markers, was employed by Tian and Noakes (1991) to evaluate the influence of several hormone treatments administered early in the postpartum period on uterine involution in ewes; they recorded involution to be complete by about 29 days after lambing. None of the treatments (progesterone, oestradiol-17β, $PGF_{2\alpha}$, oxytocin) administered shortly after lambing had an effect on the rate of uterine involution. In the USA, Akinbami *et al.* (1992) assessed uterine involution in ewes between day 30 and 50 after lambing using vascular permeability and leukocyte concentrations as markers; they concluded that the low fertility and conception rates observed in sheep mated before day 50 after lambing was probably a consequence of incomplete uterine involution.

Observations on the influence of progesterone on uterine oxytocin receptor-mediated events in the postpartum ewe recorded by workers in Aberdeen led them to suggest that early resumption of ovarian activity and the resulting progesterone dominance is detrimental to the rate of uterine involution and the re-establishment of pregnancy. Further evidence in support of this view was provided by Aitken *et al.* (1995).

Conception rates and breeding methods

There is some evidence that conception rates in out-of-season matings may be influenced by the methods used in the insemination of ewes; intrauterine insemination may enhance lamb production from ewes induced to breed during the seasonal anoestrus. In Aberdeen, for example, Aitken *et al.* (1990) used intrauterine insemination in breeding Greyface ewes 2–3 months after their spring lambings and recorded more than 80% producing lambs in November. They noted that conception and lambing rates were more than double those obtained after natural mating or using conventional cervical insemination with four times the number of viable sperm.

6.3. More Frequent Lambing Systems

6.3.1. Rowett work

Hunter (1968) reviewed attempts up to that time to increase the frequency of lambing but concluded that there was no way in which this could be achieved consistently on a flock basis with the techniques then available. Since that time, however, it has been clearly demonstrated that, given a suitable blend of sheep, feeding and management, highly acceptable results can be obtained. In particular, studies by John Robinson at the Rowett Institute in Aberdeen showed it was possible to achieve two lambings every 13 months on a flock basis and for the sheep to produce an average of twins each time, giving a remarkable figure of 3.7 lambs per ewe per year (Robinson, 1974; Robinson and Orskov, 1975). The figure of 3.5 lambs may be compared with an average of 1.28 lambs per ewe per year recorded elsewhere in the UK at that time (Anon, 1973); even in the top flocks recorded by the Meat and Livestock Commission (1.67 lambs per ewe per year), production was far below the Rowett figure.

The particular ingredients of the Rowett success story included the Finn × Dorset ewe, controlled light environment, controlled breeding (intravaginal fluorogestone acetate (FGA) pessaries), adequate and controlled nutrition and the abrupt weaning of lambs at one-month old (Fig. 6.3). From the results achieved under close supervision at the Rowett Institute, it was possible to draw up a specification for a frequent breeding system suitable for commercial exploitation in the UK (Anon, 1977).

The Scottish Agricultural College's specification dealt with a 'three lambings in two years' programme, using a two-flock system with matings in October, February and July. The suggested commercial programme was technically feasible because it had been demonstrated experimentally that light control was unnecessary. Initially, it had been felt that the regulation of the photoperiod was probably an essential ingredient, but this was later shown not to be so. Reports by Frazer *et al.* (1976) and Robinson (1979) described trials with Finn × Dorset crossbreds, using similar intravaginal progestagen and ram mating procedures to those previously employed but otherwise with normal outdoor flock management; it was concluded that the ewes breed at intervals of 7–8 months without light control if a small dose of PMSG was used in conjunction with the progestagen pessaries. This was obviously a finding of considerable practical importance, for the expense and labour involved in housing sheep for light control would otherwise be quite prohibitive.

For those interested in twice-yearly lambings, it might be noted that, even

Fig. 6.3. Finn × Dorset crossbred ewes and their lambs.

with the Finn × Dorset crossbred, attempts to lamb at six-month rather than eight-month intervals have not been successful (Speedy *et al.*, 1976), although it should be noted that this work was carried out among ewes at pasture rather than under controlled environmental/nutritional conditions.

6.3.2. French efforts

Investigators in France during the 1970s put considerable effort into developing systems of more frequent lambing (Thimonier *et al.*, 1975). Some of these efforts involved control of reproduction to the point at which lambings occurred in all months of the year, with work at weekends, at night and at public holidays largely avoided (see Fig. 6.4). Intravaginal FGA-pessaries–PMSG was employed in anovulatory ewes and pessaries alone in sheep that were cyclic. The most sophisticated system appeared to be that developed by researchers at the Nouzilly centre; this involved a series of seven flocks, each separated in their reproductive status by seven weeks; ewes which failed to conceive at the controlled oestrus in one flock were transferred to the subsequent flock for remating.

The output of lambs with this system, however, was found to be dependent on the particular type of sheep involved; Romanov × Prealpe ewes produced 301 lambs per 100 ewes per year, compared with 223 for Prealpe ewes and only

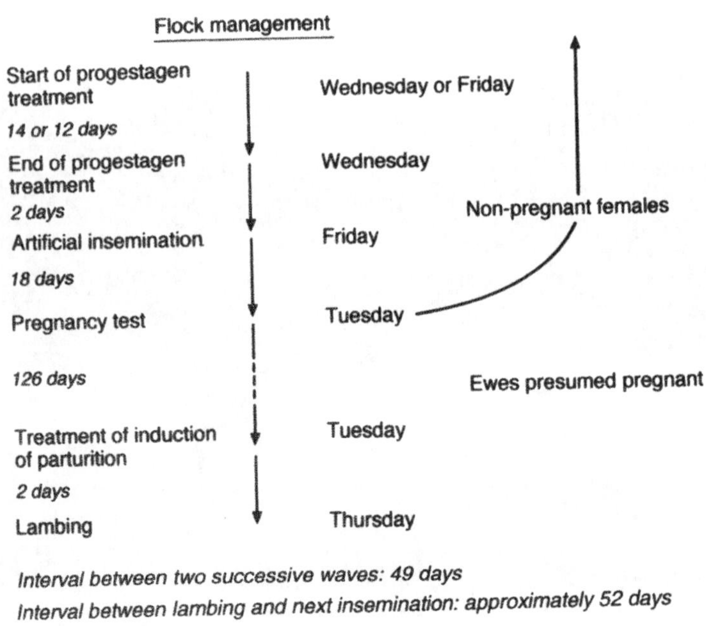

Fig. 6.4. Time-table of events in the intensive production of lambs in French sheep (from Thimonier *et al.*, 1975).

181 lambs for Ile-de-France sheep (Thimonier and Cognie, 1977). The point was well-made that sheep type was important for success.

Development of the INRA 401

One of the notable sheep success stories in France has been the development of the INRA 401 sheep breed, which is a genetic mix of the Berrichon du Cher and the prolific Russian Romanov breed. By 1995, after three decades of selection, the breed was genetically fixed and numbers had reached 40,000. Ewes of the INRA 401 breed are reputed to show high prolificacy and to be capable of lambing three times in two years.

6.3.3. Canadian efforts

Heaney *et al.* (1980) described progress in research and development in intensive lamb production at the Animal Research Institute at Ottawa in Canada, which had started in 1968. The authors dealt with what they regarded as encouraging results for an intensive lamb production system which involved total confinement; controlled breeding (intravaginal FGA and PMSG) was employed to produce three lamb crops every two years. In this system, a total of 1600 ewes was arranged in two flocks which lambed at 8-month intervals, the flocks being out of phase so that matings and lambings occurred every four months; the sheep-type was one suited to more frequent lambings. In more recent times, Shrestha *et al.* (1992) have reported on the productivity of three synthetic sheep breeds (Canadian, Outaouais and Rideau) collectively called Arcott, based on an 8-month breeding cycle; ewes have been maintained in a controlled environment (housed indoors year-round on expanded metal mesh floors in windowless barns with light controlled by time-clocks).

6.3.4. American studies

Use has been made of the Finnish Landrace in the USA in developing the Morlam crossbred at the Beltsville centre of the Department of Agriculture and the Polypay crossbred at the Dubois centre in Idaho (Hulet, 1977). Elsewhere in the 1970s, Lox *et al.* (1979) in Wisconsin examined the incidence of oestrus in different breeds of sheep to determine which breeds and management systems might afford the greatest opportunity for lambing more than once a year.

6.3.5. Israeli studies

Attempts were reported in the 1970s to increase flock profitability in Israel using more prolific ewes and an accelerated lambing frequency; this involved crossing Finn Landrace rams with the local Awassi ewe in attempts to increase its prolificacy (Goot *et al.*, 1975) and to make it more suitable for a system of more frequent breeding (Amir and Schindler, 1977). The breeding of Finn × Awassi sheep at three periods of the year (June, September, December) is dealt with in a report by Amir *et al.* (1981), who recorded changes in litter size according to the season of mating. The Israeli work was a further example of using the highly prolific Finnish Landrace sheep to introduce the capability for a longer breeding season as well as a higher litter size.

6.3.6. Egyptian studies

The breeding performance of Rahmani and Ossimi ewes bred every eight months over 18 successive mating seasons was recorded by Aboul-Naga *et al.* (1991) in Cairo. This system of accelerated lambing resulted in 42% and 55% more lambs weaned per ewe exposed to the ram for Rahmani and Ossimi ewes, respectively, than with a single breeding season each year. It was found that both ewe breeds showed oestrous activity throughout the year, although such activity was low in spring. The effects of increasing the lambing frequency on the performance of sheep production systems under the semi-arid environment of Egypt were also detailed in a report by Galal *et al.* (1993). In this, biological and economic data from a commercial flock of Barki sheep were used to evaluate the efficiency of different production systems. The authors generated six production systems from two lambing frequencies (eight and 12 months) and three breeding systems (purebreeding, production of F1 market lambs and production of market lambs from F1 ewes mated to a terminal sire). Improving management by culling non-productive animals, improving ewe : ram ratio during mating and better feeding was found to have a favourable effect on biological and financial efficiency.

6.3.7. Northern Ireland studies

Although it became clear in the 1970s that Finn × Dorset sheep, as employed in the type of intensive lamb production system described by Robinson (1979), were capable of maintaining a remarkable reproductive performance under appropriate feeding and management conditions, such prolific crossbreds were at that time limited in numbers. There was also the question of carcass quality, which is not enhanced by the Finn contribution to the fat-lamb's genotype. In Northern Ireland, work was conducted with a popular fat-lamb mother, the Greyface (Border Leicester × Blackface Mountain) in efforts to achieve three lamb crops every two years. In an investigation conducted over a four-year period, using controlled breeding techniques and weaning lambs early at one month of age, the best estimate of production was 1.82 lambs weaned/ewe/year (Foster *et al.*, 1977); this was of the same order as that found in similar studies with Greyfaces at Edinburgh (1.66 weaned lambs per ewe year^{-1}) and at Aberdeen University (1.76 weaned lambs per ewe year^{-1}).

The work in Northern Ireland showed that Suffolk, Dorset and other terminal sire breeds performed satisfactorily at all times of the year, but that conception rates, particularly in the ewe anoestrus, were variable and disappointingly low. With Finn × Dorsets, the expectation would be that 90% of the flock became pregnant to first and second services; with Greyfaces, this figure was often much lower. In fact, it was concluded that the relatively small improvement in lamb output was barely sufficient to justify the cost and labour involved.

6.3.8. South African studies

The performance of Dorper sheep under an accelerated lambing system was reported from South Africa by Schoeman and Burger (1992). In this, ewes, were exposed to three fixed 30-day mating periods per year so as to lamb in different months of the year.

6.4. Treatment Feeding and Management Considerations

As previously noted, it would appear that certain types of sheep, when provided with appropriate feeding in conjunction with early weaning, can breed satisfactorily after controlled breeding in all seasons of the year in Ireland and the UK. Reviewing the reproductive performance of sheep in frequent breeding programmes, Robinson (1979) observed that the high annual lamb output of the Finn and Finn-cross ewe was not only the result of a large litter size but also arose from their reduced lambing interval. It is known that the Finn Landrace breed has a relatively long breeding season under a wide range of environmental conditions (Wheeler and Land, 1977), a characteristic which it shares with the equally prolific Romanov breed.

It is believed to be easier to achieve an increased lambing frequency with sheep normally associated with higher than average litter sizes (e.g. Finn crossbreds; Romanov crossbreds) than with breeds, such as the Merino and its derivatives, which are associated with a low litter size, even though that particular breed is thought to have the merit of an extended breeding season. There is perhaps some support for this in the fact that Merino ewes of the high fertility Booroola strain developed in Australia continue to show breeding activity when inhibitory seasonal factors have suppressed such activity in the normal type of Merino (Bindon and Piper, 1976).

6.4.1. Progestagen–PMSG treatment

Given a suitable breed of sheep to work with, there are several considerations regarding the application of controlled breeding techniques. According to Hamilton and Lishman (1979), for instance, intravaginal progestagen sponges may not be suitable during the early postpartum period as the elimination of debris during uterine involution may be impeded; none the less, Robinson (1974) employed such sponges to good effect in Finn × Dorset crossbreds within two months of lambing and the same could be said for French researchers with their Romanov crossbreds (Thimonier *et al.*, 1975). Work by French investigators among ewes of the more usual breeds that were nursing lambs in the seasonal anoestrus seemed to indicate that it might be necessary to employ a higher dose of PMSG (at progestagen sponge removal) for lactating sheep than for those weaned at lambing or for 'dry' ewes (Thimonier *et al.*, 1968); it was suggested that this may result in a greater variation in the ovulation rate and in a greater spread in the timing of ovulation (Signoret and Cognie, 1975). It was believed that such influences on the ovulatory process might be implicated in the low fertility shown by the spring lactating ewe; fertilization rates might be lower as a result of defects in the release of oocytes. As mentioned earlier, however, studies in Ireland have not indicated a need for higher PMSG doses in the postpartum ewe.

Controlled internal drug release (CIDR) vs. progestagen pessary

Studies reported by Wallace *et al.* (1989b) have shown that the use of an intravaginal CIDR device containing progesterone, in conjunction with PMSG, was highly effective in inducing oestrous behaviour and ovulation in the early postpartum ewe during the non-breeding and breeding seasons; the synchronization technique was equally as effective at inducing oestrus on day 21 as day 35 postpartum and was independent of seasonal and lactational effects. The authors record that the structure of the CIDR device permitted fluid drainage while *in situ* and therefore may have offered some advantage over sponges in sheep undergoing uterine involution.

6.4.2. Use of melatonin and light treatment

Some studies have examined the possibility of using exogenous melatonin to stimulate early postpartum reproductive activity in spring-lambing lactating sheep. There has been evidence to show that the administration of melatonin during late gestation and the early postpartum period had no effect on milk production (Heird *et al.*, 1992). In New Mexico, Turner and Hallford (1993) recorded that, following a GnRH challenge, no more LH was released by ewes receiving melatonin than by controls. This was in agreement with data previously presented by Wheaton *et al.* (1990), who observed no effect of melatonin on LH secretion in response to GnRH in ewes treated in spring and summer. It was concluded that melatonin was not likely to be a major factor controlling return to oestrus in early postpartum spring-lambing sheep.

Embryonic mortality

According to Mauleon (1976), difficulties with the spring-lactating ewe might continue during embryonic development, with embryonic deaths occurring in the period 18–50 days after breeding. Inadequacies in the postpartum restoration of the uterus has already been mentioned as a factor which may militate against conception in sheep involved in frequent breeding programmes; a low survival rate in normal sheep embryos after transfer to the uteri of lactating ewes was noted by French workers (Cognie *et al.*, 1975).

Light manipulation to aid cyclical breeding activity

In attempts to overcome problems associated with the reproductive tract in the early months after lambing, French workers employed two inseminations and twice the normal sperm dose in breeding sheep by artificial insemination (AI). The application of photoperiodic control (to provide autumn lighting conditions at breeding), in addition to the standard FGA-sponge–PMSG oestrus induction treatment, was attempted in lactating ewes bred twice a year. Although this did not influence first service fertility (25% conceptions in April; 75% in November), non-pregnant sheep were observed to return in oestrus (Mauleon and Rougeot, 1962), something that did not occur in the absence of light control.

It should be noted that the Finn × Dorset type of ewe in a more frequent lambing system will usually return in oestrus after FGA-sponge–PMSG treatment, where she fails to conceive at the induced oestrus, and this is something that breeds of more average prolificacy (e.g. Greyface) will not readily do during the spring and summer period. Thus, Speedy and Fitzsimons (1977) did a comparison which showed conception rates of the order of 73–88% in Finn × Dorsets bred at different times of the year, whereas in Greyfaces, conception rates were high at the natural mating time in November (96%) but lower in August (55%) and still lower in February (26%); part of the problem lay in the fact that Greyfaces did not return for a second service outside their normal season. Evidence in studies conducted in late anoestrus in Ireland also suggested a relationship between readiness of the non-pregnant ewe to return to oestrus and breed type (Gordon, 1975); the higher the natural prolificacy of the breed in question, the more readily the ewe was to show evidence of cyclical breeding activity.

6.4.3. Pregnancy and lactation

Although Robinson *et al.* (1975) showed that a mean annual production of 3.5 lambs/ewe/year could be achieved in his frequent breeding programme, this was with Finn × Dorset sheep weaned at 4–6 weeks; one question of interest was whether acceptable conception rates could be maintained if the length of lactation was extended. In later studies at the Rowett Institute (Rhind *et al.*, 1977), conception rates of 100, 80, 70 and 58% to the FGA-controlled oestrus were recorded for ewes mated in December (two months after parturition) and weaned at 30, 50, 70 and 100 days of lactation, respectively; corresponding values for a second flock of sheep mated in March were 92, 83, 36 and 33%. In view of the fact that about 50% of those ewes failing to conceive at the induced oestrus did so at the second heat, it was considered possible that extending the lambing interval from seven to eight months would enable lactation to be maintained beyond one month without impairing fertility. In the system subsequently recommended for commercial use (Robinson, 1979), it was possible to take advantage of that fact; from a commercial viewpoint, it would be desirable for lambs to remain on the ewes for as long as they can, to take advantage of the mother's milk supply.

Effects of lactation

The fact that suckling and lactation need not necessarily disturb conception in the ewe is apparent from several reports dealing with autumn-lactating ewes (Gordon, 1958, 1963; Lees, 1964); in the autumn, such ewes were found to return to breeding within two months of parturition and conceive reasonably well (Fig. 6.5). Such evidence suggested that seasonal environmental factors, rather than lactation itself, were probably responsible for the particular problem of low conception rates in spring lactating sheep; even with the lactating Finn × Dorset ewe, fertility was found to be depressed to a greater extent in spring than in winter (Rhind *et al.*, 1980).

Fig. 6.5. Autumn-lambing ewes may be expected to return to oestrus spontaneously with lambs at foot.

Favourable influence of silent heats

In explaining fertility differences between autumn- and spring-lactating ewes, one consideration is the fact that autumn-lactating sheep show a silent heat prior to their first full oestrus (Mauleon and Dauzier, 1965; Restall, 1971). It seems possible that the hormonal events at this silent heat may be important in facilitating proper involution of the uterus, so that the ewe approaches full oestrus with its reproductive tract in a condition appropriate for sustaining a pregnancy. With the same sheep lambing in spring, it is unlikely that such silent heats occur in the postpartum ewe.

If the ewe is subsequently treated with progestagen–PMSG to induce mating in the early months after parturition, this may mean that she approaches the induced heat with the uterus in a relatively unprepared state. On the other hand, should such ewes show a 'repeat' oestrus (either spontaneous or induced), then they might be expected to approach that second heat period with their breeding tracts more adequately prepared. It was such reasoning that led to the development of a 'double-cycle' treatment for anoestrous sheep (Gordon, 1963), a procedure which unfortunately was too cumbersome to be commercially acceptable.

However, it may well be possible to circumvent the need for a double-cycle treatment (progestagen–PMSG + progestagen–PMSG) by simulating events that normally occur at the start of the breeding season, namely oestrogen build-up during follicle development, ovulation and formation of the corpus luteum. In artificial control, the standard progestagen–PMSG treatment may be preceded by oestrogen on the assumption that the interaction between oestrogen and subsequent progestagen treatment may improve conditions in the uterus; French efforts did find evidence of a favourable effect on fertility using a dose of 50μg oestradiol prior to the application of standard FGA–PMSG (Cognie and Pelletier, 1976).

Progestagen and PMSG dose levels

Little convincing evidence has been recorded to suggest that the dose level of progestagen and PMSG required for the postpartum ewe should be materially different from those employed in other categories of sheep. Trials which examined decreasing the dose of FGA (20mg vs. 40mg) did not result in any improvement in fertility (Mauleon, 1976); the same author noted, however, some tendency for a short progestagen treatment (six days vs. 12 days) to produce a more favourable response in ewes that showed no evidence of ovarian activity.

As previously noted, there had been suggestions that higher PMSG dose levels may be required in lactating sheep in some French studies (Thimonier *et al.*, 1968); in Australia, Evans and Robinson (1980) also suggested that more than 750IU PMSG might be required for ewes in the early postpartum period, based on evidence indicating some impairment of pituitary function in the weeks after parturition. Nevertheless, unequivocal evidence of the need for different doses of progestagen or PMSG in the early postpartum ewe has yet to be reported.

Light regimes
For those accelerated lambing systems in which sheep are to be confined indoors at all times, it is clearly possible to control the light environment precisely. It is believed that it may be important that periods of short and long daylength applied for reproduction control should be of a rhythmic nature, as they are under natural daylight conditions. If the complete production cycle (i.e. lambing to lambing) is 210 days (three lamb crops in two years) then light changes presumably would need to vary within that period as in the normal 365-day yearly pattern for sheep out of doors.

Integration of techniques
The past three decades have witnessed several useful technological advances in sheep breeding and production. These advances include the development of a practical method of administering progestagen for oestrus control in the cyclic ewe and for inducing oestrus in sheep in the non-breeding season, ability to carry out early pregnancy diagnosis at 18 days and reliable methods for inducing parturition. British, French Canadian and Australian workers have sought to put these techniques together and to employ them in systems aimed at producing lambs all the year round.

Robinson (1980) showed that it was technically feasible to produce lambs in all seasons from crossbred ewes under appropriate conditions in Australia, using the technology existing at that time. The system, as described, suffered from problems with ram matings and the early progesterone test proved to be both costly and not always available. There was also difficulty with the pregnancy test giving false positives due to embryonic mortality. It was felt that there was need for a pregnancy test which could be employed at 45 days, so that non-pregnant ewes could be identified and rebred without delay. It was concluded that the value of the system in commercial sheep farming was largely a question of costs in relation to returns from the sale of lambs; it was estimated that the cost of a viable lamb from the system was equivalent to the cost of treatment.

6.4.4. Nutritional aspects of frequent lambings

In studies at the Rowett Institute, little difficulty was experienced in maintaining the body condition of the highly productive Finn × Dorset crossbreds in an intensive lamb production programme, provided the animals were weaned at one month and well-fed for the three weeks before mating to ensure that the tissue loss in late pregnancy and early lactation was replaced before remating (Robinson, 1979). One of the important considerations during the gestation period was the pattern of feed intake; if the good body condition at mating was to be maintained until late pregnancy, then this could result in inappetance and hypoglycaemia, particularly in sheep carrying multiples.

During the 1970s, much progress was made towards a better understanding of the digestion and utilization of protein by ruminants. It became evident that for low-producing sheep, such as 'dry' ewes, pregnant ewes up to a few weeks before lambing and 'store' lambs, microbial protein will usually meet the animal's net requirements for amino-acid nitrogen. For young, fast growing lambs and ewes in the final weeks of pregnancy and in early lactation, the maximal yield of microbial protein may not necessarily meet their needs; such animals require protein supplements which, at least in part, escape degradation in the rumen.

In the frequent lambing system as described by Robinson *et al.* (1975), ewes rearing lambs were often in negative energy balance during the first month of lactation. Later work by Robinson *et al.* (1979) showed that, under such conditions, increases in the concentration of dietary crude protein stimulated the utilization of body fat and improved milk production. The response occurred within three days and was greatest with protein supplements, such as fish meal, which have a low degradability in the rumen. Using this principle, sheep in moderate condition at lambing could be stimulated to mobilize sufficient fat for the provision of energy needed in the daily production of additional milk without any detrimental effects.

Such findings were of particular interest in a frequent lambing system where a high initial growth rate in lambs was required before their early weaning. Body fat that was deposited at low cost from grass during pregnancy could be utilized after lambing when energy foods could be much more expensive. Removing the dietary protein supplement had the effect of reducing milk yield and body fat mobilization in the ewe, thus preparing the lamb for early weaning on to solid food and at the same time avoiding the detrimental effects of excessive body-fat utilization on the ewe's subsequent fertility.

The effect of body reserves on fertility and litter size in Spanish Manchega ewes managed for three lamb crops in two years on a semi-intensive system has been recorded by Molina *et al.* (1994). Body condition at mating was found to be important; ewes with a score > 3.0 had a significantly higher lambing rate (90.8%) than those with a score of < 2.0 (76.6%).

Growth hormone release blockers at weaning time

Some workers have drawn attention to the fact that in sheep, weaning or loss of lambs can result in continued milk production causing increased pressure and swelling of the mammary tissue, predisposing the tissue to mastitis. According to workers in Missouri, therapies that inhibit or reduce milk production could be useful. With this in mind, Powell and Keisler (1994) used an anticholinergic agent (methscopolamine bromide) and showed that this reduced milk yield immediately following treatment; they suggest that refinement of this approach (establishing optimum dose and timing of the agent) may represent a potential strategy for decreasing milk production in the ewe at weaning.

6.5. Lamb-Rearing Considerations

As part of frequent lambing programmes, lambs are often weaned much earlier than usual and reared on all-concentrate diets. Weaning of lambs at birth is not often regarded as an option in view of the very high feed and labour costs involved. With weaning at one month or so, the growth performance to slaughter can be high (around 350 g day^{-1}) and feed conversion rates of about 3:1 can be achieved (Robinson and Orskov, 1975). Although the carcasses of indoor-reared lambs have been a source of concern on occasions because of unacceptably soft subcutaneous fat as compared with outdoor lambs, such problems may be alleviated by modifications in the feeding techniques (use of whole barley) and probably by using genotypes that yield leaner carcasses (Orskov and Robinson, 1981).

The fact that sheep such as Finn × Dorset crossbreds are essential for success in frequent lambing systems does produce some conflict of interest when it comes to the marketing of their progeny, in view of the fact that the Finnish Landrace is not renowned for its carcass characteristics. In most situations in the UK, Suffolk rams would be used on the Finn × Dorset, although there have been attempts to employ the Texel as the terminal sire. Studies in several countries have shown that the carcasses of Texel-sired lambs contain less fat and more lean than those sired by many other breeds. In one report, Texel-cross lambs were compared with Suffolk crosses and it was shown that they had similar growth rates and feed conversion efficiencies; Texel-sired lambs, however, had significantly higher killing out percentages than those sired by the Suffolk rams (Larif and Owen, 1980).

In dealing with sheep such as the Finn × Dorset crossbred, which may not uncommonly produce litters of three and four lambs, there is the essential need to ensure the survival of such multiples. As recorded by Robinson (1981), one invaluable technique for boosting the energy supply of the small lamb, thereby preventing its demise from starvation, is the administration of colostrum (up to 30 ml kg^{-1} bodyweight of lamb) within a few minutes of birth, directly into the stomach using a catheter and syringe; the procedure was found to be fast, simple and safe and its use in supervised lambings at the Rowett Institute reduced lamb mortality to less than 2% in the frequent lambing flocks. As noted by Robinson (1990), many of the detrimental effects of maternal undernutrition on the newborn lamb can be avoided by an adequate intake of colostrum at birth.

6.5.1. Limits of lambing intervals

During the 1970s, largely due to work in Aberdeen by John Robinson and associates and in France by Thimonier and colleagues, much useful information was built up regarding more frequent lambing systems. It is clear that all-year-round production of lambs is possible given the appropriate sheep genotype (Finn × Dorset; Romanov × Berrichon du Cher) and management.

In terms of reproductive biology, the interest is in determining how prolific types of sheep are capable of an acceptable breeding performance in all seasons, whereas the more usual breed types (e.g. Suffolk crossbreds) are not. It appears to be a matter of their greater ovarian activity and the endocrine basis of that greater activity.

For the sheep farmer who may be contemplating inducing pregnancy in spring-lactating ewes (in the early postpartum period) to produce autumn as well as spring lambs, it has to be said that no effective hormonal technique can be offered. It would appear to be largely a question of factors affecting the sheep's ovarian response to progestagen–PMSG on the one hand and the resolution of inadequacies in the uterus on the other. The ultimate object of all frequent-lambing programmes is probably one in which ewes deliver twin lambs twice a year.

6.6. References

Aboul-Naga, A.M., Mansour, H., Aboul-Ela, M.B. and Almahdy, H. (1991) Breeding activity of two subtropical Egyptian sheep breeds under accelerated lambing system. *Small Ruminant Research* 4, 277–283.

Aitken, R.P., Wallace, J.M. and Robinson, J.J. (1990) A note on conception rates and litter-sizes following the intrauterine insemination of ewes at an induced oestrus during seasonal anoestrus. *Animal Production* 50, 379–382.

Aitken, R.P., Robinson, J.J. and Wallace, J.M. (1995) Is early resumption of ovarian activity detrimental to the re-establishment of pregnancy in the lactating ewe? *Biology of Reproduction* 52 (Suppl. 1), p. 107.

Akinbami, M.A., Meridith, S., Warren, J.E. Jr., Day, B.N. and Ganjam, V.K. (1992) Leukocyte concentrations and vascular permeability of the uteri in postpartum ewes. *Journal of Animal Science* 70 (Suppl. 1), p. 262.

Amir, D. and Schindler, H. (1977) Induction of oestrus and fertility of ewes at the beginning and end of the sexual season. *Hassadeh* 57, 1663–1667.

Amir, D., Schindler, H. and Genizi, A. (1981) A note on seasonal changes in litter size of Finn × Awassi ewes. *Animal Production* 32, 121–123.

Anon (1973) Sheep facts. *Meat and Livestock Commission Publication*, Milton Keynes, UK.

Anon (1977) *Technical Notes* No. 16, Publication of Scottish Agricultural Colleges.

Bekeova, E., Krajnicakova, M., Hendrishovsky, V. and Maracek, I. (1995) The effects of long-acting oxytocin, GnRH and FSH administration on the thyroxin, triodothyronin, oestradiol 17-β and progesterone levels as well as conception rates in post-partum ewes. *Animal Reproduction Science* 37, 311–323.

Bettencourt, C.M.V., Moffatt, R.F. and Keisler, D.H. (1993) Active immunization of ewes against prostaglandin F2α to control ovarian function. *Journal of Reproduction and Fertility* 97, 123–131.

Bindon, B.M. and Piper, L.R. (1976) Assessment of new and traditional techniques of selection for reproduction rate. *Proceedings of the International Congress on Sheep Breeding* (Muresk), pp. 357–371.

Call, J.W., Foote, W.C., Eckie, C.D. and Hulet, C.V. (1976) Postpartum uterine and ovarian changes and estrous behaviour from lactation effects in normal and hormone treated ewes. *Theriogenology* 6, 495–501.

Chamley, W.A., Findlay, J.K., Jonas, H., Cumming, I.A. and Goding, J.R. (1974a) Effect of pregnancy on the FSH response to synthetic gonadotrophin-releasing hormone in ewes. *Journal of Reproduction and Fertility* 37, 109–112.

Chamley, W.A., Findlay, J.K., Cumming, I.A., Buckmaster, M. and Goding, J.R. (1974b) Effect of pregnancy on the LH response to synthetic gonadotrophin-releasing hormone in the ewe. *Endocrinology* 94, 291–293.

Cognie, Y. and Pelletier, J. (1976) Preovulatory LH release and ovulation in dry and in lactating ewes after progestagen and PMSG treatment during the seasonal anoestrum. *Annales de Biologie animale Biochime Biophysique* 16, 529–536.

Cognie, Y., Hernandez-Barreto, M. and Saumande, J. (1975) Low fertility in nursing ewes during the non-breeding-season. *Annales de Biologie animale Biochimie Biophysique* 15, 329–343.

Evans, G. and Robinson, T.J. (1980) The control of fertility in sheep: endocrine and ovarian responses to progestagen–PMSG treatment in the breeding season and in anoestrus. *Journal of Agricultural Science, Cambridge* 94, 69–88.

Fitzgerald, B.P. and Cunningham, F.J. (1981) Effect of removal of lambs or treatment with bromocriptine on plasma concentrations of prolactin and FSH during the post-partum period in ewes lambing at different times during the breeding season. *Journal of Reproduction and Fertility* 61, 141–148.

Fletcher, I.C. (1971) Relationships between frequency of suckling, lamb growth and post-partum oestrous behaviour in ewes. *Animal Behaviour* 19, 108–111.

Fletcher, I.C. (1973) Effects of lactation, suckling and oxytocin on post-partum ovulation and oestrus in ewes. *Journal of Reproduction and Fertility* 33, 293–298.

Foster, W.H., McCaughey, W.J., Logan, E.F. and Irwin, D. (1977) Controlled breeding of sheep. *50th Annual Report Northern Ireland Agricultural Research Institute* pp. 19–27.

Fray, M.D., Lamming, G.E. and Haresign, W. (1995) Induction of ovulation in the acyclic postpartum ewe following continuous, low-dose subcutaneous infusion of GnRH. *Theriogenology* 43, 1019–1030.

Frazer, C., Robinson, J.J., McHattie, I. and Gill, J.C. (1976) Field studies on the reproductive performance of Finnish Landrace × Dorset Horn ewes. *Proceedings of the British Society of Animal Production* 5, 162–163.

Galal, E.S.E., Ahmed, A.M., Abdel-Aziz, A.I. and Younis, A.A. (1993) Effects of increasing lambing frequency and cross-breeding on performance of sheep production systems in semi-arid environments. *Small Ruminant Research* 10, 143–152.

Goot, H., Folman, Y., Dori, D. and Eyal, E. (1975) The Finn × Awassi cross of sheep (preliminary results). *Hassadeh* 55, 1881–1883.

Gordon, I. (1958) Studies in the extra-seasonal production of lambs. *Journal of Agricultural Science, Cambridge* 50, 152–197.

Gordon, I. (1963) The induction of pregnancy in the anoestrous ewe by hormonal therapy. *Journal of Agricultural Science, Cambridge* 60, 31–79.

Gordon, I. (1975) The use of progestagens in sheep bred by natural and artificial insemination. *Anales de Biologie animale Biochimie Biophysique* 15, 303–315.

Hamilton, C.D. and Lishman, A.W. (1979) Reducing the partum-to-mating period in autumn lactating ewes through the use of exogenous hormones. *South African Journal of Animal Science* 9, 59–63.

Heaney, D.P., Ainsworth, L., Batra, T.R., Fiser, P.S., Langford, G.A., Lee, A.J. and Hackett, A.J. (1980) Research for an intensive total confinement sheep production system. *Animal Research Institute Technical Bulletin* No. 2, Agriculture Canada, 55 pp.

Heird, C.E., Hallford, D.M., Perez-Eugia, E., Campbell, J.W. and Turner, M.L. (1992) Postpartum reproductive responses of Debouillet ewes treated with melatonin before and after lambing. *Proceedings of the Western Section American Society of Animal Science* 43, pp. 211–213.

Honmode, D. (1977) Postpartum changes in the uterus of ewes. *Animal Breeding Abstracts* 45, 384 (Abstract 3280).

Hulet, C.V. (1977) Management of reproduction in sheep. *Proceedings of Symposium on Management of Reproduction in Sheep and Goats* (Madison), Sheep Industry Development Program, pp. 119–133.

Hunter, G.L. (1968) Increasing the frequency of pregnancy in sheep. *Animal Breeding Abstracts* 36, 347–378 and 533–553.

Jenkin, G., Heap, R.B. and Symons, D.B.A. (1977) Pituitary responsiveness to synthetic LH-RH and pituitary LH content at various reproductive stages in the sheep. *Journal of Reproduction and Fertility* 49, 207–214.

Kann, G. and Martinet, J. (1975) Prolactin level and duration of post-partum aneostrus in lactating ewes. *Nature* 257, 63–64.

Kann, G., Harbert, R., Meusnier, C. and Ryniewicz, H.S. (1977) Prolactin release in response to nursing or milking stimulus in the ewe. It is mediated by thyrotrophin releasing hormone. *Annales de Biologie animale Biochimie Biophysique* 17, 441–452.

Knight, P.G., Stansfield, S.C. and Cunningham, F.J. (1990) Attenuation by an opioid agonist of the oestradiol-induced LH surge in anoestrous ewes and its reversal by naloxone. *Domestic Animal Endocrinology* 7, 165–172.

Lamming, G.E., Moseley, S.R. and McNeilly, J.R. (1974) Prolactin release in the sheep. *Journal of Reproduction and Fertility* 40, 151–168.

Larif, M.G.A. and Owen, E. (1980) A note on the growth performance and carcase composition of Texel and Suffolk-sired lambs in an intensive feeding system. *Animal Production* 30, 311–314.

Lees, J.L. (1964) Inhibitory effect of lactation on the breeding activity on the ewe. *Nature* 203, 1089.

Legan, S.J., Karsh, F.J. and Foster, D.L. (1977) The endocrine control of seasonal reproductive function in the ewe: a marked change in response to the negative feedback action of estradiol on luteinizing hormone secretion. *Endocrinology* 101, 818–825.

Lox, J., French, L.R., Chapman, A.B., Pope, A.L. and Casida, L.E. (1979) Length of breeding season for eight breed groups of sheep in Wisconsin. *Journal of Animal Science* 49, 939–942.

McDonald, M.F. and Rowson, L.E.A. (1962) Ovum transfer to lactating ewes. *Journal of Reproduction and Fertility* 4, 205–211.

McKelvey, W.A.C., Wallace, J.M., Robinson, J.J. and Aitken, R.P. (1989) Studies on increasing breeding frequency in the ewe. I. The fertilization of ova during the early post-partum period. *Animal Reproduction Science* 18, 1–12.

McNeilly, J.R., Mosely, S.R. and Lamming, G.E. (1972) Observations on the pattern of prolactin release during suckling in the ewe. *Journal of Reproduction and Fertility* 31, 487–488.

Mandiki, S.N.M., Bister, J.L. and Paquay, R. (1990) Effects of suckling mode on endocrine control of reproductive activity resumption in Texel ewes lambing in July or November. *Theriogenology* 33, 397–413.

Mandiki, S.N., Bister, J.L. and Paquay, R. (1995) Effects of progesterone treatment on ovarian and estrous activity and on LH pulsatility and PG $F_{2\alpha}$ concentration in suckling and non-suckling Texel ewes. *Small Ruminant Research* 15, 265–272.

Mauleon, P. (1976) Manipulation of the breeding cycle. *Proceedings of the International Congress on Sheep Breeding* (Muresk), pp. 310–321.

Mauleon, P. and Dauzier, L. (1965) Variations in the duration of lactation in anoestrus ewes of the Ile-de-France breed. *Annales de Biologie animale Biochimie Biophysique* 5, 131–136.

Mauleon, P. and Rougeot, J. (1962) Regulation des saisons sexuelles chez des brebis de races differentes au moyen de divers rhythmes lumeneuz. *Annales de Biologie animale Biochemie Biophysique* 2, 209–222.

Molina, A., Gallego, L., Torres, A. and Vergara, H. (1994) Effect of mating season and level of body reserves on fertility and prolificacy of Manchega ewes. *Small Ruminant Research* 14, 209–217.

Nett, T.M. (1987) Function of the hypothalmic-hypophysial axis during the post-partum period in ewes and cows. *Journal of Reproduction and Fertility* Suppl. 34, 201–213.

Newton, G.R. and Edgerton, L.A. (1989) Effects of season and lactation on luteinizing hormone secretion in postpartum ewes. *Theriogenology* 31, 885–889.

Orskov, E.R. and Robinson, J.J. (1981) The application of modern concepts of ruminant protein nutrition to sheep production systems. *Livestock Production Science* 8, 339–350.

Pelletier, J. and Thimonier J. (1975) Interactions between ovarian steroids or progestagens and LH release. *Annales de Biologie animale Biochimie Biophysique* 15, 131–146.

Peters, A.R. and Lamming, G.E. (1990) Lactational anoestrus in farm animals. *Oxford Reviews of Reproductive Biology* 12, 245–288.

Powell, M. and Keisler, D.H. (1994) A potential strategy for decreasing milk production in the ewe at weaning using a growth hormone release blocker. *Journal of Animal Science* 72 (Suppl. 1)/*Journal of Dairy Science* 77 (Suppl. 1), p. 341.

Ravault, J.P. (1976) Prolactin in the ram: seasonal variations in the concentration of blood plasma from birth until three years old. *Acta Endocrinologica* 83, 720–725.

Restall, B.J. (1971) The effect of lamb removal on reproductive activity in Dorset Horn × Merino after lambing. *Journal of Reproduction and Fertility* 24, 145–146.

Restall, B.J. and Starr, B.G. (1977) The influence of season of lambing and lactation on reproductive activity and plasma LH concentrations in Merino ewes. *Journal of Reproduction and Fertility* 49, 297–303.

Rhind, S.M., Robinson, J.J., Fraser, C. and Phillippo, M. (1977) Effects of season and lactation on ovulation rate, plasma progesterone concentrations in early pregnancy and lamb production in Finnish Landrace × Dorset Horn ewes. *Animal Production* 24 (Abstract 128).

Rhind, S.M., Robinson, J.J., Chesworth, J.M. and Crofts, R.M.J. (1980) Effects of season, lactation and plane of nutrition on prolactin concentrations in ovine plasma and the role of prolactin in the control of ewe fertility. *Journal of Reproduction and Fertility* 58, 145–152.

Robinson, J.J. (1974) Intensifying ewe productivity. *Proceedings of the British Society of Animal Production* 3, 31–40.

Robinson, J.J. (1979) Intensive systems. In: *Management and Diseases of Sheep.* Commonwealth Agricultural Bureau, Slough, pp. 431–446.

Robinson, J.J. (1981) Prenatal growth and development in the sheep and its implications for the viability of the newborn lamb. *Livestock Production Science* 8, 273–281.

Robinson, J.J. (1990) Nutrition in the reproduction of farm animals. *Nutrition Research Reviews* 3, 253–276.
Robinson, J.J. and Orskov, E.R. (1975) An integrated approach to improving the biological efficiency of sheep meat production. *World Review Animal Production* 11, 63–76.
Robinson, J.J., Fraser, C. and McHattie, I. (1975) The use of progestagens and photoperiodism in improving the reproductive rate of the ewe. *Annales de Biologie animale Biochimie Biophysique* 15, 345–352.
Robinson, J.J., McHattie, I., Calderon, C.J.F. and Thompson, J.L. (1979) Further studies on the response of lactating ewes to dietary protein. *Animal Production* 29, 257–269.
Robinson, T.J. (1959) The estrous cycle of the ewe and doe. In: Cole, H.H. and Cupps, P.T. (eds) *Reproduction in Domestic Animals*, Vol. 1. Academic Press, New York, pp. 291–333.
Robinson, T.J. (1980) Programmed year-round sheep breeding. *Australian Journal Experimental Agriculture and Animal Husbandry* 20, 667–673.
Schirar, A., Cognie, Y., Louault, F., Poulin, N., Levasseur, M.C. and Martininet, J. (1989a) Resumption of oestrus behaviour and cyclic ovarian activity in suckling and non-suckling ewes. *Journal of Reproduction and Fertility* 87, 789–794.
Schirar, A., Meusnier, C., Paly, J., Levasseur, M.C. and Martinet, J. (1989b) Resumption of ovarian activity in post-partum ewes: role of the uterus. *Animal Reproduction Science* 19, 79–89.
Schoeman, S.J. and Burger, R. (1992) Performance of Dorper sheep under an accelerated lambing system. *Small Ruminant Research* 9, 265–281.
Shrestha, J.N.B., Heaney, D.P. and Parker, R.J. (1992) Productivity of three synthetic Arcott sheep breeds and their crosses in terms of 8-month breeding cycle and artificially reared lambs. *Small Ruminant Research* 9, 283–296.
Signoret, J.P. and Cognie, Y. (1975) Determination of the moment of ovulation in ewe and sow. Influence of environment and hormonal treatment. *Annales de Biologie animale Biochimie Biophysique* 15, 205–214.
Smart, D., Sigh, I., Smith, R.F. and Dobson, H. (1994) Opioids and suckling in relation to inhibition of oestradiol-induced LH secretion in postpartum ewes. *Journal of Reproduction and Fertility* 101, 115–119.
Speedy, A.W., Black, W.J.M. and Fitzsimons, J. (1976) The performance of Finnish Landrace × Dorset Horn ewes mated every six months. *Animal Production* 22, 171.
Speedy, A.W. and Fitzsimons, J. (1977) The reproductive performance of Finnish Landrace × Dorset Horn and Border Leicester × Scottish Blackface ewes mated three times in 2 years. *Animal Production* 24, 189–196.
Theriez, M. and Molenat, G. (1975) Intensive management of sheep. Fecundity rate of ewes inseminated every 6 months as influenced by drying off immediately after parturition. *Annales Zootechnica* 24, 729–742.
Thimonier, J. and Cognie, Y. (1977) Application of control of reproduction of sheep in France. In: *Management of Reproduction in Sheep and Goats*. Sheep Industry Program, pp. 109–118.
Thimonier, J., Mauleon, P., Cognie, Y. and Ortavant, R. (1968) Induction of oestrus and pregnancy in ewes during post-partum anoestrus with the aid of vaginal sponges impregnated with fluorogestone acetate. *Annales Zootechnica* 17, 257–273.
Thimonier, J., Cognie, Y., Cornu, C., Schneberger, J. and Vernusse, G. (1975)

Intensive lamb production. *Annales de Biologie animale Biochimie Biophysique* 15, 365–367.

Tian, W. and Noakes, D.E. (1991) A radiographic method for measuring the effect of exogenous hormone therapy on uterine involution in ewes. *Veterinary Record* 129, 436–466.

Turner, M.L. and Hallford, D.M. (1993) Return to estrus and endocrine patterns in early postpartum, spring-lambing ewes treated with melatonin. *Theriogenology* 39, 1245–1256.

Van Niekerk, C.H. (1976) Limitations to female reproductive efficiency. *Proceedings of the International Congress Sheep Breeding*, Muresk, pp. 299–309.

Wallace, J.M., Robinson, J.J. and Aitken, R.P. (1989a) Does inadequate luteal function limit the establishment of pregnancy in the early post-partum ewe? *Journal of Reproduction and Fertility* 85, 229–240.

Wallace, J.M. Robinson, J.J., McKelvey, W.A.C. and Aitken, R.P. (1989b) Studies on increasing breeding frequency in the ewe. 2. The endocrine status of lactating ewes induced to ovulate 28, 35 or 42 days post-partum. *Animal Reproduction Science* 18, 271–283.

Wallace, J.M., Robinson, J.J. and Aitken, R.P. (1989c) Successful pregnancies after transfer of embryos recovered from ewes induced to ovulate 24–29 days post-partum. *Journal of Reproduction and Fertility* 86, 627–635.

Wallace, J.M., Aitken, R.P. and Cheyne, M.A. (1992a) Maternal recognition of pregnancy following blastocyst transfer to ewes induced to ovulate at 28 days post-partum. *Proceedings of the British Society Animal Production* (Winter Meeting), Paper No. 79.

Wallace, J.M., Thompson, M.G., Cheyne, M.A. and Aitken, R.P. (1992b) Oxytocin receptor concentrations, inositol phosphate turnover and prostaglandin release by endometrium from ewes induced to ovulate early post-partum. *Journal of Reproduction and Fertility* Abstract Series No. 9, p. 25.

Walton, J.S., McNeilly, J.R., McNeilly, A.S. and Cunningham, F.J. (1977) Changes in concentrations of follicle-stimulating hormone, luteinizing hormone, prolactin and progesterone in the plasma of ewes during the transition from anoestrus to breeding activity. *Journal of Endocrinology* 75, 127–136.

Wheaton, J.E., Pohl, H.A. and Windels, H.F. (1990) Effects of melatonin and progesterone administered to ewes in spring and summer. *Journal of Animal Science* 68, 923–930.

Wheeler, A.G. and Land, R.B. (1977) Seasonal variation in oestrus and ovarian activity of Finnish Landrace, Tasmanian Merino and Scottish Blackface ewes. *Animal Production* 24, 363–376.

Wright, P.J., Geytenbeek, P.E., Clarke, I.J. and Findlay, J.K. (1980) Pituitary responsiveness to LH-RH, the occurrence of oestradiol-17B-induced positive feedback and the resumption of oestrous cycles in ewes post-partum. *Journal of Reproduction and Fertility* 60, 171–176.

Wright, P.J., Geyteenbek, P.E., Clarke, I.J. and Findlay, J.K. (1981) Evidence for a change in oestradiol negative feedback and LH pulse frequency in post-partum ewes. *Journal of Reproduction and Fertility* 61, 98–102.

Induction of Multiple Births in Sheep

7

7.1. Introduction

In fat-lamb production, it is not only essential to achieve high conception rates in sheep subjected to controlled breeding procedures but important that most ewes produce twins rather than single lambs. Economic studies in lowland sheep over the years have clearly shown the importance of high fertility as a major determinant of profitability in the enterprise. A small difference in the proportion of ewes carrying multiples may make a large difference to the net income yielded by the flock.

In farming situations where the full genetic potential of a particular breed is being achieved and a further improvement in litter size is considered desirable, then the introduction of a more prolific breed, selection of ewes within a breed or the artificial control of litter size (using exogenous gonadotrophins or appropriate immunization treatments) are among several of the options available. Mention should be made of a recent book which has comprehensively reviewed and collated current knowledge of prolific sheep (Fahmy, 1996).

In the Irish Republic, in which the ewe breeding flock currently numbers 4.62 million head, some number of the lowland ewes are Galways, a breed in terms of size and lambing performance much like the Romney Marsh sheep in England. Some years ago, a survey of more than 10,000 Galways ewes on 354 farms by Daly (1966) revealed an average litter size of 1.28 (Fig. 7.1).

Litter-size limits in sheep

In terms of the limits of litter size in sheep, it is worth recording that the world lambing record is apparently held by a Swedish Landrace ewe which gave birth to and weaned a litter of eight lambs; according to Wennbom (1994), although two litters of octuplets were reported in 1991 in New Zealand and Finland, this was a world record for the number of lambs reared.

Fig. 7.1. The Galway is the major lowland sheep breed in the Irish Republic. Although its growth potential and wool characteristics are good, the weak point of the breed has been its low litter size – hence the many attempts over the years by selective breeding and other means to increase its lambing performance.

7.2. Endocrine and Ovarian Events in Multiple Ovulations

As well as the generally recognized effects of feeding, management and environment on the ovulation rate in the ewe, administration of gonadotrophins at particular stages of the oestrous cycle will induce multiple ovulations and by that means increase the number of lambs born. It is therefore of interest to examine what is known to underlie the normal variations in ovulation rate in the species.

7.2.1. Ovarian follicle populations

The total follicle population in the ewe's ovary consists of a large reserve of primordial and small preantral follicles and a much smaller number of larger vesicular follicles in their growth phase; some studies have reported a direct relationship between the number of follicles in the growth phase and ovulation rate (Cahill *et al.*, 1979). In some studies, three categories of follicles have been recognized, these being dormant, transitory and growing follicles; it is believed that recruitment of follicles in the transitory phase to the growth phase is under the control of the pituitary gonadotrophins (Dufour *et al.*, 1979). The indications are that follicles in the transitory phase acquire FSH receptors and then selectively enter the growth phase under gonadotrophin control; the recruitment of follicles from the dormant to the transitory category is still not well understood and may operate by an intraovarian mechanism independent of gonadotrophins.

Growing follicle population

In Merino sheep, some authors have estimated that three to four follicles are recruited each day into the growing follicle population and that the number of follicles finally ovulating six months later was determined by the rate of atresia occurring during this growth phase (Turnbull *et al.*, 1977). According to studies reported by Cahill and Mauleon (1980), follicle growth was very slow prior to antrum formation (130 days) but speeded up markedly in the rapid growth phase (45 days); the long growth period of the follicle indicated that follicles ovulating in the breeding season began their growth during the seasonal anoestrus, six months previously. The fact that the number of developing preantral follicles increased in anoestrus and decreased in the breeding season led to the suggestion that anoestrus may constitute an essential recovery period for the ovary.

7.2.2. Follicles and ovulation rate

Finnish Landrace and Romanov sheep have high ovulation rates and various workers have examined ovarian and oestrous activity in these breeds in comparison with other sheep. In the UK, for example, the proportion of follicles at birth in Finn × Welsh and in Finn × Blackface lambs was found to be greater than that in purebred Welsh or Blackface animals (Land, 1979); it was also recorded that later in life the duration of oestrus was greater in the Finn crossbreds than in the purebreds.

Gonadotrophin levels

Relating gonadotrophin levels to follicular growth and development in the 1970s was made difficult by the large variations found in follicle stimulating hormone (FSH) concentrations within and between animals (Findlay and Cumming, 1976) and the lack of knowledge about when the gonadotrophin-susceptible phases of follicle growth occur. At least one phase was believed to occur on days 12–14 of the oestrous cycle when gonadotrophin levels were thought to be important in determining ovulation rate at the subsequent oestrus. A comparison of FSH levels in groups of ewes, on different planes of nutrition which influenced ovulation rates, revealed relatively greater FSH levels on days 13–14 compared with day 1 (Brien *et al.*, 1976); in contrast, differences were not evident when FSH concentrations in ewes at different stages of the cycle were compared by Findlay and Cumming (1976) and elsewhere attempts to correlate the level of FSH with the ovulation rate shown by the adult ewe were not successful (Bindon *et al.*, 1975).

Evidence was available indicating that two peaks of FSH secretion occurred during the oestrous cycle, the first coincident with the preovulatory surge of luteinizing hormone (LH) and the second occurring 20–30h after the LH peak; this second FSH peak appeared to be correlated with the number of vesicular follicles present at the next heat period (Cahill and Dufour, 1979). The only characteristic of the preovulatory LH surge that could be correlated with ovulation rate appeared to be the interval between the onset of oestrus and the start of the discharge of LH (Bindon *et al.*, 1975).

Predicting ovulation rates

Circulating LH concentrations have been studied in prepubertal ewe-lambs as a possible means of making an early assessment of an individual's potential fertility (Bindon and Turner, 1974); unfortunately, it became clear that measurement of LH was complicated by fluctuations in peripheral LH concentrations throughout the day. Elsewhere, however, it was shown that selection for increased female fertility could be aided by the use of certain male characteristics, the most promising of which appeared to be testis size (Land, 1980).

7.3. Environmental and Nutritional Effects

Well-established factors known to affect litter size in sheep, such as age, breed and environmental conditions, have been discussed at length by many authors over the years (Heape, 1899; McKenzie and Terrill, 1937, Hammond, 1944). In the UK, with its considerable array of improved mutton sheep breeds and established crossbreds (Half-Breds, Greyfaces, Mule), average litter sizes, as reported in the literature, range all the way from 1.1 in Welsh Mountains, to 2.4 in the Finnish Landrace; there is ample breeding material in the country to select for flocks that provide for a wide range of litter sizes.

7.3.1. Seasonal effects

There is ample evidence from several breeds of sheep that ovulation rate increases after the commencement of the breeding season and then falls away towards the end (Johansson and Hansson, 1943; Averill, 1959; Gunn *et al.*, 1979; Mitchell *et al.*, 1996). This variation in the ovulation rate is usually reflected in a similar pattern of multiple births at lambing time which implies that fertilization rate and the incidence of embryo mortality may not be subject to seasonal variation.

The extent of the decline in ovulation rate and average litter size in the later stages of the breeding season may be influenced by body condition of the ewes, those in good condition suffering more of a decline than those in thin condition

(Newton *et al.*, 1980). It is also possible that some of the reduction in ovulation rate with advancing season may be the result of stress, particularly weather stress, since the later stages of the season tend to include the periods of worsening weather; stress has been shown to have an adverse effect on the ovulation rate (Doney *et al.*, 1976).

In the UK, as noted by Mitchell *et al.* (1995), there has been increasing interest in the development of later lambing systems for sheep which more closely match ewe nutritional requirements with herbage growth; these authors therefore examined the effect of management and season on ovulation and litter sizes in Mule ewes. Results presented by the Aberdeen workers showed that acceptable ovulation rates could be achieved throughout the period September to February but that litter sizes were reduced. These results, however, were apparently in contrast to observations which the authors had recorded in commercial flocks mated in January, in which litter size was slightly reduced compared with sheep bred in November but conception rate was high. The authors concluded that more work was required to resolve the differences between their experimental and commercial flocks. A further report by Mitchell *et al.* (1996) demonstrated that reduced fecundity in sheep mated during the latter part of the breeding season was mainly due to a reduction in ovulation rate.

7.3.2. 'Flushing' and nutritional effects

It is well established that follicle development in many mammals is affected by nutrition. This effect can either be indirect, acting by way of the hypothalamic–pituitary axis to alter the secretion of gonadotrophins, or can be direct, acting on the ovary to mediate the action of gonadotrophins on the follicle. A mechanism of action for nutrition on ovulation was proposed by Smith (1988). In this, increased nutrient intake, in particular increased protein, produces an increase in both liver size and the concentrations of hepatic microsomal enzymes. This results in an increased level of oestradiol metabolism, which in turn is reflected in increased level of FSH prior to and during luteolysis. Such increases in FSH may be responsible for a greater number of developing follicles ovulating.

Effect of nutritional signals on follicle development

The nutritional signals that may exert an action on follicle development in sheep have also been studied. In Australia, Downing *et al.* (1990) showed that the infusion of branched chain amino acids over a five-day period in the late stages of the oestrous cycle would significantly increase ovulation rate in sheep. Other studies have been discussed by Scaramuzzi (1994); according to this author, insulin-mediated uptake of glucose may reduce the capacity of the selected follicle (dominant follicle) to suppress the development of subordinate follicles, thereby increasing ovulation rate by that means.

Ovulation rate in the ewe in the usual autumn breeding season is determined by factors which operate up to the time of mating. Nutritionally, the most influential period is probably that between the previous lambing (or more accurately, the end of lactation) and the time of mating; this can be regarded as the recovery period when the sheep's reserves that had been depleted during pregnancy and lactation are replenished. There are two considerations in looking at the nutrition of the ewe during this recovery period: the first is the long-term one which largely determines the body condition and weight of the ewe at mating time; the second is the short-term 'flushing' effect operating at the time of mating.

Effect of body condition

A direct link between body condition and ovulation rate was established in sheep more than 60 years ago (Clark, 1934) but most of the reports since then have been concerned with the short-term 'flushing' effect. Coop (1966) in New Zealand was one of the first to try to define the nutritional effect more precisely by using the terms 'static' and 'dynamic' to describe it. Static effect was seen to be a matter of body condition, liveweight and size of the ewe; dynamic effect was defined as a change in liveweight during a six-week period prior to mating.

7.3.3. Body condition and ovulation rate

Body weight of the ewe at mating, representing the static effect, has been shown to influence subsequent litter size (Coop, 1966); the effect is mainly the result of differences in ovulation rate but with some involvement in the extent of embryo mortality (Edey, 1969). The body weight of the ewe has two components, basic skeletal size of the sheep on the one hand and the degree of fatness (i.e. body condition) on the other. Body condition scoring in sheep can be a useful means of assessing the flock at mating time. In Ireland, it involves giving a score on a scale ranging from 0 to 5 according to the degree of condition.

Condition is assessed by handling over and around the backbone, in the loin area immediately behind the last rib and above the kidney, using the fingers along the top and sides of the backbone (see Fig. 7.2). The ewes should be scored 6–8 weeks prior to the start of mating, so that appropriate action may be taken to get as many sheep as possible to the optimum condition (scores of 2½–3) at the time rams are introduced (Merrell, 1990). Ewes that score below 2 may need to be drawn out of the flock and given access to the best available grazing. In physiological terms, it is known that ewes in high body condition have more large, oestrogenic, ovarian follicles than ewes in low body condition (Rhind *et al.*, 1989). Studies by Rhind *et al.* (1993) examined FSH-stimulated follicle development in ewes in high and low body condition and concluded that such differences did not affect the responsiveness of the ovary to circulating concentrations of FSH.

In Australia, where flocks are large, and ewes are usually of similar genetic

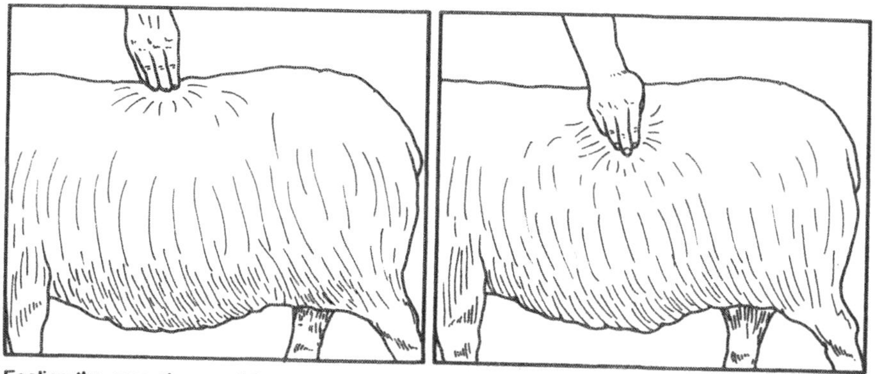

Feeling the smoothness of the ridge formed by the vertical processes of the backbone – the first step in condition scoring.

Feeling the fullness of the underside of the loin – a further step in condition scoring.

Fig. 7.2. Steps in the condition scoring of sheep (scores: 0 = emaciated; 5 = overfat).

constitution, liveweight alone has been found to be a more accurate predictor of ovulation rate than body condition (Cumming, 1977); in general, heavier ewes within a flock were found to have more ovulations than the lighter ones, showing about 2.5–3.0% increase for each 1.0kg increase in liveweight. An article by Smith (1988) also drew attention to a well-established and repeatedly documented relationship in sheep between liveweight immediately prior to mating and an increase in 2% in ovulation rate for each additional kilogram of liveweight.

'Flushing' ewes

The concept of 'flushing', the dynamic effect, has been recognized in sheep farming since at least the nineteenth century; it is generally used to describe feeding conditions in which the ewe is improving in body condition at mating time. Despite its long-standing place in sheep farming, research has still to fully elucidate the neural and hormonal mechanism which can relate factors such as mature skeletal size, tissue fat and protein reserves to ovulation rate (Doney, 1979). It is evident, for example, that there can be certain components of a ewe's diet which can markedly affect ovulation rate with little change in liveweight (Smith *et al.*, 1979). It is now accepted that the quality rather than quantity of the pasture or supplement used in flushing is an important consideration (Rattray *et al.*, 1980). In New Zealand, Smith *et al.* (1983) found the minimum flushing requirement to be three weeks of high-quality pasture feeding immediately prior to ram introduction to obtain a flushing response; this was sufficient to produce an increase of about 20% in lambs born provided most ewes conceived at first mating.

7.3.4. Lupin feeding and ovulation rate

Information on the relationship between nutrition and ovulation rate has come from Australia since the 1970s. It was found then and in later work that relatively short-term (6–9 days) supplementation of the ewe's diet with high-protein lupin grains was capable of increasing ovulation rates in sheep by some 25–30% (Oldham and Lindsay, 1984; Leury *et al.*, 1990); it was also shown that such lupin feeding could increase the number of lambs born. Some workers suggested that such short-term lupin feeding could be integrated to advantage in an oestrus synchronization programme (Cumming, 1976).

Effect of lupin feeding

The effects observed after lupin feeding have been thought by some to be the result of a significant amount of the lupin grains escaping degradation in the rumen (Nottle *et al.*, 1986) or to a higher metabolizable energy intake (Teleni *et al.*, 1984). There have also been suggestions that increases in ovulation rate in response to such short-term lupin feeding may be mediated by increases in FSH secretion late in the luteal phase (Brien *et al.*, 1976; Nottle *et al.*, 1987). Work elsewhere in Australia, however, failed to find evidence of increased FSH secretion (Ritar and Adams, 1988) which led to speculation that the higher ovulation rate stimulated by lupin feeding may be related to an increased sensitivity to FSH rather than to any increase in the rate of secretion of the gonadotrophin.

Body condition of sheep

One interesting fact, as recorded by Lindsay (1976), was that ewes did not have to change in bodyweight when supplemented with lupins to show significant changes in ovulation rate; the response was sudden and dramatic once the ewe settled into its new nutritional regimen and ceased soon after lupin supplement was withdrawn. Other work at a later stage reported by Pearse *et al.* (1994) examined the influence of body condition on ovulation rate after lupin feeding in Merino ewes; they found evidence that lupin supplementation increased ovulation rate regardless of body condition. Studies on ovulation rate and concentrations of gonadotrophic and metabolic hormones in ewes fed lupin grains have been reported by Downing *et al.* (1995); they found that insulin concentration was increased immediately after feeding and was still high 24h later. The sustained increase in insulin level suggested that an increased supply of glucose to the ovarian follicle may mediate nutritionally stimulated increases in ovulation rate. As noted earlier, there has been a suggestion that glucose may influence the way in which dominant follicles emerge (Scaramuzzi, 1994).

Practical implications of lupin feeding

Although Australian researchers have conducted numerous experiments in the 1970s on lupin feeding, the major problem has been one of determining which ewe flocks could be supplemented successfully and what causes the marked variability in lambing performance after such supplementation. In this regard,

Robertson and Hinch (1990) recorded that although ovulation rate in the first cycle was increased by 18% in ewes given lupin supplement (500 g day^{-1} for one week before breeding), the proportion of ewes returning to service in early pregnancy and showing evidence of embryonic mortality was significantly greater than controls.

Net nutritional status

Lindsay (1976) suggested that ovulation rate in ewes was related to what he termed 'net nutritional status', which was taken to be the sum of nutrients available from body reserves and those taken up daily from the digestive tract. According to this, heavy ewes given poor feeding may still show a good ovulation rate because they have a reasonable endogenous source of energy and protein. On the other hand, poor ewes temporarily well-fed will also ovulate well because of the contribution of exogenous nutrients.

'Flushing' ewe-lambs

Earlier indications that 'flushing' prior to mating had no clear effect on ovulation rate in ewe-lambs (Allen and Lamming, 1961) were not borne out by results reported by Keane (1975) in Ireland, who reported a favourable effect of such treatment.

7.3.5. Nutrition and embryo mortality

It is well established that 20–30% of sheep embryos die in the first weeks of pregnancy (Edey, 1969; Kelly, 1984); the factors responsible for such losses have still to be fully explained. It is known that progesterone plays a crucial role in maintaining pregnancy in the ewe (Denamur and Martinet, 1955) and some workers have used exogenous progesterone in attempts to improve embryo survival in sheep. In some instances, increases in pregnancy rates was achieved using progestagen supplements, although not in all. A possible explanation for such variable results was thought to be due to differences in the nutritional status of ewes during the progestagen treatment period.

There is evidence from studies in Australia that nutrition in early pregnancy and peripheral progesterone concentrations may be inversely related (Parr *et al.*, 1982; Williams and Cumming, 1982) Results from further studies reported by Parr *et al.* (1987) demonstrated that sheep fed high energy rations after mating had reduced progesterone levels and showed an increase in embryo mortality; these same authors concluded that exogenous progesterone was only likely to be effective when ewes are fed high energy rations or are in a rising nutritional state after mating.

Aberdeen studies

Elsewhere, the effect of nutrition on early embryonic mortality in sheep was examined by McKelvey and Robinson (1988) at Aberdeen; they concluded that high-plane feeding during the postmating period may adversely affect embryo survival and that this is mediated by way of a decline in plasma progesterone concentrations; low-plane feeding at this time had little effect on embryo survival. As discussed elsewhere, it is believed that high-plane feeding at mating can increase embryo loss as a result of a higher rate of metabolism decreasing progesterone concentrations. It is known that the liver is a major site of progesterone catabolism (Bedford *et al.*, 1974) and that blood flow to the liver of ewes increases with feeding (Bensadoun and Reid, 1962).

Effects of undernutrition

When undernutrition is severe, however, there can be a significant decrease is pregnancy rate in sheep, although this does not appear to be attributable to any inadequacy in corpus luteum function (Abecia *et al.*, 1994). It is also evident that undernutrition of female sheep during prenatal life and early in the postnatal period may reduce lifetime reproductive performance of these animals. Studies reported by Borwick *et al.* (1994) led to the conclusion that undernutrition of the ewe-lamb *in utero* from mating significantly retarded ovarian development in terms of oogonia differentiation.

The successful application of the more recent information on nutrition as it affects reproduction in sheep has been discussed by Robinson (1990); the author notes that this requires a recognition of the overall well-being of breeding ewes and their progeny. He instances examples of nutritional regimens that maximize a response at one physiological state but may be counterproductive to reproduction as a whole (e.g. stimulatory effect of high-plane feeding on ovulation rate, which, if sustained into early pregnancy, is detrimental to embryo survival).

7.3.6. Nutrition and fetal mortality

Work in Australia has shown that nutritional restriction during mid-pregnancy may have a significant effect on fetal mortality. A report by Kelly *et al.* (1989) showed that about 10% of multiparous Merino ewes lost one or both twin fetuses between days 30 and 95 as a result of nutritional restriction (0.3 maintenance) during that period. The authors concluded that such losses in twin-bearing ewes may need to be considered when using ultrasonic scanning to predict the number of lambs born; it also emphasized the need to confirm the viability of each fetus during ultrasound examinations of multiple fetuses.

7.3.7. Lucerne and phyto-oestrogens

Coop and Clark (1960), in a series of trials at Lincoln in New Zealand, found that barrenness in sheep was increased by 2%, multiple births decreased by 10% and the mean lambing date delayed by several days in sheep flushed and mated on lucerne in comparison with those on grass pasture; later work at Lincoln by Coop (1977) showed a depression of 20% in multiples and confirmed the existence of a real problem when lucerne was used. Smith *et al.* (1979) related the decrease in ovulation rate to the level of coumestans in the lucerne; they suggested that the coumestans may have an adverse effect on the release of FSH. Such reports and those of many others indicate the need for care in assessing the problems that may arise from the level of phyto-oestrogens in lucerne.

Adams (1990) reviewed work done over the previous 15 years on permanent infertility in ewes exposed to plant oestrogens; he recorded more than one million ewes in Australia with permanently damaged reproductive tracts after grazing oestrogenic pastures. These oestrogenic effects were evident in the absence of classical clinical 'clover disease' symptoms. It was believed that lesions resulted from an 'organizational' action of oestrogen, which caused a mild sexual transdifferentiation to occur in ewes during adult life, with the main lesion being found in the cervix.

Breed and nutrition interactions

Cumming and Findlay (1977) drew attention to the fact that certain sheep breeds have evolved a more responsive relationship between liveweight and ovulation rate than others; it has also been recorded that some strains and breeds (e.g. Booroola and Romney) may be less responsive than others to nutritional effects (Allison and Kelly, 1979). Elsewhere it was shown that certain breeds may be more responsive to nutritional effects at certain bodyweights; Scottish Blackface ewes were found to be particularly responsive between 40 and 50 kg (Gunn and Doney, 1975).

Effect of frequent blood sampling

For those involved in examining the endocrine mechanisms involved in ovulation rate in sheep, it is as well to note that ewes housed and sampled intensively to measure pulses of LH in plasma have been found to have significantly higher ovulation rates than similar sheep housed outside. This led to studies reported from Australia by Adams *et al.* (1993) in which they showed that frequent collection of blood samples from Merino ewes changed the LH pulse frequency and the mean FSH concentration; this was accompanied by a significantly altered ovulation rate.

7.4. Breed and Age Effects

7.4.1. Breed

According to Cumming and Findlay (1977), the Finnish Landrace was developed by a process of systematic selection, starting more than 70 years ago; the sheep was believed to be derived from the Mouflon unimproved breed. If this can be achieved for Finns, then improved litter size should not be impossible to achieve in other breeds with intense selection. Certainly, in Australia, using such a process of intense selection, dramatic increases in ovulation rate were achieved in one strain of Merino; the prolific Booroola strain apparently arose from a single gene mutation in one flock of Merinos (Bindon, 1984).

The Booroola Merino

Studies in Australia on the endocrine basis of the increased ovulation rate associated with the Booroola Merino suggested that this resulted from gene-specific differences in FSH secretion early in the follicular phase (Bindon *et al.*, 1991); there was no evidence that the ovaries of the Booroola ewes were especially sensitive to FSH. The authors noted that this seemed to be in contrast to data reported for the Finnish Landrace breed, where increased ovulation rate was not attributable to changes in FSH level (Adams *et al.*, 1988).

There is some evidence that different physiological and endocrinological mechanisms may be involved in explaining the high ovulation rates characteristic of breeds with high ovulation rates (Donovan and Hanrahan, 1995).

Chios, Finn and Romanov breeds

Studies in Greece reported by Avdi *et al.* (1988) ranked the Chios breed as a prolific one, similar to Finnish Landrace and Romanov sheep; in contrast to the Booroola, it appeared that the high ovulation rate was not controlled by a single gene.

It is generally accepted, however, that selection for increased litter size can be a slow process, with annual improvements of no more than about two lambs per 100 ewes (Land, 1979). It was observed by Robinson *et al.* (1977) at that time that over the previous decade, the highly prolific Finnish Landrace sheep had been imported directly from Finland by no less than 24 countries in efforts to use the breed as the means of increasing fertility and productivity of new synthetic breeds; much the same was true for the equally prolific Romanov breed which spread rapidly through various countries in Western Europe (Cornu and Cognie, 1985). In France, this breed was used in a long-term selection programme by the National Institute of Agricultural Research (INRA) to produce the INRA 401, a sheep said to be capable of achieving a 200% lambing percentage and of lambing three times in two years.

Development of new prolific breeds

A similar crossing and selection process was adopted in attempts to improve the litter size of the Awassi in the Middle East and North Africa by crossing with the moderately prolific Chios breed from Greece; another breed in North Africa regarded as useful for crossing purposes has been the D'Man. The use of breeds such as the Finn has contributed to the establishment of new synthetic breeds, such as the Cambridge (Owen, 1989), which is capable of a remarkable reproductive performance. In Ireland, imported prolific breeds have been employed in the development of synthetic breeds such as the Belclare Improver.

Finn crossbreds in the tropics

Experiences with Finn sheep in the subtropics were reported by Aboul-Naga (1989), who described crossbreeding trials conducted in various countries in the Mediterranean region. The progeny of Finn sheep crossed with local breeds inherited the high fertility of the Finn breed. However, it was recorded that pure breeding of Finn sheep gave a poorer performance in this climate than in temperate countries. Litter size of crossbred sheep increased by up to 56% and lambing was more frequent than in local breeds.

Breed differences in gonadotrophin output

Although it was shown by some workers that breed differences exist in the LH response of ewe-lambs to oestradiol challenge (Bindon *et al.*, 1974) and GnRH challenge, it was apparent that consistent individual differences did not exist within breeds which could form the basis of a selection process (Tyrrell *et al.*, 1980); it was believed that differences, if they did exist, were masked by several factors, including age, sex and season.

7.4.2. Age

Among factors that are known to have a definite effect on ovulation rate is age; ample evidence exists to clearly show that young ewes tend to have lower ovulation rates and litter sizes than mature ewes at similar liveweights (McKenzie and Terrill, 1937).

7.5. Endocrinological 'Flushing'

There are undoubtedly many sheep farming conditions in which a simple technique for increasing the twinning percentage would be of value to the farmer. As already mentioned, selective breeding, feeding and management and the use of highly prolific sheep breeds all have an important part to play in this. However, there are instances in which the hormonal induction of multiples may be considered, especially in sheep flocks that show a low twinning percentage and are not following any systematic selective breeding policy to try and increase litter size.

7.5.1. Early Work with PMSG

The hormonal induction of multiple births in sheep received much of its early attention in the former USSR, although methods and results were often difficult to interpret; a monograph by Zavadovskii (1941) in particular made it clear that PMSG treatment was being used at that time on some scale, mainly among Karakul sheep. Elsewhere, serious efforts to examine the possibility of augmenting sheep fertility by gonadotrophins date from the studies of

Robinson (1951) at Cambridge. He employed a technique, using PMSG, in which a single dose was administered during the follicular phase (day 12/13 usually) of the ewe's oestrous cycle; the procedure was effective in inducing additional ovulations, with a dose-response relationship evident over the range 500–2000 IU.

The Cambridge work formed the basis of subsequent studies reported from several countries, including Iceland (Palsson, 1956, 1962), North America (Gosset *et al.*, 1965) and the UK (Newton and Betts, 1966; Boaz and Tempest, 1975).

Elsewhere in the UK and Ireland, trials with several thousand sheep showed significant increases in litter size after PMSG doses varying from 250 to 1000 IU (see Table 7.1). Conception rate to first and later services was not affected by the practical success of the application under commercial flock conditions was very much a question of how far perinatal lamb mortality could be minimized, especially in instances of triplets, quadruplets and quintuplets. The most useful practical results were in those lowland flocks normally producing very few twins (e.g. Romney, Southdown); among breeds such as Clun Forst and Suffolk, which usually produced 50% or more twins naturally, little, if any benefit was evident. For farming conditions that are not designed to deal with triplets and higher litter numbers, the birth of such sets may prove an embarrassment rather than an asset (Fig. 7.3).

Reports of applications of this type of treatment in the former USSR, mainly among Karakul sheep, appeared in the literature during the 1950s and 1960s (Zavadovskii, 1957; Amarbaev, 1964). Elsewhere, work in Australia with high- and low-fertility strains of Merinos (Bindon *et al.*, 1971) and in New Zealand with high- and low-fertility strains of Romneys (Smith, 1976) showed

Table 7.1 Use of PMSG administered in the follicular phase of the oestrous cycle to increase litter size in sheep (from Gordon, 1983).

	Treated		Controls		
PMSG dose (IU)	No. ewes	Lambs/ewe	No. ewes	Lambs/ewe	References
500	15	1.67	15	1.47	Robinson (1951)
500	131	1.38	132	1.17	Wallace *et al.* (1954)
1000	94	1.76	103	1.07	
500	20	1.80	20	1.15	Palsson (1956)
750	20	2.05	20	1.20	
250–500	452	1.71	435	1.52	Gordon (1958a)
750–1000	595	1.89	562	1.49	
500	614	1.78	475	1.16	Palsson (1962)
5 IU lb^{-1} bodyweight	57	1.95	59	1.25	Gordon (1967a)

Figures here relate to ewes lambing to first services. In all cases, the method of treatment involved a single subcutaneous injection of PMSG in the follicular phase (days 12/13) of the ewe's natural oestrous cycle. The timing of PMSG administration was by running sterile teaser rams with the ewes beforehand.

Fig. 7.3. Endocrinological flushing of sheep – Romney Marsh ewe with quadruplets born after treatment with PMSG (from Gordon, 1958a).

that the high-fertility animals responded to a greater extent than the low ones to a given dose of PMSG.

7.5.2. Problems with PMSG in the natural cycle

From the sheep farmer's point of view, the PMSG technique involved running vasectomized teaser rams with the flock beforehand in order to work out the appropriate dates for administering the gonadotrophin. The use of sterile teasers and regular checking for heats made the technique difficult and costly to apply in normal farm practice; occasionally, there was the difficulty that ewes could not be bred at their designated time because the first oestrus was taken up with a sterile service.

For this, and other reasons, thoughts turned increasingly towards the possibility of developing a technique in which progestagen was first employed to synchronize oestrus and gonadotrophin administered independently of the oestrous cycle. It was felt that a method of augmenting fertility, simultaneously with oestrus synchronization, could permit the practical benefits of compact lambings and twinning to be usefully combined.

7.5.3. PMSG with oestrus-synchronizing treatments

Early efforts in the mid-1950s by workers in New Zealand involved multiple doses of progesterone over an 8-day period prior to administering PMSG; most ewes were in oestrus within a seven-day period and litter size was 1.5 in the PMSG group and 1.1 in the controls. In the UK, Gordon (1958a, 1963), and in Canada Howell and Woolfitt (1964), employed similar progesterone–PMSG regimes in cyclic sheep with some evidence of an increase in the incidence of multiple births. None the less, the inconvenience of administering progesterone in several doses made the procedure impracticable for commercial application and it was only with the advent of the progestagen-impregnated sponge pessary that the approach was able to receive serious consideration.

In Ireland, studies with Galway sheep showed that FGA-sponge–PMSG treatment (750 IU doses) was effective in the induction of a consistently high litter size under field conditions (Gordon, 1975; Smith, 1977); ample evidence, based on direct examination of the ovaries, was also available to show that the use of PMSG in conjunction with various progestagens (FGA/medroxyprogesterone acetate (MAP)/norgestomet) could markedly increase the number of oocytes shed at the controlled oestrus (Boland *et al.*, 1979).

Although it is only one of several options available for increasing lambing rates in sheep, there would be various opportunities for employing the progestagen–PMSG technique under Irish sheep flock conditions; it is a matter of cost and labour, relative to what is gained by the extra lambs. In New Zealand, Allison (1974) estimated that the application of the progestagen–PMSG technique could be a cheaper way of improving lamb output in low liveweight Romney ewes than by additional pre-mating feeding. In Canada, as part of their sheep research and development work, Ainsworth *et al.* (1977) did report a significant ovulatory response in ewes to a 500 IU dose of PMSG given at the time of FGA-sponge removal. Work in many countries has not produced much information to show that such a procedure can be of practical benefit.

7.5.4. Prostaglandins and PMSG

As one means of synchronizing oestrus in cyclic sheep, prostaglandin (PG) $F_{2\alpha}$ or one of its analogues has been employed, as mentioned elsewhere (see Chapter 3). Studies in which PMSG was administered at the time of the second of two luteolytic doses of prostaglandin have been reported by Boland *et al.* (1978); there was some evidence that a mild superovulatory effect could be achieved by this approach.

7.5.5. Gonadotrophins in the early luteal phase

An alternative to using a single dose of PMSG in the follicular phase of the oestrous cycle was suggested by Cahill and Dufour (1979); they recorded a response when the gonadotrophin was administered during the early luteal phase (day 2). One apparent merit of giving PMSG at day 2 rather than day 12, according to these workers, was an apparent decrease in the variability in ovulatory response; in practical terms, this might mean less likelihood of undesirably high ovulation rates and litter sizes. The treatment was based on evidence obtained by these authors showing a correlation between an FSH peak at the start of the cycle and the number of follicles ovulating at the next heat, period. However, the procedure is open to all the objections previously mentioned in dealing with follicular phase injection.

7.5.6. Use of GnRH

The use of GnRH as a means of controlling follicular development was examined in sheep by Findlay and Cumming (1976); a GnRH analogue was used by these workers to induce gonadotrophin release towards the end of the ewe's oestrous cycle. According to the evidence at the time, when administered about day 12 of the cycle, GnRH increased the mean ovulation rate by more than 20% without sheep releasing more than two oocytes. The method was, in fact, subjected to a large-scale field trial in Western Australia. Work in Ireland at the time, however, suggested that further investigations were necessary before GnRH or its analogues could be considered as useful practical aids in controlling the ovulation rate of the cyclic ewe (Quirke *et al.*, 1979).

Effects of exogenous growth hormone

In attempts to understand the mechanisms influencing ovulation rate in sheep, Gong *et al.* (1994) investigated the effect of recombinant bovine growth hormone (rBST) on follicle development; they found that such treatment significantly increased the small follicle population and concluded that rBST and/or insulin-like growth factor (IGF)-I had a stimulatory effect on the early stage of ovarian folliculogenesis.

Insulin administration

Some workers have examined the effect of insulin administration on the ovulation rate in sheep. Leury *et al.* (1990) injected Merino ewes for 10 days before oestrus with daily doses of 15IU insulin but failed to record any effect on ovulation rate.

7.5.7. Use of a 3β-hydroxysteroid dehydrogenase (3β-HSD) inhibitor (Epostane)

Treatment of animals with various competitive inhibitors of the enzyme, 3β-HSD, has been shown to reduce steroid hormone production in different reproductive tissues of various species, including the sheep (Taylor *et al.*, 1982; Jenkin *et al.*, 1984). In Scotland, Webb (1987) showed that treatment with Epostane induced a significant increase both in ovulation rate and in the number of lambs born per ewe; this appeared to be the first demonstration of an enzyme inhibitor treatment causing an increase in ovulation rate.

It was believed that Epostane treatment interfered with the hormonal feedback equilibrium between the hypothalamus–pituitary gland and the ovaries. The commercial significance of such treatment, however, depended on finding a suitable inhibitor dose to induce an increase in ovulation rate, without the detrimental side-effects of blocking oestrus and ovulation; there was also the question of finding a simple delivery system. Results presented later by Webb *et al.* (1992) confirmed that oral treatment of ewes with Epostane (twice-daily between days 10 and 15 of the oestrous cycle) had a significant effect on follicular steroidogenesis and significantly increased the number of corpora lutea per ewe. The effect on ovulation rate was evidently not caused by increased FSH secretion but may have been due to a reduction in follicular steroid activity or an alteration in the pattern of LH secretion.

7.5.8. *Effect of melatonin treatments*

The effects of melatonin treatments on the reproductive performance of sheep have been reported from several countries. In New Zealand, working with various breeds, Knight *et al.* (1992) recorded that melatonin implants could be used to increase reproductive performance; however, it was concluded that the timing of treatment and ram introduction needed to be optimized for each region and breed. In terms of long-term effects, mating 114–162 days after the start of melatonin treatment resulted in the ewes having a significantly lower pregnancy rate and fewer multiple pregnancies.

7.6. Manipulating Hormonal Feedback Effects

The normal process of ovulation, in sheep as in other mammals, depends on the balance between the stimulatory effects of pituitary gonadotrophins on developing vesicular follicles in the ovaries and the negative feedback effects of hormones such as oestradiol and inhibin (Fig. 7.4). The amount of oestradiol produced by the ovary is known to depend on the number and stage of maturation of the follicles; ewes with higher ovulation rates have more developing follicles and hence secrete more oestradiol. Cumming and Findlay (1977) suggested that the higher ovulation rates associated with the prolific breeds of sheep (Finns and Romanovs) may be the result of decreased sensitivity of the hypothalamic–pituitary axis to the negative feedback effects of oestradiol, allowing gonadotrophin levels to be maintained or enhanced at critical times to support the growth and development of a greater number of vesicular follicles. There are, however, several ways in which it has proved possible to reduce the feedback effects of ovarian steroids and peptides on the pituitary.

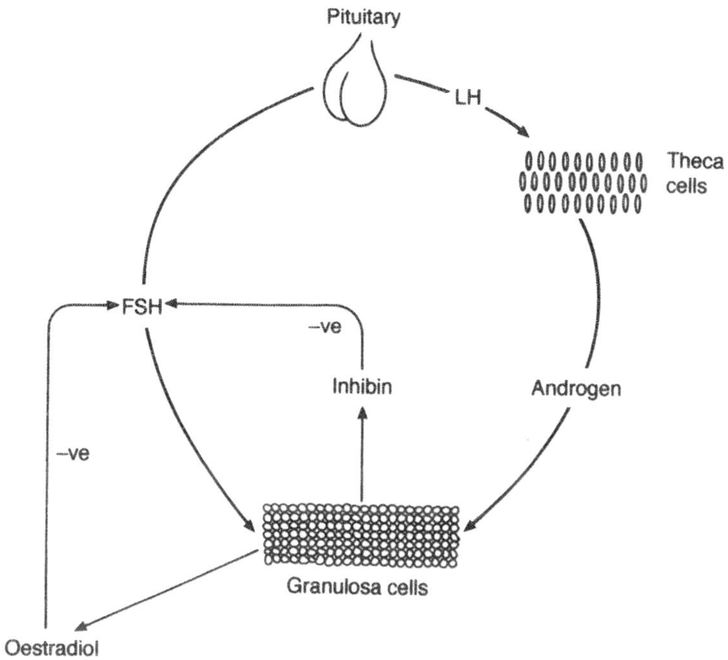

Fig. 7.4. Relationship between pituitary gonadotrophin secretion and negative feedback agents (after Henderson *et al.*, 1984).

7.6.1. Use of anti-oestrogens

One possibility is to reduce feedback action by using either an anti-oestrogen or a very weak oestrogen, which would decrease the inhibitory effects of the endogenous oestrogen produced by the ovarian follicles; the agent clomiphene citrate (an anti-oestrogen) has been one such compound used in human medicine to induce ovulation in certain instances of infertility (Cox, 1975). In Australia, Lindsay and Robinson (1970) used a dose range of 1–90mg clomiphene in ewes but at all dose levels they observed oestrogenic rather than anti-oestrogenic effects; the indications were that these doses were too high. Markedly lower doses (25μg) were employed in the efforts of Land and Scaramuzzi (1979) in sheep of several breeds; responses in the ewes were minor and variable and it was concluded that the agent has no place in controlled breeding applications. An extremely valuable alternative to the anti-oestrogen approach has been to reduce the feedback effects of ovarian oestrogens, androgens and peptides by immunizing the ewe against them.

7.6.2. Immunization against ovarian steroids

Active immunization against potentially important hormones has been widely used in research as a means of investigating their effects, and has been employed in examining a wider range of steroids, peptides and prostaglandins. The ovaries of the ewe are known to secrete at least nine different steroids, including androgens such as testosterone and androstenedione and oestrogens such as oestradiol and oestrone.

The ability of exogenous testosterone to induce oestrus, the release of LH and ovulation in ewes suggested to researchers during the 1970s and earlier that androgens as secreted by the ovary had some biological function; for example, a high proportion of ewes immunized against testosterone became anovulatory without the underlying mechanisms being evident at that time. Active immunization against oestrogens (oestradiol or oestrone) was observed to produce a castration-like effect on pituitary function (Scaramuzzi *et al.*, 1980); in this, basal levels of LH and FSH and the LH pulse frequency increased to levels only observed in ovariectomized ewes.

In New Zealand, Smith *et al.* (1981) showed that an increased ovulation rate after immunization against oestrone was followed by a 27% increase in the number of lambs that were weaned from Coopworth and Romney ewes. The results of studies reported by Scaramuzzi *et al.* (1993) showed that immunization against testosterone produced a high proportion of twins and triplets in Border Leicester × Merino ewes whereas immunity to cortisol was associated with an increased ovulation rate but reduced fertility. The latter effect may have been due to the ability of serum from ewes immunized against cortisol to bind significant amounts of circulating progesterone.

Immunization against androstenedione

Although little was known in the late 1970s of the physiological function of the androgen, androstenedione, ewes immunized against this particular ovarian steroid showed significant increases in ovulation rate (Scaramuzzi *et al.*, 1977). Such evidence was taken to indicate that androstenedione is an active regulator of ovarian activity by way of its feedback action on the hypothalamic–pituitary axis. However, in early investigations, although the ovulation rate was increased, lambing rates were often low because of an increased incidence of barren ewes in immunized flocks (Van Look *et al.*, 1978; Martin *et al.*, 1979). However, Cox *et al.* (1982) reported that problems of anoestrus and increased barrenness has been overcome by controlling the steroid antibody response to a low level by using certain adjuvants.

Development of a commercial vaccine – Fecundin

In late 1983, an anti-androstenedione compound which had been developed under an agreement between CSIRO and Glaxo Australia Pty Ltd was released commercially as Fecundin. In using Fecundin in the first year of treatment, ewes require two injections given at eight- and four-week intervals before the start of matings; in subsequent years, ewes only need a booster injection four weeks before introducing the rams.

An experiment reported by Boland *et al.* (1986) showed that in some situations, the conception rate of Fecundin-treated Merino ewes could be lowered and that this could offset the gains made in litter size; this might have been a factor in the observed variability in field results after use of the Fecundin vaccine observed by Scaramuzzi *et al.* (1983). Croker *et al.* (1987) showed that immunization against either androstenedione or oestrone produced consistent increases in the ovulation rates of Merino ewes in Western Australia. Although liveweight of the ewes at mating did not have a major effect on their performance, nutrition did have an important influence, with the better-fed pregnant ewes having more multiple births (42.4 vs. 14.0%).

Breed effects

Factors responsible for the increased ovulation rate in immunized ewes were examined by Campbell *et al.* (1991), who concluded that plasma FSH concentration was not a determinant of ovulation rate in androstenedione-immune ewes and that an increased LH concentration, or a disturbance of normal intraovarian mechanisms, may be responsible for the increased ovulation rate found in treated sheep. A review by Boland and Crosby (1993) has suggested that there may be differences in the magnitude of the androstenedione antibody response to Fecundin within breeds.

Decreased booster doses to reduce costs

Since the reports from Australia and New Zealand in the early 1980s that lambing percentages may be increased by immunization against androstenedione, using Fecundin (Scaramuzzi *et al.*, 1983), many trials have been reported from countries around the world. In Greece, Alifakiotis (1986) showed that the Fecundin vaccine caused an increase in ovulation rate in dairy breeds of sheep; this increase was not always reflected in an increase in the litter size. In Australia, workers showed that a booster dose of Fecundin lower than that recommended at that time was capable of increasing the reproductive performance of mature-age ewe flocks which had been vaccinated with the full Fecundin dose in previous years (Croker *et al.*, 1988; Sunderman, 1988); the authors noted that the 30% savings in material costs which they achieved could alter the economics of using the vaccine. Although the recommended retail price of Fecundin in Western Australia had been halved since it was first marketed in 1983, the authors felt that further reductions in cost would make it more attractive.

Fecundin examined in many countries

In New Zealand, Henderson *et al.* (1989) examined the effect of active immunization of Romney ewes with follicular fluid on the ovulation rate; their results showed that the rate could be increased but that androstenedione-based immunogens were more effective. An increased lambing performance of ewes after Fecundin treatment was recorded in India by Daroliya *et al.* (1990). In Poland, Malinowska (1989) reported increased litter sizes due to Fecundin treatment of flocks in that country. The possibilities for immunization in French breeds of sheep were discussed by Driancourt *et al.* (1990). In China, Wang *et al.* (1990) recorded increased twinning and lambing rates after Fecundin treatment of Xinjiang Finewool ewes.

Fecundin–nutrition interactions

Croker *et al.* (1990) examined the influence of vaccination with Fecundin and supplementation with lupin grains on lambing outcome in Merino ewes in Western Australia; when lupin and Fecundin treatments were combined, there was no increase in lambs born above that of Fecundin alone. A significant increase in lambs born per ewe was recorded by Willingham *et al.* (1991) in Texas in Fecundin trials conducted over a three-year period with Rambouillet ewes under range conditions; they did emphasize, however, that if such increases were to be exploited, lamb survival rate must be improved. In South America, Sirhan *et al.* (1993) recorded significantly larger litter sizes and lambing rates in Precoce sheep after Fecundin treatment in Chile.

Multi-steroid immunization

Some authors have reported on the use of multi-steroid immunizations. In Australia, for example, Wilson *et al.* (1990, 1992) reported that, in general, treatment with multi-steroid vaccines (androstenedione + testosterone + oestrone) resulted in a slightly better reproductive performance than treatment with the androstenedione vaccines.

7.6.3. Inhibin immunization in sheep

Although it had been widely accepted that the negative feedback effects of ovarian steroids was the means by which pituitary FSH secretion was regulated, in the early 1980s increasing evidence became available to show that the ovary also produces a non-steroidal compound, inhibin, which regulates FSH secretion (Channing *et al.*, 1982). Inhibin was defined as a non-steroidal compound of gonadal origin which specifically or selectively inhibits pituitary secretion of FSH. Studies by Mann *et al.* (1989) in Edinburgh led them to conclude that the rise in FSH secretion which they recorded after injecting inhibin antiserum provided strong evidence that the peptide was an important factor in the regulation of FSH production by the pituitary.

Inhibin effects

Henderson *et al.* (1984) proposed the sequence of events set out in Fig. 7.4 in explaining the way in which inhibin produces its effect. In this, after the initiation of luteal regression, progesterone starts to decrease; this releases the pituitary from the negative feedback effect of progesterone and there is a rise in LH concentration. Together with elevated FSH concentrations, the LH promotes the growth and maturation of the follicles that will eventually ovulate, which respond with an increased secretion of oestradiol. Follicular inhibin production would also be stimulated by androgens, produced by LH action on the theca interna, stimulating granulosa cell inhibin production.

While LH and oestradiol concentrations continue to rise during the follicular phase, FSH level falls during the mid-follicular phase due to the inhibitory effects of oestradiol and inhibin on the pituitary. When the sheep is actively immunized against inhibin, it is believed that this reduces the suppression of FSH during the follicular phase, thereby enabling more follicles to attain ovulatory maturity. Studies continue on the effect of inhibin on FSH concentrations in the ewe (Campbell and Scaramuzzi, 1995; Kusina *et al.*, 1995a,b), in an effort to clarify the role of the peptide in regulating the secretion of gonadotrophins.

Developing inhibin vaccines for commercial use

Much research has been directed towards developing inhibin immunization treatments for commercial use in both sheep and cattle (Glencross, 1992; Forage *et al.*, 1993). From studies conducted in the USA, Meyer *et al.* (1991) concluded that the use of a synthetic fragment of the alpha-subunit of sheep inhibin as a hapten elicited an antibody capable of increasing ovulation rate in sheep. In Ireland, the effect of immunization of ewes against the alpha-1-26 inhibin fragment was reported by Boland *et al.* (1994); the first immunization was given on day 0 and a booster injection one month later. The workers recorded evidence of a higher ovulation rate and litter size than in the untreated controls.

Studies in the UK at Reading have shown that active immunization of ewes against inhibin (bovine alpha 1–29 peptide conjugate) was capable of increasing FSH concentrations, ovulation rate and lambing rate (Fray *et al.*, 1994); they were able to demonstrate that ewes can respond to an inhibin vaccine with a sustained (at least three years) antibody response and a recurrent increase in litter size, without an attendant elevation in plasma FSH. These authors suggest that such inhibin immunization could form the basis of a practical, low input system for promoting a recurrent increase in litter size in the less fecund breeds of sheep. Other studies in that location included an examination of the effects of supplementary treatment with recombinant bovine growth hormone (rBST) on ovulation rate in inhibin-immunized Mule ewes (Tannetta *et al.*, 1995); results showed no evidence that such supplementary treatment enhanced response to inhibin immunization.

Pre-pubertal lambs

In Ireland, both low- (Suffolk cross) and high-ovulation rate (Finn × Dorset) genotypes were actively immunized against an inhibin preparation isolated from bovine follicular fluid (Morris *et al.*, 1991); in both prepubertal sheep types, substantial increases in ovulation rate were achieved.

Passive immunization

Passive immunization against an inhibin peptide has been dealt with by Kusina *et al.* (1995a,b); the antibody was injected intramuscularly 48h prior to CIDR withdrawal and increased ovulation rate in a dose-related manner.

7.7. Litters and Lamb Mortality

In sheep flocks with a high incidence of multiple births, an essential consideration is the whole question of keeping the lambs alive after they are born; there is no point in having more twins and triplets if they fail to survive. Perinatal lamb mortality is a major cause of reduced productivity in sheep, with values of 10%, 15% and even 20% deaths recorded in the literature. In Australia, as recorded by Egan (1984) at that time, it was not unusual to find that 20–25% of pregnant ewes failed to rear a lamb to marketing. Most lamb deaths occur in the first few days of life and it is recognized that the major causes are nutritional, behavioural and physiological rather than infectious (Gordon, 1967).

In general, as lamb birthweight increases within a particular breed type, mortality declines to a minimum for lambs of average or somewhat greater birthweight, and then as a result of difficult births, rises again for the high birthweights.

7.7.1. *Problems associated with multiples*

In a normal flock situation, in which the average litter size is 1.5, the proportion of litters consisting of more than two lambs would be small; with a flock averaging 2.5, on the other hand, there may well be 40% triplets and 15% quadruplets and quintuplets (Robinson *et al.*, 1977) with a consequent increase in the incidence of lamb mortality. Lees (1978), working in Wales with Clun Forest ewes, suggested that two lambs per ewe is what to aim for; he concluded, on the basis of an analysis of the literature, that the efforts of

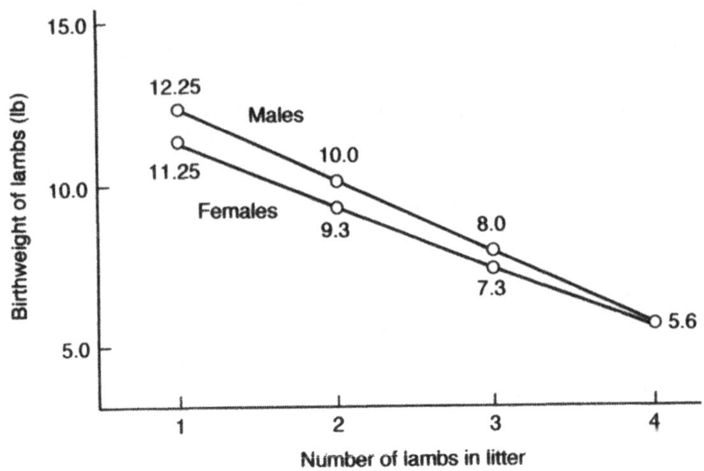

Fig. 7.5. Birthweight in Suffolk-crossbred lambs in relation to sex and litter size (from Gordon, 1958).

workers aiming at much larger litter sizes might be largely self-defeating because of the increases in lamb mortality when prolificacy exceeds two lambs per ewe. The way in which lamb birthweights decrease with increasing litter size is evident from data presented in Fig. 7.5.

In dealing with triplets and higher multiples, the number of lambs reared may actually decline as litter size goes beyond a certain point; there is also the question of the increased costs of feeding, labour and attention, as well as the reduced viability and growth rates of the lambs. It is obvious that management must be geared to deal effectively with litters if the farmer is to gain any advantage from increased litter size.

The results of workers such as Robinson (1979), dealing with highly prolific sheep, have clearly shown that lamb mortality could be kept very low if the management system was appropriate. At the Rowett Institute, it was routine practice to administer, by stomach tube (catheter and syringe) and as

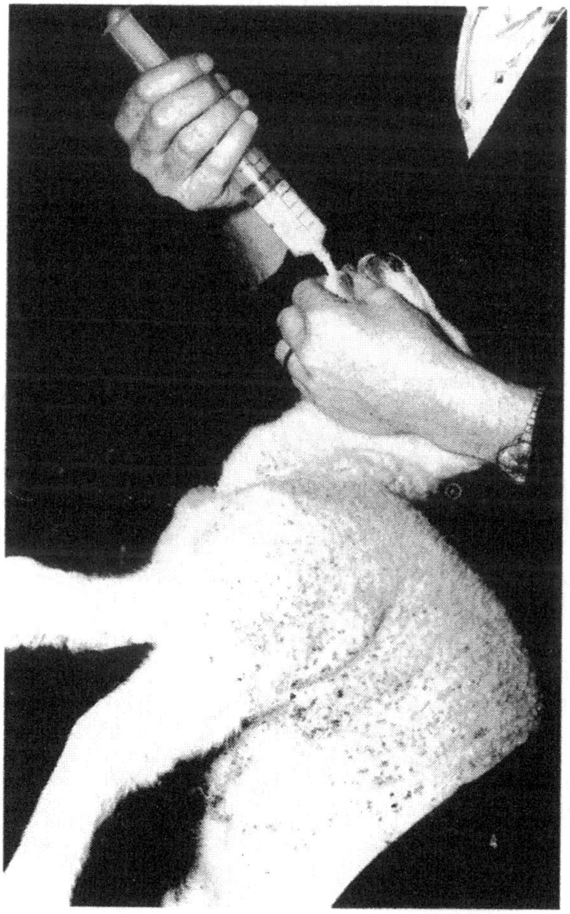

Fig. 7.6. Using the stomach-tube as a means of saving weakly lambs.

soon after birth as possible, about 60 ml of either ewe or cow colostrum to each lamb from a large litter: a quick, easy procedure which reduced perinatal mortality in such litters to a very low figure (Fig. 7.6).

It has been shown, from studies on the anatomical development of the ovine fetus, that the small lamb from a large litter is better developed than its birthweight might otherwise indicate (McDonald *et al.*, 1977); thus, it is not at risk of dying from underdevelopment *per se*, providing it gets colostrum soon after birth. Retained meconium is another situation in weakly lambs, deprived of, or receiving insufficient quantities of colostrum, which Robinson (1981) has alleviated, using a discarded stomach tube to administer an enema.

7.7.2. Large litters and ewe welfare

A large litter can result in some degree of stress in the ewe during late pregnancy. Robinson (1979) showed in the Finn × Dorset ewe that the local lamb weight produced by this prolific sheep is considerably greater in relation to her body size than that which occurs in any other farm animal. As mentioned earlier, for those sheep that are to be used in frequent breeding programmes, the ideal would probably be litters of two lambs rather than more than two. As to the ability of the ewe to deal with large litters in relation to breed, Robinson *et al.* (1977) noted some indication in published figures that lambs from the Romanov and its first crosses may have a lower mortality rate than lambs from the Finn and its crosses.

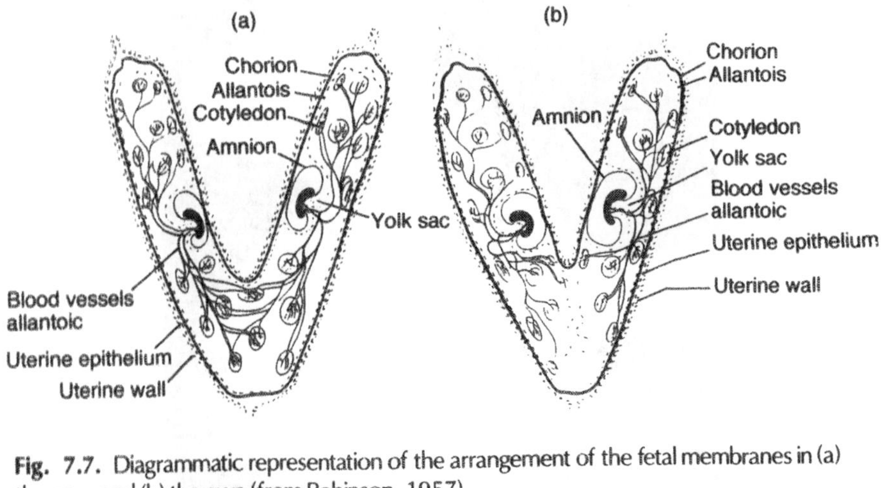

Fig. 7.7. Diagrammatic representation of the arrangement of the fetal membranes in (a) the cow, and (b) the ewe (from Robinson, 1957).

7.7.3. Freemartinism in sheep

The occurrence of the freemartin is normally thought to be restricted to cattle but it is clear that the condition can occur in sheep, and more especially in the highly prolific sheep breeds. Spedding and Dobson (1989) record two British Friesland ewes which, after clinical examination and hormone analysis, were diagnosed as freemartins. As noted by Robinson (1957), in sheep adjacent chorions but not adjacent allantoida normally fuse in instances of twin pregnancy (Fig. 7.7); with larger litter sizes (three and above), there may be a greater tendency for adjacent allantoida to fuse.

7.8. References

Abecia, J.A., Rhind, S.M. and McMillen, S.R. (1994) Effect of undernutrition on luteal function and the distribution of progesterone in endometrial tissue in ewes. *ITEA, Produccion Animal* 90A(2), 63–71.

Aboul-Naga, A.M. (1989) Some experiences with Finn sheep in the subtropics. *FAO Animal Production and Health Paper* No. 74, 136–167.

Adams, N.R. (1990) Permanent infertility in ewes exposed to plant oestrogens. *Australian Veterinary Journal* 67, 197–201.

Adams, N.R., Atkinson, S., Martin, G.B., Briegel, J.R., Boukliq, R. and Sanders, M.R. (1993) Frequent blood sampling changes the plasma concentration of LH and FSH and the ovulation rate in Merino ewes. *Journal of Reproduction and Fertility* 99, 689–694.

Adams, T.E., Quirke, J.F., Hanrahan, J.P., Adams, B.M. and Watson, J.C. (1988) Gonadotrophin secretion during the periovulatory period in Galway and Finnish Landrace ewes and Finnish Landrace ewes selected for high ovulation rate. *Journal of Reproduction and Fertility* 83, 575–584.

Ainsworth, L., Hackett, A.J., Heaney, D.P., Langford, G.A. and Peters, H.F. (1977) A multidisciplinary approach to the development of controlled breeding and intensive production systems for sheep. In: *Management of Reproduction in Sheep and Goats, Sheep Industry Development Program Symposium* (Madison), pp. 101–108.

Alifakiotis, T. (1986) Increasing ovulation rate and lambing percentage by active immunization against androstenedione in dairy sheep breeds. *Theriogenology* 25, 681–688.

Allen, D.M. and Lamming, G.E. (1961) Some effects of nutrition on the growth and sexual development of ewe lambs. *Journal of Agricultural Science, Cambridge* 57, 87–95.

Allison, A.J. (1974) Some techniques for increasing reproductive rates in sheep and their application in the industry. *Proceedings of the New Zealand Society of Animal Production* 34, 167–174.

Allison, A.J. and Kelly, R.W. (1979) Effects of differential nutrition on the incidence of oestrus and ovulation rate in Booroola × Romney and Romney ewes. *Proceedings of the New Zealand Society Animal Production* 39, 43–49.

Amarbaev, A.M. (1964) It is necessary to simplify PMS treatment of ewes. *Ovcevodstvo* 9, 29–31.

Avdi, M., Vergos, E., Alifakiotis, T., Michailidis, I., Driancourt, M.A. and Chemineau, P. (1988) Seasonal variations of oestrous behaviour and ovulation rate in Chios and

Serres ewes in Greece. *Proceedings of the 3rd World Congress in Sheep and Beef Cattle Breeding* (Paris).

Averill, R.L.W. (1959) Ovulatory activity in mature ewes in Otago. *New Zealand Journal of Agricultural Research* 2, 575–583.

Bedford, C.A., Harrison, F.A. and Heap, R.B. (1974) The splachnic, uterine, ovarian and adrenal uptake of progesterone and 20alpha-dihydroprogesterone in the pregnant and non-pregnant sheep. *Journal of Endocrinology* 62, 277–290.

Bensadoun, A. and Reid, J.T. (1962) Estimation of rate of portal blood flow in ruminants; effect of feeding, fasting and anaesthesia. *Journal of Dairy Science* 45, 540–543.

Bindon, B.M. (1984) Reproductive biology of the Booroola Merino sheep. *Australian Journal of Biological Science* 37, 163–189.

Bindon, B.M. and Turner, H.N. (1974) Plasma LH of the prepubertal lamb: a possible indicator of fecundity. *Journal of Reproduction and Fertility* 39, 85–88.

Bindon, B.M., Chang, T.S. and Turner, H.N. (1971) Ovarian response to gonadotrophin by Merino ewes selected for fecundity. *Australian Journal of Agricultural Research* 22, 809–820.

Bindon, B.M., Chang, T.S. and Turner, H.N. (1974) Genetic effects on LH release by oestradiol and Gn-RH in prepubertal lambs. *Journal of Reproduction and Fertility* 36, 477–481.

Bindon, B.M., Blanc, M.R., Pelletier, J., Terqui, M. and Thimonier, J. (1975) Preovulatory gonadotrophin and ovarian steroid changes in French sheep breeds differing in fecundity. Does FSH stimulate follicles? *Proceedings of the Australian Society of Endocrinology* 18, 64.

Bindon, B.M., Piper, L.R., Hillard, M.A., O'Shea, T. and Findlay, J.K. (1991) Endocrine basis of prolificacy in the Booroola Merino. In: *Isotope Aided Studies on Sheep and Goat Production in the Tropics* IAEA Publication, pp. 1–12.

Boaz, T.G. and Tempest, W.M. (1975) Some consequences of high flock prolificacy in an intensive grassland sheep production system. *Animal Production* 20, 219–232.

Boland, M.P. and Crosby, T.F. (1993) Fecundin: an immunological approach to enhance fertility in sheep. *Animal Reproduction Science* 33, 143–158.

Boland, M.P., Gordon, I. and Kelleher, D.K. (1978) The effect of treatment with prostaglandin analogue (ICI-80966) or progestagen (SC-9880) on ovulation and fertilization in cyclic ewes. *Journal of Agricultural Science*, Cambridge 91, 727–730.

Boland, M.P., Kelleher, D.K. and Gordon, I. (1979) Comparison of control of oestrus and ovulation in sheep by an ear implant (SC-2009) or by intravaginal sponge. *Animal Reproduction Science* 1, 275–283.

Boland, M.P., Nancarrow, C.D., Murray, J.D., Scaramuzzi, R.J., Sutton, R., Hoskinson, R.M. and Hazelton, I.G. (1986) Fertilization and early embryonic development in androstenedione-immunized Merino ewes. *Journal of Reproduction and Fertility* 78, 423–431.

Boland, M.P., Sunderland, S.J., Williams, D.H., Kane, M., Headon, D.R. and Roche, J.F. (1994) Effect of immunisation of ewes against × 1-26 inhibin fragment on antibody titres, ovulation and lambing rate. *Animal Reproduction Science* 34, 241–251.

Borwick, S.C., Rhind, S.M. and McMillen, S.R. (1994) Ovarian steroidogenesis and development in foetal Scottish Blackface ewes, undernourished in utero from conception. *Journal of Reproduction and Fertility* Abstract Series No. 14, p. 14.

Brien, F.D., Baxter, R.W., Findlay, J.K. and Cumming, I.A. (1976) Effect of lupin grain supplementation on ovulation rate and plasma follicle stimulating hormone (FSH) concentration in maiden and mature Merino ewes. *Proceedings of the Australian Society Animal Production* 11, 237–244.

Cahill, L.P. and Dufour, J. (1979) Follicular populations in the ewe under different gonadotrophin levels. *Annales de Biologie animale Biochimie Biophysique* 19, 1475–1481.

Cahill, L.P. and Mauleon, P. (1980) Influences of season, cycle and breed on follicular growth rates in sheep. *Journal of Reproduction and Fertility* 58, 321–328.

Cahill, L.P., Mariana, J.C. and Mauleon, P. (1979) Total follicular populations in ewes of high and low ovulation rates. *Journal of Reproduction and Fertility* 5, 27–36.

Campbell, B.K. and Scaramuzzi, R.J. (1995) Effect of acute immunoneutralization of inhibin in ewes during the late luteal phase of the oestrous cycle on ovarian hormone secretion and follicular development during the subsequent follicular phase. *Journal of Reproduction and Fertility* 104, 337–345.

Campbell, B.K., Scaramuzzi, R.J., Evans, G. and Downing, J.A. (1991) Increased ovulation rate in androstenedione-immune ewes is not due to elevated plasma concentrations of FSH. *Journal of Reproduction and Fertility* 91, 655–666.

Channing, C.P., Anderson, L.D., Hoover, D.J., Kolena, J., Osteen, K.G., Pomerantz, S.H. and Tanabe, K. (1982) The role of non-steroidal regulators in the control of oocyte and follicular maturation. *Recent Progress in Hormone Research* 38, 331–408.

Clark, R.T. (1934) The ovulation rate of the ewe as affected by the plane of nutrition. *Anatomical Record* 60, 125–159.

Coop, I.E. (1966) Effect of flushing on reproductive performance of ewes. *Journal of Agricultural Science, Cambridge* 67, 305–321.

Coop, I.E. (1977) Depression of lambing percentage from mating on lucerne. *Proceedings of the New Zealand Society of Animal Production* 37, pp. 149–151.

Coop, I.E. and Clark, V.R. (1960) The reproductive performance of ewes mated on lucerne. *New Zealand Journal Agricultural Research* 3, 922–933.

Cornu, C. and Cognie, T. (1985) The utilization of Romanov sheep in a system of integrated husbandry. In: Land, R.B. and Robinson, D.W. (eds) *Genetics of Reproduction in Sheep*. Butterworths, London, pp. 383–390.

Cox, L.W. (1975) Infertility: a comprehensive programme. *British Journal of Obstetrics and Gynaecology* 82, 2–6.

Cox, R.I., Wilson, P.A., Scaramuzzi, R.J., Hoskinson, R.M., George, J.M. and Bindon, B.M. (1982) *Proceedings of the Australian Society of Animal Production* 14, 511–514.

Croker, K.P., Cox, R.I., Johnson, T.J. and Wilson, P.A. (1987) The immunization of ewes against steroids as a means of increasing prolificacy in a Mediterranean environment. *Animal Reproduction Science* 13, 45–60.

Croker, K.P., Hohns, M.A., Brown, G.A. and Sunderman, F.M. (1988) Effect of reducing the boost dose of Fecundin on the reproductive performance of previously vaccinated ewes. *Australian Veterinary Journal* 65, 163–164.

Croker, K.P., Johns, M.A., Bell, S.H., Brown, G.A. and Wallace, J.F. (1990) The influence of vaccination with Fecundin and supplementation with lupin grain on the reproductive performance of Merino ewes in Western Australia. *Australian Journal of Experimental Agriculture* 30, 469–476.

Cumming, I.A. (1976) Synchronization of ovulation. *Proceedings of the International Congress on Sheep Breeding* (Muresk), pp. 430–448.

Cumming, I.A. (1977) Relationships in the sheep of ovulation rate with liveweight, breed, season and plane of nutrition. *Australian Journal Experimental Agriculture and Animal Husbandry* 17, 234–241.

Cumming, I.A. and Findlay, J.K. (1977) Evolution of ovarian function in sheep and

cattle. *Reproduction and Evolution, 4th International Symposium Biology of Reproduction* (Canberra), pp. 225–233.
Daly, P.J. (1966) *Sheep Husbandry Survey in Kilmaine, County Mayo.* An Foras Taluntais, Dublin.
Daroliya, R.K., Verma, S.K. and Chandolia, R.K. (1990) Increasing lambing performance of ewes by active immunization with polyandroalbumin. *Theriogenology* 33, 1143–1150.
Denamur, R. and Martinet, J. (1955) Effets de l'ovariectomie chez la brebis pendant la gestation. *C.r. Seanc. Soc. Biol.* 149, 2105–2107.
Doney, J.M. (1979) Nutrition and the reproductive function in female sheep. In: *The Management and Diseases of Sheep.* Commonwealth Agricultural Bureau, Slough, pp. 152–160.
Doney, J.M., Gun, R.G. and Smith, W.F. (1976) Effects of pre-mating environmental stress on oestrus and ovulation in sheep. *Journal of Agricultural Science, Cambridge* 87, 127–132.
Donovan, A. and Hanrahan, J.P. (1995) Endocrine patterns during the luteal phase of the oestrus cycle in Finn ewes genetically selected for ovulation rate and in Cambridge ewes. *Proceedings of the Irish Grassland and Animal Production Association* (21st Meeting), pp. 65–66.
Downing, J.A. Scaramuzzi, R.J. and Joss, J. (1990) Infusion of branched chain amino acids will increase ovulation rate in the ewe. *Proceedings Australian Society Animal Production* 18, 472.
Downing, J.A., Joss, J., Connell, P. and Scaramuzzi, R.J. (1995) Ovulation rate and the concentrations of gonadotrophic and metabolic hormones in ewes fed lupin grain. *Journal of Reproduction and Fertility* 103, 137–145.
Driancourt, M.A., Philipon, P., Terqui, M., Molenat, G., Mirman, B., Louault, C., Avdi, M., Folch, J. and Cognie, Y. (1990) Possibilities for immunization against steroids to improve ovulation rate and litter size in sheep and goats. *Productions Animales* 3, 31–37.
Dufour, J., Cahill, L.P. and Mauleon, P. (1979) Short and long-term effects of hypophysectomy and unilateral ovariectomy on ovarian follicular populations in sheep. *Journal of Reproduction and Fertility* 57, 301–309.
Edey, T.N. (1969) Prenatal mortality in sheep: a review. *Animal Breeding Abstracts* 37, 173–190.
Egan, A.R. (1984) Nutrition for reproduction. In: Lindsay, D.R. and Pearce, D.T. (eds) *Reproduction in Sheep.* Cambridge University Press, Cambridge, pp. 262–268.
Fahmy, M.H. (ed.) (1996) *Prolific Sheep.* CAB International, Wallingford, 560 pp.
Findlay, J.K. and Cummin, I.A. (1976) Increase in ovulation rate in sheep following administration of an LH-RH analogue. *Biology of Reproduction* 15, 115–117.
Forage, R.G., Tsonis, C.G., Brown, R.W., Hungerford, J.W., Greenwood, P.E. and Findlay, J.K. (1993) Inhibin: a novel fecundity vaccine. In: Beh, K.J. (ed.) *Animal Health and Production for the 21st Century.* CSIRO, Melbourne, pp. 122–130.
Fray, M.D., Wrathall, J.H.M. and Knight, P.G. (1994) Active immunisation against inhibin promotes a recurrent increase in litter size in sheep. *Veterinary Record* 134, 19–20.
Glencross, R.G. (1992) Effect on reproductive function in sheep and cattle of inhibin immunoneutralization. *Annales de Zootechnie* 41, 287–290.
Gong, J.G., Campbell, B.K., McBride, D. and Webb, R. (1994) The effect of recombinant bovine somatotrophin (rGH) on ovarian follicular populations in mature ewes. *Journal of Reproduction and Fertility* Abstract Series No. 13, pp. 9–10.

Gordon, I. (1958) The hormonal augmentation of fertility in the ewe during the breeding season. *Journal of Agricultural Science, Cambridge* 50, 123–151.

Gordon, I. (1963) Progesterone–PMS therapy during the breeding season. *Journal of Agricultural Science, Cambridge* 60, 31–42.

Gordon, I. (1967) Aspects of reproduction and neonatal mortality in ewe lambs and adult sheep. *Journal of the Department of Agriculture and Fisheries* (Dublin) 64, 76–130.

Gordon, I. (1975) Oestrus synchronization in sheep and its application in practice. *Proceedings of a Symposium on Detection and Control of Breeding Activity in Farm Animals*, Aberdeen University Publication, pp. 40–54.

Gordon, I. (1983) *Controlled Breeding in Farm Animals.* Pergamon Press, Oxford, pp. 181–195.

Gosset, J.W., Kiracofe, G.H., Graham, P.P. and Baker, B. (1965) Effects of equine gonadotrophin on ewe reproductivity. *Technical Bulletin Virginia Agricultural Station* No. 164.

Gunn, R.G. and Doney, J.M. (1975) The interaction of nutrition and body condition at mating on ovulation rate and early embryo mortality in Scottish Blackface ewes. *Journal of Agricultural Science, Cambridge* 79, 19–25.

Gunn, R.G., Doney, J.M. and Smith, W.F. (1979) The effect of time of mating on ovulation rate and potential lambing rate of Greyface ewes. *Animal Production* 29, 277–282.

Hammond, J. Jr (1944) On the breeding season in the sheep. *Journal of Agricultural Science, Cambridge* 34, 97–105.

Heape, W. (1899) Abortion, barrenness and fertility in sheep. *Journal of the Royal Society of England* 10, 217–220.

Henderson, K.M., Franchimont, P., Lecomte-Yerna, M.J., Charlet-Renard, Ch., Hudson, N., Ball, K. & McNatty, K.P. (1984) Ovarian inhibin: a hormone with potential to increase ovulation rate in sheep. *Proceedings of the New Zealand Society of Animal Production* 44, 29–31.

Henderson, K.M., Ellen, R.L., Weaver, A., Ball, K. and McNatty, K.P. (1989) Effect of active immunization with follicular fluid on ovulation rates in Romney ewes. *Proceedings of the New Zealand Society of Animal Production* 49, 97–101.

Howell, W.E. and Woolfitt, W.C. (1964) Hormonal control of estrus and its effect on fertility in cycling ewes. *Canadian Journal of Animal Science* 44, 195–199.

Jenkin, G., Gemmell, R.T. and Thorburn, G.D. (1984) Induction of transient functional luteolysis in cyclic sheep by a 3β-hydroxysteroid dehydroxgenase inhibitor. *Journal of Endocrinology* 100, 61–66.

Johansson, I. and Hansson, A. (1943) The sex ratio and multiple birth in sheep. *Annals Agricultural College of Sweden* 11, 145–171.

Keane, M.G. (1975) Effect of nutrition and dose level of PMS on oestrous response and ovulation rate in progestagen-treated non-cyclic Suffolk × Galway ewe lambs. *Journal of Agricultural Science, Cambridge* 84, 507–511.

Kelly, R.W. (1984) Fertilization failure and embryonic wastage. Lindsay, D.R. and Pearce, D.T. (eds) *Reproduction in Sheep.* Cambridge University Press, Cambridge, pp. 127–133.

Kelly, R.W., Wilkins, J.F. and Newnham, J.P. (1989) Fetal mortality from day 30 of pregnancy in Merino ewes offered different levels of nutrition. *Australian Journal of Experimental Agriculture* 29, 339–342.

Knight, T.W., Muir, P.D., Smith, J.F., Scales, G.H. Reid, T.C., McPhee, S.R. and Staples, L.D. (1992) Effects of long-acting melatonin implants on the reproductive performance of Corriedale, Borderdale, Romney, Coopworth and Perendale ewes in New Zealand. *New Zealand Journal of Agricultural Research* 35, 185–193.

Kusina, N.T., Meyer, R.L., Carlson, K.M. and Wheaton, J.E. (1995a) Effects of passive immunization of ewes against an inhibin-peptide on gonadotrophin levels, ovulation rate, and prolificacy. *Biology of Reproduction* 52, 878–884.

Kusina, N.T., Meyer, R.L., Carlson, K.M. and Wheaton, J.E. (1995b) Passive immunization of ewes against an inhibin-like peptide increases follicle-stimulating hormone concentrations, ovulation rate, and prolificacy in spring-mated ewes. *Journal of Animal Science* 73, 1433–1439.

Land, R.B. (1979) Genetic and physiological variation in reproductive performance. In: *Management and Diseases of Sheep.* Commonwealth Agricultural Bureau, Slough, pp. 114–123.

Land, R.B. (1980) Genetic control of ovulation rate. *Proceedings of the 9th International Congress Animal Reproduction and AI* (Madrid), 2, 63–70.

Land, R.B. and Scaramuzzi, R.J. (1979) A note on the ovulation rate of sheep following treatment with clomiphene citrate. *Animal Production* 28, 131–134.

Lees, J.L. (1978) Functional Infertility in sheep. *Veterinary Record* 102, 232–236.

Leury, B.J., Murray, P.J. and Rowe, J.B. (1990) Effect of nutrition on the response in ovulation rate in Merino ewes following short-term lupin supplementation and insulin administration. *Australian Journal of Agricultural Research* 41, 751–759.

Lindsay, D.R. (1976) The usefulness to the animal producer of research findings in nutrition on reproduction. *Proceedings of the Australian Society Animal Production* 11, pp. 217–224.

Lindsay, D.R. and Robinson, T.J. (1970) The action of clomiphene in the ewe. *Journal of Reproduction and Fertility* 23, 277–283.

McDonald, I., Wenham, G. and Robinson, J.J. (1977) Studies on reproduction in prolific ewes. 3. The development in size and shape of the foetal skeleton. *Journal of Agricultural Science, Cambridge* 89, 373–391.

McKelvey, W.A. and Robinson, J.J. (1988) The use of reciprocal embryo transfer to separate the effects of pre- and post-mating nutrition on embryo survival and growth of the ovine conceptus. *Proceedings of the 11th International Congress Animal Reproduction and AI* (Dublin) 2, Paper No. 176.

McKenzie, F.F. and Terrill, E. (1937) Estrus, ovulation and related phenomena in the ewe. *Research Bulletin of the Missouri Agricultural Experiment Station* No,. 264.

Malinowska, T. (1989) The use of Fecundin to improve fertility in sheep. *Przeglad Hodowlany* 57(16), 30–31.

Mann, G.E., Campbell, B.K., McNeilly, A.S. and Baird, D.T. (1989) Passively immunizing ewes against inhibin during the luteal phase of the oestrous cycle raises the plasma concentration of FSH. *Journal of Endocrinology* 123, 383–391.

Martin, G.B., Scaramuzzi, R.J., Cox, R.I. and Gheradi, P.B. (1979) Effects of active immunization against androstenedione or oestrone on oestrus, ovulation and lambing in Merino ewes. *Australian Journal of Experimental Agriculture and Animal Husbandry* 19, 673–678.

Merrell, B.G. (1990) The effect of duration of flushing period and stocking rate on the reproductive performance of Scottish Blackface ewes. In: *New Developments in Sheep Production* (BSAP Symposium, Malvern), pp. 138–141.

Meyer, R.L., Carlson, K.M., Rivier, J. and Wheaton, J.E. (1991) Antiserum to an inhibin alpha-chain peptide neutralizes inhibin bioactivity and increases ovulation rate in sheep. *Journal of Animal Science* 69, 747–754.

Mitchell, L.M., King, M.E., Aitken, R.P. and Wallace, J.M. (1995) The effect of

management history and time of mating on ovulation and lambing rates in Mule ewes. *Proceedings of the British Society of Animal Science* (Winter Meeting), Paper 95.
Mitchell, L.M., King, M.E., Aitken, R.P. and Wallace, J.M. (1996) Effect of mating season and body condition on ovulation, fertilization and pregnancy rates in crossbred ewes. *Theriogenology* 45, 293.
Morris, D.G., McDermott, M.G. and Sreenan, J.M. (1991) Effect of immunizing prepubertal lambs of low and high ovulation rate genotypes with inhibin partially purified from bovine follicular fluid. *Theriogenology* 35, 339–350.
Newton, J.E. and Betts, J.E. (1966) Factors affecting litter-size in the Scotch Half-Bred ewe. I. Treatment with PMS and progesterone. *Journal of Reproduction and Fertility* 12, 167–172.
Newton, J.E., Betts, J.E. and Wulde, R. (1980) The effect of body condition and time of mating on the reproductive performance of Masham ewes. *Animal Production* 30, 253–260.
Nottle, M.B., Armstrong, D.T., Setchell, B.P. and Seamark, R.F. (1986) Lupin supplementation, FSH secretion and ovulation rate in the ewe. *Proceedings of the Australian Society of Reproductive Biology* 18, 49.
Nottle, M.B., Setchell, B.P. and Seamark, R.F. (1987) Short-term supplementation with lupin gain increases FSH in the ovariectomized, oestradiol-implanted ewe. *Proceedings of the Australian Society Reproductive Biology* 19, 37.
Oldham, C.M. and Lindsay, D.R. (1984) The minimum period of intake of lupin grain required by ewes to increase their ovulation rate when grazing dry summer pasture. In: Lindsay, D.R. and Pearce, D.T. (eds) *Reproduction in Sheep*. Cambridge University Press, Cambridge, pp. 274–275.
Owen, J.B. (1969) *Sheep Production; View of the Future*. Seale-Hayne Agricultural College Publication, Newton Abbot, Devon.
Palsson, H. (1956) Augmentation of fertility of Iceland ewes. *Proceedings of the 3rd International Congress of Reproduction and AI* (Cambridge) 1, 112–115.
Palsson, H. (1962) Augmentation of fertility in Iceland ewes with PMS in successive years. *Journal of Reproduction and Fertility* 3, 55–60.
Parr, R.A., Cumming, I.A. and Clarke, I.J. (1982) Effects of maternal nutrition and plasma progesterone concentrations on survival and growth of the sheep embryo in early gestation. *Journal of Agricultural Science, Cambridge* 98, 39–46.
Parr, R.A., Davis, I.F., Fairclough, R.J. and Miles, M.A. (1987) Overfeeding during early pregnancy reduces peripheral progesterone concentration and pregnancy rate in sheep. *Journal of Reproduction and Fertility* 80, 317–320.
Pearse, B.H.G., McMeniman, N.P. and Gardner, I.A. (1994) Influence of body condition on ovulatory response to lupin (*Lupinus angustifolius*) supplementation of sheep. *Small Ruminant Research* 13, 27–32.
Quirke, J.F., Jennings, J.J., Hanrahan, J.P. and Gosling, J.P. (1979) Oestrus, time of ovulation, ovulation rate and conception rate in progestagen-treated ewes given Gn-RH analogues and gonadotrophins. *Journal of Reproduction and Fertility* 56, 479–488.
Rattray, P.V., Jagusch, K.T., Smith, J.F., Winn, G.W. and Maclean, K.S. (1980) Flushing responses from heavy and light ewes. *Proceedings of the New Zealand Society of Animal Production* 40, pp. 34–37.
Rhind, S.M., McMillen, S.R., McKelvey, W.A.C., Rodriguez-Herrejon, F.F. and Mcneilly, A.S. (1989) Effect of the body condition of ewes in the secretion of LH and FSH and the pituitary response to gonadotrophin-releasing hormone. *Journal of Endocrinology* 120, 497–502.

Rhind, S.M., Goddard, P.J., McMillen, S.R. and McNeilly, A.S. (1993) FSH-stimulated follicle development in ewes in high and low body condition and chronically treated with gonadotrophin-releasing hormone agonist. *Journal of Reproduction and Fertility* 97, 451–456.

Ritar, A.J. and Adams, N.R. (1988) Increased ovulation rate, but not FSH or LH concentrations in ewes supplemented with lupin grain. *Proceedings of the Australian Society for Animal Production* 17, 310–313.

Robertson, J.A. and Hinch, G.N. (1990) The effect of lupin feeding on embryo mortality. *Proceedings of the Australian Society of Animal Production* 18, 544.

Robinson, J.J. (1979) Intensive systems. In: *Management and Diseases of Sheep.* Commonwealth Agricultural Bureau, Slough, pp. 431–446.

Robinson, J.J. (1981) Prenatal growth and development in the sheep and its implications for the viability of the newborn lamb. *Livestock Production Science* 8, 273–281.

Robinson, J.J. (1990) Nutrition in the reproduction of farm animals. *Nutrition Research Reviews* 3, 253–276.

Robinson, J.J., Fraser, C. and McHattie, I. (1977) Development of systems for lambing sheep more frequently than once per year. Technical Publication US Feeds Grains Council, 5–33.

Robinson, T.J. (1951) The augmentation of fertility by gonadotrophin treatment of the ewe in the normal breeding season. *Journal of Agricultural Science, Cambridge* 41, 6–38.

Robinson, T.J. (1957) Pregnancy. *Progress in the Physiology of Farm Animals.* Hammond, J. (ed.) Butterworths, London, Vol. 3, pp. 793–904.

Scaramuzzi, R.J. (1994) The role of metabolic hormones in mediating nutritional influences on ovarian function. *Journal of Reproduction and Fertility*, Abstract Series No. 14, p. 2.

Scaramuzzi, R.J., Davidson, W.G. and Van Look, P.F.A. (1977) Increasing ovulation rate in sheep by active immunisation against an ovarian steroid androstenedione. *Nature* 269, 817–818.

Scaramuzzi, R.J., Martensz, N.D. and Van Look, P.F.A. (1980) Ovarian morphology and the concentration of steroids, and of gonadotrophins during the breeding season in ewes actively immunized against oestradiol-17β or oestrone. *Journal of Reproduction and Fertility* 59, 303–310.

Scaramuzzi, R.J., Gelhard, H., Beels, C.M., Hoskinson, R.M. and Cox, R.I. (1983) Increasing lambing percentages through immunization against steroid hormones. *Wool Technology and Sheep Breeding* 31, 87–97.

Scaramuzzi, R.J., Hoskinson, R.M. and Cogni, Y. (1993) The reproductive performance of Border Leicester × Merino ewes immunized against testosterone and cortisol. *Animal Reproduction Science* 34, 55–68.

Sirhan, A.L., Manterola, B.H., Cerda, A.D. and Mira, J.J. (1993) Reproductive performance of Precoce ewes immunized with Fecundin and flushed. *Advances en Produccion Animal* 18, 39–53.

Smith, J.F. (1976) Selection of fertility and response to PMSG in Romney ewes. *Proceedings of the New Zealand Society of Animal Production* 36, pp. 247–251.

Smith, J.F. (1988) Nutrition and ovulation rate in the ewe. *Australian Journal of Biological Science* 41, 27–36.

Smith, J.F., Jagusch, K.T., Brunswick, L.F.C. and Kelly, R.W. (1979) Coumestans in lucerne and ovulation in ewes. *New Zealand Journal of Agricultural Research* 22, 411–416.

Smith, J.F., Cox, R.I. McGowan, T.L., Wilson, P.A. and Hoskinson, R.M. (1981) Increasing the ovulation rate in ewes by immunization. *Proceedings of the New Zealand Society of Animal Production* 41, 193–197.

Smith, J.F., Jagusch, K.T. and Farquhar, P.A. (1983) The effect of duration and timing of flushing on the ovulation rate of ewes. *Proceedings of the New Zealand Society Animal Production* 43, 13–16.

Smith, P.A. (1977) Studies in the artificial insemination of sheep. Ph.D Thesis, National University of Ireland, Dublin.

Spedding, R.N. and Dobson, H. (1989) Diagnosis of freemartinism in sheep. *Veterinary Record* 123, 18–19.

Sunderman, F.M. (1988) Effect of reducing the boost dose of Fecundin on the reproductive performance of previously vaccinated ewes. *Australian Veterinary Journal* 65, 163–164.

Tannetta, D., Fray, M.D., Wrathall, J.H.M., Bleach, E.C.L., Glencross, R.G. and Knight, P.G. (1995) Effects of supplementary treatment with bovine growth hormone (bST) on hormonal and ovulatory responses to inhibin immunization in the ewe. *Journal of Reproduction and Fertility*, Abstract Series No. 15, p. 22.

Taylor, M.S., Webb, R., Mitchell, M.D. and Robinson, J.S. (1982) Effect of progesterone withdrawal in sheep during late pregnancy. *Journal of Endocrinology* 92, 185–193.

Telini, E., Rowe, J.B. and Croker, K.P. (1984) Ovulation rates in ewes: the role of energy-yielding substrates. In: Lindsay, D.R. and Pearce, D.T. (eds) *Reproduction in the Sheep.* Cambridge University Press, Cambridge, pp. 277–278.

Turnbull, K.E., Braden, A.W.H. and Mattner, P.E. (1977) The pattern of follicular growth and atresia in the ovine ovary. *Australian Journal of Biological Science* 30, 229–241.

Tyrrell, R.N. Starr, B.G., Restall, B.J. and Donnelly, J.B. (1980) Repeatability of LH responses by lambs to monthly challenge with synthetic gonadotrophin releasing hormone (GnRH). *Animal Reproduction Science* 3, 155–160.

Van Look, P.F.A., Clarke, I.J., Davidson, W.G. and Scaramuzzi, R.J. (1978) Ovulation and lambing rates in ewes actively immunized against androstenedione. *Journal of Reproduction and Fertility* 53, 129–130.

Wallace, L.R., Lambourne, L.J. and Sinclair, D.P. (1954) Effect of PMS on the reproductive performance of Romney ewes. *New Zealand Journal of Scientific and Technical Agriculture* 55, 421–425.

Wang, F.R., Li, W.P., Ding, P.J. and Li, D.W. (1990) Study on Fecundin-AAC for improving reproduction in ewes. *Chinese Journal of Animal Science* 26(2), 3–5.

Webb, R. (1987) Increasing ovulation rate and lambing rate in sheep by treatment with a steroid enzyme inhibitor. *Journal of Reproduction and Fertility* 79, 231–240.

Webb, R., Baxter, G., McBride, D. and McNeilly, A.S. (1992) 3β-hydroxysteroid dehydrogenase inhibitor reduces ovarian steroid production but increases ovulation rate in the ewe: interactions with gonadotrophins and inhibin. *Journal of Endocrinology* 134, 115–125.

Wennbom, I. (1994) World lambing record. *Farskotsel* 74(5), 28.

Williams, A.H. and Cumming, A.I. (1982) Inverse relationship between concentration of progesterone and nutrition in ewes. *Journal of Agricultural Science*, Cambridge 98, 517–522.

Willingham, T.D., Shelton, M. and Thompson, P. (1991) The efficacy of fecundin in improving reproduction in Rambouillet ewes. *Sheep Research Journal* 7, 11–15.

Wilson, P.A., Cox, R.I., Wong, M.S.F. and Paull, D.R. (1990) Multi-steroid immunization – a practical treatment to increase fecundity in Merino ewes. *Proceedings of the Australian Society of Animal Production* 18, 432–435.

Wilson, P.A., Cox, R.I., Wong, M.S.F. and Paull, D.R. (1992) Effect of repeated annual multi-steroid immunisation on the reproductive performance of fine-wool Merino sheep. *Proceedings of the Australian Society Animal Production* 19, 188–191.

Zavadovskii, M.M. (1941) Hormonal stimulation of multi-foetation in sheep. *Agiz. Seljhozgiz, Moscow.*

Zavadovskii, M.M. (1957) The effect of treatment with PMS on Karakul ewes, and on the embryonic development and quality of newborn lambs. *Trud. vsesjuz. nauc-issled Inst. Ziovotn* 21, 167–179.

8 Pregnancy Testing in Sheep

8.1. Introduction

A practical and economic method for the early diagnosis of pregnancy in sheep was sought for some time; it was a question of practical importance in sheep farming as well as being of scientific interest in the study of reproduction and fertility in the species. There are several practical advantages in knowing in advance whether sheep are pregnant or not:

1. it can eliminate the cost and labour involved in providing expensive supplementary feed to barren sheep;
2. barren ewes can be culled and sold earlier than otherwise would be possible;
3. meal feeding and attention can be restricted to pregnant sheep as they approach lambing and not wasted on the barren animals.

There is the additional consideration of determining whether the pregnant ewe is carrying a single or multiple fetus(es); this permits the farmer to direct feed and attention to where it is needed most.

8.1.1. Use of the marking harness

The usual method of checking for early pregnancy in sheep in the autumn breeding season is by noting whether the ewe returns in oestrus after a previous breeding. To aid in detecting oestrous sheep, the farmer can employ raddle paste (or block) or a marking harness with crayons (Fig. 8.1), based on the device first described by Radford *et al.* (1960). In changing the colour of crayons, it is important to do this at 14-day rather than at longer intervals; some ewes can show oestrus after an interval of only two weeks and this would be missed if the colour change is left to 16 days.

In ewes that are bred after controlled breeding, where rams have to cover some number of sheep in a relatively short period, it may be advisable to use raddle paste (powder and animal oil) rather than the harness.

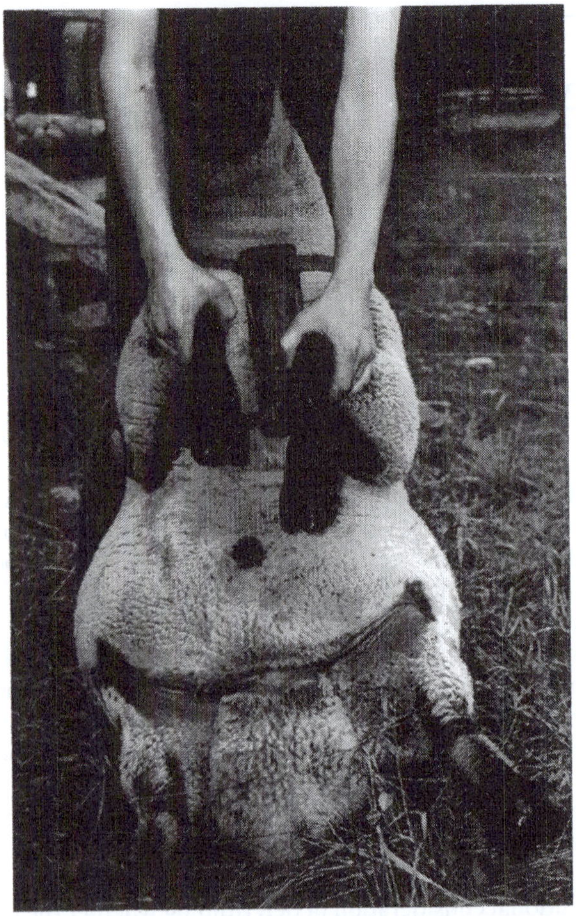

Fig. 8.1. Use of the marking harness in the identification of ewes that fail to become pregnant. Australian workers were the first to come up with the ram marking harness. An alternative – and cheaper – option is to use a mixture of ochre and oil (neat's foot) as a raddle paste.

Method unreliable during sheep anoestrus

In the ewe anoestrus, pregnancy diagnosis on the basis of non-return of oestrus is not always possible and an alternative method may have to be employed. It is a matter of how close to the start of the breeding season the first matings have occurred. For farmers involved in more frequent lambing systems, which may mean some out-of-season matings in early anoestrus, the question of separating out ewes that are non-pregnant in the early weeks after mating and making arrangements for their rebreeding is likely to be important; there is also the need to identify the pregnant ewes as they approach the time for expensive meal feeding in late gestation.

8.1.2. Distinguishing singles from twins

As already noted in the previous chapter (see Section 7.7), the incidence of perinatal mortality in sheep is related to size and birthweight of the lamb; this, in turn, is influenced by litter size and the particular nutritional regime enjoyed by the ewe in late pregnancy (Robinson, 1990). Provision of appropriate amounts of supplementary feed to ewes carrying multiple fetuses can ensure that these reach the maximum size at birth and therefore ensure their survival. On the other hand, reducing the mount of feed provided to single-bearing ewes may help to control birthweights and assist in minimizing lamb losses arising from difficult births (dystocia).

For farmers breeding their own flock replacements, if ewes can be grouped before lambing into those that are carrying twins and triplets and those carrying single lambs, then subsequent selection for natural twinning ability of the ewes could be made that much easier in flocks where no record of individual ewes is maintained. It is worth mentioning that, in carrying out pregnancy testing, it is clearly a great advantage to be dealing with a group of ewes all of which are at known stages of gestation; this may well be possible, where the sheep have been bred after oestrus control measures.

Range of pregnancy testing methods available

A useful review of the literature on pregnancy testing in sheep was the report of Richardson (1972); this author described 24 methods which had been employed up to that time and gave first-hand information on 17 which she had herself checked out. Literature was subsequently reviewed by several other authors in the 1980s (Memon and Ott, 1980; Plant, 1980) and 1990s (Ishwar, 1995). In the present discussion, attention is mainly confined to those methods which may be regarded as sufficiently accurate and feasible to be employed in commercial sheep farming; in such a context, techniques must be simple, inexpensive and accurate.

8.2. Physiology and Endocrinology of Early Pregnancy in the Ewe

Pregnancy in the ewe starts at the instant an oocyte is fertilized and continues through until the fetus, fluids and membranes are expelled at the time of parturition. Prenatal life can be divided into two periods: that covering the development of the embryo and that dealing with the fetus. Although the longest period is that of the fetus, the first period is usually the most critical for the life of the developing organism. During the first period, the growing sheep embryo becomes increasingly dependent on the uterine environment and its secretions for its survival and growth; in turn, the uterine environment undergoes various modifications to provide for the needs of the embryo.

8.2.1. Rapid growth and elongation of the blastocyst

In the embryonic period, the sheep embryo is basically spherical between day 1 and day 10 and then elongates to a characteristic filamentous form by day 12 and day 14 (see Fig. 8.2). By the 15th day of gestation, the sheep blastocyst has grown sufficiently in size to come into contact with the inner surface of the whole of the pregnant horn of the uterus, and by this time discrete areas of trophoblast are closely attached to the caruncular epithelium (Steven and Morriss, 1975). Where there is only a single conceptus, the trophoblast extends into the contralateral horn of the uterus; where there are multiple embryos, they do not overlap but remain aligned end to end.

According to data provided by Sviatko *et al.* (1993) the morphology of the sheep blastocyst at day 13 of pregnancy is related to the concentration of plasma progesterone during formation of the corpus luteum (days 2–6 of the oestrous cycle) but not during the luteal phase (days 7–13). The corpus luteum of the sheep is maintained throughout pregnancy, but all the factors involved in the maintenance of luteal function are not understood. Studies in the USA reported by Jablonka-Shariff *et al.* (1994) showed that ovine corpora lutea from early pregnancy produce fibroblast growth factors; they suggested that these growth factors may play a role in luteal cell proliferation or turnover and may thereby contribute to the maintenance of luteal function.

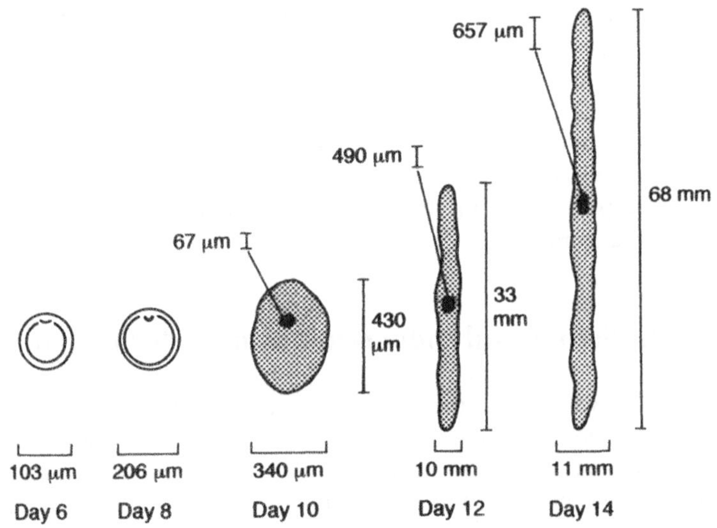

Fig. 8.2. Diagrammatic representation of the growth and development of the sheep embryo.

8.2.2. Growth factors and the sheep embryo

Development of the sheep conceptus involves a series of processes that include cell proliferation, differentiation, migration and invasion. It is clear that there are some number of secretory proteins released by the sheep conceptuses which provide biochemical communication between embryo and mother (Roberts *et al.*, 1994). Much evidence also now exists to support the view that polypeptide growth factors play a central role in the regulation of embryo development. In France, for example, Gharib-Hamrouche *et al.* (1995) identified transforming growth factor-alpha (TGF-α) expressed by ovine embryo and trophoblast at the attachment stage (day 15); they also established that the endometrium is a source of the same growth factor. Data reported by Harbison *et al.* (1995) have demonstrated that the ovine embryo starts to produce immuno- and bioactive TGF-β when attachment is initiated; these authors suggest that because TGF-β neutralizes interferon's (IFN's) antiviral activity and affects expression of cell adhesion molecules, production of this growth factor may determine the fate of sheep embryos during the period of peri-implantation.

8.2.3. Development of placentomes

The non-pregnant sheep uterus contains some 60–150 endometrial thickenings, termed 'caruncles', which are the potential sites for attachment of the allanto-chorion. Attachment takes place at about day 30 after fertilization and usually occupies 70–80% of available caruncles, depending on litter size and other factors. The points of attachment develop into placentomes, a compound structure consisting of the fetal cotyledon and the maternal caruncle. The placentomes provide the main sites of anchorage and of gas and nutrient exchange for the fetus throughout the remainder of gestation. The presence of the embryos prevents luteal regression in the pregnant sheep; a functional corpus luteum is required for at least 60 days of the five-month pregnancy period (Bazer and First, 1983). After that time placental progesterone production is adequate to allow the maintenance of pregnancy.

The placenta

The vascular allanto-chorionic placenta in the ewe is established gradually, with a consequent bringing closer together of fetal and maternal blood vessels. A period of rapid growth of the placenta then occurs, followed in turn by the period of maximum growth of the fetus during which most of the nutrients and metabolic end-products are transferred across the placenta. Finally, there is a period of autolysis of the placenta in late pregnancy, when the fetus draws not only on maternal nutrients but also on the products of autolysis. In the sheep, which usually has one or two young, there is a good deal of fetal fluid remaining at the time of parturition, in contrast to what is found in litter-bearing animals, such as the pig, in which fetal fluids are much less.

8.2.4. Pregnancy recognition in the ewe

It is now well accepted that normal pregnancy in the ewe depends upon the early ovine embryo signalling its presence to the maternal system, a process termed maternal recognition of pregnancy (see Fig. 8.3). It is clear that maternal recognition of pregnancy involves the immune, vascular and endocrine systems; a fuller understanding of the complex relationships between the conceptus and its maternal environment is required if methods are to be developed to minimize the incidence of embryo mortality (Findlay, 1984).

The ewe is one of those mammalian species in which females have uterine-dependent ovarian oestrous cycles; in the absence of a functional endometrium, there is no local endogenous source of prostaglandin (PG) $F_{2\alpha}$, the uterine luteolytic agent that brings about the structural and functional demise of the corpus luteum. The endocrinology responsible for the uterine production of luteolytic amounts of PG is gradually becoming better understood (Ashworth and Bazer, 1989; Bazer *et al.*, 1991; Spencer *et al.*, 1995a,b). It is known that in the cyclic ewe, luteal cells release oxytocin in a pulsatile manner during late dioestrus; oxytocin then binds to its endometrial receptors and initiates luteolytic pulses of $PGF_{2\alpha}$, which is released from the endometrium between days 15 and 17 of the cycle.

Actions of IFN

IFN, also termed interferon tau or ovine trophoblast protein-1, is a Type I IFN secreted by trophoblastic cells of the expanding sheep embryo between days 10 and 21 of pregnancy; this IFN is primarily responsible for inhibiting uterine production of luteolytic amounts of PG, although the mechanisms involved remain unknown (Ott *et al.*, 1995). During early pregnancy, the numbers of endometrial receptors for oxytocin are significantly lower than in the cyclic ewe

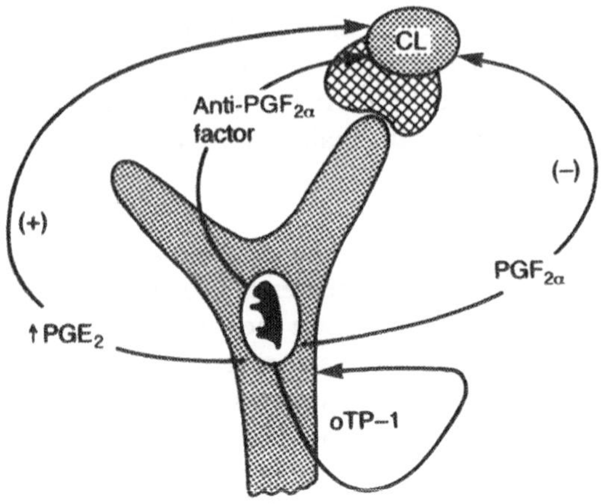

Fig. 8.3. Factors in the maternal recognition of pregnancy in the ewe (from Niswender *et al.*, 1994).

and stimulatory effects of exogenous oxytocin on uterine production of $PGF_{2\alpha}$ are correspondingly reduced or absent. Exogenous oestrogens can, through a uterine-dependent mechanism, stimulate synthesis of endometrial receptors for oxytocin and production of PG, an effect which is significantly attenuated during early pregnancy. It is believed that IFN may exert its effect by inhibiting synthesis of oxytocin receptors. A report by Spencer *et al.* (1995b) proposed that the antiluteolytic effects of ovine IFN are to repress endometrial oestradiol receptor-gene expression, thereby inhibiting formation of oxytocin receptors which prevents the generation of $PGF_{2\alpha}$ pulses.

There are some conditions in sheep in which the maternal recognition of the early embryo may be seriously disturbed. In the early postpartum ewe, for example, Wallace *et al.* (1995) have shown evidence implying that suboptimal conceptus growth rates and secretion of IFN, which result in an inability to regulate endometrial oxytocin receptor-mediated $PGF_{2\alpha}$ secretion, may be central to pregnancy failure in the early postpartum ewe.

Luteal protective agents

It is known that the concentration of PGE_2 in the utero-ovarian blood system increases on days 13 and 14 and it is believed that this prostaglandin may be a luteal protective agent. According to Bazer *et al.* (1991), PGF_2 may initiate and/or accelerate depletion of luteal oxytocin before endometrial receptors for oxytocin are maximal; this may explain pulses of oxytocin with lower amplitude in pregnant ewes. However, as observed by Niswender *et al.* (1994), definitive studies on the role of PGF_2 in maternal recognition of pregnancy in sheep have yet to be reported. Another factor that may be important in sheep is a luteal protective protein(s), secreted by the early ovine embryo, that apparently antagonizes the action of $PGF_{2\alpha}$ (Wiltbank *et al.*, 1992)

8.3. Radiographic Techniques

Radiography has been successfully used in farm animal research programmes for some time; in studies on frequent lambing systems in sheep at the Rowett Institute in Scotland, the technique was used routinely in obtaining an accurate diagnosis of pregnancy and fetal numbers (Wenham and Robinson, 1972). Earlier studies elsewhere had reported accurate X-ray diagnosis in the ewe after the 55th day.

In Australia, there have been those who maintained that if X-ray diagnoses were available in commercial flocks, feed resources available in late pregnancy could be used more efficiently (Rizzoli *et al.*, 1976); they demonstrated that with specially designed handling equipment it was technically feasible to use radiography in sheep on a farm scale to determine fetal numbers with a high degree of accuracy during the fourth month of gestation (90% accuracy at 100–120 days in diagnosing twin pregnancies). Using such a unit under farm conditions, the Australians were able to deal with 400–600 ewes per day; the hope was that the twin-bearing ewes, with the greater nutritional requirements in late pregnancy, could be fed at a more appropriate level and that lambs would be saved by this and by preferential husbandry at lambing.

In New Zealand, Grace *et al.* (1989) reported multiple pregnancy diagnosis in 4700 sheep in 21 flocks using a real-time ultrasonic body scanner or X-rays (video-fluoroscopy) and checked the accuracy of the two methods at lambing time. The percentage of ewes accurately diagnosed for number of lambs by ultrasonics (19 flocks) was 96.1–100% and by X-ray (13 flocks) 94.3–99.6%. The maximum number of ewes that could be tested by each method was 200 for scanning and 250 per hour for X-rays. However, the cost of the X-ray system was calculated as being about 12 times that of ultrasonics. Clearly although technically the two systems gave the same results, the ultrasonic scanner is by far the cheaper.

8.4. Ultrasonic Techniques

Extensive use has been made of ultrasonics in human medicine in dealing with various aspects of pregnancy and assisted reproduction. At this point in time, a wide range of ultrasonic equipment has been used in sheep for pregnancy testing, ranging from simple pulse-echo devices to those capable of providing an instantaneous image of the uterine contents. In farm animals, including sheep, early work involved the use of either the Doppler technique (detection of the fetal pulse) or the A-mode (amplitude-depth) technique (detection of the fluid-filled uterus).

8.4.1. Fetal pulse detection

The Doppler technique utilized ultrasound at frequencies similar to the A-mode method. In this approach, the ultrasound, which was transmitted into the ewe and received back by the transducer after meeting rapidly moving particles (i.e. blood in the fetal heart and umbilical vessels), was slightly shifted in frequency; such sound was converted into an audible signal by the equipment, whereas sound returning from motionless structures had exactly the same frequency as transmitted sound and was not heard.

The Doppler method as used in sheep was reported by various workers in Ireland (Keane, 1969), the UK (Richardson, 1972), the USA (Hulet, 1969), France (Bosc, 1971) and New Zealand (Allison, 1971). The ewe was examined in either a sitting or standing position; the surface of the transducer was smeared with oil (to ensure good contact) and placed on the bare area of skin close to the udder. Such external application of the Doppler technique achieved a high accuracy of diagnosis when employed during the second half of pregnancy. A Japanese study reported by Fukui *et al.* (1986) compared the accuracy at days 80–95 of gestation by the Doppler method and real-time ultrasound scanning; unexpectedly, the Doppler method proved to be the more useful for multiple pregnancy diagnosis.

8.4.2. Rectal probe Doppler

The intra-rectal Doppler technique was reported to be more accurate and enabled determinations of pregnancy in sheep to be made sooner than with an external transducer. According to work reported by Lindahl (1972), covering more than 2000 ewes, pregnancy could be determined at mid-gestation with an accuracy of 90% or more. In the UK Deas (1977) dealt with 1396 sheep by the intra-rectal technique, examining them at 31–40, 41–60, 61–80, 81–100 and 100–120 days after mating; the percentages of correct positive diagnoses were 58, 80, 88, 96 and 97, respectively.

Detection of fluid-filled uterus

Work reported some years ago demonstrated the examination of ewes by A-mode sound to be quick, convenient and simple (Lindahl, 1966; Meredith and Madani, 1980). The ewe was usually dealt with in the standing position; the transducer was smeared with oil and placed on the bare skin of the belly about 50mm in front of the udder on the ewe's right side. When the narrow beam of ultrasound met tissue which had a different acoustic value (e.g. fluid-filled, pregnant uterus), it was reflected at the boundary of the object; the echoes were received by the transducer and converted into signals which were amplified and displayed on a cathode-ray screen or in some other visual form.

The general experience was that the ewe needed to be into the second half of pregnancy if a highly accurate diagnosis was to be made; BonDurant (1980) recorded an accuracy of 91% in sheep greater than 65 days' pregnant and 35% in ewes examined before that time. The speed of testing was much greater than when using Doppler instruments; with good handling facilities, three operators were easily able to test about 160 sheep an hour (Wroth and McCallum, 1979).

8.4.3. Real-time ultrasonics and detection of multiples

A commercial animal 'scanner' for determining backfat and loin areas in pigs and rib-eye areas in cattle was first reported as a possibility in pregnancy testing in sheep by Stouffer *et al.* (1969); subsequently, Lindahl (1976) used such equipment to generate a two-dimensional image of uterine contents. Abdominal scanning as close as possible to the udder resulted in 100% detection of pregnant ewes when carried out after mid-term (day 70 or beyond); an accuracy of 84% was recorded in distinguishing between single and multiple fetuses, but this was not regarded as sufficiently high to warrant the expense and labour involved.

At that time, Plant (1980), with reference to Australian conditions, mentioned one disadvantage of such ultrasonic scanners as the high initial cost of such equipment and the need for maintenance and recharging of batteries if they were to be used in isolated areas where mains electricity was not available. Real-time ultrasonic systems were used transabdominally in sheep and were recorded as being reliable in determining pregnancy and fetal numbers from day 50 after breeding (Fowler and Wilkins, 1984). In the UK, it was clearly shown in the mid-1980s that a high accuracy of diagnosis could be achieved between days 50 and 100 of gestation (White *et al.*, 1984).

According to Johns (1986), although the cost of equipment had shown a three-fold reduction in Australia over a three-year period, the disadvantages at that time still included the high labour costs, which arose from the fact that the ewe had to be thoroughly prepared for examination. However, for sheep farmers looking for an accurate answer on single- and multiple-bearing ewes, the scanner was to prove very successful (Buckrell *et al.*, 1986; Buckrell, 1988). In Spain, Blasco *et al.* (1989) diagnosed pregnancy in 646 Aragon ewes and recorded that the greatest accuracy was achieved for sheep tested on days 30–60 of gestation; position of the fetus and the position of the ewe during the test had a significant effect on accuracy for ewes tested on days 28–51 of gestation.

Fetal losses

Various studies have noted instances in which early diagnosis by scanning may not always be accurate because of fetal losses. Wilkins *et al.* (1982) observed a low level of fetal loss in mid to late pregnancy (60–90 days). Studies reported by Kelly *et al.* (1989), working with multiparous Merino ewes, led them to conclude that fetal losses in ewes bearing twins during periods of nutritional restriction may need to be taken into account when using real-time ultrasonics in predicting the number of lambs; they recorded about 10% of twin-bearing ewes losing one or both fetuses between days 30 and 95 of gestation after feed was restricted. The same authors also emphasized the need to confirm the viability of each fetus during ultrasound scanning.

Real-time ultrasonic scanning has been used by Australian workers to determine the reproductive potential of Merino sheep in New South Wales (NSW), Kilgour (1992) recorded results of commercial scanning in 47,648 autumn-joined and 7846 spring-joined sheep; the author estimated that sheep in NSW were achieving only two-thirds of their potential.

In Denmark, Wootton (1993) recorded that the use of real-time ultrasonic scanning at 80–90 days after mating resulted in pregnancy diagnosis with 100% accuracy and that litter size could be predicted with 98% accuracy. Alan *et al.* (1994) in Turkey have recorded an accuracy rate of 97.8% in diagnosing pregnancy in ewes with a 5 MHz real-time ultrasonic scanner; ewes in which pregnancy was earlier than day 25 were misdiagnosed as negative by the procedure. The authors concluded that such scanning could be usefully employed to avoid the economic losses caused by the slaughter of pregnant sheep.

8.4.4. Estimating the time of conception by scanning

Real-time ultrasonic scanning can be used to estimate the week of conception in sheep, the diagnosis being based on the size of the head and body of the fetus. Working with Merino ewes in Australia, Johns (1993) recorded that he could correctly estimate the week of conception for 71% of the ewes when the examination was conducted 92 days after ram introduction to the flock. In terms of being accurate within 2 weeks, the diagnoses were correct for 91% of the sheep.

8.4.5. Scanning in early pregnancy

The accuracy of real-time ultrasonic scanning in early pregnancy has been examined by several workers. Clearly, the ability to predict pregnancy by day 20 in ewes with a high degree of accuracy would provide an excellent opportunity to study embryonic and/or fetal mortality. In the USA, Garcia *et al.* (1993) showed that pregnancy could be diagnosed by detection of an embryo and by embryonic heartbeat during days 21–34 of gestation; the accuracy was low (52%) on days 17–19, but reached 85% on days 32 and 34. The ability to detect non-pregnant ewes was 80% on days 21–23 and reached 98% by days 32–34. The authors note that before day 24 of gestation, diagnosis in many instances was based on the appearance of the uterine lumen and location of the uterus in relation to the bladder, rather than on detection of the conceptus.

Elsewhere in the USA, Schrick and Inskeep (1993) used transrectal ultrasound scanning to demonstrate that this technique was rapid and accurate for diagnosing early pregnancy. A 7.5 MHz transducer was employed, with the ewes in dorsal recumbency in a tilting squeeze chute. Extraembryonic fluid and membranes were observed in the uterine horns ipsilateral to corpora lutea by day 15 of gestation in all ewes subsequently diagnosed pregnant. Heartbeats within the embryonic vesicles were first detected on days 18 and 19; an accurate count of embryos was not possible until day 25.

8.5. Progesterone and Other Hormone Assays

Early pregnancy diagnosis in the ewe, based on plasma progesterone levels at a certain time after mating, was reported in the late 1960s and early 1970s by workers in several countries, including Ireland (McDonnell, 1974). As in cattle, the test is based on the principle that plasma progesterone concentrations are much lower in the cyclic ewe at oestrus than in the pregnant sheep. In France, early progesterone testing in sheep was employed as a complimentary technique to controlled breeding, especially when such breeding had occurred during the ewe anoestrus. In such work, Thimonier *et al.* (1977) measured progesterone levels at day 18 after breeding; almost all ewes diagnosed non-pregnant failed to lamb, whereas only 84% of those diagnosed pregnant actually gave birth. Robinson (1980), using the 18-day progesterone test in frequent lambing systems, was one who expressed concern at the number of false positives, presumably arising from early embryo mortality; as in cattle, it is essential to regard the progesterone test as no more than an accurate method of detecting non-pregnancy in the early weeks after mating.

In Turkey, Alacam *et al.* (1988) recorded the diagnosis of pregnancy on the basis of progesterone values as determined by radioimmunoassay (RIA); at 15–17 days after mating, a correct diagnosis was made in 85 and 86% of pregnant and non-pregnant ewes, respectively. In Pakistan, Rao *et al.* (1990) concluded that a progesterone level of 3.10 ng ml^{-1} or above on the 16th day after mating was indicative of pregnancy in sheep, using an enzyme immunoassay technique.

8.5.1. Progesterone in milk

Efforts in Israel by Shemesh *et al.* (1979) to base progesterone determinations on milk rather than blood plasma proved successful in dealing with Awassi milking sheep during the natural breeding season (92–100% accuracy in detecting non-pregnancy). During the seasonal anoestrus, however, the accuracy in diagnosing pregnancy was found to be unusually low and it was thought that this may have been due to a higher level of a particular protein in the milk which interfered with the assay.

Ewes at unknown stages of gestation

A progesterone-testing method applicable to situations in which the mating dates for individual ewes was unknown was described by Tyrrell *et al.* (1980); the test, which gave a very accurate diagnosis, involved taking three blood samples from the ewe over a 12-day period. Such procedures would usually be limited to research applications.

8.5.2. Progesterone levels and litter size

Placental production of progesterone is known to increase markedly between days 70–100 of gestation in the ewe, after the placenta has reached its maximum size (Thorburn *et al.*, 1977). In the ewe, unlike the goat, it is possible to make some estimate of fetal number by quantitative progesterone determinations because steroid level is related to litter size. In the UK, Gadsby *et al.* (1972) attempted to classify litter size in prolific sheep by measuring progesterone concentration between days 91 and 105 of gestation; the success rate of 65% was too low for the procedure to have any practical relevance.

Use of on-farm progesterone tests

The use of on-farm tests to evaluate progesterone in sheep blood was reported by Lewis and Young (1992) in the USA; samples were collected from ewes on days 0, 10 and 15 of the oestrous cycle and assayed by RIA or a commercially available progesterone test (Target test). Results showed that the on-farm progesterone test was 100% accurate for detecting ewes with a functional corpus luteum and could be performed simply and rapidly.

8.5.3. Oestrogen tests

Oestrogens are synthesized in the placenta and are present either in the fetal blood or in the maternal circulation. Total oestrogens in the maternal peripheral circulation are known to increase as pregnancy progresses in the ewe. In France, Thimonier *et al.* (1977) tested ewes at 100–110 days of gestation and determined pregnancy with an accuracy of 99%. On the basis of other studies in the same country, it had become evident that there was a relationship between total lamb weight at birth and the concentration of total oestrogens in the maternal blood of sheep. Thimonier *et al.* (1977) found, however, that determining litter size with any reasonable degree of accuracy required prior knowledge of the maternal oestrogen pattern for the particular breed or crossbred examined. Although the same authors found no evidence of a seasonal variation in total oestrogen levels, they did consider that nutritional status of the dam and the genotype of the fetuses themselves may affect maternal oestrogen concentrations.

Studies reported by Worsfold *et al.* (1986) in the UK showed that oestrone sulphate is the major detectable oestrogen found in peripheral plasma of the pregnant ewe and becomes increasingly apparent from around 70 days after mating; levels of unconjugated oestrogens, such as oestrone, remain low until the final days of pregnancy, when they rise markedly. The same authors noted that although oestrone sulphate level varied according to the number of live fetuses, the considerable variation between individual ewes made it unsatisfactory as a predictive tool.

8.5.4. Ovine placental lactogen

Placental lactogenic activity in sheep was demonstrated and an ovine placental lactogen (oPL) purified and characterized some time ago (Chan *et al.*, 1978); the hormone appeared to have mammogenic, lactogenic and growth promoting activities and for this reason was also known as ovine chorionic somatomammotrophin (Kelly *et al.*, 1974). Thimonier *et al.* (1977) in France and Robertson *et al.* (1980a) in Canada described pregnancy tests for sheep based on the detection of this pregnancy-specific agent in the blood after 80 and 57 days of gestation, respectively. Further studies by Robertson *et al.* (1980b) confirmed that an extremely accurate diagnosis of pregnancy and non-pregnancy could be achieved when blood samples were taken later than day 55 of gestation; the incidence of fetal loss beyond that stage of gestation was recorded as being extremely low (< 3%).

In the matter of relating oPL concentrations to litter size, studies reported by Taylor *et al.* (1980) recorded the mean concentrations of the hormone to be 718 ± 227, 1387 ± 160 and 1510 ± 459ng ml^{-1} for single, twin and triplet pregnancies, respectively. However, peak concentrations of oPL were observed at days 130–139 of gestation, which was too late to be of any practical interest.

8.6. Other Pregnancy Detection Methods

8.6.1. *Manual examinations*

Several techniques have been used in late pregnancy in attempts to sort out pregnant from non-pregnant animals. Pratt and Hopkins (1975) reported accuracies of 80–95% in ewes examined by abdominal palpation at 90–130 days of gestation. According to West (1986), this examination may be aided by withholding feed and water for 12–24h beforehand. With the method, the sheep was normally restrained in a sitting position, one hand was placed against the left wide of the ewe's abdomen, the other side being palpated using the fingertips. According to Plant (1980), the fetus could be felt as a floating body that was pushed away and then returned to the fingertips. Although the method is simple and relatively fast (up to 200 ewes per hour), the number of lambs cannot be determined with any accuracy.

8.6.2. *Rectal–abdominal palpation*

A rectal–abdominal palpation technique for pregnancy diagnosis in sheep was first described by Hulet (1972). The method was based on detecting the enlarged pregnant uterus by means of a probe inserted in the rectum. In the hands of an experienced operator, the procedure was reliable after mid-pregnancy and it was possible to deal with 120 or more sheep per year (Plant, 1980). Although the rectal probe was easy to make, was inexpensive and required no maintenance, in the hands of inexperienced personnel it could give rise to serious welfare concern.

8.6.3. *Vaginal biopsy*

The vaginal biopsy method was described by Richardson (1972). It is based on the principle that cell layers in the vaginal wall of the pregnant ewe differ from those that are found in non-pregnant sheep. Non-pregnant ewes have more than ten layers of polygonal and squamous cells whereas pregnant ewes have fewer layers of cells that are usually cuboidal. Sections of vaginal mucosa are taken from the vaginal wall just anterior to the urethral orifice. The stratified squamous epithelium of the non-pregnant ewe is gradually replaced during early pregnancy by cell layers that tend to be cuboidal in shape and which show changes in the nuclei and cytoplasm. The technique apparently had an accuracy of 90% in ewes pregnant for more than 60 days. Although the method was quick and easy to perform, it gave no prediction of single- or multiple-bearing ewes and the fact that the biopsy sample had to be processed in some distant laboratory made it quite unsuitable for commercial use.

8.6.4. Immunological tests

Antigens associated with pregnancy in the ewe were described by Australian workers in the 1970s and elsewhere in that country progress was made towards the characterization of these antigens (Clarke *et al.*, 1980). It was evident that the antigens could be detected in the maternal circulation of the pregnant ewe as early as 24h after mating; it appeared that a factor was released into the maternal circulation soon after fertilization and that it may act by modifying lymphocyte activity so that the early embryo was protected from rejection by the maternal tissues. However, the rosette inhibition test employed to detect the 'early-pregnancy-factor' was too complex for thoughts of routine use.

8.6.5. Pregnancy-specific proteins

The economic benefit of pregnancy detection in ewe-lambs by RIA of pregnancy-specific protein B (PSPB) in blood was studied in work reported by Packham *et al.* (1989) from Idaho. In this, lambs were exposed to a 65-day breeding season, after which blood samples were collected within 30 days and sera assayed for PSPB. Ewe-lambs detected non-pregnant were sold within two weeks of blood sampling and made much higher prices ($106–$112) than untested ewe-lambs sold at a later time when it became evident they were non-pregnant ($26). The workers concluded that a substantial economic advantage resulted from such early pregnancy detecting.

8.6.6. Twin detection by PSPB

A quantitative RIA was developed by Willard *et al.* (1995) in the USA to assay serum concentrations of PSPB during and after pregnancy in ewes carrying single and twin fetuses; the authors recorded that PSPB became detectable 20 days after mating and increased steadily in concentration to day 30. During the period 60–120 days after mating, the PSPB test was 78% correct in predicting twins; this would not be regarded as sufficiently high to be of practical value.

8.7. References

Alan, M., Timurkan, H. and Gulyuz, F. (1994) Pregnancy diagnosis by realtime ultrasonagraphy in ewes. *Turk Veterinerlik ve Hayvancilik Dergisi* 18, 161–163.

Alacam, E., Dinc, D.A., Guler, M., Eroz, S. and Sezer, A.N. (1988) Use of a radioimmunoassay method for early pregnancy diagnosis in ewes after oestrus synchronization with MAP, PMS and GnRH. *Veteriner Fakultesi Dergisi, Selcuk Universitesi* 4(1), 91–98.

Allison, A.J. (1971) Ultrasonics for pregnancy detection. *New Zealand Journal of Agriculture* 123, 25–30.

Ashworth, C.J. and Bazer, F.W. (1989) Interrelationships of proteins secreted by the ovine conceptus and endometrium during the periattachment period. *Animal Reproduction Science* 20, 117–130.

Bazer, F.W. and First, N.L. (1983) Pregnancy and parturition. *Journal of Animal Science* 57 (Suppl. 2), 425–431.

Bazer, F.W., Thatcher, W.W., Hansen, P.J., Mirando, M.A., Ott, T.L. and Plante, C. (1991) Physiological mechanisms of pregnancy recognition in ruminants. *Journal of Reproduction and Fertility* Suppl. 43, 39–47.

Blasco, I., Folch, J. and Echegoyen, E. (1989) Early pregnancy diagnosis and determination of the number of foetuses in sheep by means of ultrasonics. *ITEA, Informacion Tecnica, Economica Agraria* 20(82), 22–31.

BonDurant, R.H. (1980) Pregnancy diagnosis in sheep and goats: field tests with an ultrasound unit. *California Veterinarian* 34, 26–28.

Bosc, M.J. (1971) A study of pregnancy diagnosis in the ewe based on ultrasonics and the doppler effect. *Annales de Zootechnie* 20, 107–110.

Buckrell, B.C. (1988) Applications of ultrasonography in reproduction in sheep and goats. *Theriogenology* 29, 71–84.

Buckrell, B.C., Bonnett, B.N. and Johnson, W.H. (1986) The use of real-time ultrasound rectally for early pregnancy diagnosis in sheep. *Theriogenology* 25, 665–673.

Chan, J.S.D., Robertson, H.A. and Friesen, H.G. (1978) Maternal and fetal concentration of ovine placental lactogen measured by RIA. *Endocrinology* 102, 1606–1613.

Clarke, F.M., Morton, H., Rolfe, B.E. and Clunie, G.J.A. (1980) Partial characterization of early pregnancy factor in the sheep. *Journal of Reproductive Immunology* 2, 151–162.

Deas, D.W. (1977) Pregnancy diagnosis in the ewe by an ultrasonic rectal probe. *Veterinary Record* 101, 113–115.

Findlay, J.K. (1984) Maternal recognition of pregnancy. In: Lindsay, D.R. and Pearce, D.T. (eds) *Reproduction in Sheep*. Cambridge University Press, Cambridge, pp. 105–111.

Fowler, D.G. and Wilkins, J.P. (1984) Diagnosis of pregnancy and number of fetuses in sheep by real time ultrasound imagining. I. Effects of number of fetuses, stage of gestation, operator, and breed of ewe on accuracy of diagnosis. *Livestock Production Science* 11, 437–450.

Fukui, Y., Kobayashi, M., Tsubaki, M., Tetsuka, M., Shimoda, K. and Ono, H. (1986) Comparison of two ultrasonic methods for multiple pregnancy diagnosis in sheep and indicators of multiple pregnant ewes in the blood. *Animal Reproduction Science* 11, 25–33.

Gadsby, J.E., Heap, R.B., Powell, D.G. and Walters, D.E. (1972) Diagnosis of pregnancy and of the number of fetuses in sheep from plasma progesterone concentrations. *Veterinary Record* 90, 339–342.

Garcia, A., Neary, M.K., Kelly, G.R. and Pierson, R.A. (1993) Accuracy of ultrasonography in early pregnancy diagnosis in the ewe. *Theriogenology* 39, 847–861.

Gharib-Hamrouche, N., Chene, N. and Martal, J. (1995) Comparative expression of TGF-alpha and EGF genes in the ovine conceptus and uterine endometrium in the peri-implantation period. *Reproduction, Nutrition and Development* 35, 291–303.

Grace, N.D., Beach, A.D., Quinlivan, T.D. and Ward, B. (1989) Multiple pregnancy diagnosis of using real time ultrasonic body scanner and video-fluoroscopy systems. *Proceedings of the New Zealand Society of Animal Production* 49, 107–111.

Harbison, L.A., Tamura, K., Khan, S., Christenson, R.K. and Imakawa, K. (1995) Conceptus production of immuno- and bioactive transforming growth factor beta (TGF-beta) during peri-implantation period. *Biology of Reproduction* 52, p. 94.

Hulet, C.V. (1969) Pregnancy diagnosis in the ewe using an ultrasonic doppler instrument. *Journal of Animal Science* 28, 44–47.

Hulet, C.V. (1972) A rectal-abdominal palpation technique for diagnosing pregnancy in the ewe. *Journal of Animal Science* 35, 814–819.

Ishwar, A.K. (1995) Pregnancy diagnosis in sheep and goats: a review. *Small Ruminant Research* 17, 37–44.

Jablonka-Shariff, A., Grazui-Bilska, A.T., Reynolds, L.P. and Redmer, D.A. (1994) Cellular proliferation and fibroblast growth factors (FGF) of the corpus luteum (CL) during early pregnancy in ewes. *Journal of Animal Science* 72 (Suppl. 1)/ *Journal of Dairy Science* 77 (Suppl. 1), p. 283.

Johns, M.A. (1986) Pregnancy diagnosis using ultrasound. *Journal of Agriculture of Western Australia* 27(1), 32–35.

Johns, M.A. (1993) Estimation of the week of conception in Merino ewes using real-time ultrasonic imaging. *Australian Journal of Experimental Agriculture* 33, 839–841.

Keane, M.G. (1969) Pregnancy diagnosis in the sheep by an ultrasonic method. *Irish Veterinary Journal* 23, 194–196.

Kelly, P.A. Robertson, H.A. and Friesen, H.G. (1974) Temporal pattern of placental lactogen and progesterone secretion in sheep. *Nature* 248, 435–437.

Kelly, R.W., Wilkins, J.F. and Newnham, J.P. (1989) Fetal mortality from day 30 of pregnancy in Merino ewes offered different levels of nutrition. *Australian Journal of Experimental Agriculture* 29, 339–342.

Kilgour, R.J. (1992) Lambing potential and mortality in Merino sheep as ascertained by ultrasonography. *Australian Journal of Experimental Agriculture* 32, 311–313.

Lewis, G.S. and Young, J.N. (1992) Use of on-farm tests to evaluate progesterone in sheep blood. *Sheep Research Journal* 8, 63–65.

Lindahl, I.L. (1966) Detection of pregnancy in sheep by means of ultrasound. *Nature* 212, 642–643.

Lindahl, I.L. (1972) Early pregnancy detection in ewes by intra-rectal reflection echo ultrasound. *Journal of Animal Science* 34, 772–775.

Lindahl, I.L. (1976) Pregnancy diagnosis in ewes by ultrasonic scanning. *Journal of Animal Science* 43, 1135–1140.

McDonnell, H. (1974) A pregnancy test using progesterone binding protein by a competitive binding technique: a preliminary report of its application to ewes. *Irish Veterinary Journal* 28, 1–10.

Memon, M.A. and Ott, R.S. (1980) Methods of pregnancy diagnosis in sheep and goats. *Cornell Veterinarian* 70, 226–231.

Meredith, M.J. and Madani, M.O.K. (1980) The detection of pregnancy in sheep by

A-mode ultrasound. *British Veterinary Journal* 136, 325–330.

Niswender, G.D., Juengel, J.L., McGuire, W.J., Belifore, C.J. and Wiltbanck, M.C. (1994) Luteal function: the estrous cycle and early pregnancy. *Biology of Reproduction* 50, 239–247.

Ott, T.L., Wiley, A.A., Spencer, T.E., Bertol, F.F. and Bazer, F.W. (1995) Uterine expression of interferon-induced Mx in cyclic and pregnant ewes. *Biology of Reproduction* 52 (Suppl. 1), p. 143.

Packham, J.H., Mitchell, L.A., Smith, G.W., Withers, R.V. and Sasser, R.G. (1989) Economic importance of pregnancy detection in ewe lambs. *Proceedings of the Western Section of the American Society of Animal Science and Canadian Society of Animal Science* 40, 317–319.

Plant, J.W. (1980) Pregnancy diagnosis in the ewe. *World Animal Review* 36, 44–47.

Pratt, M.S. and Hopkins, P.S. (1975) The diagnosis of pregnancy in sheep by abdominal palpation. *Australian Veterinary Journal* 49, 378–380.

Radford, H.M., Watson, R.H. and Wood, G.C. (1960) A crayon and associated harness for the detection of mating under field conditions. *Australian Veterinary Journal* 36, 57–66.

Rao, K.M., Jabbar, M.A. and Naz, N.A. (1990) Early pregnancy diagnosis in the ewes based on plasma progesterone level. *Pakistan Veterinary Journal* 10, 76–77.

Richardson, C. (1972) Pregnancy diagnosis in the ewe. A review. *Veterinary Record* 90, 264–275.

Rizzoli, D.J., Winfield, C.G., Howard, T.J. and England, I.K.H. (1976) Diagnosis of multiple pregnancy in ewes on a field scale. *Journal of Agricultural Science, Cambridge* 87, 67–77.

Roberts, R.M., Kramer, K.K., Xie, S., Duffy, J., Negal, R.J. and Wooding, F.B.P. (1994) Identification of secretory proteins released by preimplantation embryos of the sheep. In: Mastroianni *et al.* (eds) *Gamete and Embryo Quality*. Parthenon Publishing, London, pp. 209–223.

Robertson, H.A., Chan, J.S.D., Hackett, A.J., Marcus, G.J. and Friesen, G.H. (1980a) Diagnosis of pregnancy in the ewe at mid-gestation. *Animal Reproduction Science* 3, 69–71.

Robertson, H.A., Chan, J.S.D. and Friesen, H.G. (1980b) The use of a pregnancy-specific antigen, chorionic somatomammotrophin, as an indicator of pregnancy in sheep. *Journal of Reproduction and Fertility* 58, 279–281.

Robinson, J.J. (1990) Nutrition in the reproduction of farm animals. *Nutrition Research Reviews* 3, 253–276.

Robinson, T.J. (1980) Programmed year-round sheep breeding. *Australian Journal of Experimental Agriculture and Animal Husbandry* 20, 667–673.

Schrick, F.N. and Inskeep, E.K. (1993) Determination of early pregnancy in ewes utilizing transrectal ultrasonography. *Theriogenology* 40, 295–306.

Shemesh, M., Ayalon, N. and Mazor, T. (1979) Early pregnancy diagnosis in the ewe, based on milk progesterone levels. *Journal of Reproduction and Fertility* 56, 301–304.

Spencer, T.E., Mirando, M.A. Ogle, T.F. and Bazer, F.W. (1995a) Ovine interferon tau (oIF-τ) regulates endometrial receptors for estrogen (ER) and oxytocin (OTR), but not progesterone (PR). *Journal of Animal Science* 73 (Suppl. 1), p. 2.

Spencer, T.E., Becker, W.C., George, P., Mirando, M.A., Ogle, T.F. and Bazer, F.W. (1995b) Ovine interferon tau regulates expression of endometrial receptors for estrogen and oxytocin but not progesterone. *Biology of Reproduction* 53, 732–745.

Steven, D. and Morriss, G. (1975) Development of the foetal membranes. In: Steven,

D.H. (ed.) *Comparative Placentation*. Academic Press, London, pp. 58–86.
Stouffer, J.R., White, W.R.G., Hogue, D.E. and Hunt, G.L. (1969) Ultrasonic scanner for detection of single or multiple pregnancy in sheep. *Journal of Animal Science* 29, 104.
Sviatko, M.B., Cardenas, H., McClure, K.E. and Pope, W.F. (1993) The effect of small doses of progesterone on blastocyst morphology in sheep. *Sheep Research Journal* 9, 119–124.
Taylor, M.J., Jenkin, G., Robinson, J.S., Thorburn, G.D., Friesen, H.G. and Han, J.S.D. (1980) Concentrations of placental lactogen in chronically catheterized ewes and fetuses in late pregnancy. *Journal of Endocrinology* 85, 27–34.
Thimonier, J., Bosc, M., Djiane, J., Martal, J. and Terqui, M. (1977) Hormonal diagnosis of pregnancy and number of foetuses in sheep and goats. In: *Management of Reproduction in Sheep and Goats, Sheep Industry Development Program Symposium* (Madison), pp. 79–88.
Thorburn, G.D., Challis, J.R. and Currie, W.B. (1977) Control of parturition in domestic animals. *Biology of Reproduction* 16, 18–27.
Tyrrell, R.N., Gleeson, A.R., Peter, D.A. and Connell, P.J. (1980) Early identification of non-pregnant and pregnant ewes in the field using circulating progesterone concentration. *Animal Reproduction Science* 3, 149–153.
Wallace, J.M., Aitken, R.P. and Cheyne, M.A. (1995) Conceptus interferon in uterine flush, endometrial concentrations of oxytocin receptors and prostaglandin $F_{2\alpha}$ release in vitro after transfer of conceptuses to ewes induced to ovulate at 28 days postpartum. *Journal of Reproduction and Fertility* 103, 299–305.
Wenham, G. and Robinson, J.J. (1972) Radiographic pregnancy diagnosis in sheep. *Journal of Agricultural Science, Cambridge* 78, 233–238.
White, I.R., Russel, A.J.F. and Fowler, D.G. (1984) Real time ultrasonic scanning in the diagnosis of pregnancy and the determination of fetal numbers in sheep. *Veterinary Record* 115, 140.
Wilkins, J.F., Fowler, D.G., Piper, L.R. and Bindon, B.M. (1982) Observations on litter-size and reproductive wastage using ultrasonic scanning. *Proceedings of the Australian Society of Animal Production* 14, p. 637.
Willard, J.M., White, D.R., Wesson, C.A.R., Stellflug, J., Sasser, R.G. (1995) Detection of fetal twins in sheep using a radioimmunoassay for pregnancy-specific protein B. *Journal of Animal Science* 73, 960–966.
West, D.M. (1986) Pregnancy diagnosis in the ewe. In: Morrow, D.A. (ed.) *Current Therapy in Theriogenology*. W.B. Saunders, Philadelphia, pp. 850–852.
Wiltbank, M.C., Wiepz, G.J., Knickerbocker, J.J., Belfiore, C. and Niswender, G.D. (1992) Proteins secreted from the early ovine conceptus block the action of prostaglandin $F_{2\alpha}$ on large luteal cells. *Biology of Reproduction* 46, 475–482.
Wootton, D. (1993) Scanning of pregnant ewes. *Tidsskrift for Dansk Fareavl* 58, 8–9.
Worsfold, A.I., Chamings, R.J. and Booth, J.M. (1986) Measurement of oestrone sulphate in sheep plasma as a possible indicator of pregnancy and the number of viable fetuses present. *British Veterinary Journal* 142, 195–197.
Wroth, R.H. and McCallum, M.J. (1979) Diagnosing pregnancy in sheep – the 'Scanopreg'. *Journal of Agriculture* (Western Australia) 20, 85.

9 Control of Lambing

9.1. Introduction

The more widespread application of controlled breeding techniques, by which a flock of ewes can be bred by natural or artificial insemination as a group rather than as individuals, has made the feasibility of using controlled lambing techniques in practical farming much easier. Clearly, if most ewes in a farmer's flock are at an identical stage of gestation, it is relatively simple to gather the sheep on a given day in late pregnancy and administer an agent which will initiate parturition.

As a follow-on to controlled matings, especially in flocks where a very high conception rate to first service occurred, controlled lambings by an appropriate induction agent may occasionally have practical appeal. In view of the fact that lambs are often born during the inclement weather of late winter and many in night time, controlled lambings could ensure the shepherd being on the spot while the sheep are giving birth, thereby saving more of the lambs and improving labour efficiency during the lambing season.

9.1.1. Practical advantages of controlled lambings

As previously noted (Chapter 6), some intensive lamb production systems have aimed at programmed lambing of artificially-bred oestrus-synchronized ewes during mid-week by initiating parturition in the last week of gestation; by such means, it was possible to ensure a minimum of weekend and holiday work.

In the context of making the most efficient use of labour, ultra-compact synchronized lambing may prove to be an economic proposition in many practical situations. It means that in flocks kept for fat-lamb production, oestrus-synchronized ewes which would normally lamb down within the space of 7–10 days can be further aligned to give birth within a 48h period, with most lambs being born during daytime. Induced parturition in sheep may have some unusual applications; in Karakul flocks, for example, in which the pelt of the unborn fetus is of primary economic importance, premature induction rather than Caesarian section may be employed to advantage.

9.2. Duration of Pregnancy in Sheep

It is well established that the duration of pregnancy in sheep can vary according to several factors; these include the number of lambs in the conceptus, the sex of the lambs, the sire breed, the breed of ewe and its age. Some authors have distinguished certain categories of sheep in terms of the length of their gestation period. These include:

1. the early-maturing improved meat breeds (e.g. Southdown, Suffolk, Hampshire, Dorset Horn) with gestation periods varying from 144 to 147 days;
2. the slow-maturing fine-wool breeds (e.g. Merino, Rambouillet) with periods averaging 149–151 days; and
3. crossbred long-wool breeds (e.g. Columbia, Corriedale) with periods in the intermediate range.

It is generally accepted that the duration of gestation within a particular breed is extremely stable (Forbes, 1967).

In the UK, dealing in the main with long-wool mutton breeds such as the Romney Marsh, an average gestation period of 147.8 days for 515 ewes was recorded (Gordon, 1958a); in Ireland, lowland breeds of ewes, mainly Galways and their crosses, have shown average gestation periods of 148.1 days in single-bearing sheep and 147.0 in twin-bearers (Gordon, 1967). Other authors with other breeds have also recorded about a day's difference in gestation length between single and twin pregnancies. When larger litters are included, there is an obvious decrease in gestation length with increasing litter size, as shown in Table 9.1 for Clun Forest ewes in the UK. In Nigeria, Osinowo *et al.* (1994) has argued that selection for large litter size would be accompanied by a reduction in gestation length.

9.2.1. Season and gestation length

In the UK, there was some suggestion that summer pregnancies were of shorter duration than those occurring in the usual autumn/winter period (Gordon,

Table 9.1. Duration of gestation in relation to litter size in Clun Forest sheep (from Gordon, 1983).

	Single	Twin	Triplet	Quadruplet
No. of ewes	176	164	31	10
Gestation length				
Range	138–153	142–152	142–152	140–150
Mean	147.5	146.8	146.7	144.8

1958b); ewes lambing in the autumn after a summer pregnancy carried their lambs on average for 144.7 days whereas those lambing in spring carried their young for 147.8 days. Elsewhere, it was reported that under a decreasing daylength pattern and low environmental temperatures during late pregnancy, gestation length in Finn-cross ewes was shorter than under the opposite environmental conditions (Amir *et al.*, 1980).

9.2.2. Diurnal distribution of lambing

Although many shepherds may hold the view that the majority of lambings occur at night, reports in the literature regarding the diurnal distribution of lambings in sheep have not revealed any clear pattern. In New Zealand, Wallace (1949a) found that the peak incidence of lambings occurred between 00.00 and 04.00h. A study of Awassi ewes showed lambing occurring at random throughout a 24h period (Younis and El-Gabory, 1978). In Australia, George (1969) found that significantly more Merino ewes lambed at night, whereas more Dorset Horn ewes lambed during the day; the author attempted to reconcile this with the origins of the two breeds.

In the USA, Lindahl (1964) suggested that physical and metabolic activity of ewes could be factors influencing the onset of lambing, since peak times of parturition were preceded by feedings at 0800 and 1400h. The lowest incidence of lambings was observed by Sharafeldin *et al.* (1971) during the period of meal feeding and the indications were that imminent lambings were postponed as a result of increased adrenal activity at the time of feeding.

Effect of feeding and movement

The physical characteristics of diet are known to be important in determining the rate of food intake in sheep and can influence the time ewes spend ruminating. For such reasons, Hudgens *et al.* (1986) in the USA examined the association between form of roughage and time of lambing. However, they recorded that neither long-stem lucerne hay nor finely-chopped lucerne haylage influenced the time the ewes gave birth. They found that peak lambings appeared to correspond to a lull or reduction in routine activities in the sheep barn; feeding time *per se* did not influence the time at which ewes gave birth.

Some authors have examined the effect of sheep movements on the onset of lambings. In New Zealand, for example, Bray and Burton (1988) concluded that moving ewes to a new paddock in the evening had no advantage over moving ewes in the morning in inducing the sheep to lamb in daylight. In Australia, however, Alexander *et al.* (1993a) did find that the birth process was delayed when pregnant Merino sheep were moved slowly from one paddock to another; the same workers recorded a peak in the incidence of lambings during the five-hour period 0900–1400h, when 28% of the sheep lambed.

9.2.3. Sheep under constant illumination

Under intensive management, ewes are typically housed in artificially illuminated pens for the entire lambing period and are under continuous observation. In Aberdeen, Mitchell (1995) investigated the diurnal distribution of lambings under such conditions. Although a previous report had suggested that under such housed conditions, the timing of meal feeding might influence the timing of parturition (Burgess, 1994), no such association was found by the Aberdeen worker. Her results showed that the timing of parturition in intensively managed ewes was: (i) evenly distributed throughout a 24-h period; and (ii) was not influenced by age of dam and litter size.

9.3. Physiology and Endocrinology of Late Pregnancy and Parturition

9.3.1. Prenatal effects on the lamb fetus

Various reports have dealt with the way in which the environment may exert an influence on the lamb during its prenatal life. In the USA, for example, results presented by Stark and Daniel (1989) demonstrated that, despite the fact that the fetal lamb is sequestered from the direct effects of the environment, exposure of the pregnant ewe to constant light disrupts the fetal pacemaker that generates the circadian rhythm of vasopressin in the cerebrospinal fluid. Elsewhere in the USA, Ebling *et al.* (1989) found that serum prolactin concentration in lambs at birth reflected the photoperiodic treatment of their dam. They concluded that the fetal lamb receives and responds to information about daylength *in utero*, and begins developing a seasonal photoperiodic history before birth. In Michigan, Herbosa *et al.* (1995) concluded that a sex difference in the reproductive response to photoperiod exists in prepubertal sheep and that this was due to the action of testicular androgens during fetal development. In Ireland, Sunderland *et al.* (1995) found evidence suggesting that the developing fetal lamb is unable to utilize information about prevailing photoperiod received before birth and that the sequence of changing photoperiod after birth is a more important photoperiodic determinant of puberty in ewe-lambs.

9.3.2. Mid- to late-pregnancy effects

During the second half of pregnancy in the ewe, growth of the fetus predominates, fetal weight increasing twofold during the final month of the gestation period. Heat stress in mid-pregnancy was shown by McCrabb *et al.* (1993) to have a significant effect on placental growth and to influence the birthweight of the lamb; such results were consistent with previous reports in sheep showing that fetal growth retardation was primarily due to reduced placental function. According to Bell (1984), the mechanism by which heat stress reduces placental growth is not known.

Lambings in housed sheep

In Northern Ireland, Black and Chestnutt (1990) examined the influence of shearing regime and grass silage quality on lambings in housed sheep. They showed an average increase in twin-lamb birthweight from ewes shorn at least six weeks before lambing over that of unshorn lambs to be about 1 kg per lamb with no significant effect on ewe liveweight change. They also showed that unshorn ewes had a significantly shorter gestation length (by two days) than those with the wool removed. Elsewhere in the UK, Symonds *et al.* (1990) reported that the shearing of housed ewes prior to lambing enhanced the growth rate of their lambs by as much as 20% during the first 30 days after birth. It was postulated that metabolic adaptations in recently shorn ewes in a cold environment may result in endocrine changes likely to increase the partitioning of nutrients towards milk production. In South Africa, Cloete *et al.* (1994) also found evidence that shearing a few weeks before lambing improved lamb birthweight (by 5%); these workers concluded that shearing prior to lambing was likely to be a cost-effective method of improving the efficiency of winter-lambing flocks in that country.

9.3.3. Initiation of parturition

The late 1960s and early 1970s witnessed a marked increase in knowledge of factors involved in the hormonal regulation of parturition in mammals and much of this was due to studies conducted with the sheep as the experimental model. The classic experiments of Liggins and associates in New Zealand did much to establish the role of the fetal pituitary–adrenal axis in the initiation of parturition in the sheep; ablation of the fetal pituitary was found to result in an indefinite prolongation of pregnancy, whereas infusion of synthetic adrenocorticotrophin (ACTH) or glucocorticoids into the fetus led to premature delivery. It is now well recognized that birth in the sheep is triggered by increasing activity of the fetal hypothalamic–pituitary–adrenal axis characterized by strongly pulsatile ACTH secretion (Howe and Brooks, 1995).

Fetal adrenal activity

The observations of Comline and Silver (1961) that the fetal adrenals markedly increased in size in the final two weeks of gestation and those of Bassett and Thorburn (1969) showing that the plasma concentration and secretion rate of corticosteroids, such as cortisol, also increased at that time, agreed with the view that activation of the fetal pituitary–adrenal axis initiated parturition. It was not clear, however, whether increased cortisol concentrations were due to an increase in the trophic stimulation of the fetal adrenal or to a maturational change in the responsiveness of the fetal adrenals in the presence of a basal level of ACTH. There were lines of evidence indicating that an increase in adrenal sensitivity probably did play a role in the initiation of parturition, although there was doubt about the cause of such changes (Thornburn, 1978).

9.3.4. Production of cortisol

Although stimulation of the fetal adrenal to produce cortisol was originally thought to be a question of ACTH from the fetal pituitary, it subsequently became apparent that other substances, perhaps of pituitary origin, were responsible for the initial stimulation of cortisol secretion by the fetal adrenals and for increasing the sensitivity of the adrenals to ACTH (First, 1979); certainly, the final surge of cortisol was believed to be the result of ACTH action. It was believed that in the ewe, this cortisol stimulated placental enzymes (17-alpha-hydroxylase) which were responsible for the conversion of placental progesterone into oestrogen; this placental oestrogen then acted on a uterus which was no longer under progesterone influence, causing the synthesis of prostaglandin (PG) $F_{2\alpha}$ and PGE_2, both of which are powerful stimulants of uterine activity.

As uterine contractions continue under the influence of PGs, the fetus advances into the relaxed cervix and the anterior vagina, which triggers the reflex release of oxytocin from the posterior pituitary; the action of this peptide strengthens the uterine contractions.

Placental lactogen

It had been suggested that placental lactogen (oPL) may play an important role in the control of pregnancy in the sheep by inhibiting the synthesis of PG and that the rise in fetal cortisol before parturition may switch off oPL secretion (Thornburn, 1978).

Changes in body temperature

Studies reported by Winfield and Makin (1975) showed that body temperature of the ewe drops about 0.5°C during the final 48h before lambing; a decrease to below 39.2°C was proposed as a possible method for selecting which ewes are due to lamb within the next two days.

9.3.5. Changes in the cervix

In the early part of pregnancy in the ewe, as the uterus grows to accommodate the developing conceptus, the cervix forms an essential mechanical barrier. At lambing, however, the cervix becomes soft and distensible to allow the passage of the fetus through the birth-canal. It is known that an active ripening process occurs within the cervix prior to the onset of labour. According to a review by Rice *et al.* (1984), it was uncertain at that time whether a relaxin-type hormone was involved in the process of ovine cervical ripening. Results of a study by Gazal *et al.* (1993) did, however, reveal that an antepartum relaxin surge occurs in sheep four days before normal parturition.

9.4. Induction by the Use of Steroids

There are two ways in which the timing of lambing may be influenced by exogenous hormones: it may be a matter of prolonging gestation or shortening it. In regard to shortening the gestation period, there are known limits to how far this may be taken; Dawes and Parry (1965) concluded, on the basis of their evidence, that lambs of gestational age less than 95% of normal were not of normal viability. In practical terms, if the lambs are to survive, an induction treatment should not be applied earlier than about one week before the average date for lambing in the sheep in question (Fig. 9.1).

Although the administration of progesterone, with its well-known inhibitory action on uterine motility, might possibly be employed to delay parturition, there are few reports of using this approach in controlling lambings. There was one small-scale instance in which oral doses of progestagen (50 mg MAP daily) were employed in the last few days of the gestation to control the initiation of parturition (Garm and Nedkvitne, 1968); most ewes lambed 48h after the last dose of progestagen without evidence of adverse side-effects.

However, most of the reports in the literature have been concerned with using agents to shorten rather than to prolong the gestation period.

Fig. 9.1. Very compact lambings are possible in flocks bred after oestrus control. One farmer in Gloucestershire in the UK with a flock of some 60 Finn × Dorset ewes planned to have all lambings confined to Easter Monday in one particular year – and that is exactly how it turned out. Induced lambings would only have practical appeal at a certain range of flock sizes – and only when a high conception rate to the controlled oestrus had been achieved – entirely possible with Finn × Dorset sheep.

9.4.1. Use of synthetic corticosteroids

As noted earlier, the perfusion of the fetal lamb with ACTH or with cortisol can result in premature parturition. After finding that cortisol was effective in this way, highly potent corticosteroid analogues with glucocorticoid or mineralocorticoid activity were employed as induction agents. Dexamethasone, a glucocorticoid with about 25 times greater potency than cortisol, was demonstrated to be an effective induction agent (Bosc, 1972; Harrison, 1982); elsewhere studies have employed flumethasone (Rommereim and Slyter, 1981) or betamethasone (Lucas and Notman, 1974) as alternative glucocorticoids of extremely high potency.

It has been assumed that the injection of synthetic corticosteroids can simulate one step in the normal sequence of events which occurs at birth in the sheep. There is no evidence that dexamethasone directly stimulates uterine motility in the preparturient ewe (Prud'Homme and Bosc, 1977); there is, instead, a sharp increase in oestrogen levels (Bosc *et al.*, 1977) and a decrease in progesterone concentration (Fylling, 1971). The mechanism by which the exogenous glucocorticoid leads to increased oestrogen concentrations and decreased progesterone levels is presumably one involving the action of placental enzymes.

Initial inhibitory effect of corticosteroid

It is known that the fetal cortisol level is markedly depressed after administering dexamethasone to the ewe in late pregnancy; this inhibition lasts for less than 24h, the initial fetal cortisol levels having returned by that time and thereafter rising to higher concentrations than in controls (Bosc and Fevre, 1974). In view of evidence that the secretion of cortisol by the fetal adrenals depends on their stage of maturation at the time of treatment, the effect of a single dose of corticosteroid, in bringing about the subsequent surge of fetal cortisol, cannot be expected until a week or so ahead of the normal lambing date.

The inhibition of fetal cortisol secretion for some hours after administering the induction agent is believed to be responsible for the well-recognized lull of about one day before lambings commence after treatment (Bosc, 1972; Joyce, 1974). However, for those sheep that have already embarked on the parturition process, lambings can be expected to go ahead; it is known that labour normally commences about 12h before the expulsion of the fetus, so that ewes already started into labour at the time of corticosteroid injection would be expected to deliver on schedule.

It might be as well to note that certain environmental effects can influence the time of day at which the ewe gives birth. It has been said that although the lamb fetus decides the day on which it is to be born, the mother decides the hour. Such mechanisms would have an obvious survival value among sheep in the wild state. With domesticated sheep, a relationship between the distribution of lambings throughout the day and certain farm routines has been mentioned (Sharafeldin *et al.*, 1971); the lowest incidence of lambings was observed during the period of meal feeding and the indications were that imminent lambings were postponed as a result of increasing adrenal activity at the time of feeding.

9.4.2. Corticosteroid doses and responses

The usual dose level of dexamethasone employed as an induction agent is 15–20mg; the more potent flumethasone is used at a dose level of 2mg. The actual day of the gestation period when the agent is given can be expected to vary with the gestation length of the breed in question; generally, it would be 4–5 days ahead of the mean gestation length for the breed. Even the time of day when the corticosteroid is injected may be a consideration; in some studies, ewes treated in the evening (20.00h) lambed sooner and over a shorter time period than those treated at 08.00 on the same day (Bosc, 1972). On the other hand, it would not appear that parity of the ewe, litter size or sex of the lamb influences response to the corticosteroid (Bosc *et al.*, 1977).

It may be worth quoting an example of how the induction may be expected to operate in a normal commercial-type situation, taken from work at the Rowett Institute in Aberdeen. In this, Finn × Dorset ewes were bred after oestrus synchronization on a Thursday; on Saturday evening, at 21.00h, 142 days later, they were each given a dose of 15mg dexamethasone to induce lambing, which started in the early morning of the following Monday and finished in the evening of the Tuesday (Robinson, 1979).

Timing of lambings after induction treatment

In summary, the use of dexamethasone results in a characteristic delay of 24–36h with a peak over the next 36h and lambings virtually complete by 72h. It should be noted that although early evidence supported the view that the viability and growth rates of lambs and the health and subsequent fertility of ewes after corticoid induction was normal (Bosc *et al.*, 1977), there have been reports in which dexamethasone treatment increased the interval from parturition to placental expulsion and the percentage of ewes retaining the fetal membranes > 6h after parturition (Rubianes *et al.*, 1991). In the UK, however, Peters and Dent (1992) found the induction of lambing (dexamethasone on day 143) to be safe for ewes and lambs without effects on lambing difficulty or retention of fetal membranes, although the interval between lambing and placental expulsion was slightly longer than normal in some instances.

Corticoids in conjunction with other agents

Some investigators have sought to decrease the variance of lambing time in sheep by using clenbuterol and oxytocin in association with a corticosteroid (Kiesling and Meredith, 1991); results showed that oxytocin treatment could reduce the variability in lambing time in clenbuterol-treated ewes.

9.4.3. Oestrogen as the induction agent

It is well established for many mammals that the level of oestrogen in the maternal circulation increases as pregnancy progresses. In the sheep, the rise in oestrogen concentration that occurs prior to parturition has been well documented

Table 9.2. Induction of parturition (day 143/144) by various agents (from Boland *et al.*, 1982).

	Oestradiol benzoate (20 mg)	Dexamethasone (16 mg)	Prostaglandin (15 mg)	Saline
Ewes	39	41	40	39
Interval to parturition (h)	38.6	44.2	83.5	82.9

(Challis, 1971; Bedford *et al.*, 1972; Obst and Seamark, 1972); it appears that this increase occurs for a few days prior to lambing with the major rise recorded within 48h of parturition. As previously noted, there is clear evidence that the last major increase in circulating oestrogen is an important part of the hormonal events associated with the initiation of lambing.

Early reports showing that parturition could be induced in sheep by oestrogens included those of Liggins *et al.* (1973) using the synthetic oestrogen, diethyl-stilboestrol (DES). Studies with the natural steroid showed that oestradiol benzoate (ODB) at the 15–20mg dose level, administered in the last week of gestation, was also effective in the induction of parturition (Cahill *et al.*, 1976; Bosc *et al.*, 1977; Robinson, 1980; Konig, 1982; Rawlings *et al.*, 1983; Rawlings and Howell, 1988). Both dexamethasone and ODB were used effectively in induction studies in Ireland (Table 9.2).

Where the stage of gestation is accurately known, it has been suggested, on the basis of comparative evidence, that the use of ODB to synchronize lambings may be preferable to dexamethasone (Robinson, 1980); with 20mg ODB doses, lambing of 125/128 ewes was virtually completed within 48h with very few lambs lost at parturition and none in the following 72h. It should be noted, however, that in some oestradiol induction studies, there have been sheep which did not respond (Cahill *et al.*, 1976); these ewes apparently suffered a higher incidence of dystocia and there was a greater perinatal mortality. As shown by Rawlings and Howell (1988), this may have been due to the particular ODB dose level chosen.

Effect on milk yields

A positive effect of oestradiol on lactation has been observed in some of these studies (Bosc *et al.*, 1977), increased milk yield being reflected in greater weight gains by the lambs in the early weeks of life. In view of reports of dystocia and higher lamb mortality in some instances, it was considered necessary to evaluate ODB treatment further. According to studies in Canada by Rawlings and Howell (1988), not only was ODB found to be a safe treatment for inducing lambing but it could also be employed to increase the number of ewes lambing during daylight hours. It should be noted that they found a single injection of 2mg ODB given in the last week of gestation (day 142) effective; increasing the ODB dose to 15mg resulted in a 50% incidence of dystocia, with considerable fetal malpresentation.

The Canadian workers regarded 2mg ODB to be a simple and effective treatment in a commercial setting.

9.5. Induction Using Other Agents

9.5.1. Prostaglandins

New Zealand workers in the early 1970s produced evidence implicating $PGF_{2\alpha}$ as a factor in normal parturition in the sheep. Although there is a large increase in PG in the maternal utero-ovarian venous plasma of the ewe during the final 24h of gestation (Challis *et al.*, 1976), PGs have proved relatively ineffective in initiating parturition earlier than about a week from full-term (Oakes *et al.*, 1971). In one comparison between corticoid (2mg flumethasone) and a normal luteolytic dose of PG (15mg $PGF_{2\alpha}$) administered on day 141 of gestation, 89% and 33% of ewes, respectively, delivered lambs within 72h (Harman and Slyter, 1980); in other studies in the USA reported by Barta *et al.* (1980), cloprostenol (250µg) was used without effect as an induction agent.

9.5.2. Use of Epostane

Where pregnancy is maintained solely by the corpus luteum (e.g. pigs and goats), parturition can be induced near term by $PGF_{2\alpha}$ or one of its analogues. In the horse, where the source of progesterone is placental, oxytocin is more effective in inducing labour than PGs. In the sheep, however, neither PG nor oxytocin will consistently induce labour before full-term. In this species, where pregnancy is also maintained by placental progesterone, parturition is normally triggered by a prepartum rise in fetal cortisol. With the development of agents such as Epostane, which prevent progesterone synthesis by inhibiting 3-beta-hydroxysteroid dehydrogenase, the possibilities for induction of labour in the sheep were widened. In the UK, Silver (1987) found an intramuscular injection of 50mg Epostane (in 0.5ml dimethyl sulphoxide (DMSO)) to be effective in terms of successful induction and neonatal viability. The author concluded that his findings opened up new prospects for controlled lambing times, as well as providing a useful, safe and easy method for induction of labour in problem cases during late pregnancy.

9.5.3. RU 486 (Mifepristone)

Epostane is an anti-progestin compound which exerts its effect by inhibiting the 3-beta hydroxysteroid dehydrogenase enzyme system, thereby blocking the *in vivo* synthesis of progesterone. Another avenue of progesterone blockade involves the use of an appropriate receptor blocking agent. In the early 1980s, researchers at Roussel Uclaf developed a novel steroidal antiprogestagen, RU 38486 (later abbreviated to RU 486) which was marketed in France under the name Mifepristone. Although RU 486 binds to progesterone receptors, it has little progesterone-like activity and prevents the progestational effects required to maintain pregnancy.

Results reported by Gazal *et al.* (1993) in the USA indicated that RU 486 can precisely control the time of parturition in sheep in late pregnancy without detrimental effects on dystocia, retention of fetal membranes or delayed postpartum fertility. In sheep, placental progesterone synthesis accounts for up to 80% of circulating progesterone in the last two-thirds of pregnancy and it is not clear how RU 486 compromises placental production of the steroid.

9.6. Agents for Terminating Pregnancy in Sheep

There may be occasions (mismating due to rams breaking into flocks during the breeding season) when it becomes necessary to terminate early pregnancy in the sheep.

9.6.1. Prostaglandin $F_{2\alpha}$

Although it is possible to use $PGF_{2\alpha}$, or one of its analogues, to terminate pregnancy in cattle in the early months of gestation, the results were found to be much less clear-cut in sheep (Inskeep *et al.*, 1975; Pratt *et al.*, 1977). There was a suggestion that the failure of ewes to respond may be due to an anti-luteolytic effect arising from the conceptus which is capable of overcoming the action of PG (Inskeep *et al.*, 1975; Pratt *et al.*, 1977).

Further evidence along these lines was provided in the report of Tyrrell *et al.* (1981), who showed that a single dose of 125 μg cloprostenol was not always effective in terminating pregnancy in ewes during the first trimester of pregnancy; the authors suggested that a second dose of PG should be given after a seven-day interval to ewes predicted to be carrying multiples, on the assumption that the anti-luteolytic effect of the embryos was related to the number present. It is now clear that several factors may influence the response of the corpus luteum of early pregnancy in the ewe to luteolytic treatments, including the stage of pregnancy, the repetition of the treatment, the dose of PG and the number of embryos.

There is a transitory period of refractoriness to luteolytic treatment, developing at the time of maternal recognition, peaking at day 15 or 16 and declining after day 21. A report by Audicana and Harvey (1993) reported that, in general, cloprostenol was more effective than $PG\text{-}F_{2\alpha}$ because there was less refractoriness on day 20 and none on days 26–30 to this analogue. The fact that the sheep's placenta can maintain pregnancy on its own after 55 days means that PG treatment after that time can no longer be guaranteed to terminate pregnancy.

9.6.2. Oestrogen in terminating late pregnancy

The use of oestrogen as the agent for synchronizing lambings in sheep could be hazardous if the farmer is not quite certain of their conception dates. The action of oestradiol, employed in doses varying from 10 to 40mg in the final trimester of pregnancy, was effective in inducing abortion within three days in 40–70% of ewes injected between day 126 and 130 of gestation (Restall *et al.*, 1976). Reference has already been made to the studies of Dawes and Parry (1965), which led those workers to conclude that lambs of a gestational age less than 95% of normal show poor viability. However, a report by Rawlings and Howell (1988) noted that responsiveness to a single 2mg dose of ODB was restricted to the near-term ewe, thereby allowing the delivery of viable lambs.

In using corticosteroids, rather than oestrogens, there may be the safeguard that the treatment is unlikely to result in the delivery of lambs through inadvertently administering the agent to sheep several weeks away from full-term.

9.7. Neonatal Mortality in Sheep

Delivery of the lamb is but one of several coordinated events occurring at parturition. Other events include lung maturation, closing of the ductus arteriosus in the lamb; in the ewe, there is the ripening of the cervix, development of maternal behaviour, milk formation, milk ejection and separation of the fetal membranes from the uterine wall. Treatments for the control of lambing should obviously not result in a failure in any of the several mechanisms which normally operate at parturition.

9.7.1. Incidence of lamb deaths

As noted in a previous chapter (Section 7.7), under some flock conditions, up to 20% of lambs may fail to survive the early weeks of life (Fig. 9.2). Looking at the story in sheep flocks in Ireland and the UK, it has been recorded that perinatal lamb losses may account for 15% or more of the lambs born, though in well-managed flocks the losses can be as low as 5–10% (Gordon, 1958a, 1967; Kilkenny and Read, 1974).

Elsewhere and in earlier times, an analysis of data covering 25 years by Bell (1947) in the USA had shown a 20% loss of lambs up to two months of age. In Australia, Moule and Jackson (1949) recorded a 13% mortality rate in Merino lambs in the first two weeks of life; a similar figure was provided by Wallace (1949b) for New Zealand Romney lambs in their first month. In Wales, Williams (1954) recorded a lamb mortality incidence of 6–10%. There was general agreement among authors that much of this lamb mortality was avoidable by improvements in the management and feeding of the lambing

Fig. 9.2. Lamb deaths, many of which could be avoided, impose a severe penalty on overall flock productivity. Meat and Livestock Commission (MLC) data for the UK show that the average rate of lamb deaths across all flocks was 12% in 1995.

flock. It is rare for the death of the lamb to occur prior to the start of parturition; the generally accepted incidence of antenatal death is about 2% (Haughey, 1991). Survival of the live-born lamb depends mainly on its ability to withstand environmental stress, cold and starvation.

As previously mentioned, lambs at both low and high birthweights may be at a disadvantage; in the one case, it would be because the lamb's energy resources may not be adequate and in the other because of difficulties in the birth process. Management should obviously be geared towards ensuring that all lambs are as close as possible to the optimum birthweight for the breed in question. A review article by Alexander (1984) notes that mortality in sheep is typically related to birthweight by a 'U'-shaped curve; mortality is likely to be lowest at a weight between 3 and 5 kg, depending on breed and age of ewe. Multiple births usually have the disadvantage of lower than optimum birthweights and it is obviously beneficial for ewes to be separated during the final trimester of pregnancy and fed according to the number of lambs they are carrying.

Maternal behaviour in sheep

Distress and mortality of lambs due to poor quality maternal care, is an important welfare consideration. Thus, South African workers identify mismothering as playing a major role in deaths from three days to weaning (Haughey, 1991; Cloete, 1992). The effect of age and parity on maternal behaviour in single-bearing Merino ewes was examined by Alexander *et al.* (1993b). They found the duration of labour was significantly longer for first-parity ewes than for sheep in higher parities; lamb mortality (up to 3–4 days of age) was also significantly higher in primiparous than in multiparous ewes. Primiparous ewes showed a higher tendency to desert their lambs than multiparous ewes and a smaller proportion of primiparous ewes stood to be suckled.

At Cambridge, research has been aimed at establishing the neurohormonal factors that control the induction of maternal behaviour and the formation of the selective ewe–lamb bond (Keverne *et al.*, 1992). It is clear that maternal behaviour in the ewe is induced by signals from the vagina and cervix during labour contractions, provoking neurochemical changes in the brain. Release of oxytocin within the brain has been shown to be directly related to the induction of maternal behaviour. The Cambridge workers were able to demonstrate that oxytocin is released within the medial preoptic area at parturition and during suckling.

Experiments designed to develop methods for fostering orphan and triplet lambs also showed that the administration of 2–3 min of mechanical stimulation of the vagina and cervix induced ewes which had formed selective bonds with their lambs to accept orphan lambs, even 24 h postpartum; this technique had a > 80% success rate and caused the sheep minimal discomfort.

In earlier work in Australia, Poindron *et al.* (1984) reviewed knowledge of mechanisms controlling maternal behaviour in the sheep and suggested various methods that might be employed in that country in dealing with the fostering of lambs (see Table 9.3).

9.7.3. Use of anoestrous ewes as foster mothers

Cambridge workers were the first to show that anoestrous ewes could be used as foster mothers (Keverne *et al.*, 1992). They described the method and suggested that it was economical enough to allow its use in commercial practice. In this, lactation was induced over a six-week period using vaginal sponges containing oestradiol and progesterone. When these lactating ewes were given 2–3 min of mechanical stimulation of the vagina and cervix, they immediately fostered orphan lambs and formed selective bonds with them, even though some had never had lambs of their own previously; the ewes proved to be good mothers and reared their adopted lambs through until weaning.

Compact lambings

Part of the reason for interest in compact matings and lambings is in being able to reduce lamb mortality by the application of well-proven management techniques for ensuring survival of the lambs. The use of the stomach tube for feeding weakly lambs and cross-fostering crates (lamb adopters) for fostering triplet lambs on to ewes producing singles are among such techniques. Overnight housing and shelter can also be expected to increase the number of lambs reared by way of reductions in losses arising from starvation and exposure.

Table 9.3. Possible fostering methods in sheep in relation to the physiological control of maternal behaviour (modified from Poindron *et al.*, 1984).

Technique	Type of ewes used	Advantages of technique	Disadvantages of technique
(A) Vaginal stimulation in parturient ewes	Parturient ewes with single lambs	Cheapest and simplest. Suitable for flocks with a high degree of synchrony	Need for high supervision at lambing. Foster ewes unknown in advance. Period for fostering limited to a few hours. Tends to produce two lambs reared per lambing ewe
(B) Anosmia before lambing	Parturient ewes with single lambs	Greater duration of fostering period than in (A). Characteristics of lambs of little importance in success of fostering. Highest rate of fostering success	Single bearing ewes must be known before lambing. Tends to produce 2 lambs reared per lambing ewe

9.7.4. Difficult parturition

Some authors have reported on the incidence and causes of difficult parturition in sheep (defined as cases requiring assistance at birth); a paper by Grommers *et al.* (1985) dealt with the way in which the presentation of the lamb at birth influenced the incidence of such difficulties.

9.8. References

Alexander, G. (1984) Constraints to lamb survival. In: Lindsay, D.R. and Pearce, D.T. (eds) *Reproduction in Sheep*. Cambridge University Press, Cambridge, pp. 199–209.

Alexander, G., Bradley, L.R. and Stevens, D. (1993a) Effect of age and parity on maternal behaviour in single-bearing Merino ewes. *Australian Journal of Experimental Agriculture* 33, 721–728.

Alexander, G., Stevens, D., Baker, P. and Bradley, L.R. (1993b) The timing of birth in grazing Merino sheep. *Australian Journal of Experimental Agriculture* 33, 557–560.

Amir, D., Genizi, A. and Schindler, H. (1980) Seasonal and other changes in the gestation duration. *Journal of Agricultural Science, Cambridge* 95, 47–49.

Audicana, L. and Harvey, M.J.A. (1993) Termination of early pregnancy in sheep with dinoprost or cloprostenol: comparison of two commercial preparations. *Veterinary Record* 133, 574–576.

Barta, M., Wallace, A.K., Humes, E., Williams, J.C. and Godke, R.A. (1980) Attempts to induce parturition in domestic ewes and goats using cloprostenol (ICI 80996). *Journal of Animal Science* 50 (Suppl. 1), 257.

Basett, J.M. and Thorburn, G.D. (1969) Fetal plasma corticosteroids and the initiation of parturition in the sheep. *Journal of Endocrinology* 44, 285–286.

Bedford, C.A., Harrison, F.A. and Heap, R.B. (1972) The metabolic clearance rate and production rate of progesterone and the conversion of progesterone to 20-alpha hydroxpregn-4-en-3-one in the sheep. *Journal of Endocrinology* 55, 105–118.

Bell, A.W. (1984) Factors controlling placental and foetal growth and their effects on future production. In: Lindsay, D.R. and Pearce, D.T. (eds) *Reproduction in Sheep.* Cambridge University Press, Cambridge, pp. 144–152.

Bell, D.S. (1947) Dead lambs do tell tales. *The Sheepman* 17, 466.

Black, H.J. and Chestnutt, D.M.B. (1990) Influence of shearing regime and grass silage quality on the performance of pregnant ewes. *Animal Production* 51, 573–582.

Boland, M.P., Crosby, T.F. and Gordon I. (1982) Induction of lambing: comparison of the effects of prostaglandin, oestradiol and dexamethasone. *Journal of Agricultural Science, Cambridge* 98, 391–394.

Bosc, M.J. (1972) The induction of synchronization of lambing with the aid of dexamethasone. *Journal of Reproduction and Fertility* 28, 347–357.

Bosc, M.J., De Louis, C. and Terqui, M. (1977) Control of the time of parturition of sheep and goats. In: *Management of Reproduction in Sheep and Goats. Sheep Industry Program Symposium* (Madison), pp. 89–100.

Bosc, M.J. and Fevre, J. (1974) Etude du mode d'action de la dexamethasone utilisee pour induire l'agnelage chez la brebis. *Comptes Rendus de l'Académie des Science* (Paris), Series D 278, 315–318.

Bray, A.R. and Burton, R.N. (1988) Effect of time of day at paddock change on the proportion of ewes giving birth in daytime. *New Zealand Journal of Agricultural Research* 31, 129–131.

Burgess, S. (1994) Lambing patterns. *The Sheep Farmer* (April 1994) p. 15.

Cahill, L.P., Knee, B.W. and Lawson, R.A.S. (1976) Induction of parturition in ewes with a single injection of oestradiol benzoate. *Theriogenology* 5, 289–294.

Challis, J.R.G., Dilley, S.R., Robinson, J.S. and Thorburn, G.D. (1976) Prostaglandins in the circulation of the foetal lamb. *Prostaglandins* 11, 1041.

Challis, R.R.C. (1971) Sharp increase in free circulating oestrogens immediately before parturition in sheep. *Nature* 229, 208–209.

Cloete, S.W.P. (1992) Observations on litter size, parturition and maternal behaviour in relation to lamb mortality in fecund Dormer and South African Mutton Merino ewes. *South African Animal Science* 22, 214–221.

Cloete, S.W.P., Niekerk, van F.E. and Van der Merwe, G.D. (1994) The effect of shearing pregnant ewes prior to a winter-lambing season on ewe and lamb performance in the southern Cape. *South African Journal of Animal Science* 24 (4), 140–142.

Comline, R.S. and Silver, M. (1961) The release of adrenaline and nor-adrenaline from the adrenal glands of the foetal sheep. *Journal of Physiology* (London) 156, 424–444.

Dawes, G.S. and Parry, H.B. (1965) Premature delivery and survival of lambs. *Nature* 207, 330.

Ebling, F.J.P., Wood, R.I., Suttie, J.M., Adel, T.E. and Foster, D.L. (1989) Prenatal photoperiod influences neonatal prolactin secretion in the sheep. *Endocrinology* 125, 384–391.

First, N.L. (1979) Mechanisms controlling parturition in farm animals. In Hawk, H. (ed.) *Animal Reproduction Symposium* No. 3. Allenheld, Osmun, pp. 215–257.

Forbes, J.W. (1967) Factors affecting the gestation length in sheep. *Journal of Agricultural Science, Cambridge* 68, 191–194.

Fylling, P. (1971) Premature parturition following dexamethasone administration to pregnant ewes. *Acta Endocrinologica* 66, 289–295.

Garm, O. and Nedkvitne, O. (1968) Synchronization of parturition in ewe groups of the Norwegian Dalabreed. *Proceedings of the 6th International Congress on Reproduction and AI* (Paris), Resumes, p. 273.

Gazal, O.S., Li, Y., Schwabe, C. and Anderson, L.L. (1993) Attenuation of antepartum relax in surge and induction of parturition by antiprogesterone RU 486 in sheep. *Journal of Reproduction and Fertility* 97, 233–240.

George, J.M. (1969) Variation in the time of parturition of Merino and Dorset Horn ewes. *Journal of Agricultural Science, Cambridge* 73, 295.

Gordon, I. (1958a) Hormonal augmentation of fertility in the ewe during the breeding season. *Journal of Agricultural Science, Cambridge* 50, 123–151.

Gordon, I. (1958b) Studies in the extra-seasonal production of lambs. *Journal of Agricultural Science, Cambridge* 50, 152–197.

Gordon, I. (1967) Aspects of reproduction and neonatal mortality in ewe lambs and adult sheep. *Journal of the Department of Agriculture* (Dublin) 64, 76–127.

Gordon, I. (1983) *Controlled Breeding in Farm Animals*. Pergamon Press, Oxford, pp. 181–195.

Grommers, F.J., Elving, L. and Van Eldik, P. (1985) Parturition difficulties in sheep. *Animal Reproduction Science* 9, 365–374.

Harman, E.L. and Slyter, A.L. (1980) Induction of parturition in the ewe. *Journal of Animal Science* 50, 391–393.

Harrison, F.A. (1982) Dexamethasone-induced parturition in sheep. *British Veterinary Journal* 138, 402.

Haughey, K.G. (1991) Perinatal lamb mortality – its investigation, causes and control. *Journal of South African Veterinary Association* 62, 78.

Herbosa, C.G., Wood, R.I. and Foster, D.L. (1995) Prenatal androgens modify the reproductive response to photoperiod in the developing sheep. *Biology of Reproduction* 52, 163–169.

Howe, D.C. and Brooks, A.N. (1995) Differential coupling of *N*-methyl-D-Asparate (NMDA) receptor to adrenocorticotrophin (ACTH) and growth hormone (GH) secretion in the late gestation ovine fetus. *Journal of Reproduction and Fertility* Abstract Series No. 15, p. 39.

Hudgens, R.E., Albright, J.L. and Pennington, J.A. (1986) Influence of feeding time and diet on time of parturition in multiparous ewes. *Journal of Animal Science* 63, 1036–1040.

Inskeep, E.K., Smutny, W.J. and Butcher, R.L. (1975) Effects of intrafolicular injections of prostaglandins in non-pregnant and pregnant ewes. *Journal of Animal Science* 41, 1098–1104.

Joyce, M.J.B. (1974) The use of dexamethasone to induce parturition in ewes. *Irish Veterinary Journal* 28, 127–131.

Keverne, E.B., Kendrick, K.M. and da Costa, A.P. (1992) Maternal behaviour in sheep. *Report Institute of Animal Physiology and Genetics Research (1990–1991)*, Cambridge, pp. 138–139.

Kiesling, D.O. and Meridith, S. (1991) Decreasing the variance of lambing time in crossbred ewes using flumethasone, clenbuterol and oxytocin. *Theriogenology* 36, 999–1008.

Kilkenny, J.B. and Read, J.L. (1974) British sheep production economics. *Livestock Production Science* 1, 165–178.

Konig, C.D.W. (1982) Artificial control of parturition. *Proceedings of the British Sheep Veterinary Society* 6, p. 7.

Liggins, G.C., Fairclough, R.J., Grieves, S.A., Kendall, J.Z. and Knoz, B.S. (1973) The mechanism of initiation of parturition in the ewe. *Recent Progress in Hormone Research* 29, 111–159.

Lindahl, I.L. (1964) Time of parturition in ewes. *Animal Behaviour* 12, 231.

Lucas, J.M.S. and Notman, A. (1974) The use of corticosteroids to synchronize parturition in sheep. *British Veterinary Journal* 130, 1–5.

McCrabb, G.J., McDonald, B.J. and Hennoste, L.M. (1993) Heat stress during mid-pregnancy in sheep and the consequences for placental and fetal growth. *Journal of Agricultural Science, Cambridge* 120, 265–271.

Mitchell, L.M. (1995) Timing of parturition in intensively managed ewes. *Proceedings of the British Society of Animal Science* (Winter Meeting), Paper 96.

Moule, G.R. and Jackson, M.N.S. (1949) Studies on lamb mortality. *Queensland Agriculture Journal* 69, 235.

Oakes, G., Mofid, M., Brinkman, C.R. and Assali, N.S. (1971) Insensitivity of the sheep to prostaglandins. *Proceedings of the Society of Experimental Biology and Medicine* 142, p. 194.

Obst, J.M. and Seamark, R.F. (1972) Plasma oestrogen concentrations in ewes during parturition. *Journal of Reproduction and Fertility* 28, 161–162.

Osinowo, O.A., Abubakar, B.Y. and Trimnell, A.R. (1994) Genetic and phenotypic relationships between gestation length, litter size and litter birth weight in Yankasa sheep. *Animal Reproduction Science* 34, 111–118.

Peters, A.R. and Dent, C.N. (1992) Induction of parturition in sheep using dexamethasone. *Veterinary Record* 131, 128–129.

Poindron, P., Neindre, Le P. and Levy, F. (1984) Maternal behaviour in sheep and its physiological control. In: Lindsay, D.R. and Pearce, D.T. (eds) *Reproduction in Sheep*. Cambridge University Press, Cambridge, pp. 191–198.

Pratt, B.R., Butcher, R.L. and Inskeep, E.K. (1977) Antiluteolytic effect of the conceptus and of PG E2 in ewes. *Journal of Animal Science* 46, 784–791.

Prud'Homme, N. and Bosc, M.J. (1977) Uterine activity in the ewe before, during and after spontaneous parturition or after dexamethasone priming. *Annales de Biologie animale Biochimie Biophysique* 17, 9–19.

Rawlings, N.C. and Howell, W.E. (1988) The use of estradiol benzoate to manage lambing period in ewes bred at synchronized estrus. *Journal of Animal Science* 66, 851–854.

Rawlings, N.C., Jeffcoate, I.A., Savage, N.C., Steuart, D.M.K. and Steuart, L.H.M. (1983) The effect of season and technique on synchronized and induced estrus and the induction of lambing in the ewe in a commercial setting. *Theriogenology* 19, 665.

Restall, B.J., Herdegen, J. and Carberry, P. (1976) Induction of parturition in sheep using oestradiol benzoate. *Australian Journal of Experimental Agriculture and Animal Husbandry* 16, 462–466.

Rice, G.E., Leach Harper, C.M., Hooper, S. and Thorburn, G.D. (1984) Endocrinology of pregnancy and parturition. In: Lindsay, D.R. and Pearce, D.T. (eds)

Reproduction in Sheep. Cambridge University Press, Cambridge, 165–173.
Robinson, J.J. (1979) Intensive systems. In: *Management and Diseases of Sheep*. Commonwealth Agricultural Bureau, Slough, pp. 431–446.
Robinson, T.J. (1980) Programmed year-round sheep breeding. *Australian Journal of Experimental Agriculture and Animal Husbandry* 20, 667–673.
Rommereim, D.N. and Slyter, A.L. (1981) Effect of day of gestation in induction of lambing with flumethasone. *Journal of Animal Science* 53, 564.
Rubianes, E., Rodas, E., Benech A., Carrau, A. and Ferreira, A. (1991) Lambing and placental expulsion time after dexamethasone-induced and Corriedale and Polwarth. *Theriogenology* 36, 329–334.
Sharafeldin, M.A., Ragab, M.T. and Kandeel, A.A. (1971) Behaviour of ewes during parturition. *Journal of Agricultural Science, Cambridge* 76, 419–422.
Silver, M. (1987) Successful induction of labour in sheep. *Veterinary Record* 120, 299–300.
Stark, R.I. and Daniel, S.S. (1989) Circadian rhythm of vasopressin levels in cerebrospinal fluid of the fetus: effect of continuous light. *Endocrinology* 124, 3095–3101.
Sunderland, S.J., O'Callaghan, D., Boland, M.P. and Roche, J.F. (1995) Effect of photoperiod before and after birth on puberty in ewe lambs. *Biology of Reproduction* 53, 1178–1182.
Symonds, M.E., Bryant, M.J. and Lomax, M.A. (1990) Metabolic adaptation during lactation in winter-shorn sheep. *Journal of Agricultural Science, Cambridge* 114, 201.
Thorburn, G.D. (1978) Hormonal control of parturition in sheep and goat. *Seminar in Perinatology* 2, 235–245.
Tyrrell, R.N., Lane, J.G., Nancarrow, C.D. and Connell, P.J. (1981) Termination of early pregnancy in ewes by use of a prostaglandin analogue and subsequent fertility. *Australian Veterinary Journal* 57, 76–78.
Wallace, L.R. (1949a) Observations of lambing behaviour in ewes. *Proceedings of the New Zealand Society of Animal Production* 9, pp. 85–96.
Wallace, L.R. (1949b) Parturition in ewes and lamb mortality. *Massey Agricultural College, Sheep Farming Annual* 2, 5.
Williams, S.M. (1954) Fertility in Clun Forest sheep. *Journal of Agricultural Science, Cambridge* 45, 202–228.
Winfield, C.G. and Makin, A.W. (1975) Prediction of the onset of parturition in sheep from observations of rectal temperature changes. *Livestock Production Science* 2, 393.
Younis, A.A. and El-Gabory, I.A.H. (1978) On the diurnal variation in lambing and time for placenta expulsion in Awassi ewes. *Journal of Agricultural Science, Cambridge* 91, 757.

10 Embryo Transfer and Associated Techniques in Sheep

10.1. Introduction

Although small-scale studies in sheep embryo transfer (ET) were first reported more than 60 years ago (Warwick *et al.*, 1934), results were usually disappointing with no report showing more than a limited proportion of recipient ewes producing lambs after transfer. It was not until the work of Tim Rowson and his co-workers at Cambridge that the possibilities of the technique were clearly demonstrated; Averill (1958) showed, for example, that 80% of sheep embryos transferred to recipients developed into viable lambs. The methods employed in the mid-1950s by Rowson continued to be used as the standard procedure for some 30 years, before serious attempts were made to develop non-surgical techniques. In expert hands and used for research, the Cambridge procedure proved to be valuable; for commercial applications, however, the surgical approach led to concern in several countries.

10.1.1. Need to develop non-surgical methods

In the context of employing non-surgical methods, it should be noted that serious concern has been expressed in several countries about the use of surgical methods. Already in the UK, government sponsored committees have produced reports which, if implemented, would effectively outlaw the routine use of ET in sheep. Elsewhere, a committee appointed by the Norwegian Ministry of Agriculture in 1994 has decreed that for ethical reasons, further trials on ET in pigs, sheep and goats should not be allowed in Norway until suitable non-surgical techniques have been developed (Anon, 1994).

10.1.2. Practical merit of sheep ET

One obvious area of commercial interest has been employing ET to expand the population of a particular breed which is in demand; in Ireland, for example, the technique was used at one time in attempts to increase the number of high quality Texel sheep (Fig. 10.1). A further consideration with sheep ET is in importing and exporting sheep in the form of frozen embryos rather than as animals on the hoof. According to Sakul *et al.* (1992), the freezing of sheep embryos can contribute to genetic improvement efforts by facilitating transfer of embryos among countries while reducing the risk of disease transmission. Then, there is the potential of the technique to increase the efficiency of breeding improvement programmes (e.g. multiplied ovulation and embryo transfer; MOET).

MOET programmes in sheep

Theoretical studies suggest that rates of genetic progress in sheep could be increased by 100% through use of MOET to increase selection intensity among ewes and reduce female generation intervals. There has been a slow uptake of this technology in sheep breed improvement programmes due to: (i) the need for surgical intervention in the recovery and transfer of embryos; and (ii) the poor repeatability of the superovulatory response.

Fig. 10.1. Texel ewes as donors in an embryo transfer programme.

Variability in superovulatory responses

However, during the 1980s, as detailed below, the first of these problems was adequately resolved by development of minimally invasive techniques for embryo recovery and transfer by laparoscopy. The problem of variability in superovulatory response still remains. In the UK, results presented by Haresign *et al.* (1994a,b) have shown that MOET technology could be effectively applied as part of a breed improvement programme with Scottish Blackface and Welsh Mountain ewes maintained in harsh hill environments. The relatively poor body condition and declining nutritional status of the sheep over the winter months did not adversely affect the response of donor ewes or the ability of recipient ewes to conceive. The authors suggested that this may have not been due to chance.

10.1.3. Nutritional effects on embryo survival

Haresign *et al.* (1994a,b) pointed to previous studies which had shown that high levels of feeding around the time of ovulation reduced progesterone concentrations and impaired development of embryos recovered from superovulated sheep (McEvoy *et al.*, 1993; Creed *et al.*, 1994). Other work reported by McEvoy *et al.* (1995a,b) showed that the provision of additional supplementary progesterone to ewes on a high plane of feeding around time of ovulation elevated plasma progesterone levels and significantly enhanced subsequent embryo development.

Prepubertal donors

In passing, it might be noted that in New Zealand, Rangel-Santos *et al.* (1991) evaluated the feasibility of a juvenile MOET scheme in sheep but concluded from their results that superovulatory treatments did not produce satisfactory responses in prepubertal sheep. In Wales, however, Wolf and McDougall (1994) found that they achieved similar superovulatory rates with 8-month-old ewe-lambs as with older sheep. There is also evidence, as noted in a later section, that sheep embryos can be generated from oocytes recovered from young lambs and processed in the laboratory by way of *in vitro* techniques (Armstrong *et al.*, 1994; Earl *et al.*, 1995).

10.1.4. Sheep ET and endangered breeds

The use of ET in dealing with endangered sheep species is another valuable way in which the technique can be employed. Rorie *et al.* (1994) drew attention to the possible uses of interspecific and intergeneric ETs in the preservation of rare or endangered sheep breeds. In Italy, Ledda *et al.* (1995) has reported on the recovery of embryos from superovulated Mouflon ewes (*Ovis gmelini musimon*) and their survival after transfer into domestic sheep; gestation length was longer in ewes carrying Mouflon lambs than in those carrying sheep embryos (155 vs. 148 days). In Texas, Flores-Foxworth *et al.* (1995) demonstrated that when using compatible species, *in vitro* maturation/fertilization technology combined with interspecific embryo transfer could result in the birth of lambs and the propagation of an exotic ovine species (Red sheep: *Ovis oreintalis gmelini*).

10.1.5. ET in research

For those researchers investigating many aspects of reproductive biology, as well as those concerned more specifically with sheep reproduction, ET technology is likely to remain of considerable importance. The first documented sheep × goat chimera produced via blastocyst injection was reported by Roth *et al.* (1989), this was only possible using ET techniques. A further example would be the study of Wallace *et al.* (1993), who utilized ET to show that a high plane of nutrition immediately after ovulation did not influence the secretory dialogue between the rapidly expanding conceptus and the endometrium.

10.2. Superovulation Techniques

In the sheep, the induction of superovulation follows much the same lines as those employed in cattle; a follicle-stimulating preparation is administered either towards the end of the sheep's normal oestrous cycle (days 11–13) or around the end of a progestagen treatment which is employed to control oestrus (Fig. 10.2). Unlike the donor cow, the seasonally breeding ewe may not always be showing oestrous cycles; for that reason, progestagen treatment preceding the gonadotrophin part of the regimen may be more usual in dealing with the sheep.

In the 1990s, when real-time ultrasonic equipment could be employed to visualize follicular events in the ruminant ovary, several reports have dealt with sheep. In Germany, Riesenberg *et al.* (1995) demonstrated that scanning of ovarian structures can be used in sheep to follow ovarian dynamics during superovulation. In the same country, Kaulfuss *et al.* (1995) successfully used real-time ultrasonics to check the results of superovulation in their ET programme, so that non-responding or poorly-responding donors could be identified at an early stage.

10.2.1. Pregnant mare serum gonadotrophin (PMSG)

The most widely used gonadotrophin for superovulation in sheep has been PMSG; anterior pituitary preparations from pigs, sheep, cattle and horses have also been used. PMSG resembled the pituitary gonadotrophins and human chorionic gonadotrophin (hCG); they all consist of two chemically dissimilar

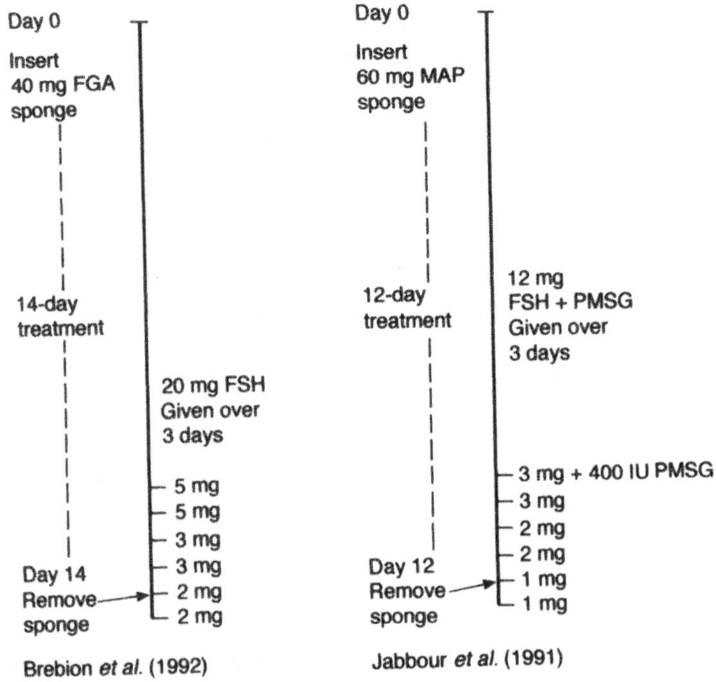

Fig. 10.2. Hormonal protocols used in the superovulation of sheep.

alpha- and beta-subunits, with the beta-subunit being responsible for both the follicle stimulating hormone (FSH)-like and luteinizing hormone (LH)-like activities of the intact molecule. Bindon and Piper (1982) described the PMSG molecule as being very nearly the complete gonadotrophin; it is able to induce follicle growth, oestrogen production, ovulation, luteinization and progesterone synthesis. Sialic acid makes up part of the PMSG molecule and is known to influence liver degradation and hence the plasma half-life of the gonadotrophin; PMSG is known to have a longer half-life *in vivo* than other pituitary and placental gonadotrophins.

Ovarian response to PMSG

In naturally cyclic ewes, PMSG can induce a dose-related ovarian response; Averill (1958) reported an increase in the average number of ovulations from 2.8 to 9.1 as the dose of PMSG increased from 700IU to 1300IU given on day 12/13 of the oestrous cycle. However, the number of large follicles that fail to rupture increases as the dose-level of PMSG rises; about 2000IU would be regarded as the highest permissible dose of the gonadotrophin.

10.2.2. Use of anti-PMSG

Although PMSG has been widely employed as a superovulation-inducing agent, the hormone suffers from the disadvantage of prolonged action, which may result in a second postovulatory wave of follicle growth (Bouters *et al.*, 1983; Monniaux *et al.*, 1984) together with high levels of steroid production by follicles. Such adverse effects of the gonadotrophin can be minimized if oestrogen concentrations are reduced by neutralizing PMSG after initial follicle stimulation; this can be attempted by injection of anti-PMSG after an appropriate interval (Bindon and Piper, 1977; Jabbour and Evans, 1991b). In Italy, Martemucci *et al.* (1995) have shown that addition of monoclonal anti-PMSG to their superovulatory treatment of 1000IU PMSG improved ovarian response and the yield of transferable embryos when carried out at fixed times following the onset of oestrus (12–24h after onset).

10.2.3. Pituitary extracts vs. PMSG

Of the various anterior pituitary preparations employed in superovulation in sheep, horse anterior pituitary (HAP) has probably been most frequently used, generally with reasonable success when given in the form of several consecutive daily injections in the follicular phase of the natural cycle. The HAP extract was effective when given by Moore and Shelton (1962a) as three equal daily doses beginning on day 12 of the oestrous cycle. Subsequently, Moore and Shelton (1964) recorded a linear relationship between the dose of HAP and the mean number of corpora lutea; the average number of ovulations increased from four to 11 as the total dose of the extract increased from 60 to 135mg.

With high PMSG dose levels, as already noted, there may be a tendency towards increased numbers of unruptured luteinized follicles in sheep; this was not the case with HAP. Optimum responses to the horse preparation, as observed by Moore and Shelton (1964), were in ewes that came in oestrus 24–48h after the end of the treatment, when an average of more than nine embryos per ewe was obtained with less variability than found with PMSG.

10.2.4. Commercial pituitary preparations

During the 1980s, commercial pituitary FSH preparations (e.g. FSH-P) became available for use in ET programmes (see Table 10.1); in using FSH-P, a series of injections were required to elicit optimal ovarian response. The response was shown to be affected by the dose of FSH-P administered (Ryan *et al.*, 1984; Smith, 1984). In sheep, FSH-P was reported to be a superior gonadotrophin to PMSG in terms of fertilization rates and numbers of embryos recovered when cervical insemination was performed (Evans *et al.*, 1984). The substitution of FSH-P for PMSG was shown to enhance the

Table 10.1 Commercial preparations of FSH available for superovulation (after Baril *et al.*, 1993).

Origin	Commercial name	Firm
Ovine	EMBRYO-S	Embryo Plus; Australia
Ovine	OVAGEN	Immuno-chemicals Products; New Zealand
Porcine	FSH-P	Schering; USA
Porcine	FOLLTROPIN	Vetrepharm; Canada
Porcine	STIMUFOL	Rhône-Mérieux; France
Porcine	SUPER-OV	Ausa International; USA

effectiveness of superovulation treatments in providing good quality embryos in studies reported by Torrie *et al.* (1987).

Dose-response relationship with FSH preparations

It was believed that improved embryo quality was due mainly to the lower degree of hyperstimulation and less prolonged elevation of oestradiol concentrations during the pre- and postovulatory periods after FSH-P treatment. There is evidence suggesting that although FSH-P increases the size of follicles, neither the steroidogenic efficiency nor the pattern of steroid secretion is changed (Moor *et al.*, 1985); in contrast, PMSG increases not only follicle size but also the steroidogenic efficiency and pattern of steroid secretion. The fact that treatment with FSH-P improved embryo yields, especially when cervical insemination was used in breeding donor sheep, suggested that the gonadotrophin did not markedly alter sperm transport in the reproductive tract. The effect of FSH-P dose on the number of ovulations and embryos in superovulated Targhee ewes was determined by Smith *et al.* (1984); results are shown in Table 10.2.

Table 10.2 Effect of FSH dose level on the number of ovulations and embryos recovered from donor sheep (from Smith, 1984).

FSH* (mg Armour)	No. ovulations	No. embryos
0	1.9 ± 2.5	1.1 ± 2.5
12.0	6.7 ± 3.8	1.3 ± 3.9
14.5	6.7 ± 3.8	4.3 ± 3.9
17.0	8.4 ± 1.7	5.9 ± 1.7
19.5	12.1 ± 1.6	7.7 ± 1.6
22.0	14.4 ± 1.6	8.8 ± 1.6
30.0	10.7 ± 1.7	7.9 ± 1.7

*FSH-P Burns Biotec; 79 Targhee ewes used in study.

Ovine vs. porcine preparations

Two commercial FSH preparations were employed by Dingwall *et al.* (1991) in attempts to increase the superovulation rate in Suffolk ewes used in an MOET programme; the Ovagen preparation appeared to have the potential to induce superovulatory responses (ovulation rates of 13.1 and 11.3 in the months of August and October, respectively) which were at a level and a consistency required for the breeding improvement programme. Later studies in Scotland reported by Fernie *et al.* (1994) also showed that the superovulatory response to Ovagen was markedly superior to porcine FSH; the same authors found that variation in response was not due to factors such as season, breed and age of sheep.

Irish studies

In Ireland, Crosby (1993) recorded that ewes treated with Folltropin-V showed a significantly higher ovulatory response and more transferable embryos than sheep receiving Super-Ov. In the same country, Boland *et al.* (1995) recorded that ovine FSH (Ovagen) was superior to porcine FSH under the particular superovulatory regimen employed.

10.2.5. Use of human menopausal gonadotrophin (HMG)

Among FSH-type preparations employed by Scudamore *et al.* (1991a,b) in superovulation treatments was human menopausal gonadotrophin (HMG) which is extensively employed in human assisted reproduction in stimulating the growth of additional follicles. The Aberdeen workers recorded the mean ovulation rate of ewes treated with HMG to be comparable with those treated with porcine FSH (7.7±2.3 vs. 7.7±0.8). In Brazil, Alvarez and Feitoza (1991) used HMG or PMSG to superovulate Polwarth ewes; they recorded a significantly higher number of good quality embryos recovered per ewe with the HMG (3.0 vs. 0.8). Other work in that country reported by Stefani *et al.* (1991) showed that HMG induced significantly more ovulations per ovary than did FSH-P (see Table 10.3).

Table 10.3 Superovulatory responses in a comparison of FSH-P and HMG (from Stefani *et al.* 1991).

		Corpora lutea		Embryo/ova recovered	
Group	No. of donors	No.	Mean	No.	%
FSH-P	32	225	7[a]	53	24
HMG	17	208	12[b]	60	29

a, b ($P > 0.05$).

10.2.6. Single-dose FSH treatments

Various workers have attempted to simplify multiple FSH injections to a single administration of the gonadotrophin to save time and labour. The response of ewes given a single injection of FSH at two different times of the synchronized cycle (at sponge withdrawal or 36h before withdrawal) was reported by Meinecke-Tillmann *et al.* (1992); rather surprisingly, the ovulation rate was much the same as with multiple dose treatments. A further report by Meinecke-Tillmann *et al.* (1993) supported their findings by showing that a single injection of 25mg FSH on sponge withdrawal was a more efficient method of superovulation than administering the gonadotrophin over a four-day period.

Use of PVP in formulation

Work reported by Lopez-Sebastian *et al.* (1993) in the USA showed that FSH-P diluted in propylene glycol and administered in a single dose on day 13 of the oestrous cycle significantly increased the ovulation rate in comparison to controls (5.5 vs. 1.5); however, neither delayed absorption nor an augmentation effect accounted for the benefit of this agent. In Italy, Dattena *et al.* (1994) concluded from studies conducted with anoestrous sheep that a single intramuscular injection of porcine FSH (dissolved in 30% polyvynlpyrrolidone) resulted in superovulation similar to that with multiple injections of FSH-P or the injection of FSH-P in combination with PMSG.

How gonadotrophins achieve superovulation

In terms of how the FSH preparation achieves its superovulatory effect, Jablonka-Shariff *et al.* (1994) found evidence that FSH-P treatment in the ewe's follicular phase of the cycle decreased the incidence of atresia and induced recruitment and growth of follicles, as shown by increased numbers of medium and large vesicular follicles. Earlier work by Driancourt and Fry (1992) examined the effect of superovulation with porcine FSH or PMSG on growth and maturation of the ovulatory follicles in sheep; they concluded that PMSG increased ovulation rate by (i) recruiting small follicles, (ii) causing up to a threefold increase in follicular growth rate, and (iii) altering the size distribution of the largest follicles at oestrus but not by reversing atresia.

10.2.7. FSH-P in combination with PMSG

It was found that one of the disadvantages of using FSH-P alone was that a proportion of ewes treated failed to show evidence of any superovulatory response, regardless of the total dose or dose regimen used (Eppleston *et al.*, 1984). It was possible to overcome this problem by using a combination of FSH-P and PMSG in the superovulatory treatment. Ryan *et al.* (1984) and Jabbour *et al.* (1991) reported that a higher proportion of ewes showed a superovulatory response when PMSG was used in conjunction with FSH-P than when FSH-P was administered alone. The use of moderate doses of PMSG in conjunction with FSH-P appeared to be superior to employing either gonadotrophin alone in obtaining viable embryos. There was also evidence that an injection of gonadotrophin releasing hormone (GnRH) 24h after progestagen-sponge withdrawal in PMSG + FSH-treated ewes could increase the superovulatory response; Jabbour *et al.* (1991) demonstrated a beneficial effect of GnRH on ovulation rate during the autumn (15.8 vs. 11.8).

Variable responses with combination

Studies by Ryan *et al.* (1991) in Australia involved the treatment of Merino ewes with PMSG (0, 800 and 1600IU) plus FSH (0, 12 or 18mg FSH); compared with FSH alone, 800IU PMSG plus 12 or 16mg FSH significantly increased the percentage of superovulating ewes from 74 to 99%. In that same year, Jabbour and Evans (1991c) reported results suggesting that combined treatment of PMSG and FSH-P was suitable for superovulation of Merino ewes. Not all reports showed the combined gonadotrophin treatment to be of benefit; in Hungary, Cseh and Seregi (1993) compared the efficiency of PMSG alone and PMSG + FSH in combination and found no significant differences, either in ovulation rate (8.7 vs. 8.6) or in transferable quality embryos (5.5 vs. 6.2). In Ireland, Boland *et al.* (1993) similarly recorded no advantage in using a combined PMSG + FSH treatment in comparison with FSH alone.

10.2.8. Prostaglandin and gonadotrophin

The use of $PGF_{2\alpha}$ or one of its analogues has been employed as part of some superovulation regimens. The analogue cloprostenol was employed to induce luteolysis after ovarian stimulation in sheep at Cambridge (Trounson *et al.*, 1976); in this, 100µg cloprostenol was administered to sheep treated 24–72h previously with PMSG, the regimen resulting in most ewes coming in oestrus within 36h. However, according to Willadsen (1979a), a major drawback to the use of prostaglandin in PMSG-treated ewes was the high incidence of premature regression of the corpora lutea formed as a result of superovulation. This premature regression apparently occurred between days 5 and 7 in some 50% or more of ewes.

10.2.9. Premature luteal regression

Although it appeared that the uterus was implicated, the exact reasons for the demise of the corpora lutea were unknown at that time. The ewes returned in oestrus after regression of the corpora lutea and embryos were expelled from the uterus. For that reason, it was necessary not to postpone collection of embryos beyond day 4 or alternatively that sheep should receive progestagen support until the day of collection (Willadsen, 1979a).

Although a similar phenomenon of premature corpus luteum regression has been recorded in cattle, the timing of the regression and its low incidence do not pose problems.

Whyman and Moore (1980) examined the use of the analogue cloprostenol, administered at various intervals after giving PMSG on day 12 of the natural cycle, but such treatment showed no advantage over that in which prostaglandin was not employed.

In the USA, Schiewe *et al* (1990a) used HMG in serial doses totalling 1350IU to superovulate ewes in the mid-luteal phase of the cycle; prostaglandin $F_{2\alpha}$ was administered 36h after the initial injections. The workers recorded that embryo recovery was compromised by premature luteal regression.

10.2.10. Progestagen and gonadotrophin

Many of the superovulation treatments used in sheep attempt to combine ovarian stimulation with control of oestrus. As noted previously, superovulation and oestrus at a predetermined time can be induced with FSH preparations or PMSG given in conjunction with progestagens. In such treatments, PMSG was usually given as a single injection 48h before (Jabbour and Evans, 1991a), 24h before (Larsson *et al.*, 1991; Rangel-Santos *et al.*, 1991; Burgman, 1993), or at the time of sponge pessary removal (Gatica and Correa, 1993; Samartzi *et al.*, 1995).

Pituitary FSH preparations were usually given in the form of several daily injections, the final one of which was timed for 12h after the termination of the progestagen treatment (Jabbour *et al.*, 1991; Brebion *et al.*, 1992). Superovulated donors could usually be expected to exhibit oestrus some 24–48h after terminating the progestagen treatment. Studies in Ireland by Boland *et al.* (1995) indicated that an optimum superovulatory response in sheep was achieved by continuing FSH injections until the time of oestrus, rather than terminating gonadotrophin treatment at the time of progestagen sponge withdrawal.

In Germany, Nellenschulte and Niemann (1992) synchronized ewes by intravaginal progestagen and superovulated the ewes using 16mg pig FSH or 1500IU PMSG; there was no significant difference between the gonadotrophins in the number of corpora lutea recorded (4.7 + 4.6). In South Africa, Burgman (1993) reported that the superovulatory response to 1000IU PMSG, given 24h prior to progestagen withdrawal on days 13, 14, 15 and 16 of intravaginal treatment, did not differ significantly.

Progestagen dose effects

Some authors have reported effects on embryo quality arising from the dose of progestagen employed in the intravaginal sponge. In Aberdeen, Scudamore *et al.* (1992) compared two doses of fluorogestone acetate in sponges (30 and 40mg) and found evidence that the level of progestagen priming prior to superovulation affected embryo viability; survival rates for 3-day embryos were 58.3% and 75% for the 30- and 40-mg FGA treatments, respectively.

10.2.11. Use of the controlled internal drug release device

Studies reported by Thompson *et al.* (1990) in New Zealand suggested that abnormal follicular development occurred when superovulatory gonadotrophin treatment was given during oestrus synchronization with a single CIDR device. It was found that ovarian response with the CIDR device was similar to that in progestagen-sponge-treated sheep when two devices were employed. In Aberdeen, Scudamore *et al.* (1993) recorded significantly higher progesterone concentrations in ewes treated with three CIDR devices than in those treated with one (7.3 vs. 3.3 ng ml^{-1}).

Effect of progesterone levels

Further studies by Scudamore *et al.* (1994) examined the effect of method of oestrus synchronization on the superovulatory response of ewes; their results indicated that for superovulation with pFSH, a higher level of steroid priming (e.g. 30 mg FGA + 400 mg progesterone) gave better results than the steroid level supplied by a CIDR device or a 30 mg FGA pessary. In New Zealand, Campbell *et al.* (1994) reported results suggesting that it may be necessary to use two CIDR devices to maintain a sufficiently long period of high progesterone and that the day of the cycle on which the CIDR device is inserted may be important in achieving maximum superovulatory response in yearling ewes.

10.2.12. Need for ovulating hormone

There appears to be no clear justification for administering an ovulating hormone as part of a superovulation regimen in the belief that this may increase the ovulation rate. Wright *et al.* (1980) administered an LH preparation in the early hours of oestrus to ewes treated with progestagen and an FSH preparation without observing any effect. The general view would be that the sheep possesses sufficient endogenous LH to cope with the needs of superovulation.

10.2.13. Pretreatment with GnRH

The combined administration of GnRH agonists and gonadotrophins was considered to be a significant advance in ovarian stimulation for IVF purposes in human assisted reproduction. In some protocols, which are based on the suppression of ovarian activity before the initiation of gonadotrophin treatment, the agonist is administered for at least 14 days. Workers in France have shown evidence of increased embryo production in superovulated Lacaune ewes pretreated for two weeks with a GnRH agonist (40 μg day^{-1}, Buserelin); they recorded the average number of transferable embryos to increase from 2.6 in controls to 7.2 in ewes which had been infused continuously with the agonist over a 15-day period (Briois *et al.*, 1992).

In Australia, Evans *et al.* (1994) also reported results demonstrating that treatment with a GnRH agonist (via a subcutaneous minipump) prior to FSH produced viable embryos. Unlike in humans, however, where the intranasal administration of GnRH can often be left to the patient, administration of the agent in sheep presents problems of cost and labour that would clearly limit its commercial application.

10.2.14. Repeated superovulation

One of the requirements of a successful superovulation and ET technique is that donor females should be capable of responding to gonadotrophin treatment on several occasions. In Australia, Moore and Shelton (1962b) induced superovulation at intervals of one year in sheep without any significant decrease in the superovulatory response. Palsson (1962), working with a once-yearly treatment with PMSG in Icelandic sheep over a 3-year period, likewise found no evidence of a reduced response. In Ireland, studies with PMSG (Lynch, 1968) and HAP (Boland, 1973) showed that sheep could be superovulated on three occasions in a 6–9 month period, including treatments both during the ewe anoestrus as well as in the breeding season.

In Australia, a study reported by Gherardi and Martin (1978) examined the responsiveness of Merino ewes given PMSG (1000 IU) regularly every cycle for 12 months; results clearly showed that the sheep did not become progressively less responsive to the gonadotrophin. Such evidence proved encouraging to those in sheep ET who were to attempt superovulation on several occasions. Reports from Japan (Fukui *et al.*, 1985) and France (Torres and Sevellec, 1987) supported the view that superovulatory treatment could be repeated without this resulting in a decreased ovulation rate. It did not, however, solve the major problem associated with repeated surgery, that of a build-up of adhesions; the general rule at that time was that three surgical interventions were possible, with pregnancies intervening between the successive attempts. However, animal welfare considerations as well as technical reasons made it essential during the 1970s and 1980s that non-surgical methods should be developed both for the recovery and transfer of sheep embryos.

Effects on uterus

The collection and transfer of sheep embryos by laparoscopy was described by Nellenschulte and Niemann (1992); they recorded that repeated embryo recovery was possible up to three times a month, with a recovery rate of approximately 75–80%. They noted that a limiting factor for continued laparoscopic embryo collection was the postoperative protrusion of the endometrium at the puncture wounds of the uterine walls. In South Africa, Steyn *et al.* (1993) compared fertility of Merino ewes after embryo collection by laparoscopy or surgical intervention; they found that the surgical technique adversely affected ewe fertility during subsequent breeding seasons.

10.3. Factors Affecting Superovulatory Response

10.3.1. Breed effects

It is known that sheep of high fecundity are more responsive to PMSG treatment than animals of low fecundity, Cahill and Dufour (1979) showed that high fecundity ewes have more follicles in the growth phase than low fecundity animals and thus had more follicles responsive to PMSG. Smith (1976) observed that Romney ewes selected from high fecundity flocks had a greater number of ovulations over the range of PMSG doses used compared with ewes from low fecundity flocks. Bindon *et al.* (1971) recorded the ovarian response in Merino ewes selected for a high incidence of multiple births to be three times greater than in ewes selected for a low incidence of twin births. Such results appeared to indicate that the ovaries of sheep selected for high fecundity might be more sensitive to gonadotrophins.

Certainly, Merino ewes derived from Booroola crosses carrying the *F* gene appear to be more sensitive to exogenous gonadotrophins (Piper *et al.*, 1982; Kelly *et al.*, 1983). In one report, Booroola Merinos responded to 1000IU PMSG with an average of about 12 ovulations, whereas control Merinos showed about seven ovulations (Bindon *et al.*, 1986). However, it has not been demonstrated that the ovaries of prolific sheep breeds are more sensitive to gonadotrophins other than PMSG; there were some indications, for example, that Booroola Merinos may not be more sensitive to commercial preparations of FSH (Bindon *et al.*, 1986).

French studies

The effect of breed on superovulatory response to a 16mg FSH-P dose was indicated in studies reported by Torres and Cognie (1984) in France with Prealpes ewes (9.0 ovulations) when compared to those of Cognie *et al.* (1986) with Ile-de-France ewes (18.6 ovulations).

10.3.2. Season

Working in Spain with Manchega ewes, Lopez-Sebastian *et al.* (1990) recorded that a dose of 16mg FSH-P in decreasing doses yielded a mean of 4.2 viable embryos, regardless of time of year when the ewes were treated. Studies reported by Greaney *et al.* (1991) dealt with ewes of five breeds of sheep that were either superovulated in the breeding season or during the anoestrus; ovulation rate was found to be significantly higher in the non-breeding than in the breeding season (7.8 vs. 6.6) but all other parameters were similar. In Japan, Fukui *et al.* (1994) recorded a significantly lower ovulation rate during spring than in ewes superovulated during the autumn (8.4 vs. 14.7); the developmental capacity of the embryos did not differ significantly between seasons.

10.3.3. Nutritional effects

There are several studies which have shown that the level of peri- and postovulatory progesterone in ewes can markedly influence embryo survival in sheep. In view of the inverse relationship known to exist between the plasma concentration of progesterone and food intake in sheep (Parr, 1992; Creed *et al.*, 1994), a study was carried out by McEvoy *et al.* (1993) to determine the effect of altering feed intake on progesterone concentrations during the preovulatory priming phase and on the subsequent viability of embryos collected from superovulated ewes. It was shown that high levels of feeding which reduced preovulatory progesterone concentrations to below 3ng ml^{-1}, the threshold apparently required to prevent subsequent embryo losses, also led to a significant decrease in the yield of viable embryos. The Aberdeen workers suggested that conventional wisdom regarding the maintenance of animals on a high plane of feeding prior to ovulation may need to be reconsidered.

Implications for MOET sheep programmes

Later work reported by McEvoy *et al.* (1995b) led them to conclude that their findings could have important implications for embryo survival in genetic selection programmes involving MOET; they showed that 'flushing' the superovulated ewe before mating did not increase ovulation rate; to the contrary, by adversely affecting preovulatory progesterone levels, high plane feeding potentially incurs serious penalties in terms of embryo development and survival. As noted elsewhere, the avoidance of embryo losses when high preovulatory feeding regimes are employed might be achieved via the use of exogenous progesterone treatments (McEvoy *et al.*, 1995a).

10.3.4. Body condition effects

Results reported by Jabbour *et al.* (1991) demonstrated a beneficial effect of supplementary lupin grain feeding in reducing the incidence of premature luteal regression in superovulated Merino ewes. In Ireland, results reported by Boland *et al.* (1993) indicated that body condition score of ewes had a significant effect on ovulatory response and embryo yield. In terms of the effects of protein nutrition on the performance of donor and recipient sheep, a study by Bishonga *et al.* (1994) in Aberdeen indicated that high levels of rumen-degradable protein in the form of added urea were associated with reductions in both the rate of embryo recovery and the percentage of pregnancies established after autotransfer; it was believed that high levels of plasma urea and ammonia may have had adverse effects on early embryo survival.

Unusual effects

Further studies reported by Madibela *et al.* (1995) led them to suggest that the higher lamb birthweights previously recorded by Bishonga *et al.* (1994) may have had their origin in upregulated metabolic activity. According to the Aberdeen workers, although no direct link is known to exist between precocious early embryo development and eventual neonatal oversize, there is a growing body of circumstantial evidence to support the view that disruption of early developmental events, by means ranging from invasive nuclear transfer procedures to dietary-mediated changes in the embryo's micro-environment, may not only increase mortality rates, but may occasionally stimulate gigantism *in utero.*

Age of ewe

The influence of age in donor sheep used in MOET breeding improvement programmes has been examined by some workers. In the UK, Wolf and Mylne (1994) reported that ovulation rates and embryo recovery were similar for aged Texel ewes and yearling sheep, but embryo quality and survival tended to be lower in the younger donors. The trend towards lower embryo quality and survival in the younger sheep was consistent with data presented by Dingwall *et al.* (1993).

Response and follicle stage

As in cows, there appears to be some evidence in sheep that the presence of a large follicle at the time of gonadotrophin administration decreases the superovulatory response in the ewe. In Uruguay, Rubianes *et al.* (1995) studied the superovulatory response of anoestrous ewes to conventional treatment with PMSG and a progestagen sponge (MAP/FGA); a qualitatively greater superovulatory response was recorded in ewes in the absence of a large follicle at the time of gonadotrophin administration.

10.3.5. Prepubertal lambs as a source of embryos

Mansour (1959), working at Cambridge, was one of the first to attempt the induction of ovulation in very young prepubertal ewe-lambs; he treated animals varying in age from 1 to 22 weeks with PMSG and hCG, but observed an absence of ovulation in the lambs until they were 16 weeks or more of age. Elsewhere, however, Land and McGovern (1968) and Worthington and Kennedy (1979) did achieve ovulations at earlier ages (6–9 week-old lambs) with PMSG and hCG. None the less, even when superovulation was induced and embryos recovered, there has been evidence showing that prepubertal ewe-lambs did not always yield embryos with the same potential for continued development as do older sheep; this problem is also evident in prepubertal cattle.

Embryos from prepubertal sheep

Studies in which sheep embryos were cultured *in vitro* or transferred to adult recipients (Quirke and Hanrahan, 1977) provided evidence of problems. According to experiences recorded by Armstrong and Evans (1983), superovulation failure and poor fertilization limited the yield of embryos obtained from donor sheep less than one year of age. It was also apparent that questions of embryo normality often arise in prepubertal lambs that are not found in more mature sheep.

The potential of more recently developed superovulatory procedures (progestagen/Ovagen/prostaglandin), with intrauterine insemination as the method of breeding, were evaluated by Wolf and McDougall (1994) in ewe-lambs. The authors concluded that superovulation rates in 8-month ewe-lambs can be similar to those found in older sheep and that the quality of embryos was acceptable. In New Zealand, Campbell and McDonald (1995) have reported data suggesting that treating ewe-lambs at or near puberty, either early or late in an oestrous cycle, will not affect the response to synchronization and superovulation treatment, as measured by follicle characteristics, LH surge and number of ovulations.

10.3.6. Use of GnRH

The use of GnRH was examined by Walker *et al.* (1989) as a means of increasing the efficacy of embryo collection; treatment consistently improved fertilization rates and the number of embryos collected per ewe was enhanced when compared with untreated controls. In the timing of GnRH, 24h after progestagen treatment was the preferred time in ewes treated with PMSG; 36h was preferred in ewes treated with FSH. These results modified a previous report by Walker *et al.* (1986) which had indicated that GnRH administered 24h after sponge removal synchronized the time of ovulation, irrespective of the gonadotrophin used in superovulation. In more recent times, a highly effective ovarian stimulation protocol used by Earl *et al.* (1995) in 8- to 9-week-old lambs also included treatment with GnRH 24h after sponge withdrawal.

10.3.7. Use of melatonin to enhance response

It is well established that the use of melatonin in ewes can influence ovulatory response. Some workers have examined the effect of melatonin in conjunction with FSH treatment on superovulation in ewes. A study by Joly *et al.* (1993) suggested that melatonin treatment was able to enhance superovulatory treatments in sheep during late anoestrus but not in the breeding season (see Table 10.4); in this, Ile-de-France ewes received melatonin implants (36mg) 30–50 days before breeding. In Aberdeen, Robinson *et al.* (1994) found no evidence to suggest that long-term melatonin treatment of anoestrous recipient ewes was beneficial in an ET programme; neither was evidence found of a

Table 10.4 Comparative ovarian response to superovulation after melatonin treatment during two periods (from Joly *et al.*, 1993).

	Outside of breeding season (May–June)		Early breeding season (July–August)	
	CL	Fol. of 2mm diameter	CL	Fol. of 2mm diameter
Group 1 (melatonin)	$12.5^a \pm 5.4$	$7.0^a \pm 4.2$	9.5 ± 3.9	$7.8^a \pm 2.6$
Group 2 (control)	$8.2^b \pm 2.6$	$11.4^b \pm 6.1$	10.2 ± 3.0	$10.8^b \pm 3.5$

[a, b]: mean number of corpora lutea (CL) and follicles (Fol.) per ewe within columns with different superscripts differ ($P < 0.05$).

beneficial effect of melatonin treatment of the donor ewe on the potential viability of her embryos.

10.3.8. *Effect of growth hormone treatment*

There is evidence, in cattle, that treatment with recombinant bovine somatotrophin (BST) can influence follicle development and the incidence of twin-ovulations. In the USA, Eckery *et al.* (1994) administered BST to ewes from day 5 of the cycle for 13 days; superovulatory treatment consisted of a single dose of PMSG followed by twice-daily injections of FSH for four days. No increase in ovulation rate or in the number of small follicles was evident.

10.4. Breeding the Donor Ewe

10.4.1. *Artificial insemination after laparotomy*

Regardless of the superovulation regimen employed, fertilization failure has been a frequent occurrence in superovulated sheep; according to Willadsen (1979a), fertilization after mating was the exception rather than the rule with donors having more than about 10 ovulations. However, it was evident that high fertilization rates could be achieved in superovulated sheep by the direct deposition of semen into the uterine horns (Trounson and Moore, 1974a; Boland and Gordon, 1978). Although fertilization rates in excess of 90% could be achieved after breeding donor ewes by intrauterine inseminations, the subsequent embryo recovery rate could be markedly decreased. For such reasons, Willadsen (1979a) suggested that surgical artificial insemination (AI) should be done as soon as possible in oestrus and as gently as possible, without touching the ovaries or oviducts. It was clear that embryos produced by uterine insemination were viable and capable of normal development when transferred to recipients (Killeen and Moore, 1971).

10.4.2. Intrauterine AI by laparoscopy

The use of laparoscopy rather than laparotomy in carrying out intrauterine AI has been reported by many workers. In progestagen-treated, oestrus-synchronized, superovulated ewes, the insemination is usually carried out 40–60h after pessary withdrawal (Robinson *et al.*, 1989a; Dingwall *et al.*, 1991; Stefani *et al.*, 1991; Maxwell *et al.*, 1993; Haresign *et al.*, 1994a,b; Boland *et al.*, 1993, 1995; Scudamore *et al.*, 1994; McEvoy *et al.*, 1995a).

Timing of AI

In Aberdeen, Robinson *et al.* (1989a) reported results which led them to suggest adoption of intrauterine insemination 60h after progestagen withdrawal to maximize fertilization and embryo recovery rates in superovulated ewes. In a later report from the same laboratory, however, Scudamore *et al.* (1991b) noted that the proportion of transferable quality embryos was improved by insemination at 48h rather than 60h after pessary withdrawal (100 vs. 35.4%). In the USA, Rexroad and Powell (1991) concluded that intrauterine AI at 40h or later resulted in optimum fertilization and embryo recovery rates.

The relative merits of inseminating into the oviduct or uterus have also been examined. In Australia, Maxwell *et al.* (1993) recorded significantly higher fertilization rates after oviductal than after intrauterine insemination.

One advantage of breeding by AI, especially by intrauterine insemination, was that ewes which did not exhibit oestrus after the superovulation treatment were bred as well as those that did show heat; Moore (1982) mentioned that some 10–15% of donors usually fail to come in oestrus after superovulation but that they were capable of yielding normal embryos after surgical AI.

10.4.3. Intrauterine AI in combination with natural service

In some reports, natural service has been augmented by intrauterine insemination in an effort to improve the efficiency of embryo production. Results of work by Haresign *et al.* (1994a,b), in a MOET breeding improvement programme with superovulated Scottish Blackface and Welsh Mountain ewes, indicated that intrauterine AI at 46h after pessary withdrawal (ewes bred at the onset of oestrus by hand-mating) increased fertilization rate and the quality of the embryos recovered.

10.4.4. Conventional AI and natural service

Using orthodox artificial or natural insemination procedures to breed donor sheep, the usual practice was to mate by hand service or perform cervical inseminations at 12h intervals during the heat period; oestrus could be expected to be more prolonged than normal, presumably the result of the increased oestrogen concentrations arising from the stimulated ovaries.

Effect of ram on embryo quality

It might be noted that the ram used in matings may influence the yield of viable embryos. Back in the 1960s, Newton and Betts (1968) presented data showing that one particular Suffolk ram sired significantly more lambs per superovulated ewe lambing than any of the other five rams employed in their study. In sheep IVF studies, it is also apparent that significant ram differences exist in the quality of embryos produced (Fukui *et al.*, 1988a).

10.5. Embryo Recovery and Handling Procedures

For many years, sheep embryos were obtained from the reproductive tract using the flushing methods first described by Hunter *et al.* (1955). In this, collection was carried out under general anaesthesia with the ovaries, oviducts and uterus exposed by a mid-ventral incision. The oviducts were cannulated via the fimbriated end and part or all of the tract flushed by gently expressing the recovery medium along the uterine horns and through the fallopian tubes. In the sheep, the embryo, at the 8- to 16-cell stage, enters the uterus around the third to fourth day after the end of oestrus; however, regardless of the timing of collection, flushing of medium back through the oviducts can result in a high rate of embryo recovery. In New Zealand, Tervit and Havik (1976) employed a flushing technique which involved inserting a urethral catheter into the lumen of the uterine horn, inflating the catheter cuff and then flushing the horn with medium injected near the tip.

10.5.1. Laparoscopy for embryo recovery

Workers at Aberdeen reported on embryo recovery on days 5 and 6 by laparoscopy in ewes that had been subjected to a range of superovulatory treatments (Scudamore *et al.*, 1991a); they reported recovering 52% of the ova shed, of which 87% were embryos. As noted above, Nellenschulte and Niemann (1992) in Germany showed that repeated embryo recovery was possible with laparoscopy and reported an embryo recovery rate of approximately 75%.

10.5.2. Transcervical recovery of sheep embryos

Attempts have been made by some workers to recover embryos from ewes via the cervix, despite its complex structure. Clearly, the introduction of non-surgical methods for embryo recovery is essential on animal welfare grounds. Studies in South Africa by Barry *et al.* (1990) were among the first to show that the cervices of maiden and adult ewes could be passed for embryo collection by 'ripening' the cervix with PGE_2 and oestradiol without detrimental effect to the embryo. In Germany, Grupp (1991) similarly applied intracervical treatment of superovulated ewes with PGE_2, (with or without oestradiol benzoate) and reported transcervical embryo recovery in 80–90% of ewes; the author suggested that hormonal cervical dilation followed by transcervical embryo recovery may provide a rapid and non-traumatic method of embryo recovery in sheep.

In Scotland, Mylne *et al.* (1992) attempted transcervical embryo recovery, manipulating a fine catheter through a cervix dilated by treatment with PGE_2 and oestradiol; the animals were anaesthetized and in dorsal recumbency. The workers recorded a significant difference in the success of cervical penetration between multiparous ewes (46%) and maiden ewes (5%). It was concluded that further work was needed to refine the technique to achieve consistent cervical dilation.

The availability of real-time ultrasonic scanning equipment permits some new approaches to the introduction of catheters through the tortuous cervical canal in sheep. Gobel *et al.* (1995) attempted transcervical embryo collection in anaesthetized German Merino ewes, using transrectal ultrasonics to guide a catheter into the uterus with the animal restrained in dorsal recumbency. When the cervix was passed, 100ml phosphate-buffered saline (PBS) (in 5 × 20ml volumes) was infused into each uterine horn and recovered by spontaneous backflow. Although non-surgical, such forms of recovery procedure would have very limited application in commercial practice.

10.5.3. Detecting superovulated donors prior to embryo recovery

Various attempts have been made to determine whether donor ewes subjected to superovulation treatments have responded and are worth considering for recovery attempts. Ivkov *et al.* (1994) recorded a significant increase in progesterone level on the fourth day after oestrus in donor ewes compared with recipients (4.21 vs. 1.9ng ml^{-1}); they suggested that by measuring the level of progesterone, it would be possible to determine the effectiveness of superovulation before attempting recovery.

10.5.4. Premature luteal regression in superovulated sheep

One consideration in recovering embryos from superovulated ewes is the early regression of the corpora lutea, which can negatively influence embryo recovery rate (Ledda *et al.*, 1995). In Australia, Ryan *et al.* (1991) recorded a seasonal component to the incidence of sheep with regressed corpora lutea (autumn, 37%; spring, 21%); a similar finding was reported by Jabbour *et al.* (1991) in the same country.

10.5.5. Media for embryo recovery

Different forms of media, ranging from complex tissue culture media (e.g. Medium 199; Ham's F-10) to simple balanced salt solutions enriched with sheep serum, or serum albumin of sheep or cattle origin, have been used for the collection and holding of sheep embryos. The Cambridge early work generally involved the use of sheep serum with trace amounts of antibiotics added (Hunter *et al.*, 1955; Averill and Rowson, 1958). With the development of more effective methods for culturing sheep embryos, serum was wholly or partially replaced by bicarbonate or phosphate-buffered solutions enriched with either 2–3% sheep or cattle serum albumin or 10–20% sheep serum. For collection and transfer, with media exposed to air, the phosphate-buffered solutions (e.g. Dulbecco's phosphate-buffered saline) proved to be particularly useful; the composition of this medium is listed in Table 10.5.

As a general rule, sheep embryos should be transferred as soon as possible after collection; between collection and transfer, the embryos can be stored at room temperature for several hours provided precautions are taken to avoid

Table 10.5 Composition of Dulbecco's phosphate-buffered saline (Whittingham modification) (from Whittingham, 1971).

NaCl	8.0 g		
KCl	0.2 g		
Na_2HPO_4	1.15 g		
KH_2PO_4	0.20 g		
$CaCl_2$	0.10 g	or $CaCl_2.6H_2O$	0.1974 g
$MgCl_2.6H_2O$	0.10 g		
Supplemented with:			
Na pyruvate	0.036 g		
Glucose	1.0 g		
BSA	4.0 g	Bovine serum albumin (Fraction V) (Sigma)	
Penicillin	0.060 g	or Kanamycin sulphate (Sigma) 25 mg	
Streptomycin	0.050 g		
Phenol red	1.0 ml		
All in 1 litre of double distilled deionized water			

contamination of the medium during storage. According to Torres and Cognie (1984), lowering the storage temperature from 37° to 20°C improved embryo survival; they mention the possibility that maintaining metabolism at 37°C had an adverse effect on nutrient reserves.

10.5.6. Evaluation of sheep embryos

The normal process of fertilization and early development of sheep embryos is now well understood. The first cleavage occurs about 15–18h after the oocyte has been penetrated by the sperm; the second cleavage occurs some 12h later to give the four-cell embryo by 48h after fertilization (day 3 after oestrus onset). From then on, blastomeres usually cleave every 16–24h, with 4–6-cell embryos being typical of day 4 and 24–32-cell morulae typical of day 5. By day 6, most embryos have developed into compacted late morulae or into early blastocysts; by day 7, the majority are blastocysts and expanding blastocysts. The blastocyst hatches from the zona pellucida on day 8/9 and it may be difficult to distinguish embryos from other cell aggregates that are flushed from the uterus at that time.

Usually, day 3 to day 7 embryos have been used in transfers; relatively underdeveloped embryos (for their age) are probably best left out of transfer attempts (Moore, 1976). Morphological abnormalities in sheep embryos have been commented on by various investigators (Averill, 1958; Tervit and McDonald, 1969) but their influence on embryo survival has not always been clear. Killeen and Moore (1971) noted that sheep embryos possessing one or more anucleate cells were still capable of developing normally; they suggested that these embryos should be regarded as atypical rather than abnormal. According to Moore (1982), retarded sheep embryos rarely show continued development in culture, indicating that they are not viable.

10.5.7. Short-term storage of sheep embryos

As noted above, the use of sterile sheep blood serum (usually with trace amounts of antibiotics added) as the recovery and storage medium for embryos, together with transfers only to those recipients closely synchronized with their respective donors, probably accounted for much of the success achieved by Cambridge workers in sheep ET during the 1950s.

Although it was shown to be possible, more than 40 years ago, to ship rabbit embryos in blood serum held at 10°C from one continent to another and achieve acceptable survival rates (Marden and Chang, 1952), the embryos of farm animals proved much less amenable to such treatment. However, a fortunate discovery in the mid-1950s by Cambridge workers showed that sheep embryos, at an early stage of cleavage, could be stored for several days in the rabbit oviduct and retain their ability to develop into lambs on re-transfer to recipient sheep (Averill *et al.*, 1955); this subsequently enabled sheep embryos to be transported between England and South Africa and between Australia and New Zealand (Adams *et al.*, 1961; Welch, 1969). Later studies at Cambridge showed that in preserving early-cleavage embryos destined for re-transfer, the rabbit oviduct could provide a satisfactory environment for up to five days (Lawson *et al.*, 1972).

Longer-term storage

A variety of culture media have been used over the years in an effort to support development of the early sheep embryo. These included synthetic oviduct fluid (SOF), which was based on the chemical composition of the oviductal fluid; culture was carried out in a low oxygen atmosphere (Tervit *et al.*, 1972; Tervit and Rowson, 1974). During the 1980s, research in the sheep showed that it is possible for sheep oviductal cells, maintained in an appropriate culture medium (e.g. M-199), to support early embryo development (Gandolfi and Moor, 1987).

Storage at reduced temperatures

Storing sheep embryos at a reduced temperature may have advantages, such as avoiding the need to freeze-thaw embryos. Reduced temperature can inhibit continued cleavage, and it may permit synchronization of donor and recipient ewes to be based on the age of the embryo at the start of storage. In sheep, as in cattle, there is evidence that early-cleavage stage embryos (2–16-cell) are more sensitive to cooling to 0°C than morulae stages (Willadsen *et al.*, 1976), although there have been reports suggesting that early sheep embryos could be stored at temperatures varying from 0 to 13°C for several days (Averill and Rowson, 1959; Kardymowicz, 1972; Renard *et al.*, 1976).

10.6. Freezing the Sheep Embryo

Although commercial interest in the low temperature storage of sheep embryos has been very limited in comparison to that shown in cattle, from the research viewpoint, the sheep has often been used as an experimental animal in testing freeze-thaw methodology because of its low cost, relative to the cow. The first lamb born after transfer of a frozen-thawed embryo was born in Cambridge (Willadsen *et al.*, 1974) and this was followed by similar success in Australia (Moore and Bilton, 1976) and in Poland (Smorag *et al.*, 1978). Among the cryoprotectants examined at that time were dimethylsulphoxide (DMSO) and glycerol.

10.6.1. Cryoprotectants used

At Cambridge, Willadsen *et al.* (1976) successfully stored sheep embryos in media containing 1.5 M DMSO; Australian workers also reported lambs from embryos frozen in media containing 1.0–2.0 M DMSO (Bilton and Moore,

1976). Where glycerol was used, it was employed at a concentration of 1.4 M and some reports indicated that this agent was preferable to DMSO (Willadsen, 1980b). A study by Smith *et al.* (1994) compared three methods of stepwise cryodilution using glycerol as the cryoprotectant; they found no difference between the one- and three-step method of cryodilution, but that a five-step method may have adverse effects on embryo survival. Among the more unusual cryoprotectants employed in the freezing of sheep embryos is methanol, used at the 3.0 M concentration (Czlonkowska *et al.* 1991).

Phosphate-buffered saline has been the basis of media employed in the freezing as well as for the storage of sheep embryos for the few hours that may elapse from the time of collection until freezing. Cambridge studies indicated that even when only half the normal cell number was present at the early blastocyst stage (after embryo-splitting), a normal lamb could still be produced after freezing and thawing (Willadsen, 1980a).

10.6.2. Use of ethylene glycol

In a review of embryo freezing methods in sheep, Brebion *et al.* (1992) refer to the use of 1.5 M ethylene glycol as the cryoprotectant and 20% fetal calf serum in the medium; according to these authors, 85–90% of sheep embryos survived freezing and 65% survived after transfer to recipient ewes. The use of ethylene glycol in the freezing of sheep embryos was also reported by McGinnis *et al.* (1993); they examined one-step vs. two-step addition and removal of ethylene glycol. The one-step addition involved placement of embryos directly into 1.5 M ethylene glycol, whereas the two-step addition used an intermediate 10-min exposure to 0.75 M ethylene glycol. Similarly, the one-step removal involved direct placement of thawed embryos into 1.0 sucrose, and the two-step procedure included a 10-min exposure to 0.25 M sucrose before placement in 1.0 M sucrose. No difference was observed in embryo survival rate after culture for 96h between the cryoprotectant addition and removal methods. These results were validated by a limited number of transfers in which embryo survival of thawed embryos (73%) was similar to that with fresh embryos (74%). In Canada, Songasen *et al.* (1995) reported that the survival rate of sheep embryos frozen in ethylene glycol, propylene glycol and DMSO was 76.9, 62.5 and 55.6%, respectively, based on the *in vitro* development rate to hatched blastocysts.

10.6.3. Vitrification of sheep embryos

More than a decade ago, Rall and Fahy (1985) described an innovation in the cryopreservation of mammalian embryos; they devised a mixture of solutes (dimethylsulphoxide, acetamide, propylene glycol and polyethylene glycol) that permitted vitrification at practical cooling rates. Vitrification is a process of solidification in which crystalline ice does not separate and consequently the solutes are not concentrated; there is simply a gross increase in viscosity, producing a solid, glassy state. It had already been shown that embryos may vitrify internally when DMSO alone was used as a cryoprotectant; in the new method, vitrification occurred outside the cells as well.

Advantages of vitrification

With vitrification, cooling rate became relatively unimportant but exposure to the vitrifying media had to be brief in order to avoid toxicity; warming had to be rapid in order to prevent crystallization as the temperature returned to normal. Sheep embryos have been subjected to vitrification (mixture of glycerol and propanediol) and subsequent transfer of thawed embryos resulted in the birth of lambs (Gajda *et al.*, 1989). In the USA, the efficacy of conventional freezing or vitrification of sheep embryos was examined by Schiewe *et al.* (1990b); these authors showed that sheep embryos were able to survive their simple and rapid vitrification procedure.

Cryoprotectants used

In Australia, Szell *et al.* (1990) used 25% glycerol and 25% propylene glycol in their vitrification attempts; embryo survival rates varied from 36% for late morulae to 70% for late blastocysts and the authors recorded the birth of lambs after using the procedure. In the same country, other studies gave results with day 6 embryos comparable with those found with conventional freeze-thaw procedures (Ali and Shelton, 1993); the use of ethylene glycol as the major permeating cryoprotectant in the vitrification medium was believed to be an important factor in explaining the improved success rate. Naitana *et al.* (1994) in Italy also found that a high pregnancy rate could be achieved after vitrification of sheep embryos using glycerol and ethylene glycol as cryoprotectants.

Results provided by Szell and Windsor (1994) showed their best results with embryos exposed to a vitrification mixture of 3.5 M glycerol + 3.5 M propylene glycol; the survival rate was similar to that obtained using more complex traditional procedures. As in cattle, vitrification is viewed by some in sheep as a possible means of eliminating the need for expensive, programmable cryogenic units in freeze-storing sheep embryos under field conditions.

10.7. Recipient Management and Embryo Transfer Techniques

10.7.1. Synchronizing donors and recipients

It is well accepted that the occurrence of oestrus in donor and recipient ewes must be closely synchronized if the survival rate of transferred embryos is to be optimized (Fig. 10.3). In Australia, Shelton and Moore (1966) examined synchronization requirements in the sheep, transferring embryos to recipients in oestrus 48h before to 48h after their respective donors; optimum results, in terms of pregnancy rates and embryo survival, were found among recipients in heat 12h before to 12h after donors. Rowson and Moor (1966) confirmed and extended these observations, recording that 75% of their recipients became pregnant when oestrus was exactly synchronized; a difference of as much as +2 days was tolerated reasonably well but with a difference of +3 days, only 8% of recipient ewes became pregnant. At a later date, Wilmut and Sales (1981) in Edinburgh showed that sheep embryos transferred into an advanced asynchronous recipient failed to survive because their development was so modified that they failed to inhibit luteolysis. In a study dealing with variables that influence the outcome of ET in sheep, Alabart *et al.* (1995) in Spain showed that one of the main factors contributing to a high fertility rate was the degree of synchrony between donors and recipients.

10.7.2. Selecting recipients on the basis of progesterone levels

Among various problems in sheep embryo transfer is the need for more efficient selection of ewes as potential recipients; such selection might be one means of improving the success rate of transfers. In sheep, unlike cattle, it is not possible to determine the status of the corpus luteum by palpation. In the former Yugoslavia, Inkov *et al.* (1995) examined the possibility of selecting recipients on the basis of measured concentrations of serum progesterone on the fourth day after oestrus; the best pregnancy rate was obtained in ewes showing levels of more than 3 ng ml^{-1} progesterone.

Fig. 10.3. Pure-bred Texel lamb with its Suffolk-cross foster mother.

10.7.3. Oestrus control in recipients

Early studies in sheep ET, in which daily doses of progesterone were administered to control oestrus in recipient sheep, indicated that this form of control did not adversely affect pregnancy and embryo survival rates (Hunter *et al.*, 1955). The same was found to be true when progestagens such as FGA and MAP were administered to recipient ewes by the intravaginal route (Crosby *et al.*, 1980). At Cambridge, Willadsen (1979a), reported that practically all prostaglandin-treated recipients (125 μg cloprostenol at 10–12-day intervals) came in oestrus 24–48 h after the second dose of the agent and that this was a satisfactory method of oestrus control in such sheep; however, in view of difficulties reported elsewhere (see Chapter 3) with prostaglandins in oestrus control, progesterone or progestagens may be the preferred option.

10.7.4. Surgical procedures

Transfer of embryos by surgical intervention was usually conducted under general anaesthesia with the ewe suitably restrained; access to the reproductive tract to effect transfer was generally by a mid-ventral incision and there was little advantage in exploring alternative surgical approaches. By virtue of its small size and ease of handling in comparison with pigs and cattle, the ewe permitted many transfers to be carried out within a short time period. The developmental stages at which embryos or oocyte have been transferred have ranged all the way from follicular oocytes (transferred to mated recipients) and zygotes to elongated blastocysts on day 12 of the cycle. The effect of age of embryo, number of embryos transferred and the site of transfer have been reported on by several investigators and covered in various works (Moore, 1982; Baril *et al.*, 1993). Embryos recovered about four days after oestrus are usually transferred to the uterus. The standard procedure was to transfer two embryos to each recipient (one to each uterine horn), although single embryo transfers can give acceptable pregnancy rates (Crosby *et al.*, 1980).

10.7.5. Transfer of embryos by laparoscopy

The significance of improved ET techniques for sheep breeding, using fibre optics, was reviewed by Robinson *et al.* (1989b) in Aberdeen. An efficient and rapid non-surgical method for use in transferring early embryos in sheep, which was efficient for 2-day and 7-day embryos, was described by Vallet *et al.* (1989) in France. In Germany, Nellenschulte and Niemann (1992) have reported on the laparoscopic transfer of fresh and thawed embryos.

10.7.6. Transcervical embryo transfer

A report from Canada by Buckrell *et al.* (1993) dealt with the use of a proven technique for transcervical sheep AI (Guelph System for Transcervical AI) for transferring day 7 embryos to recipient ewes. The results indicated that the cervix could be penetrated successfully and in a reasonable period of time (average of 3.17 min); lambing outcome, however, suggested a need for further development of the transfer technique.

10.7.7. Animal health considerations

One of the advantages of using embryos rather than sheep on the hoof in movements between countries and continents is in minimizing the risks of disease transmission. However, the embryos need to be examined and washed to minimize the risk of contamination by pathogens. In the USA, Riddell *et al.* (1989) examined the question of whether *Brucella abortus* would adhere to the zona pellucida of ovine embryo and to evaluate the efficiency of embryo washing procedures; the authors concluded that measures additional to washing (antibiotics) may be needed to ensure that sheep embryos are free of pathogens.

10.8. Producing Sheep Embryos by *In Vitro* Techniques

The *in vitro* fertilization (IVF) of ovulated oocytes in sheep goes back to the 1950s, when Dauzier and Thibault (1959) and Thibault and Dauzier (1961) presented reports of their work in France; however, it was little more than a decade ago that the first IVF lamb was born in Japan (Hanada, 1985). Although it was demonstrated, more than a quarter of a century ago, in the UK and Ireland (see Table 10.6), that sheep oocytes released from ovarian follicles readily undergo meiotic maturation *in vitro* (Edwards, 1965; Crosby and Gordon, 1971), it soon became evident from work in Cambridge that such nuclear maturation was not necessarily synonymous with cytoplasmic maturation (Moor and Trounson, 1977; Warnes *et al.*, 1977).

Culture of intact ovarian follicles

In contrast to consistent failure to induce normal cytoplasmic maturation in sheep oocytes released from vesicular follicles, the Cambridge workers showed that oocytes matured *in vitro* within intact preovulatory follicles subsequently developed normally to term (Moor and Trounson, 1977); the same work also showed that the potential for complete oocyte maturation *in vitro* was not restricted to the preovulatory follicle but could be induced in sheep oocyte from non-atretic and atretic follicles obtained at any stage of the cycle.

Table 10.6 Maturation stages in sheep oocytes cultured *in vitro* (*in vitro* maturation) (from Crosby and Gordon, 1971).

Maturation stages	Culture period (h) 2	4	6	8	10	12	14	16	18	20	22	24	25–27	28–30	48	Totals
Vesicular nucleus	11	7	3	1	14	1	5	1	2	1	4	6	14	2	–	72
Diakinesis chromatin	1	3	5	5	5	3	2	0	2	0	2	0	3	0	–	31
Pro-metaphase	–	–	8	3	1	1	1	3	4	4	9	5	8	0	–	47
Metaphase plate	–	–	6	8	8	11	21	14	0	4	4	14	35	11	3	139
Late anaphase	–	–	–	–	–	–	2	2	9	1	0	2	7	–	–	23
Teleophase	–	–	–	–	–	–	1	3	7	7	3	4	5	–	–	30
Pro-metaphase-II + P. body	–	–	–	–	–	–	–	–	5	–	1	0	0	–	–	6
Metaphase-II + P. body	–	–	–	–	–	–	–	–	2	–	5	7	40	11	3	68
All stages	12	10	22	17	28	16	32	23	31	17	28	38	112	24	6	416

Successful maturation of sheep oocytes in vitro

Normal nuclear and cytoplasmic maturation of the sheep oocyte, according the Moor and Trounson (1977), was regulated by both gonadotrophins and steroids. It was the work of Staigmiller and Moor (1984) in Cambridge that was the first to show that, under appropriate *in vitro* maturation (IVM) conditions, sheep oocytes were capable of acquiring full developmental competence. Oocyte recovery was by follicle dissection and Staigmiller and Moor (1984) employed M-199 supplemented with 10% fetal calf serum, gonadotrophins (FSH/LH), oestradiol and additional granulosa cells; oocytes were cultured at 37°C for 24h in a culture system providing gentle agitation. Developmental competence was assessed by transfer of oocytes to inseminated recipient ewes.

Birth of IVMF-derived lambs

The first report of lambs being born from oocytes matured and fertilized *in vitro* was that of Cheng *et al* (1986) in Cambridge. These results were subsequently extended by Crozet *et al.* (1987) in France; in their studies, the early sheep embryos were usually cultured in the rabbit oviduct prior to transfer to the sheep.

10.8.1. Laboratory production of sheep embryos

In the IVM of sheep oocytes, Wahid *et al.* (1991) showed the value of morphological evaluation of oocytes prior to maturation. Later, the same workers reported that high rates of nuclear maturation could be achieved in the absence of hormone supplementation (Wahid *et al.*, 1992). Elsewhere, Galli and Moor (1991) had earlier drawn attention to results with sheep indicating that the addition of steroids during IVM enhanced neither cell cycle progression nor developmental potential of sheep oocytes. The same workers reported that sheep oocytes undergoing IVM in the absence of gonadotrophins had higher viability than many of those cultured in the presence of gonadotrophin. In Australia, O'Brien *et al.* (1994) likewise concluded that there was no advantage in using hormones (FSH/LH and oestradiol) in the sheep IVM medium.

Sperm capacitation

In the capacitation of ram spermatozoa, work with sheep and pigs by Cheng *et al.* (1986) led them to conclude that there were basic differences between the two species in the capacitation requirements. Other work in Cambridge, reported by Fukui *et al.* (1988) made it clear that individual rams affected not only the fertilization rate but also the ability of zygotes to cleave after fertilization. The authors concluded that selection of rams with high fertilization ability could be a key factor determining the success of IVF work in sheep. Earlier work by Newton and Betts (1968) with superovulated ewes had also indicated the importance of the individual ram used in natural service in influencing fertilization and embryo survival rates.

In vitro *fertilization medium*

In the former Czechoslovakia, Slavik and Fulka (1991) and Slavik *et al.* (1992) described the use of M-199 enriched with heparin as their sheep fertilization medium. In these studies, high fertilization and cleavage rates of *in vitro* matured sheep oocytes were achieved, but transfer of early-stage embryos resulted in a low birth-rate in recipient ewes (five lambs from 34 embryos). In Australia, Walker *et al.* (1994) concluded from their studies that the use of synthetic oviductal fluid (SOF) plus 2% oestrous sheep serum both for swim-up and IVF was an acceptable procedure. In Germany, Mocker (1994) using a modified Brackett and Oliphant (BO) medium plus heparin, reported high fertilization rates, both with fresh and frozen ram semen.

In terms of the non-surgical transfer of *in vitro* produced sheep embryos, Cseh *et al.* (1995) have reported results in which they showed that the laparascopic technique was suitable for the transfer of such embryos.

10.8.2. Oocyte recovery from live sheep

The recovery of sheep oocytes and the production of embryos from them may be of interest to those involved in conventional sheep ET applications for breeding improvement purposes. An evaluation of MOET in Suffolk sheep by Dingwall *et al.* (1993) noted, for example, that current ET rates in this species fell some way short of what was required for an effective MOET programme

(Smith, 1988). As with cattle, the repeated recovery of oocytes from live sheep may be of good commercial interest in the development of effective breeding improvement programmes. In this regard, it is possible that sexed sperm can eventually be employed in sheep IVF. One report from Argentina has described the *in vitro* maturation and fertilization (IVM/IVF) of oocytes collected from Merino-cross ewes by way of a laparoscopic follicular aspiration procedure (Baldassarre *et al.*, 1994); data showed such oocytes to be as capable of maturation and fertilization as abattoir derived oocytes.

In Texas, Flores-Foxworth *et al.* (1995) found the laparoscopic oocyte aspiration procedure to be relatively simple, as well as being less traumatic than normal embryo recovery procedures in dealing with donor Red sheep (*Ovis orientalis gmelini*) in work aimed at producing Red sheep lambs from domestic sheep (*Ovis aries*).

Embryos from prepubertal sheep

Using embryos generated by IVF of oocytes recovered from young lambs by laparotomy, and transferred to adult recipient ewes, Armstrong *et al.* (1994) showed that a minimum of four lambs was attainable per donor lamb, with prospects for considerably more with further improvement of techniques. In contrast to such IVF results, 2–3-month lambs subjected to superovulation treatments and inseminated laparoscopically yielded no embryo; this failure was attributed to immaturity of the lamb's reproductive tract rather than to lack of ovulatory response or developmental incompetence of oocytes. In a further report, the Australian IVF methods were modified to obtain a higher yield of mature oocytes from young lambs (Earl *et al.*, 1995); results indicated that 10–15 blastocysts could be obtained from each oocyte collection from 8–9-week-old gonadotrophin-treated donor breeding stock.

10.8.3. Use of abattoir sheep ovaries

For a valuable pedigree ewe, at the end of its productive life, the recovery of the ovaries at slaughter, with subsequent IVF of oocytes, may be an additional means of producing high-quality lambs.

10.9. Embryo Splitting and Cloning in Sheep

10.9.1. Incidence of identical twins in sheep

The normal incidence of monozygotic twins in sheep is believed to be extremely low; Johansson and Hansson (1943), examining extensive data on the sex ratio and multiple-births in sheep, could find no statistical evidence of their occurrence. Elsewhere, in Australia and New Zealand, again on the basis of analysing sheep lambing data, other workers also concluded that identicals were either absent or extremely rare. Studies in prenatal physiology in sheep did suggest, however, that such twins may occur on occasions; Cohrs (1934) and Henning (1937), working with abattoir materials, both observed twins in ewes possessing a single corpus luteum and concluded that they were monozygotic. At a later date, Rowson and Moor (1964), in examining 424 sheep embryos ranging from early blastocysts (day 6/7) to elongated blastocysts (day 14), recorded four with two embryonic areas.

10.9.2. Embryo splitting

The artificial production of identical twinning in a species such as the sheep, where such twins rarely occur, may have advantages, particularly in certain areas of research. An attempt to produce identical twins by mechanically dividing 4–7-day-old sheep embryos was reported by Trounson and Moore (1974b) in Australia; although some 25% of the demi-embryos developed during culture into apparently normal blastocysts, only two lambs were born from 19 blastocysts which were transferred to recipient ewes. A major step forward in the production of monozygotic twins came with the report of Willadsen (1979b), who developed a technique in Cambridge for producing identicals in sheep involving the microsurgical separation of the blastomeres of two-cell embryos (Fig. 10.4), their insertion into foreign zonae pellucidae, embedding in a protective cylinder of agar and cultured in the ligated sheep oviduct (see Fig. 10.5).

The viability of late morulae and early blastocysts produced by such means was about 50%, the lower than normal survival rate being attributed primarily to mechanical damage resulting from the manipulative procedure by which the agar was removed from the embryo before transfer to the recipient. In a later report (Willadsen, 1980a), the same author showed that two blastomeres of a four-cell embryo and four blastomeres of an eight-cell embryo developed into blastocysts which were as normal as those produced by isolated blastomeres from two-cell embryos; pregnancy and embryo survival rates were as high on this occasion as those recorded after the transfer of ordinary sheep embryos.

Splitting embryos at a later stage

Although the Cambridge technique was valuable for research purposes, it did require the use of an intermediate host and was tedious to perform. For such reasons, work elsewhere examined the development of sheep demi-embryos without the use of an intermediate host (Gatica *et al.*, 1984); sheep embryos were bisected at the morula stage and transferred to recipients immediately. Pregnancy rate and embryo survival rate proved to be much the same as found with non-manipulated embryos (see Table 10.7). In France, studies reported by Chesne *et al.* (1987) showed that monozygotic twin pairs could be obtained by splitting embryos at days 8–10; there was a tendency for the success rate to increase with day of collection (see Table 10.8). The importance of the

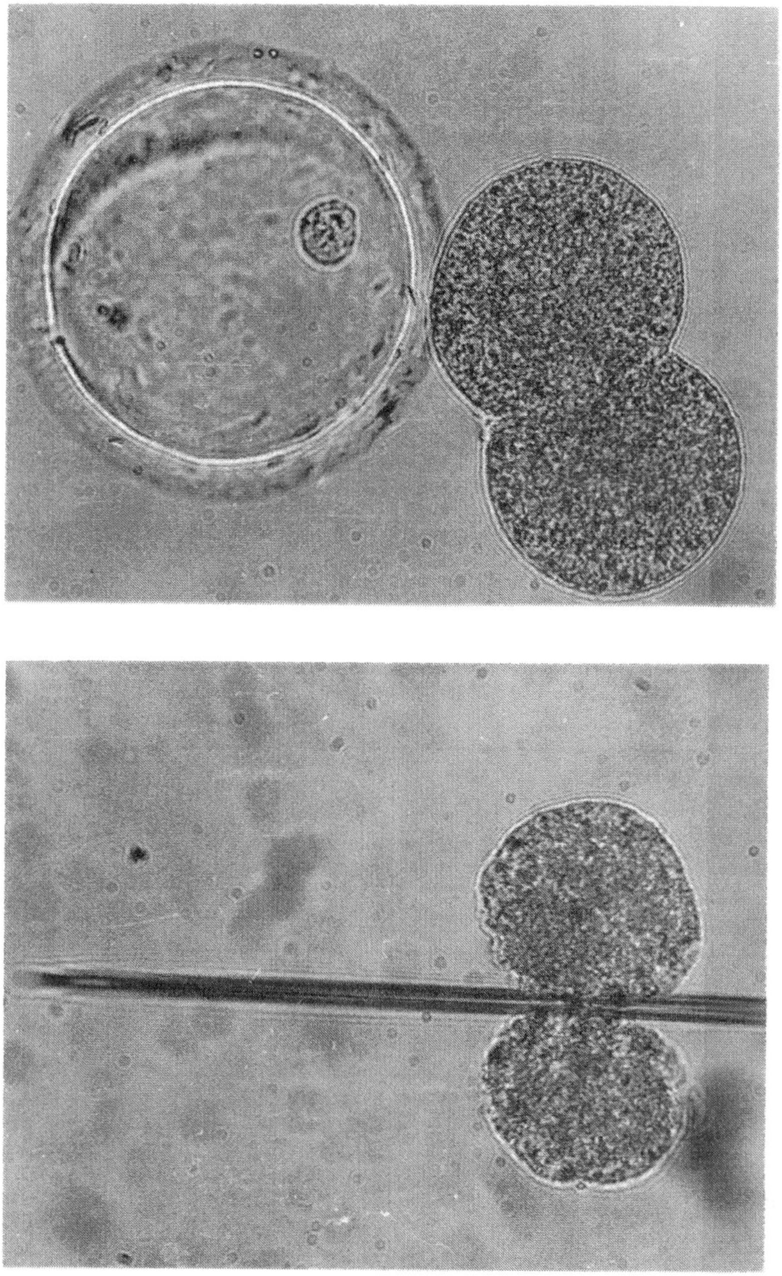

Fig. 10.4. Splitting at the early (two-cell) and later (morula) stages of embryo development in the sheep (from Gatica, 1988).

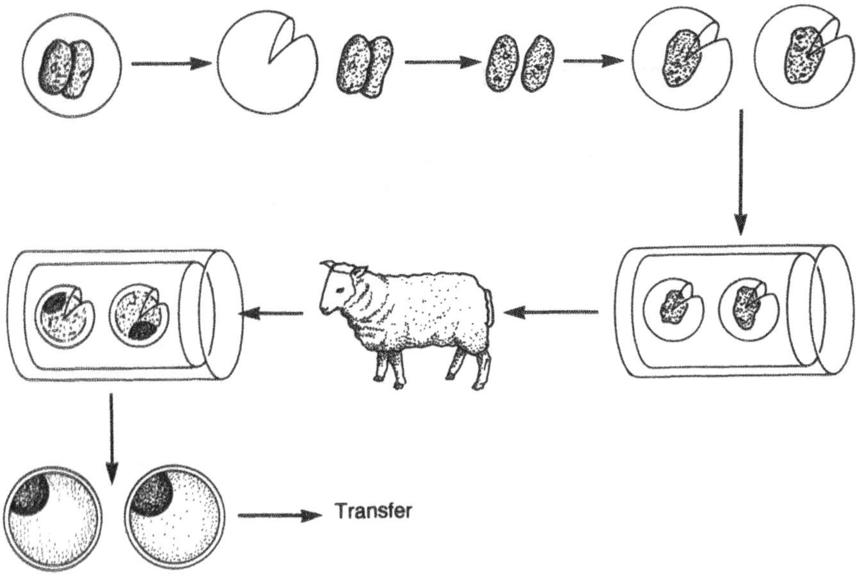

Fig. 10.5. Production of identical twin embryos in sheep (after Willadsen, 1979).

Table 10.7 Lambing outcome after transfer of bisected embryos to recipient ewes (from Gatica *et al.*, 1984).

No. of embryos bisected	17
No. of halves transferred	34
No. of recipient ewes	17
No. of ewes lambing	9 (52.9%)
No. of ewes producing twins	7 (77.8%)
No. of ewes producing singles	2 (22.2%)

selection of the best quality embryos for splitting has been stressed by Szell *et al.* (1994); they demonstrated that the survival rate decreased when Grade-2 rather than Grade-1 embryos were employed.

10.9.3. Cloning in sheep

Although embryo splitting in sheep and other farm mammals has been possible since the late 1970s, the technique has several obvious limitations as a means of increasing the number of lambs of a specific genotype. The viability of embryos is low when the embryonic mass is divided into four parts. The technology of cloning by nuclear transfer (NT) was first successfully employed in sheep at Cambridge by Willadsen (1986) and further developed in the species by Smith and Wilmut (1989) in Edinburgh. In Australia, McLaughlin

Table 10.8 Results of embryo splitting and transfer in sheep in relation to day of embryo recovery (from Chesne *et al.*, 1987).

	Day of recovery		
	Day 8	Day 9	Day 10
No. of embryos bisected	12*	11	16
No. of recipient ewes	12	11	16
No. of ewes lambing (%)	7 (58)	8 (73)	13 (81)
No. of sets of identical twins	5	5	8
No. of singlets	2	3	5
Total number of lambs	12	13	21
Efficiency†	100%	118%	131%

*Includes four embryos with an intact zona pellucida.
†No. of lambs born per 100 embryos bisected and transferred.

et al. (1990) reported the birth of identical twin and quadruplet Merino lambs by way of nuclear transfer. The methods employed by the Roslin workers in cloning by nuclear transfer are shown diagrammatically in Fig. 10.6.

Unlike in cattle, those who have used *in vitro* matured oocytes as cytoplasts have not found them to be as effective as those matured *in vivo* (McLaughlin *et al.*, 1991). However, more progress has been made in sheep, using embryonic stem cells as nuclear donors than in any of the other farm mammals. In Edinburgh, for example, Campbell *et al.* (1995) assessed the totipotency of embryonic disc cells from day 9 sheep embryos by nuclear transfer; they showed that they were totipotent and that such totipotency was maintained in culture for three passages. Further work reported by the same group has been able to demonstrate for the first time the development to term of a sheep embryo reconstructed by nuclear transfer from an established cultured cell line (Campbell *et al.*, 1996a,b).

10.10. Sex Diagnosis of Sheep Embryos

The general principles applicable to the sexing of cattle embryos are applicable to sheep. Bredbacka and Peippo (1992) are among authors who have reported sex diagnosis in sheep using a polymerase chain reaction-based sex determination assay for sheep embryos. It is also likely that sperm separation via flow cytometry could be employed to yield sexed sperm for use in IVF applications.

10.11. Production of Transgenic Sheep

The transgenic modification of sheep was first reported by Hammer *et al.* (1985) in the USA. At this point in time, Edinburgh in Scotland has become

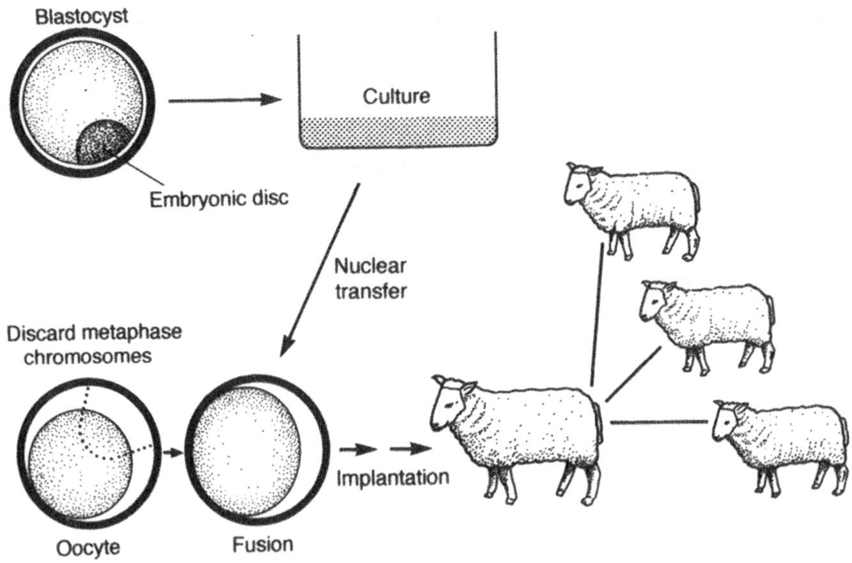

Fig. 10.6. Producing lambs by nuclear transfer using donor nuclei from permanent cell lines (from Solter, 1996). Edinburgh workers have provided the first evidence of live mammalian offspring following nuclear transfer from an established cell line. In this work, embryonic discs were cultured on a feeder layer and permanent cell lines established. Enucleated recipient oocytes were produced by removing a small amount of cytoplasm containing the metaphase plate and fused with single cells from the totipotent cell line.

a well-recognized world centre of transgenic research in this species (Cherfas, 1990); among the successes recorded is the first sheep to carry human genes and the first demonstration that genetic defects can be corrected in an animal and the correction passed on to the offspring. As noted by Ward *et al.* (1984), it is important to develop dialogue between researchers who understand the physiology of sheep and molecular biologists with the skills required for the isolation and modification of genes. Progress in gene transfer research in sheep has been covered in various research papers from several countries (Rexroad and Wall, 1987; Walker *et al.*, 1990; Gol'dman *et al.*, 1992; Gogolevskii *et al.*, 1993) and review articles (Ebert and Schindler, 1993; Pursel and Rexroad, 1993). The recent significant progress made in the production of embryonic stem cells (Campbell *et al.*, 1996a,b) should increase the efficiency of gene transfer in this species. The potential for commercial applications of transgenic technology in sheep has been discussed by various authors (Niemann *et al.*, 1994; Nancarrow and Ward, 1995).

10.12. References

Adams, C.E., Rowson, L.E.A., Hunter, G.L. and Bishop, G.P. (1961) Long distance transport of sheep ova. *Proceedings of the 4th International Congress of Animal Reproduction* (The Hague) 2, 381–382.

Alabart, J.L., Folch, J., Fernandez-Arias, A., Ramon, J.P., Garbayo, A. and Cocero, M.J. (1995) Screening of some variables influencing the results of embryo transfer in the ewe. I. Five-day-old embryos. *Theriogenology* 44, 1011–1026.

Ali, J. and Shelton, J.N. (1993) Successful vitrification of day-6 sheep embryos. *Journal of Reproduction and Fertility* 99, 65–70.

Alvarez, R.H. and Feitoza, A.S.L. (1991) Transfer of sheep embryos under field conditions. *Boletim de Industria Animal* 48, 7–11.

Anon (1994) Embryo transfer in small ruminants. *Norsk Veterinaertidsskrift* 106, 750–753.

Armstrong, D.T. and Evans, G. (1983) Factors influencing success of embryo transfer in sheep and goats. *Theriogenology* 19, 31–42.

Armstrong, D.T., Irvine, B. and Earl, C.R. (1994) In vitro fertilization (IVF) of follicular oocytes from juvenile lambs and their developmental competence in vitro and in vivo. *Biology of Reproduction* 50 (Suppl. 1), 189.

Averill, R.L.W. (1958) The production of living sheep eggs. *Journal of Agricultural Science, Cambridge* 50, 17–33.

Averill, R.L.W. and Rowson, L.E.A. (1958) Ovum transfer in sheep. *Journal of Endocrinology* 16, 326–336.

Averill, R.L.W. and Rowson, L.E.A. (1959) Attempts at storage of sheep ova at low temperatures. *Journal of Agricultural Science, Cambridge* 52, 392–395.

Averill, R.L.W., Adams, C.E. and Rowson, L.E.A. (1955) Transfer of mammalian ova between species. *Nature* 176, 167.

Baldassarre, H., De Matos, D.G., Furnus, C.C., Castro, T.E. and Fischer, E.I.C. (1994) Technique for efficient recovery of sheep oocytes by laparoscopic folliculocentesis. *Animal Reproduction Science* 35, 154–150.

Baril, G., Brebion, P. and Chesne, P. (1993) Manuel de formation pratique pour la transplantation embtryonnaire chez la brebis et la chevre. *FAO Publication* No. 115.

Barry, D.M., Van Niekerk, C.H., Rust, J. and Van der Walt, T. (1990) Cervical embryo collection in sheep after 'ripening' of the cervix with prostaglandin E2 and estradiol. *Theriogenology* 33, 190.

Bilton, R.J. and Moore, N.W. (1976) Effects of ice seeding and of freezing and thawing rate on the development of sheep embryos stored at –196°C. *Theriogenology* 6, 635.

Bindon, B.M. and Piper, L.R. (1977) Induction of ovulation in sheep and cattle by injections of PMSG and ovine anti-PMSG immune serum. *Theriogenology* 8, 171.

Bindon, B.M. and Piper, L.R. (1982) Physiological basis of ovarian response to PMSG in sheep and cattle. In: *Embryo Transfer in Cattle, Sheep and Goats.* Australian Society Reproductive Biology, Canberra, pp. 1–5.

Bindon, B.M., Ch'ang, T.S. and Turner, H.N. (1971) Ovarian response to gonadotrophin by Merino ewes selected for fecundity. *Australian Journal of Agricultural Research* 22, 809–820.

Bindon, B.M., Piper, L.R., Cahil, L.P., Driancourt, M.A. and O'Shea, T. (1986) Genetic and hormonal factors affecting superovulation. *Theriogenology* 25, 53–70.

Bishonga, C., Robinson, J.J., McEvoy, T.G., Aitken, R.P., Findlay, P.A. and Robertson,

I. (1994) The effects of excess rumen degradable protein in ewes on ovulation rate, fertilization and embryo survival in vivo and during in vitro culture. *Proceedings of the British Society of Animal Production* (Winter Meeting), Paper No. 81.
Boland, M.P. (1973) Studies related to egg transfer in sheep. PhD Thesis, National University of Ireland, Dublin.
Boland, M.P. and Gordon, I. (1978) Recovery and fertilization of eggs following natural service and uterine insemination in the Galway ewe. *Irish Veterinary Journal* 32, 123–125.
Boland, M.P., O'Doherty, J.V. and Crosby, T.F. (1993) Superovulation in sheep influenced by FSH and body condition. *Proceedings of the 9th Meeting European Embryo Transfer Association* (Lyon), p. 158.
Boland, M.P., Kelly, P., Crosby, T.F. and Roche, J.F. (1995) The effect of type and number of FSH injections on superovulation in ewes. *Proceedings of the British Society of Animal Science* (Winter Meeting), Paper No. 57.
Bouters, R., Moyaert, I., Corijn, M. and Vandeplasche, M. (1983) The use of a PMSG antiserum in superovulated cattle endocrinological changes and effects on timing of ovulation. *Zuchthygiene* 18, 172–177.
Brebion, P., Baril, G., Cognie, Y. and Vallet, J.C. (1992) Embryo transfer in sheep and goats. *Annales de Zootechnie* 41, 331–339.
Bredbacka, P. and Peippo, J. (1992) Sex diagnosis of ovine and bovine embryos by enzymatic amplification and digestion of DNA from the ZFY/ZFX locus. *Agricultural Science in Finland* 1(2), 233–238.
Briois, M., Belloc, J.P., Brebion, P., Guerin, Y. and Cognie, Y. (1992) Increased embryo production in superovulated elite Lacaune ewes pretreated with a GnRH agonist. *Proceedings of the 8th Meeting of the European Embryo Transfer Association* (Lyon), p. 132.
Buckrell, B.C., Gartley, C.J., Buschbeck, C., Jordan, P. and Walton, J.W. (1993) Evaluation of a transcervical AI technique for transferring embryos in sheep. *Theriogenology* 39, 197.
Burgman, J.M. (1993) Effect of delayed sponge withdrawal on the superovulatory response to pregnant mare serum gonadotrophin and on embryo recovery in sheep. *South African Journal of Animal Science* 23, 111–112.
Cahil, L.P. and Dufour, J. (1979) Folicular populations in the ewe under different gonadotrophin levels. *Annales de Biologie animale Biochimie Biophysique* 19, 1475–1481.
Campbell, J.W. and McDonald, M.F. (1995) Effects of stage of the estrous cycle at initiation of synchronization and superovulation on ovrian characteristics in ewe lambs. *Journal of Animal Science* 73 (Suppl. 1), 304.
Campbell, J.W., McDonald, M.F. and Wickham, G.A. (1994) Hormonal and ovarian responses in Romney ewe hoggets after synchronization and superovulation treatment. *Proceedings of the New Zealand Society of Animal Production* 54, pp. 239–242.
Campbell, K., McWhir, J., Ritchie, B. and Wilmut, I. (1995) Production of live lambs following nuclear transfer of cultured embryonic disc cells. *Theriogenology* 43, 181.
Campbell, K.H.S., McWhir, J., Ritchie, W.A. and Wilmut, I. (1996a) Live lambs by nuclear transfer from an established cell line. *Theriogenology* 45, 286.
Campbell, K.H.S., McWhir, J., Ritchie, W.A. and Wilmut, I. (1996b) Sheep cloned by nuclear transfer from a cultured cell line. *Nature* 380, 64–66.
Cheng, W.T.K., Moor, R.M. and Polge, C. (1986) In vitro fertilization of pig and sheep oocytes matured in vivo and in vitro. *Theriogenology* 25, 146.

Cherfas, J. (1990) Molecular biology lies down with the lamb. *Science* 249, 124–126.

Chesne, P., Colas, G., Cognie, Y., Guerin, Y. and Sevellec, C. (1987) Lamb production using superovulation, embryo bisection and transfer. *Theriogenology* 27, 751–757.

Cohrs, P. (1934) Uniovular twins in the sheep and pig and binovular unifollicular twins in the sheep. *Zeitschrift Anatomische* 102, 584–593.

Cognie, Y., Chupin, D. and Saumande, J. (1986) The effect of modifying FSH/LH ratio during the superovulatory treatment in ewes. *Theriogenology* 25, 148.

Creed, J., McEvoy, T.G., Robinson, J.J., Aitken, R.P., Palmer, R. M. and Robertson, I. (1994) The effect of pre-ovulatory nutrition on the subsequent development of superovulated sheep ova in an in vitro culture system. *Proceedings of the British Society of Animal Production* (Winter Meeting), Paper No. 82.

Crosby, T.F. (1993) Superovulation in sheep: the effects of pFSH type and ewe breed. *Theriogenology* 39, 205.

Crosby, T.F. and Gordon, I. (1971) II. Timing of nuclear maturation in oocytes cultured in growth medium. *Journal of Agricultural Science, Cambridge* 76, 373–374.

Crosby, T.F., Boland, M.P., El-Kamali, A.A. and Gordon, I. (1980) Superovulation in the ewe using HAP. *Theriogenology* 13, 92.

Crozet, N., Huneau, D., De Smedt, V., Theron, M.-C., Szollosi, D., Torres, S. and Sevellec, C. (1987) In vitro fertilization with normal development in the sheep. *Gamete Research* 19, 291–303.

Cseh, S. and Seregi, J. (1993) Practical experiences with sheep embryo transfer. *Theriogenology* 39, 207.

Cseh, S., Besenfeldr, U., Treuer, A., Brem, G. and Seregi, J. (1995) Successful laparoscopic implantation of sheep embryos produced by in vitro fertilization of oocytes. *Proceedings of the 11th Meeting of the European Embryo Transfer Association* (Hanover), p. 152.

Czlonkowska, M., Papis, K., Guszkiewicz, A., Kossakowski, M. and Eysymont, U. (1991) Freezing of sheep embryos in 3.0 M methanol. *Cryo-letters* 12, 11–16.

Dattena, M., Vespignani, S., Branca, A., Gallus, M., Ledda, S., Naitana, S. and Cappai, P. (1994) Superovulatory response and quality of embryos recovered from anestrus ewes after a single injection of porcine FSH dissolved in polyvinylpyrrolidone. *Theriogenology* 42, 235–239.

Dauzier, L. and Thibault, C. (1959) Donnees nouvelles fur la fecondation in vitro de l'oeuf de la Lapine et de la brebis. *Comptes Rendus Academy of Science, Paris* 248, 2655–2656.

Dingwall, W.S., Fernie, K., Fitzsimons, J. and McKelvey, W.W.C. (1991) MOET in Suffolks: increasing the superovulation rate. *Proceedings of the British Society of Animal Production* (Winter Meeting), Paper No. 173.

Dingwall, W.S., McKelvey, W.A.C., Mylne, J. and Simms, G. (1993) An evaluation of MOET in Suffolk sheep. *Proceedings of the British Society of Animal Production* (Winter Meeting), Paper No. 103.

Driancourt, M.A. and Fry, R.C. (1992) Effect of superovulation with pFSH or PMSG on growth and maturation of the ovulatory follicles in sheep. *Animal Reproduction Science* 27, 279–292.

Earl, C.R., Irvine, B.J., Kelly, J.M., Rowe, J.P. and Armstrong, D.T. (1995) Ovarian stimulation protocols for oocyte collection and in vitro embryo production from 8 to 9 week old lambs. *Theriogenology* 43, 203.

Ebert, K.M. and Schindler, J.E.S. (1993) Transgenic farm animals: progress report. *Theriogenology* 39, 121–135.

Eckery, D.C., Moeller, C.L., Nett, T.M. and Sawyer, H.R. (1994) Recombinant bovine somatotropin does not improve superovulatory response in sheep. *Journal of Animal Science* 72, 2425–2430.

Edwards, R.G. (1965) Maturation in vitro of mouse, sheep, cow, pig, rhesus monkey and human ovarian oocytes. *Nature* 206, 349.

Eppleston, J., Bilton, R.J. and Moore, N.W. (1984) Effect of FSH dose and treatment regime on ovulatory response in sheep. *Proceedings of the Australian Society of Reproductive Biology* 16, 68.

Evans, G., Holland, M.K., Nottle, H.B., Sharpe, P.H. and Armstrong, D.T. (1984) Production of embryos in sheep using FSH preparations and laparoscopic intrauterine insemination. In: Lindsay, D.R. and Pearce, D.T. (eds) *Reproduction in Sheep*. Cambridge University Press, Cambridge, pp. 313–315.

Evans, G., Brooks, J., Struthers, W. and McNeilly, A.S. (1994) Superovulation and embryo recovery in ewes treated with gonadotrophin-releasing hormone agonist and purified follicle-stimulating hormone. *Reproduction, Fertility and Development* 6, 247–252.

Fernie, K., Dingwall, W.S., McKelvey, W.A.C. and Fitzsimons, J. (1994) Superovulation in the ewe: the effects of source of gonadotrophin, season, breed and age. *Proceedings of the British Society of Animal Production* (Winter Meeting), Paper No. 59.

Flores-Foxworth, G., Coonrod, S.A., Moreno, J.F., Byrd, S.R., Kraemer, D.C. and Westhusin, M. (1995) Interspecific transfer of IVM IVF-derived Red sheep (*Ovis orientalis gmelini*) embryos to domestic sheep (*Ovis areies*). *Theriogenology* 44, 681–690.

Fukui, Y., Kano, H., Kobayashi, M., Tetsura, M. and Ono, H. (1985) Response to repeated superovulation treatment in the ewe. *Japanese Journal of Animal Reproduction* 31, 155–157.

Fukui, Y., Glew, A.M., Gandolfi, F. and Moor, R.M. (1988a) Ram-specific effects on in vitro fertilization and cleavage of sheep oocytes matured in vitro. *Journal of Reproduction and Fertility* 82, 337–340.

Fukui, Y., Glew, A.M. Gandolfi, F. and Moor, R.M. (1988b) In vitro culture of sheep oocytes matured and fertilized in vitro. *Theriogenology* 29, 883–891.

Fukui, Y., Tashiro, Y., Kimura, H. and Miyamoto, A. (1994) Effects of progestagen treatment and season on superovulatory responses of ewes and developmental capacity of early embryos recovered. *Journal of Reproduction and Development* 40, 251–257.

Galli, C. and Moor, R.M. (1991) Gonadotrophin requirements for the in vitro maturation of sheep oocytes and their subsequent embryonic development. *Theriogenology* 35, 1083–1093.

Gandolfi, F. and Moor, R.M. (1987) Stimulation of early embryonic development in sheep by co-culture with oviduct epithelial cells. *Journal of Reproduction and Fertility* 81, 23–28.

Gatica, R. (1988) Studies in the micromanipulation of farm animal embryos. PhD Thesis, National University of Ireland, Dublin, 171 pp.

Gatica, R. and Correa, J.E. (1993) Manipulation of chimaeric sheep and goat embryos. *Agro Sur* 21, 102–108.

Gatica, R., Boland, M.P., Crosby, T.F. and Gordon, I. (1984) Micromanipulation of sheep morulae to produce monozygotic twins. *Theriogenology* 21, 555–560.

Gherardi, P.G. and Martin, G.B. (1978) The effect of multiple injections of pregnant mares serum gonadotrophin on the ovarian activity of Merino ewes. *Proceedings of*

the Australian Society of Animal Production 12, p. 260.
Gobel, W., Meinecke-Tillmann, S. and Meinecke, B. (1995) Transcervical embryo collection in small ruminants controlled by transrectal ultrasonography. Preliminary results. *Proceedings of the 11th Meeting of the European Embryo Transfer Association* (Hanover), p. 178.
Gogolevskii, P.A., Gol'dman, I.L., Baksheev, Y.D., Kadulin, S.G., Kasymov, K., Yazykov, A.A. and Ernst, L.K. (1993) Cytological problems of transgenic sheep raising. *Russian Agricultural Sciences* 51(8), 18–28.
Gol'dman, I.L., Baksheev, E.D., Gogolevskii, P.A., Kadulin, S.G., Yazykov, A.A., Novak, V.S., Dvoryanchikov, G.A., Rud'ko, N.P., Zhadanov, A.B. and Ernst, L.K. (1992) Advanced biotechnology for obtaining transgenic sheep. *Soviet Agricultural Sciences* 51(10), 1–7.
Greaney, K.B., Mcdonald, M.F., Vivanco, H.A. and Tervit, H.R. (1991) Out-of-season embryo transfer in five breeds of imported sheep. *Proceedings of the New Zealand Society Animal Production* 51, 129–131.
Grupp, T. (1991) Investigations on dilation of the cervix and transcervical embryo recovery in ewes and goats. Thesis, Ludwig-Maximilians-Universtitat Munchen, Germany, 144 pp.
Hammer, R.E., Pursel, V.G., Rexroad, C.E., Wall, R.J., Bolt, D.J., Ebert, K.M., Palmiter, R.D. and Brinster, R.L. (1985) Production of transgenic rabbits, sheep and pigs by microinjection. *Nature* 315, 680–685.
Hanada, A. (1985) A successful case of a lamb born by transfer of in vitro fertilized eggs in sheep. *Annual Research Report of the National Institute of Animal Husbandry*, pp. 49–51.
Haresign, W., Merrell, B. and Richards, R.I.W.A. (1994a) Multiple ovulation and embryo transfer in hill ewes: effects of mating system on embryo quality, and its relationship with pregnancy rates. *Proceedings of the British Society of Animal Production* (Winter Meeting), Paper No. 87.
Haresign, W., Merrell, B. and Richards, R.I.W.A. (1994b) Multiple ovulation and embryo transfer in hill ewes. *Proceedings of the British Scoiety of Animal Production* (Winter Meeting), Paper No. 88.
Henning, W.L. (1937) A double pregnancy with a single corpus luteum. *Journal of Heredity* 28, 61–62.
Hunter, G.L., Adams, C.E. and Rowson, L.E.A. (1955) Interbreed ovum transfer in sheep. *Journal of Agricultural Science, Cambridge* 46, 143–149.
Ivkov, V., Veselinovic, S., Micic, R., Veselinovic, S. and Medic, D. (1994) Detection of superovulation in donor ewes according to level of serum progesterone. *Proceedings of the 10th Meeting of the European Embryo Transfer Association* (Lyon), p. 188.
Ivkov, V., Vesselinovic, S., Micic, R., Vesselinovic Snezana and Medic, D. (1995) Levels of serum progesterone in recipient ewes before embryo transfer as an indication of subsequent fertility. *Proceedings of the 11th Meeting of the European Embryo Transfer Association* (Hanover), p. 192.
Jabbour, H.N. and Evans, G. (1991a) Ovarian and endocrine responses of Merino ewes to treatment with PMSG and/or FSH-P. *Animal Reproduction Science* 26, 93–106.
Jabbour, H.N. and Evans, G. (1991b) Ovarian and endocrine reponses of Merino ewes following treatment with PMSG and GnRH or PMSG antiserum. *Animal Reproduction Science* 24, 259–270.
Jabbour, H.N. and Evans, G. (1991c) Superovulation of Merino ewes with an ovine pituitary follicle stimulating hormone extract. *Reproduction, Fertility and Development* 3, 561–569.

Jabbour, H.N., Ryan, J.P., Evans, G. and Maxwell, W.M.C. (1991) Effects of season, GnRH administration and lupin supplementation on the ovarian and endocrine responses of Merino ewes treated with PMSG and FSH-P to induce superovulation. *Reproduction, Fertility and Development* 3, 699–707.

Jablonka-Shariff, A., Fricke, P.M., Grazul-Bulska, A.T., Reynolds, L.P. and Redmer, D.A. (1994) Size, number, cellular proliferation and atresia of gonadotropin-induced follicles in ewes. *Biology of Reproduction* 51, 531–540.

Johansson, I. and Hansson, A. (1943) The sex ratio and multiple births in sheep. *Annals Agricultural College of Sweden* 11, 145.

Joly, T., Nibart, M., Chemineau, P. and Thibier, M. (1993) Use of melatonin implant to enhance ovarian response of superovulated ewes. *Proceedings of the 9th Meeting of the European Embryo Transfer Association* (Lyon), p. 216.

Kardymowicz, O. (1972) Successful in vitro storage of fertilized sheep ova for ten days. *Proceedings of the 7th International Congress of Animal Reproduction and AI* (Munich) 1, pp. 499–502.

Kaulfuss, K.H., Brabnt, S., Blume, K. and May, J. (1995) The improvement of embryo transfer programmes in sheep by examination of the ovary response in superovulated donor ewes by transrectal real-time ultrasound. *Deutsche Tierarztliche Wochenschrift* 102, 208–212.

Kelly, R.W., Owens, J.L., Crosbie, S.F., McNatty, K.P. and Judson, N. (1983) Influence of Booroola Merino genotype on the responsiveness of ewes to pregnant mares serum gonadotrophin, luteral tissue weights and peripheral progesterone concentration. *Animal Reproduction Science* 6, 199–207.

Killeen, I.D. and Moore, N.W. (1971) The morphological appearance and development of sheep ova fertilized by surgical insemination. *Journal of Reproduction and Fertility* 24, 63–70.

Land, R.B. and McGovern, P.T. (1968) Ovulation and fertilization in the lamb. *Journal of Reproduction and Fertility* 15, 325–327.

Larsson, B., Gustafsson, A., Nasholm, A. and Bjurstrom, L. (1991) A programme for oestrus synchronization and embryo transfer in sheep. *Reproduction in Domestic Animals* 26, 301–308.

Lawson, R.A.S., Adams, G.E. and Rowson, L.E.A. (1972) The development of sheep eggs in the rabbit oviduct and their viability after re-transfer to ewes. *Journal of Reproduction and Fertility* 29, 105–116.

Ledda, S., Naitana, S., Loi, P., Dattena, M., Gallus, M., Braca, A. and Cappai, P. (1995) Embryo recovery from superovulated mouflons (*Ovis gmelini musimon*) and viability after transfer into domestic sheep. *Animal Reproduction Science* 39, 109–117.

Lopez-Sebastian, A., Cognie, Y., Cocero, M.J., De La Fuente, J. and Poulin, N. (1990) Effect of season and duration of FSH treatment on embryo production in sheep. *Theriogenology* 34, 175–180.

Lopez-Sebastian, A., Gomez-Brunet, A., Lishman, A.W., Johnson, S.K. and Inskeep, E.K. (1993) Modification by propylene glycol of ovulation rate in ewes in response to a single injection of FSH. *Journal of Reproduction and Fertility* 99, 437–442.

Lynch, J.J. (1968) Superovulation and egg transfer in sheep. Master of Agricultural Science Thesis, University College, Dublin.

McEvoy, T.G., Robinson, J.J., Aitken, R.P., Kyle, C.E. and Robertson, I.S. (1993) The effect of feeding level during a 12-day progesterone-priming period on the viability of embryos collected from superovulated ewes. *Proceedings of the British Society of Animal Production* (Winter Meeting), Paper No. 57.

McEvoy, T.G., Robinson, J.J., Aitken, R.P., Findlay, P.A. and Robertson, I.S. (1995a) Relationship between pre-ovulatory feed intake, progesterone priming and the subsequent in vitro development of ova collected from superovulated ewes. *Theriogenology* 43, 276.

McEvoy, T.G., Robinson, J.J., Aitken, R.P., Findlay, P.A., Palmer, R.M. and Robertson, I.S. (1995b) Dietary-induced suppression of pre-ovulatory progesterone concentrations in superovulated ewes impairs the subsequent in vivo and in vitro development of their ova. *Animal Reproduction Science* 39, 89–107.

McGinnis, L.K., Duplantis, S.C. Jr. and Youngs, C.R. (1993) Cryopreservation of sheep embryos using ethylene glycol. *Animal Reproduction Science* 30, 273–280.

McLaughlin, K.J., Davies, L. and Seamark, R.F. (1990) In vitro embryo culture in the production of identical Merino lambs by nuclear transplantation. *Reproduction, Fertility and Development* 2, 619–622.

McLaughlin, K.J., Pugh, A.P., Logan, K. and Tervit, H.R. (1991) Assessment of oocyte source for ovine nuclear transfer. *Theriogenology* 35, 240.

Madibela, O.R., McEvoy, T.G., Robinson, J.J., Findlay, P.A., Aitken, R.P. and Robertson, I.S. (1995) Excess rumen degradable protein influences the rate of development and glucose metabolism of fertilized sheep ova. *Proceedings of the British Society of Animal Production* (Winter Meeting), Paper No. 98.

Mansour, A.M. (1959) The hormonal control of ovulation in the immature lamb. *Journal of Agricultural Science, Cambridge* 52, 87–94.

Marden, W.G.R. and Chang, M.C. (1952) The aerial transport of mammalian ova for transplantation. *Science* 115, 705.

Martemucci, G., D'Alessandro, A., Toteda, F., Facciolongo, A.M. and Gambacorta, M. (1995) Embryo production and endocrine response in ewes superovulated with PMSG, with or without monoclonal anti-PMSG administered at different times. *Theriogenology* 44, 691–703.

Maxwell, W.M.C., Evans, G., Rhodes, S.L., Hillard, M.A. and Bindon, B.M. (1993) Fertility of superovulated ewes after intrauterine or oviductal insemination with low numbers of fresh or frozen-thawed spermatozoa. *Reproduction, Fertility and Development* 5, 57–63.

Meinecke-Tillmann, S., Lewalski, H. and Meinecke, B. (1992) Superovulatory reaction of ewes after a single injection of FSH given at two different times of the synchronized cycle. *Proceedings of the 8th Meeting European Embryo Transfer Association* (Lyon), p. 192.

Meinecke-Tillman, S., Lewaslski, H. and Meinecke, B. (1993) Induction of superovulation in Merinolandschaf ewes after single or multiple FSH injections. *Reproduction in Domestic Animals* 28, 433–440.

Mocker, S. (1994) In vitro fertilization in sheep. Investigations on the effects of heparin, caffeine and glucose on in vitro sperm capacitation. Thesis, University of Munich, 120 pp.

Monniaux, D., Mariana, J.C. and Gibson, W.R. (1984) Action of PMSG on follicular populations in the heifer. *Journal of Reproduction and Fertility* 70, 243–253.

Moor, R.M. and Trounson, A.O. (1977) Hormonal and follicular factors affecting maturation of sheep oocytes in vitro and their subsequent developmental capacity. *Journal of Reproduction and Fertility* 49, 101–109.

Moor, R.M., Osbourne, J.C. and Crosby, I.M. (1985) Gonadotrophin-induced abnormalities in sheep oocytes after superovulation. *Journal of Reproduction and Fertility* 74, 167–172.

Moore, N.W. (1976) Culture, storage and transfer of sheep embryos. *Proceedings of the*

International Congress on Sheep Breeding, Muresk, pp. 495–499.
Moore, N.W. (1982) Egg transfer in the sheep and goat. Chapter 7. In: Adams, C.E. (ed.) *Mammalian Egg Transfer*. CRC Press, Boca Raton, Florida, pp. 119–133.
Moore, N.W. and Shelton, J.N. (1962a) Oestrous and ovarian response of the ewe to a horse anterior pituitary extract. *Nature* 194, 1283–1284.
Moore, N.W. and Shelton, J.N. (1962b) The application of the technique of egg transfer to sheep breeding. *Australian Journal of Agricultural Research* 13, 718–724.
Moore, N.W. and Shelton, J.N. (1964) Response of the ewe to a horse anterior pituitary extract. *Journal of Reproduction and Fertility* 7, 79–87.
Moore, N.W. and Bilton, R.J. (1976) Storage, culture and transfer of embryos of domestic animals. *Proceedings of the 8th International Congress of Animal Reproduction and AI* (Kracow), 3, 306.
Mylne, M.J.A., McKelvey, W.A.C., Fernie, K. and Matthews, K. (1992) Use of a transcervical technique for embryo recovery in sheep. *Veterinary Record* 130, 450–451.
Naitana, S., Dattena, M., Branca, A., Ledda, S., Loi, P. and Cappai, P. (1994) High viability rate of vitrified ovine embryos. *Proceedings of the 10th Meeting of the European Embryo Transfer Association* (Lyon), 224.
Nancarrow, C.D. and Ward, K.A. (1995) The commercial and agricultural applications of animal transgenesis. *Molecular Biotechnology* 4, 167–178.
Nellenschulte, E. and Niemann, H. (1992) Collection and transfer of ovine embryos by laparoscopy. *Animal Reproduction Science* 27, 293–304.
Newton, J.E. and Betts, J.E. (1968) The effect of superovulation, synchronization and the ram on ewe litter size. *Proceedings of the 6th International Congress on Reproduction and AI* (Paris), Resumes, 289.
O'Brien, J.K., Rhodes, S.L., Maxwell, W.M.C. and Evans, G. (1994) Hormonal requirements for in vitro maturation of sheep oocytes. *Theriogenology* 41, 266.
Palsson, H. (1962) Hormonal augmentation of fertility in sheep by PMS. *Journal of Reproduction and Fertility* 3, 55–63.
Parr, R.A. (1992) Nutrition–progesterone interactions. *Reproduction, Fertility and Development* 4, 297–300.
Piper, L.R., Bindon, B.M., Curtis, Y.M., Cheers, M.A. and Nethery, R.D. (1982) Response to PMSG in Merino and Booroola Merino crosses. *Proceedings of the Australian Society of Reproductive Biology* 14, p. 82.
Pursel, V.G. and Rexroad, C.E. (1993) Recent progress in the transgenic modification of swine and sheep. *Molecular Reproduction and Development* 36, 251–254.
Quirke, J.F. and Hanrahan, J.P. (1977) Comparison of the survival in the uteri of adult ewes of cleaved ova from adult ewes and ewe lambs. *Journal of Reproduction and Fertility* 51, 487–489.
Rall, W.F. and Fahy, G.M. (1985) Ice-free cryopreservation of mouse embryos at –196°C by vitrification. *Nature* 313, 573–575.
Rangel-Santos, R., McDonald, M.F. and Wickham, G.A. (1991) Evaluation of the feasibility of a juvenile MOET scheme in sheep. *Proceedings New Zealand Society Animal Production* 51, pp. 139–142.
Renard, J.P., Winterberger-Torres, S. and Du Mesnil du Buisson, F. (1976) Storage of ewe and cow eggs at 10°C. In: Rowson, L.E.A. (ed.) *Egg Transfer in Cattle*. EEC Symposium (Cambridge), pp. 165–171.
Rexroad, C.E. Jr. and Wall, R.J. (1987) Development of one-cell fertilized sheep ova following microinjection into pronuclei. *Theriogenology* 27, 611–619.
Rexroad, C.E. Jr. and Powell, A.M. (1991) FSH injections and intrauterine insemina-

tion in protocols for superovulation of ewes. *Journal of Animal Science* 69, 246–251.

Riddell, M.G., Stringfellow, D.A., Wolfe, D.F. and Galik, P.K. (1989) In vitro exposure of ovine ova to *Brucella abortus*. *Theriogenology* 31, 895–901.

Riesenberg, S., Lewalski, H., Meinecke-Tillmann, S. and Meinecke, B. (1995) Ultrasonic documentation of follicular dynamics following different superovulatory regimens in small ruminants – preliminary results. *Proceedings of the 11th Meeting European Embryo Transfer Association* (Hanover), p. 234.

Robinson, J.J., Wallace, J.M. and Aitken, R.P. (1989a) Fertilization and ovum recovery rates in superovulated ewes following cervical insemination or laparoscopic intrauterine insemination or laparoscopic intrauterine insemination at different times after progestagen withdrawal and in one or both uterine horns. *Journal of Reproduction and Fertility* 87, 771–782.

Robinson, J.J., Wallace, J. and Atiken, R. (1989b) Significance of improved embryo transfer techniques for sheep breeding. *Agriculture and Food Research Council News* 16–17.

Robinson, J.J., McEvoy, T.G., Aitken, R.P. and Robertson, I. (1994) Melatonin treatment of donor and recipient ewes in an embryo transfer programme: effects on ovulation rate and embryo survival during normal seasonal anoestrus. *Proceedings of the British Society of Animal Production* (Winter Meeting), Paper No. 78.

Rorie, R.W., Pool, S.H., Prichard, J.F., Betteridge, K.J. and Godke, R.A. (1994) A simplified procedure for making reconstituted blastocysts for interspecific and intergeneric transfer. *Veterinary Record* 135, 186–187.

Roth, T.L., Anderson, G.B., BonDurant, R.H. and Pashen, R.L. (1989) Survival of sheep × goat hybrid inner cell masses after injection into ovine embryos. *Biology of Reproduction* 41, 675–682.

Rowson, L.E.A. and Moor, R.M. (1964) Occurrence and development of identical twins in sheep. *Nature* 201, 521–522.

Rowson, L.E.A. and Moor, R.M. (1966) Embryo transfer in the sheep: the significance of synchronizing oestrus in donor and recipient animal. *Journal of Reproduction and Fertility* 11, 207–212.

Rubianes, E., Ibarra, D., Ungerfeld, R., Carbajal, B. and Castro, T. De (1995) Superovulatory response in anestrous ewes is affected by the presence of a large follicle. *Theriogenology* 43, 465–472.

Ryan, J.P., Bilton, R.J. and Hunton, J.R. (1984) Superovulation of ewes with a combination of PMSG and FSH-P. In: Lindsay, D.R. and Pearce, D.T. (eds) *Reproduction in Sheep*. Cambridge University Press, Cambridge, pp. 338–341.

Ryan, J.P., Hunton, J.R. and Maxwell, W.M.C. (1991) Increased production of sheep embryos following superovulation of Merino ewes with a combination of pregnant mare serum gonadotropin and follicle stimulating hormone. *Reproduction, Fertility and Development* 3, 551–560.

Sakul, H., Bradford, G.E., Bondurant, R.H., Anderson, G.B. and Donahue, S.E. (1992) Cryopreservation of embryos as a means of germ plasm conservation in sheep. *Journal of Animal Science* 70 (Suppl. 1), Abstract 26.

Samartzi, F., Boscos, C., Vainas, E. and Tsakalof, P. (1995) Superovulatory response of Chios sheep to PMSG during spring and autumn. *Animal Reproduction Science* 39, 215–222.

Schiewe, M.C., Howard, J.G., Goodrowe, K.L., Stuart, L.D. and Wildt, D.E. (1990a) Human menopausal gonadotropin induces ovulation in sheep, but embryo recovery after prostaglandin $F_{2\alpha}$ synchronization is compromised by premature

luteal regression. *Theriogenology* 34, 469–486.
Schiewe, M.C., Rall, W.F., Stuart, O.D. and Wildt, D.E. (1990b) In situ straw dilution of ovine embryos cryopreserved by conventional freezing or vitrification. *Theriogenology* 33, 321.
Scudamore, C.L., Robinson, J.J., Aitken, R.P., Kennedy, D.J., Ireland, S. and Robertson, I.S. (1991a) Laparoscopy for intrauterine insemination and embryo recovery in superovulated ewes at a commercial embryo transfer unit. *Theriogenology* 35, 329–337.
Scudamore, C.L., Robinson, J.J. and Aitken, R.P. (1991b) The effect of timing of laparoscopic insemination in superovulated ewes, with or without sedation, on the recovery of embryos, their stage of development and subsequent viability. *Theriogenology* 35, 907–914.
Scudamore, C.L., Robinson, J.J., Aitken, R.P. and Robertson, I.S. (1992) A comparison of two doses of fluorogestone acetate in pessaries on the quality of embryos recovered from superovulated ewes. *Theriogenology* 37, 445–456.
Scudamore, C.L., McEvoy, T.G., Aitken, R.P., Robinson, J.J. and Robertson, I.S. (1993) The effect of two different levels of progesterone priming on the response of ewes to superovulation. *Theriogenology* 39, 433–442.
Scudamore, C.L., Robinson, J.J., Aitken, R.P. and Robertson, I.S. (1994) The effect of method of oestrous synchronization on the response of ewes to superovulation with porcine follicle stimulating hormone. *Animal Reproduction Science* 34, 127–133.
Solter, D. (1996) Lambing by nuclear transfer. *Nature* 380, 24–25.
Songasen, N., Buckrell, B.C., Plante, C. and Leibo, S.P. (1995) In vitro and in vivo survival of cryopreserved sheep embryos. *Cryobiology* 32, 78–91.
Shelton, J.N. and Moore, N.W. (1966) Survival of fertilized eggs transferred to ewes after progesterone treatment. *Journal of Reproduction and Fertility* 11, 149–151.
Slavik, T. and Fulka, J. (1991) Pregnancies after transfer of sheep embryos produced from oocytes matured and fertilized in vitro. *Folia Biologica-Praha* 37, 94–100.
Slavik, T., Fulka, J. and Goll, I. (1992) Pregnancy rate after the transfer of sheep embryos originated from randomly chosen oocytes matured and fertilized in vitro. *Theriogenology* 38, 749–756.
Smith, C.L. (1984) Dose effect of follicle stimulating hormone for superovulation of crossbred Targhee ewes. *Theriogenology* 21, 262.
Smith, J.F. (1976) Selection for fertility and response to PMSG in Romney ewes. *Proceedings of the New Zealand Society of Animal Production* 36, pp. 247–251.
Smith, C. (1988) Genetic improvement of livestock, using nucleus breeding units. *World Animal Review* 65, 2–10.
Smith, L.C. and Wilmut, I. (1989) Influence of nuclear and cytoplasmic activity on the development in vivo of sheep embryos after nuclear transplantation. *Biology of Reproduction* 40, 1027–1035.
Smith, C.L., Peter, A.T. and Appell, K.M. (1994) Effects of stepwise cryodilution prior to freezing and stepwise post-thaw rehydration on viability of ovine embryos. *Theriogenology* 41, 1267–1271.
Smorag, Z., Wierzbowski, S. and Wierzchos, E. (1978) Results of transplanting 3-, 6- and 7-day old frozen sheep embryos. *Bulletin Academy of Polish Science* 26, 273–275.
Staigmiller, R.B. and Moor, R.M. (1984) Effect of follicle cells on the maturation and

developmental competence of ovine oocytes matured outside the follicle. *Gamete Research* 9, 221–229.

Stefani, J.S., Palma, M.D.C. and Roodrigues, J.L. (1991) Ovine superovulation response to HMG. *Proceedings of the 7th Meeting European Embryo Transfer Association* (Cambridge), p. 210.

Steyn, M.C., Morgenthal, J.C. and Barry, D.M. (1993) The effect of embryo collection technique on subsequent fertility in SA Mutton Merino ewes. *Theriogenology* 39, 317.

Szell, A.Z. and Windsor, D.P. (1994) Survival of vitrified sheep embryos in vitro and in vivo. *Theriogenology* 42, 881–889.

Szell, A., MacLeod, I.M., Windsor, D.P. and Kelly, R.W. (1994) Production of identical twin lambs by embryo splitting. *Theriogenology* 41, 1643–1652.

Tervit, H.R. and Havik, P.G. (1976) A modified technique for flushing ova from the sheep uterus. *New Zealand Veterinary Journal* 24, 138–140.

Tervit, H.R. and McDonald, M.F. (1969) Culture and transplantation of sheep ova. *New Zealand Journal of Agricultural Research* 12, 313.

Tervit, H.R. and Rowson, L.E.A. (1974) Birth of lambs after culture of sheep ova in vitro for up to 6 days. *Journal of Reproduction and Fertility* 38, 177–179.

Tervit, H.R., Whittingham, D.G. and Rowson, L.E.A. (1972) Successful culture in vitro of sheep and cattle ova. *Journal of Reproduction and Fertility* 30, 493–497.

Thibault, C. and Dauzier, L. (1961) Analyse des conditions de la fecondation in vitro de l'oeuf de la lapine. *Annales de Biologie animale Biochimie Biophysique* 1, 277.

Thompson, J.G.E., Simpson, A.C., James, R.W. and Tervit, H.R. (1990) The application of progesterone-containing CIDR devices to superovulated ewes. *Theriogenology* 33, 1297–1304.

Torres, S. and Cognie, Y. (1984) Superovulation and egg transfer in the ewe. *Reproduction, Nutrition and Development* 24, 623–631.

Torres, S. and Sevellec, C. (1987) Repeated superovulation and surgical recovery of embryos in the ewe. *Reproduction, Nutrition and Development* 27, 859–863.

Torrie, S., Cognie, Y. and Colas, G. (1987) Transfer of superovulated sheep embryos obtained with different FSH-P. *Theriogenology* 27, 407–419.

Trounson, A.O. and Moore, N.W. (1974a) Fertilization in the ewe following multiple ovulation and uterine insemination. *Australian Journal of Biological Science* 27, 301–304.

Trounson, A.O. and Moore, N.W. (1974b) Attempts to produce identical offspring in the sheep by mechanical division of the ovum. *Australian Journal of Biological Science* 27, 505–510.

Trounson, A.O., Willadsen, S.M. and Moor, R.M. (1976) Effect of prostaglandin analogue Cloprostenol and oestrus, ovulation and embryonic viability in sheep. *Journal of Agricultural Science, Cambridge* 86, 609–611.

Vallett, J.C., Folch, J., Poulin, N. and Cognie, Y. (1989) Surgical or laparoscopic embryo transfer in sheep before or after the first cleavages. *Proceedings of the 5th Meeting European of the Embryo Transfer Association* (Lyon), p. 188.

Wahid, H., Gordon, I., Sharif, H., Lonergan, P., Monaghan, P. and Gallagher, M. (1991) Development of ovine blastocysts following maturation, fertilization and culture of oocytes in vitro. *Proceedings of the 7th Meeting of the European Transfer Association* (Cambridge), p. 214.

Wahid, H., Monaghan, P. and Gordon, I. (1992) In vitro maturation (IVM) of the sheep follicular oocyte. *Journal of Reproduction and Fertility* Abstract Series No. 9, p. 52.

Walker, S.K., Smith, D.H. and Seamark, R.F. (1986) Timing of multiple ovulations in the ewe after treatment with FSH or PMSG with and without GnRH. *Journal of Reproduction and Fertility* 77, 135–142.

Walker, S.K., Smith, D.H., Frensham, A., Ashman, R.J. and Seamark, R.F. (1989) The use of synthetic gonadotropin releasing hormone treatment in the collection of sheep embryos. *Theriogenology* 31, 741–752.

Walker, S.K., Heard, T.M., Verma, P.J., Rogers, G.E., Bowden, C.S., Sivaprasad, A.V., McLaughlin, K.J. and Seamark, R.F. (1990) In vitro assessment of the viability of sheep zygotes after pronuclear microinjection. *Reproduction, Fertility and Development* 2, 633–640.

Walker, S.K., Hill, J.L., Bee, C.A. and Warnes, D.M. (1994) Improving the rate of production of sheep embryos using in vitro maturation and fertilization. *Theriogenology* 41, 330.

Wallace, J.M., Aitken, R.P. and Cheyne, M.A. (1993) Post-ovulation nutritional status in ewes does not influence early conceptus development in vivo or luteotrophic protein secretion in vitro. *Proceedings of the British Society Animal Production* (Winter Meeting), Paper No. 58.

Ward, K.A., Murray, J.D., Nancarrow, C.D., Boland, M.P. and Sutton, R. (1984) The role of embryo gene transfer in sheep breeding programmes. In: Lindsay, D.R. and Pearce, D.T. (eds) Reproduction in Sheep. Cambridge University Press, Cambridge, pp. 279–285.

Warnes, G.N., Moor, R.N. and Johnson, M.H. (1977) Changes in protein synthesis during maturation of sheep oocytes in vivo and in vitro. *Journal of Reproduction and Fertility* 49, 331–335.

Warwick, B.L., Berry, R.O. and Horlacher, W.R. (1934) Results of mating rams to angora female goats. *Proceedings of the 27th Annual Meeting of the American Society of Animal Production*, pp. 225–227.

Welch, R.A.S. (1969) Transport of sheep ova in rabbits. *Proceedings of the New Zealand Society of Animal Production* 29, 87–94.

Whittingham, D.G. (1971) Survival of mouse embryos after freezing and thawing. *Nature* (London) 233, 125.

Whyman, D. and Moore, R.W. (1980) Effects of PMSG and the prostaglandin $F_{2\alpha}$ analogue, Cloprostenol, on superovulation, fertilization and egg transport in the ewe. *Journal of Reproduction and Fertility* 60, 267–272.

Willadsen, S.M. (1979a) Embryo transplantation in sheep. In: *Management and Diseases of Sheep.* Commonwealth Agricultural Bureaux, Slough, pp. 69–85.

Willadsen, S.M. (1979b) A method for culture of micromanipulated sheep embryos and its use to produce monozygotic twins. *Nature* 277, 289–300.

Willadsen, S.M. (1980a) The viability of early cleavage stages containing half the normal number of blastomeres in the sheep. *Journal of Reproduction and Fertility* 59, 357–362.

Willadsen, S.M. (1980b) Deep freezing of embryos in the large domestic species. *Proceedings of the 9th International Congress of Animal Reproduction and AI* (Madrid), 255–261.

Willadsen, S.M. (1986) Nuclear transplantation in sheep embryos. *Nature* 320, 63–66.

Willadsen, S.M., Polge, C., Rowson, L.E.A. and Moor, R.M. (1974) Preservation of sheep embryos in liquid nitrogen. *Crybiology* 11, 560.

Willadsen, S.M., Polge, C., Rowson, L.E.A. and Moor, R.M. (1976) Deep freezing of sheep embryos. *Journal of Reproduction and Fertility* 46, 151–154.

Wilmut, I. and Sales, D.I. (1981) Effect of an asynchronous environment on embryonic development in sheep. *Journal of Reproduction and Fertility* 61, 179–184.

Wolf, B.T. and McDougal, I. (1994) Comparison of MOET in Texel ewe lambs and yearling ewes. *Proceedings of the British Society of Animal Production* (Winter Meeting), Paper No. 85.

Wolf, B.T. and Mylne, M.J.A. (1994) Influence of age of donor ewe on MOET in Texel sheep. *Proceedings of the British Society of Animal Production* (Winter Meeting), Paper No. 86.

Worthington, C.A. and Kennedy, J.P. (1979) Ovarian response to exogenous hormones in six-week old lambs. *Australian Journal of Biological Science* 32, 91–95.

Wright, R.W. Jr, Bondioli, K.R., Grammer, J.C., Kuzan, F. and Menino, A.R. Jr (1980) FSH or FSH+LH superovulation in ewes following estrus synchronization with medroxyprogesterone acetate pessaries. *Journal of Animal Science* 51 (Suppl. 1), 339.

11 Breeding Sheep at Younger Ages

11.1. Introduction

For those formulating proposals for improving the productivity of lowland sheep, breeding the ewe as early as possible would be an obvious consideration. Farmers who are attempting to improve their efficiency and to counter rising production costs are likely to adopt early breeding practices more readily if they feel the outcome is likely to be successful. The worldwide advantages and disadvantages of breeding sheep as ewe-lambs was reviewed more than 20 years ago by Dyrmundsson (1973); at that time, the author observed that for this practice to be acceptable to farmers, the ewe-lamb's reproductive performance must be satisfactory without this adversely affecting subsequent lifetime performance.

Increasing lifetime performance

It is certainly possible that lifetime lamb production can be increased by reducing the unproductive periods in a breeding ewe's lifetime. One of the most unproductive periods is from weaning until first breeding. In the USA, Hulet *et al.* (1969) demonstrated that ewe-lambs reaching puberty during the first breeding season possess a higher production potential than other ewe-lambs, even if they are not bred in their first year of life. Certainly, in lowland flocks, in countries such as Ireland, under conditions of good feeding and management, there is every reason for arguing that ewes should be selected, managed and bred to lamb first at one year of age. The obvious advantages of breeding 7–8-month-old ewe-lambs, to give birth at about a year, include reduced maintenance costs before the start of reproduction, a shortened generation interval that results in more rapid genetic gains from selection and increased lifetime production.

However, the success rate in breeding ewe-lambs can vary markedly, according to bodyweight and size, according to breed type and age and according to the time of year chosen for breeding. The fact is that sheep farmers throughout the world have usually bred ewes for the first time at the yearling stage. There is, nevertheless, growing interest in sexual and reproductive performance in ewe-lambs, particularly in relation to systems of intensified sheep production.

Controlled reproduction

Controlled reproduction can be applied to ewe-lambs at 7–10 months of age to advance their breeding season by several weeks, to synchronize oestrus in those that have already reached puberty and to induce oestrus in lambs that may not have mated at all during their first year of life (Fig. 11.1). There is even the possibility of delaying the induction of oestrus until the sheep is one year or more of age and getting the first lambs born in the early autumn when the sheep is some 18 months old.

11.1.1. Puberty and sexual maturity

Puberty in the ewe-lamb has been defined as the time of first ovulation and first oestrus, and occurs only during the breeding season (Adams and Steiner, 1988). As will be noted later, however, the first ovulation precedes first oestrus by two or three weeks in this species. Although sexual maturity is a term

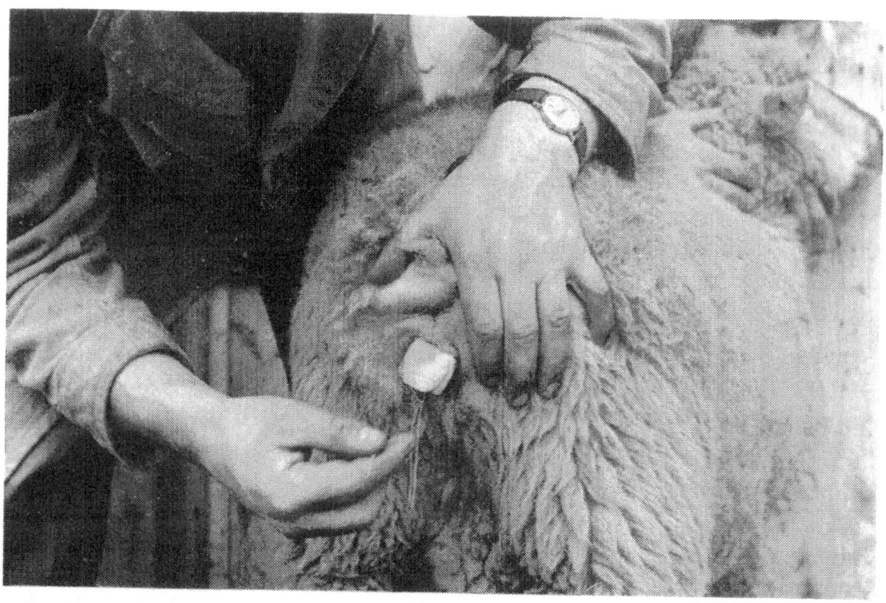

Fig. 11.1. Provided the ewe-lamb is sufficiently well grown, the use of the intravaginal sponge, usually with a dose of PMSG at the end of treatment, can be extremely useful in bringing these young females into oestrus. Care must always be taken during the insertion and withdrawal of sponges because of the relatively smaller size of the vagina compared with adult sheep – and these are not tasks to be undertaken by inexperienced personnel.

Fig. 11.2. Autumn lambing at 18 months of age after hormone treatment in the spring (from Gordon, 1958).

occasionally used as an alternative to puberty, it should be noted that the terms should relate to two different states in sheep: puberty is the time at which reproduction first becomes possible whereas sexual maturity is not reached until the animal expresses its full reproductive potential. In ewe-lambs such a distinction is important, since ewes do not acquire their full reproductive capacity until the adult stage is reached.

11.2. Physiology and Endocrinology of Puberty in Sheep

11.2.1. Endocrine events

Puberty is the process whereby the young female sheep becomes capable of spontaneous ovulation and a fertile mating; in sheep, such events occur as a consequence of activation of the gonadotrophin surge mechanism by the positive (stimulatory) feedback action of oestradiol. In sheep, competency to respond to oestradiol positive feedback becomes established within a few weeks of birth and the magnitude of the luteinizing hormone (LH) discharge in response to exogenous oestradiol is similar to that of the adult sheep by 27 weeks of age (Chu *et al.*, 1979). The evidence is that many of the endocrine mechanisms are capable of operating long before they are normally called on to function in the ewe-lamb.

LH release

In contrast to the inactivity of the mechanism governing LH surge in the prepubertal lamb, that regulating tonic LH secretion is relatively active throughout the prepubertal period. Although tonic LH production in the ewe-lamb is characterized by pulsatile releases of the gonadotrophin from the pituitary this apparently does not apply to follicle stimulating hormone (FSH) (Foster *et al.*, 1975); the indications are that the mechanisms which regulate the tonic secretion of LH and FSH in the growing ewe-lamb differ appreciably. Evidence in the 1970s suggested that tonic LH levels increased from the early weeks after birth (Hanrahan *et al.*, 1977). As noted by Adams and Steiner (1988), in some sheep, pulsatile release of LH starts at 4 weeks of age and is observed in virtually all animals by 9–11 weeks; the pulsatile mode of release results in mean LH levels that exceed the baseline values found in the adult ewe.

Tonic LH secretion occurs in the form of pulsatile releases, the pulse rate being less than one per hour in the developing lamb. Puberty is the time during the ewe-lamb's development when hourly LH pulses (LH pulse frequency) are first permitted to occur, a situation which appears to be due to a reduction in the negative feedback action of ovarian oestradiol (Foster and Ryan, 1979); what was not clear for some time was the nature of the stimulus for the critical decrease in the responsiveness of the ewe-lamb to oestradiol feedback.

11.2.2. Sexual differentiation of the gonadotrophin releasing hormone neurosecretory apparatus

It is now known that an important mechanism determining the onset of puberty in sheep is the increase in secretion of GnRH, the neuropeptide that controls the release of the pituitary gonadotrophins. According to Bucholtz and Foster (1994), during prenatal development in the sheep, the GnRH neurosecretory apparatus becomes sexually differentiated. In the ewe-lamb, high frequency GnRH secretion at puberty normally starts in response to postnatal photoperiodic cues (long-day to short-day length in the autumn). In the ram-lamb, however, which is differentiated to be photoperiodically insensitive, the pubertal onset of high frequency GnRH secretion is normally timed by postnatal growth cues.

Endogenous opioid mechanisms

Studies in the USA have examined the endogenous opioid regulation of pulsatile LH secretion as a factor influencing puberty in sheep. Ebling *et al.* (1989a) concluded that although endogenous opioid mechanisms are an important inhibitory mechanism controlling pulsatile LH in the growing ewe-lamb, changes in opioid inhibition are unlikely to underlie the decrease in sensitivity to steroid negative feedback and increase in LH pulse frequency.

11.2.3. Endocrine events at puberty

It is believed that there is a sustained rise, which occurs in the space of a few days, in the tonic LH baseline at the onset of puberty, due to an increase in the LH pulse frequency to about one per hour; this results in one or more follicles developing towards the preovulatory stage and in a steady increase in oestradiol production, which eventually activates the preovulatory LH surge mechanism.

As noted earlier, it is believed that puberty occurs because of a marked decrease in the response of the hypothalamic–pituitary axis to the negative feedback action of oestradiol on tonic LH secretion. There is evidence, for example, showing that removing the ovaries of the young lamb increases the LH pulse frequency to once per hour (Foster *et al.*, 1975); other studies showed that artificially producing such a rapid LH pulse rate (by administering LH hourly) will result in an LH surge and ovulation (Foster and Ryan, 1979).

Hypersensitivity to oestradiol

Such data were taken to indicate that the young lamb is readily capable of producing the hourly pulses of LH if inhibitory ovarian steroids are removed and to show that the ovaries are capable of producing oestradiol in amounts sufficient to invoke the LH surge, if they are artificially exposed to more frequent LH stimulation. The similarity of puberty in the ewe-lamb and the onset of the breeding season in adult sheep, in which there is evidence that the season starts because of a marked reduction in response to the inhibitory feedback effect of oestradiol on tonic LH (Legan *et al.*, 1977) suggested to the Michigan workers that hypersensitivity to oestradiol feedback on LH secretion may be the final common mechanism at work in both the prepubertal ewe-lamb and the anoestrous ewe.

The gonadostat theory of puberty

According to this view, the ewe-lamb becomes less inhibited by oestradiol as it grows older, this eventually enabling the LH pulse frequency to occur at a level sufficient to cause follicle development, oestrogen production and the preovulatory LH surge which leads to the first ovulation and the initiation of puberty. The ewe-lamb is believed to conform to the gonadostat theory of puberty (Ramirez and McCann, 1963) by exhibiting a change in frequency (increase) of pulsatile LH secretion in response to increasing oestradiol concentrations during development; circulating oestradiol concentrations that are able to suppress gonadotrophin secretion in prepubertal females are less effective in gonadotrophin suppression after puberty.

Events preceding first ovulations

A study by Ryan *et al.* (1991) provided evidence that the first gonadotrophin surge in the ewe-lamb during maturation is invariably preceded by a sustained rise in circulating oestradiol. In view of the evidence showing that the lamb exhibits a peripubertal decrease in the ability of oestradiol to inhibit the hypothalamic–pituitary axis regulating LH secretion (Foster and Ryan, 1979), it appears that this important change in feedback effect allows a stimulatory rise in gonadotrophic secretion to occur, resulting in the final endocrine cascade of events leading to the first ovulation.

Although oestradiol inhibits the secretion of LH in the prepubertal ewe-lamb, it has been shown that this suppression is not absolute; it is known that LH pulses occur and that there are weekly variations in their frequency. Studies reported by Friedman *et al.* (1992) indicated that a dynamic relationship exists between the ovaries and the neuroendocrine system before puberty. They suggested that fluctuations in LH pulse frequency may arise from changes in the amount of oestradiol secreted by developing and regressing ovarian follicles.

11.2.4. Growth hormone effects

Puberty in the ewe-lamb is accompanied by an increased frequency of LH pulses, and during normal development this is preceded by a decline in growth hormone. Some work has examined the possibility that growth hormone may act as a metabolic signal from the brain to affect the secretion of LH and that the decline in growth hormone controls the onset of puberty (Suttie *et al.*, 1991); however, no evidence was found to support the hypothesis that a decrease in growth hormone secretion is required for the attainment of puberty.

11.2.5. First ovulations and corpora lutea

Although, to the farmer, puberty is indicated by the ewe-lamb exhibiting its first oestrus, in endocrinological terms this is by no means the first important event occurring at that time. The probability is that two preliminary ovarian cycles may actually have preceded the first heat period (Oyedipe *et al.*, 1986); in the preceding cycle, there has been the well-established normal luteal phase after a silent heat and before that another short cycle. The initial short cycle is less than half the length of the normal cycle and is apparently initiated by the first LH surge. This sequence of events is not unique to ewe-lambs at puberty; it is now well-established that short cycles may occur before the commencement of cyclical breeding activity in some proportion of adult ewes after ram introduction in the ewe anoestrus (Oldham and Martin, 1979).

During their transition to cyclicity at puberty, ewe-lambs must establish a luteolytic mechanism, and it is evident that inadequacies during this transitional phase may often lead to premature luteolysis and luteal regression. According to Batten *et al.* (1995), in the prepubertal ewe-lamb, the animals possess the prerequisites of a luteolytic mechanism, in having a dormant population of endometrial oxytocin receptors in which progesterone can induce oxytocin stimulated prostaglandin $F_{2\alpha}$ release.

11.2.6. Oestrus in the ewe-lamb

It is well accepted that the duration of oestrus is shorter in ewe-lambs than in adult sheep (Hafez, 1952; Dyrmundsson, 1973). In more recent times, Loftsson and Dyrmundson (1990) reported the duration of oestrus in Icelandic ewe-lambs as 31h and in ewes as 47h; neither body condition nor synchronization of oestrus influenced its duration.

In terms of hormonal events at oestrus, a comparison of prolactin, LH and progesterone concentrations during oestrus and early pregnancy in Clun ewe-lambs and adult sheep by Davies and Beck (1993) led them to suggest that the different profiles of these hormones found in the two groups may contribute to the lower reproductive performance of the ewe-lambs.

11.3. Environmental and Other Effects and Puberty

On the basis of evidence reviewed up to the early 1970s, Dyrmundsson (1973) concluded that there is no fixed age, bodyweight or time of year at which ewe-lambs experience their first heat period, this being a result of the complex interaction between these factors and the time of birth. Certainly, the reproductive performance of ewe-lambs, in terms of oestrous response, conception rate and litter size, differs markedly from that of the adult sheep.

11.3.1. Daylight environment

There is plenty of evidence showing that in certain breeds a proportion of ewe-lambs may fail to attain oestrus (i.e. show oestrus) before the changing daylength environment in late winter inhibits sexual activity. Studies in Cambridge suggested that lambs born late in the lambing season were unlikely to exhibit oestrus in their first autumn (Hammond, 1944; Hafez, 1952). As well as that, lambs with a low rate of growth during the summer months had an increased chance of remaining prepubertal until their second autumn season (Hafez, 1952; Allen and Lamming, 1961).

In ewe-lambs, photoperiod times the transition to puberty. Studies in the USA have indicated that in the spring-born lamb, decreasing daylengths after the summer solstice provide the seasonal cue that times puberty to the autumn of the first year (Foster *et al.*, 1988a). When ewe-lambs were reared under a constant short photoperiod, puberty was delayed for at least half a year (Yellon and Foster, 1985). The direction of change in photoperiod is also known to be important; Michigan workers have shown that exposure of ewe-lambs to long days, followed by an abrupt transfer to short days, induced a more rapid onset of puberty than a photoperiod sequence in the opposite direction (Foster *et al.*, 1988b).

It is known that specific melatonin binding sites are present in the fetal lamb from an early age. It is also evident that the fetal lamb receives information about ambient photoperiod from the transfer of maternal melatonin across the placenta and that photoperiodic information received by the lamb in prenatal life affects postnatal neuroendocrine function in the lamb (Zemdegs *et al.*, 1988; Ebling *et al.*, 1989b). However, studies on the influence of prenatal photoperiod and the timing of puberty in the Suffolk ewe-lamb indicated that this occurred early in postnatal life and not before birth (Herbosa and Foster, 1992; Herbosa *et al.*, 1994).

11.3.2. Conceptions and silent heats

In those ewe-lambs that do come in oestrus and mate, the percentage conceiving can often be markedly below the 92% or so figure expected in adult sheep of the same breed; the barrenness rate in the young animals is commonly within the 20–40% range (Gordon, 1967; Dyrmundsson, 1973; Forrest and Bichard, 1974; Keane, 1974; Edey *et al.*, 1978). Female lambs which do attain puberty in their first year of life may show a high incidence of silent heats; a high frequency of multiple cycle intervals during the breeding season of ewe-lambs at Cambridge was interpreted by Hafez (1952) as evidence of silent heats.

'Flushing' and the ewe-lamb

There has been no general agreement that nutritional 'flushing' prior to mating is capable of increasing the ovulation rate in ewe-lambs (Allen and Lamming, 1961; Southam *et al.*, 1971). The sheep farmer may, however, be more interested in avoiding twins rather than promoting them in this category of young sheep (Dyrundsson, 1981).

11.3.3. Breed effects

Several reports have suggested that genetic factors can contribute to the variable reproductive performance of ewe-lambs. Laster *et al.* (1972) conducted a study involving 19 genetic groups and recorded lambing results varying from 33% for purebred Corriedales to 100% in Finn crossbreds (percentage lambing of those bred); Rambouillet crossbreds and Finn crosses reproduced significantly better than did purebred lambs from a range of domestic breeds. These American workers found that the performance of Finn crosses exceeded that of any other sheep examined.

The Romanov is a breed well known for its exceptionally high ovulation rate and its early age of puberty (Ricordeau *et al.*, 1990). In South Africa, Boshoff *et al.* (1975) showed that Romanov crosses exhibited oestrus significantly earlier than Karakul ewes. At a later time, in the same country, Greeff *et al.* (1993) sought to determine to what extent early puberty and high ovulation rate could be enhanced by increasing the proportion of Romanov genes in crosses with the Dorper breed; they recorded that the infusion of

Romanov genes into a population via crossbreeding advanced the onset of puberty, increased the ovulation rate and decreased ewe mass.

11.3.4. Bodyweight effects

The effect of increasing liveweight in the adult ewe on its reproductive performance is well recorded; in the young ewe-lamb, bodyweight is of even greater significance because the occurrence of puberty is likely to be dependent on the animal attaining a certain critical liveweight in its first autumn. Most reports agree that reproductive performance in the ewe-lamb improves with increasing liveweight; much of this improvement apparently resulted from an increase in the proportion of ewe-lambs that came into oestrus.

In general, first oestrus in ewe-lambs is attained at weights varying from 50 to 70% of adult bodyweight (Hafez, 1952; Dyrmundsson, 1973). However, the liveweight may also depend on the season; in Irish Suffolk-cross lambs, for instance, the threshold of bodyweight at puberty declined from 44 kg in early October to 33 kg in late December (Keane, 1974). Lightweight ewe-lambs are not a good prospect for breeding because they will attain puberty late in the year or may not attain puberty at all, in which case they are precluded entirely from mating and lambing.

11.3.5. Age of lamb

An increase in the ewe-lamb's age at mating has been found by some to result in a significant increase in conception and lambing rates (Laster *et al.*, 1972); Keane (1974), on the other hand, found that date of birth, within the January to early April period, had no influence on the reproductive performance of ewe-lambs which were similar in liveweight at the start of the breeding season.

11.3.6. Season of birth

The season of birth can markedly influence the age at which puberty occurs. Michigan workers reported that in lambs born out of their natural birth season, environmental factors can delay puberty by delaying the reduction in negative feedback responsiveness to oestrogen until the developing female enters an appropriate season for reproduction.

11.3.7. Ram effect

Although well-recorded in adult sheep, evidence of the ram's effect on the attainment of puberty in ewe-lambs is much more limited. Dyrmundsson and Lees (1972) did record, however, that the sudden introduction of rams to ewe-lambs in the normal period of transition from their prepubertal to pubertal condition resulted in a high degree of synchronization of first matings. In Ireland, O'Riordan and Hanrahan (1989) reported that introduction of the ram in early October advanced first oestrus in ewe-lambs by two weeks (compared with ewe-lambs first exposed to the ram in late October). In the UK, Al-Mauly *et al.* (1991) examined the effect of introducing rams on the pulsatile release of LH and the onset of puberty in ewe-lambs; they recorded that the mean pulse frequency of LH secretion was significantly increased by the introduction of rams in mid-September and mid-October.

11.3.8. Effect of temperature

Little information is to be found in the literature on the direct effect of temperature on sexual development in ewe-lambs. There was some evidence in Wales that removal of the fleece towards the end of anoestrus may bring forward the start of breeding activity in adult sheep (Lees, 1967); after autumn shearing, however, there was no clear effect on the onset of puberty among ewe-lambs in that country (Dyrmundsson and Lees, 1972). In Ireland, O'Doherty and Crosby (1991) were able to show that premating shearing of ewe-lambs significantly enhanced their conception rate after controlled breeding.

11.3.9. Exposure to electromagnetic fields

Considerable concern is often expressed about the impact of animal agriculture on the quality of the environment. Less attention has been paid to the impact of man-made environmental effects on animals. In the USA, Lee *et al.* (1993) reported results suggesting that chronic exposure of developing ewe-lambs to the electromagnetic field from a 500-kV transmission line did not affect the circadian pattern of melatonin secretion or the mechanisms involved in the onset of puberty.

11.4. Subfertility in Ewe-Lambs

There is much evidence clearly showing that fertility is more variable and lower in ewe-lambs than in adult sheep. Conception rates as low as 16% in Merinos (Watson and Gamble, 1961) and as high as 76% in Clun sheep (Dyrmundsson and Lees, 1972) have been recorded. In Ireland, work by Keane (1974) with Suffolk-cross ewe-lambs showed conception rates varying from 37 to 58%. Unlike adult sheep, ewe-lambs that do not conceive immediately often fail to return to service and this contributes to the reduced reproductive performance (Gordon, 1967; Keane, 1974). The poor fertility, it should be noted, applies to ewe-lambs that mate after controlled breeding treatments as it does to those that breed naturally.

11.4.1. Causes of infertility

Oestrus without ovulation (anovulatory oestrus), although regarded as rare in adult sheep, has been recorded in normal and hormone-treated ewe-lambs. Edey *et al.* (1978) found an incidence of anovulatory oestrus of between 7 and 33% in groups of Merino and Perendale lambs in Australia; in Ireland, Quirke (1979a,b) observed a similar phenomenon in 7% of Galway ewe-lambs. Further studies by Quirke *et al.* (1981) suggested that reduced fertility in Galway ewe-lambs was unlikely to be the result of an unfavourable relationship between the timing of ovulation and behavioural oestrus; they recorded the duration of oestrus to be 30h and that ovulation occurred in most lambs around the end of oestrus, as observed in adult sheep (Holst and Braden, 1972).

11.4.2. Fertilization rates

Several authors attempted to estimate the fertilization rate in ewe-lambs as part of their investigations into the problem of reduced reproductive performance. Figures for untreated ewe-lambs varied from 77 to 90% (Allen and Lamming, 1961; Keane, 1974; Hamra and Bryant, 1979). Although it was possible that the particular behavioural responses of ewe-lambs to the ram may be a factor resulting in fertilization failure (Edey *et al.*, 1978) and in some instances it has been possible to attribute fertilization failure to lack of insemination (Quirke, 1981), the general conclusion was that fertilization failure was unlikely to be the source of the major difference in conception and pregnancy rates between ewe-lambs and adult sheep.

There has been some suggestion that sperm transport may be a problem in ewe-lamb fertility. In Montana, Lane *et al.* (1991) found evidence that sperm require more than 2 h to reach the oviducts of ewe-lambs mated at the first or third oestrus; they note that this may explain the lower pregnancy rate of such sheep compared with mature ewes. However, further studies by Lane *et al.* (1993) led them to conclude that sperm transport through, and the distribution of sperm within the reproductive tract, may not be major limiting factors affecting fertility during pubertal transition in ewe-lambs.

11.4.3. Embryo mortality

With fertilization rates high and lambing rates low, the indications were that the level of embryo mortality contributed substantially to the lower reproductive performance of ewe-lambs; Quirke (1979b) estimated an embryo survival rate of 37% in one Irish study involving 556 ewe-lambs, as against the 75% or so recorded for adult sheep in the literature. The factors involved were by no means clear, but evidence strongly indicated that the problem lay with the embryos themselves rather than with unfavourable uterine conditions (Quirke and Hanrahan, 1977); embryo transfer studies showed that ewe-lamb and adult ewe uteri were equally capable of supporting normal adult sheep embryos.

The effect of long daylength on embryo survival and growth in ewe-lambs was studied by Beck and Davies (1994a); although the significance of the results is unclear, they recorded that embryo viability and growth was reduced under a long photoperiod (16h light:8h darkness).

11.4.4. Oestrogenic effects in ewe-lambs

Although those who have examined various characteristics of the oestrous cycle in the ewe-lamb and adult sheep have found them to be very similar (Smith *et al.*, 1977), there was some evidence in Irish work of a different pattern of oestrogen secretion by the preovulatory ovarian follicles (Quirke *et al.*, 1981). This was apparently supported in earlier data reported by Trounson *et al.* (1977) from studies involving the culture of isolated follicles *in vitro*. The indications were that conditions in the developing follicle may be related to the reduced fertility of ewe-lambs.

In Wales, Beck and Davies (1994b) conducted a study to determine the effect on ewe-lamb fertility of mating at puberty or the third oestrus or after premating oestrogen (25μg oestradiol benzoate) and progestagen therapy, designed to mimic the changes that occur in plasma oestrogen and progesterone concentrations during the first, second and third oestrous cycles. They recorded results showing that fertility was significantly improved in ewe-lambs mated at the third oestrus or after progestagen–oestrogen treatment designed to simulate three oestrous cycles.

11.5. Hormonal Induction of Puberty in Sheep

Although there is a considerable literature now available on the control and induction of oestrus in adult sheep, there have been much fewer reports dealing with the use of controlling breeding in the ewe-lamb. More than 20 years ago, Dyrmundsson (1973) reviewed papers published up to that time and concluded that there was much variability in the proportion of ewe-lambs which responded successfully to treatment; hormone treatments were generally more effective when applied close to the time of the natural onset of breeding activity. As observed by Quirke (1984), dealing with Irish conditions, controlled breeding may be particularly valuable in ewe-lamb breeding as foreknowledge of the lambing date permits close supervision of lambing which is necessary to ensure that lamb losses are not excessive.

11.5.1. USA studies

In the USA, work reported by Foote and Matthews (1969) clearly showed the possibility of inducing puberty by exogenous steroid and gonadotrophin treatment; elsewhere, however, there were those who pointed to the need for the sheep at one year of age to be an adequate size to bear normal healthy young and to produce enough milk to nourish them (Hulet, 1977).

11.5.2. French studies

In France where controlled breeding has been used on some scale in ewe-lambs, the sheep treated with the FGA-sponge–PMSG regimen must be older than 7 months and heavier than 60–65% of their adult weight; there is clearly little merit in inducing oestrus and ovulation among sheep that are incapable of rearing lambs after birth because of small size and general unsuitability. The French have employed an FGA-impregnated pessary which was specifically designed for ewe-lambs.

11.5.3. Irish studies

In certain sheep breeds, and this would apply to the Galway in Ireland, because of their genetic constitution, there can be a failure of many well-grown ewe-lambs to exhibit oestrus in their first year of life (Gordon, 1967); there is also the fact that conception rates at the first mating often proved to be less than 50%. The use of the progestagen-impregnated intravaginal sponge (FGA or MAP) and PMSG (400–500 IU) has been employed to induce oestrus in such animals, as well as in Suffolk-cross and other sheep (Keane, 1974; Quirke, 1979a).

Proper placement of the progestagen sponge within the vagina may occasionally be impeded by the presence of a muscular constriction; forcing either the sponge or speculum past this constriction may cause tissue damage in the vagina which can render the subsequent removal of the device either difficult or impossible (1.7% impossible in Keane, 1974; 0.8% in Quirke, 1979a). However, care should be taken to avoid treating such sheep and there is no merit in simply depositing the sponge in the posterior vagina. In New Zealand, Ch'ang *et al.* (1968) recorded a 26% loss rate in Romney ewe-lambs which was apparently due to a failure to lodge the device in the anterior vagina in many instances.

PMSG in combination with the sponge

With ewe-lambs, it was often considered necessary that PMSG should be routinely used, because of the uncertainty of knowing whether the sheep were prepubertal or cyclic. With intravaginal progestagen and PMSG, 90% or more of the ewe-lambs were expected to mate within 2–3 days of sponge withdrawal.

The interval between progestagen withdrawal and oestrus onset was found to be rather longer in ewe-lambs than in adult sheep (Quirke, 1979a) but the heat period was of the same duration as in older animals. Ovulation without oestrus (3%) and oestrus without ovulation (7%) was recorded in Galway ewe-lambs in some Irish studies. It was also evident that although progestagen–PMSG treatment can induce oestrus, it will not improve fertility in ewe-lambs.

A trial was undertaken by Crosby and Murray (1988) in which they investigated the effect of substituting the 'ram effect' for PMSG in an artificial insemination (AI) programme with ewe-lambs (Table 11.1). The authors found that joining teaser rams with the progestagen-treated ewe-lambs was as cheap and effective in inducing pregnancy as the use of PMSG, when set-time AI (insemination at 52h post-treatment) was the method of breeding.

As well as using intravaginal sponges, earlier Irish studies included the use of subcutaneous progesterone implants (375mg Sil-Estrus); the implant was one means of circumventing the occasional difficulty experienced in sponge insertion, but had its own problems, in terms of time and care required in using this approach.

11.5.4. Ram effect in progestagen-treated ewe-lambs

In Ireland, O'Riordan and Hanrahan (1989) employed intravaginal progestagen sponges, inserted for 12 days prior to the introduction of the ram at the end of October, as an induction treatment; they recorded no effect on the percentage of ewe-lambs mated and the pregnancy rate, compared with untreated control lambs. The effect of progestagen type, PMSG dosage and time of ram introduction on reproductive performance in ewe-lambs was reported by O'Doherty and Crosby (1990); PMSG did not influence the percentage of young ewes lambing in this work.

Table 11.1 The effect of PMSG and teaser rams on lambing rate to first service in ewe-lambs (from Crosby and Murray, 1988).

	Teaser rams		No teaser rams	
	500IU PMSG	No PMSG	500IU PMSG	No PMSG
No. treated	34	34	35	35
No. lambing (%)	17	19	20	9
Litter size	1.35	1.32	1.30	1.22
Proportion of pregnant animals lambing to first service	73.9	61.3	71.4	31.0

11.5.5. Effect of premating shearing

In Ireland, lambs weighing a minimum of 40kg were shorn a month ahead of treatment with intravaginal progestagen sponges (FGA/MAP). After sponge removal, ewes were bred to fertile rams at a ram:ewe ratio of 1.6. O'Doherty and Crosby (1991) recorded no significant difference in reproductive performance between the two progestagen treatments, but premating shearing significantly increased conception rate and reduced the percentage of barren ewes. In Canada, Kemp *et al.* (1991) sheared ewe-lambs at three months of age and three months later exposed them to vasectomized rams; shearing did not influence age at puberty in this approach. In Germany, Schlolaut (1992) reported on the effect of shearing ewe-lambs in September and breeding them after progestagen-sponge treatment in October; they recorded no significant differences between FGA-treated and MAP-treated ewes in lambing outcome but shorn ewe-lambs had slightly higher conception rates. In agreement with O'Doherty and Crosby (1991), the German author recorded a longer gestation period (by 1.24 days) in the shorn animals.

Further studies in Ireland by O'Riordan and Hanrahan (1994) recorded an increased conception rate in shorn lambs in one year but the opposite effect in the following year; the authors concluded that shearing did not necessarily always improve reproductive performance.

11.5.6. Intrauterine AI in breeding autumn-born ewe-lambs

In the USA, Stellflug *et al.* (1993) used progestagen (MAP) sponges with 400IU PMSG in preparing autumn-born Polypay and Targhee ewe-lambs for breeding in spring. For Targhee ewe-lambs, the authors recorded that AI by laparoscopy resulted in a significantly better lambing outcome than using rams in natural service; in the Polypay ewe-lambs, however, the opposite was true.

11.5.7. Breeding at one-year old

Getting the young sheep to accept the ram in its first autumn of life is not the only opportunity that exists for earlier breeding. The availability of progestagen–PMSG treatments permits sheep to be bred in spring at one year of age and to give birth for the first time in the late summer or early autumn (Gordon, 1958); this has the advantage of the animal being that much older at the time of its first lambing (Fig. 11.2). As noted elsewhere (Chapter 6), the experience with adult sheep lambing well within their natural breeding season is for the ewes to return to the ram and conceive with lambs at foot.

11.6. Light Manipulation and Melatonin Treatment

In some commercial sheep farming conditions, light treatments have been employed in achieving early breeding in one-year-old ewe-lambs. In the USA, for example, poor pregnancy rates in ewe-lambs used in autumn-lambing flocks has been a problem. A study by Hanson and Slyter (1995) evaluated the effect of extending the daily photoperiod from December until February on conception in one-year-old ewe-lambs. Extended light treatment, providing 18h light per day through that period resulted in a significantly higher percentage of light-treated sheep lambing after an April/May breeding (54.8 vs. 27.9%). The authors concluded that such extended light treatments could benefit producers breeding ewe-lambs for autumn lambings.

In the USA, Perez-Eguia and Hallford (1994) examined the reproductive and endocrine characteristics of ewe-lambs after short- (30 days) or long-term (60 days) administration of melatonin before the autumn breeding season; melatonin did not significantly affect reproductive performance.

11.7. Lamb Mortality Considerations

Lamb mortality, especially in the perinatal period, is known to be substantially greater in the offspring of ewe-lambs than in that of older sheep. Losses may be particularly high among the twin-born lambs because of low birthweight and associated lack of vitality (Dyrmundsson, 1973). On the other hand, in sheep that are carrying singles, generous feeding of the immature ewe during the late stages of pregnancy may result in difficult births due to fetal oversize (Laster *et al.*, 1972). It would seem that proper feeding in late pregnancy may be even more critical in the ewe-lamb than in the adult sheep.

Evidence suggesting that the pattern of nutrient utilization of pregnant ewe-lambs differs from that of adult ewes was reported by workers in the 1970s. Although the maintenance of net bodyweight during the final trimester of pregnancy may be a reliable indicator of nutritional adequacy in adult sheep, there may be the need to ensure that the net maternal weight of the young ewe increased during this period if there is not to be a reduction in lamb birthweight.

11.7.1. Shepherding skills

It is essential that shepherds have a good standard of training in the skills required to minimize lamb mortality. The shepherd needs to be capable of assessing quickly whether intervention is necessary in the lambing process. It is likely that premature intervention may be as harmful as delaying too long; unskilled intervention may be less helpful than no assistance at all.

11.7.2. Mothering qualities of young sheep

Ewe-lambs which give birth at a year of age acquire their mothering experience earlier and usually with the advantage of having to care for only single lambs; the sheep have the chance to become better mothers sooner and before they have the responsibility of larger litters. Although some ewe-lambs may well exhibit poor mothering qualities after lambing, yearlings previously bred as lambs tend to be more reliable breeders, better mothers and to have fewer lambing problems.

11.8. References

Adams, L.A. and Steiner, R.A. (1988) Puberty. In: Clarke, J.R. (ed.) *Oxford Reviews Reproductive Biology* Vol. 10. Oxford University Press, Oxford, pp. 1–52.

Allen, D.M. and Lamming, G.E. (1961) Some effects of nutrition on the growth and sexual development of ewe lambs. *Journal of Agricultural Science, Cambridge* 57, 87–95.

Al-Mauly, N.Z.N., Bryant, M.J. and Cunningham, F.J. (1991) Effect of the introduction of rams on the pulsatile release of luteinizing hormone and the onset of reproductive activity in ewe lambs. *Animal Production* 53, 209–214.

Batten, M., Scholey, D. and Lamming, G.E. (1995) Endometrial oxytocin receptor concentrations and activity in prepubertal lambs and calves. *Journal of Reproduction and Fertility* Abstract Series No. 15, p. 58.

Beck, N.F.G. and Davies, B. (1994a) The effect of long day length on embryo survival and growth in ewe lambs. *Journal of Reproduction and Fertility* Abstract Series No. 13, p. 42.

Beck, N.F.G. and Davies, M.C.G. (1994b) The effect of stage of breeding season or pre-mating oestrogen and progestagen therapy on fertility in ewe lambs. *Animal Production* 59, 429–434.

Boshoff, D.A., Burger, F.J.L. and Cronje, J.A. (1975) Sexual activity of Romanov-Karakul crosses under extensive conditions. *South African Journal of Animal Science* 5, 91.

Bucholtz, D.C. and Foster, D.L. (1994) Timing of puberty in sheep: new concepts and conceptual challenges. *Journal of Animal Science* 72 (Suppl.1)/*Journal of Dairy Science* 77 (Suppl.1), p. 121.

Ch'ang, T.S., McDonald M.F. and Wong. E.D. (1968) Induction of oestrus and ovulation in Romney ewe hoggets with a progestagen. *New Zealand Journal Agricultural Research* 11, 525–532.

Chu, T.J., Edey, T.N. and Findlay, J.K. (1979) Pituitary response of prepubertal lambs to oestradiol-17B. *Australian Journal Biological Science* 32, 463–467.

Crosby, T.F. and Murray, B.F. (1988) A comparison of PMSG and teaser rams on reproductive performance in ewe lambs. *Proceedings of the 11th International Congress of Animal Reproduction and AI* (Dublin), 4, 429.

Davies, M.C.G. and Beck, N.F.G. (1993) A comparison of plasma prolactin, LH and progesterone concentrations during oestrus and early pregnancy in ewe lambs and ewes. *Animal Production* 57, 281–286.

Dyrmundsson, O.R. (1973) Puberty and early reproductive performance in sheep. *Animal Breeding Abstracts* 41, 273–280.

Dyrmundsson, O.R. (1981) Natural factors affecting puberty and reproductive performance in ewe lambs: a review. *Livestock Production Science* 8, 55–65.

Dyrmundsson, O.R. and Lees, J.L. (1972) Attainment of puberty and reproductive performance in Clun Forest ewe lambs. *Journal of Agricultural Science, Cambridge* 78, 39–45.

Ebling, F.J.P., Schwartz, M.L. and Foster, D.L. (1989a) Endogenous opioid regulation of pulsatile luteinizing hormone secretion during sexual maturation in the female sheep. *Endocrinology* 125, 369–383.

Ebling, F.J.P., Wood, R.I., Suttie, J.M., Adel, T.E. and Foster, D.L. (1989b) Prenatal photoperiod influences neonatal prolactin secretion in the sheep. *Endocrinology* 125, 384–391.

Edey, T.N., Kilgour, R. and Bremner, K. (1978) Sexual behaviour and reproductive performance of ewe lambs at and after puberty. *Journal of Agricultural Science, Cambridge* 90, 83–91.

Foote, W.C. and Matthews, D.H. (1969) Hormonal induction of precocious puberty and related phenomena in the ewe. *Journal of Animal Science* 29, 189.

Forrest, P. and Bichard, N. (1974) Analysis of production records from a lowland flock. 2. Flock statistics and reproductive performance. *Animal Production* 19, 25–32.

Foster, D. and Ryan, K. (1979) Mechanisms governing onset of ovarian cyclicity at puberty in the lamb. *Annales de Biologie animale Biochimie Biophysique* 19, 1369–1380.

Foster, D.L., Lemons, J.A., Jaffe, R.B. and Niswender, G.D. (1975) Sequential patterns of circulating luteinizing hormone and follicle-stimulating hormone in female sheep from early post-natal life through the first estrous cycles. *Endocrinology* 97, 985–993.

Foster, D.L., Ebling, F.J.P. and Claypool, L.E. (1988a) Timing of puberty by photoperiod. *Reproduction, Nutrition and Development* 28, 349–364.

Foster, D.L., Yellon, S.M., Ebling, F.J.P. and Claypool, L.E. (1988b) Are ambient short-day cues necessary to puberty in a short day breeder? *Biology of Reproduction* 38, 821–829.

Friedman, C.R., Anson, H.I. and Manning, J.M. (1992) Acute and chronic inhibitory feedback of physiological concentrations of estradiol on LH pulse frequency in the prepubertal female sheep. *Biology Reproduction* 46 (Suppl. 1), p. 80.

Gordon, I. (1958) Studies in the extra-seasonal production of lambs. *Journal of Agricultural Science, Cambridge* 50, 152–197.

Gordon, I. (1967) Aspects of reproduction and neonatal mortality in ewe lambs and adult sheep. *Journal of the Department of Agriculture and Fisheries* (Dublin) 64, 76–127.

Greeff, J.C., Langenhoven, J. and Wyma, G.A. (1993) Puberty and ovulation rate of Romanov, Dorper, and their crosses during the first breeding season. *South African Journal of Animal Science* 23, 113–115.

Hafez, E.S.E. (1952) Studies on the breeding season and reproduction of the ewe. *Journal of Agricultural Science, Cambridge* 42, 189–265.

Hammond, J. Jr. (1994) On the breeding season in sheep. *Journal of Agricultural Science, Cambridge* 34, 97–105.

Hamra, A.N. and Bryant, M.J. (1979) Reproductive performance during mating and early pregnancy in young female sheep. *Animal Production* 28, 235–243.

Hanrahan, J.P., Quirke, J.F. and Gosling, J.P. (1977) Genetic and non-genetic effects on plasma LH concentrations in lambs at 4 and 8 weeks of age. *Journal of Reproduction and Fertility* 51, 343–349.

Hanson, D.J. and Slyter, A.L. (1995) The effect of extended light exposure on growth and reproductive performance of crossbred ewe lambs. *Journal of Animal Science* 73 (Suppl. 1), p. 245.

Herbosa, C.G. and Foster, D.L. (1992) Prenatal photoperiod and the timing of puberty in the female lamb. *Biology Reproduction* 46 (Suppl. 1), p. 60.

Herbosa, C.G., Wood, R.I., I'Anson, H. and Foster, D.L. (1994) Prenatal photoperiod and the timing of puberty in the female lamb. *Biology of Reproduction* 50, 1367–1376.

Holst, P.J. and Braden, A.W.H. (1972) Ovum transport in the ewe. *Australian Journal of Biological Science* 25, 167–173.

Hulet, C.V. (1977) Management of reproduction in sheep. In: *Management of Reproduction in Sheep and Goats.* Sheep Industry Development Program Symposium (Madison), pp. 119–133.

Hulet, C.V., Wiggins, E.L. and Ercanbrack, S.K. (1969) Estrus in range lambs and its relationship to lifetime reproductive performance. *Journal of Animal Science* 28, 246.

Keane, M.G. (1974) Effect of bodyweight on attainment of puberty and reproductive performance in Suffolk × Galway ewe lambs. *Irish Journal of Agricultural Research* 13, 263–274.

Kemp, R.A., Lane, S.F. and Berger, Y.M. (1991) Effects of shearing and prebreeding ram exposure on days to first mark and pregnancy rate of ewe lambs. *Canadian Journal of Animal Science* 71, 905–907.

Lane, M.A., Berardinelli, J.G., Cardenas, H. and Staigmiller, R. (1991) Sperm transport may be a problem in fertility of ewe lambs. *Montana AgResearch* 8(1), 33–37.

Lane, M.A., Berardinelli, J.G., Cardenas, H. and Staigmiller, R.B. (1993) Sperm transport and distribution during the pubertal transition in ewe lambs. *Journal of Animal Science* 71, 707–713.

Laster, D.B., Glimp, H.A. and Dickerson, G.E. (1972) Factors affecting reproduction in ewe lambs. *Journal of Animal Science* 35, 79–83.

Lee, J.M.Jr., Stormshak, F., Thompson, J.M., Thinesen, P., Painter, L.J., Olenchek, E.G., Hess, D.L., Forbes, R. and Foster, D.L. (1993) Melatonin secretion and puberty in female lambs exposed to environmental electric and magnetic fields. *Biology of Reproduction* 49, 857–864.

Lees, J.L. (1967) Effect of time of shearing on the onset of breeding activity in the ewe. *Nature* 214, 743–744.

Legan, S.J., Karsh, F.J. and Foster, D.L. (1977) The endocrine control of seasonal reproductive function in the ewe: a marked change in response to negative feedback action of estradiol on LH secretion. *Endocrinology* 101, 800.

Loftsson, E. and Dyrmundsson, O.R. (1990) Duration of oestrus in Icelandic ewes and ewe lambs. *Buvisindi* 4, 71–76.

O'Doherty, J.V. and Crosby, T.F. (1990) The effect of progestagen type, PMSG dosage and time of ram introduction on reproductive performance in ewe lambs. *Theriogenology* 33, 1279–1286.

O'Doherty, J.V. and Crosby, T.F. (1991) Effect of premating shearing on reproductive performance in ewe lambs after oestrus synchronization. *Journal of Agricultural Science, Cambridge* 116, 135–138.

Oldham, C.M. and Martin, G.B. (1979) Stimulation of seasonally anovular Merino ewes by rams. II. Premature regression of ram-induced corpora lutea. *Animal Reproduction Science* 1, 291–295.

O'Riordan, E.G. and Hanrahan, J.P. (1989) Advancing first estrus in ewe lambs. *Farm and Food Research* 20, 25–27.

O'Riordan, E.G. and Hanrahan, J.P. (1994) Effects of autumn shearing of spring-born ewe lambs on liveweight and reproductive performance. *Irish Journal of Agricultural and Food Research* 33, 131–139.

Oyedipe, E.O., Pathiraja, N., Edquist, L.E. and Buvanendran, V. (1986) Onset of puberty and estrous cycle phenomena in Yankasa ewes as monitored by plasma progesterone concentrations. *Animal Reproduction Science* 12, 195–199.

Perez-Eguia, A. and Hallford, D.M. (1994) Reproductive and endocrine characteristics of ewe lambs after short- or long-term administration of melatonin. *Sheep and Goat Research Journal* 10, 194–200.

Quirke, J.F. (1979a) Control of reproduction in adult ewes and ewe lambs and estimation of reproductive wastage in ewe lambs following treatment with progestagen impregnated sponges and PMSG. *Livestock Production Science* 6, 295–305.

Quirke, J.F. (1979b) Oestrus, ovulation, fertilization and early embryo mortality in progestagen-PMSG treated Galway ewe lambs. *Irish Journal of Agricultural Research* 18, 1–11.

Quirke, J.F. (1981) Regulation of puberty and reproduction in female lambs: a review. *Livestock Production Science* 8, 37–53.

Quirke, J.F. (1984) Breeding management, early season breeding and breeding from ewe lambs. In: *Sheep Production* Handbook Series No.20. An Foras Taluntais, Dublin, pp. 17–20.

Quirke, J.F. and Hanrahan, J.P. (1977) Comparison of the survival in the uteri of adult ewes of cleaved ova from adult ewes and ewe lambs. *Journal of Reproduction and Fertility* 51, 487–489.

Quirke, J.F., Hanrahan, J.P. and Gosling, J.P. (1981) Duration of oestrus, ovulation rate, time of ovulation and plasma LH, total oestrogen and progesterone in Galway adult ewes and ewe lambs. *Journal of Reproduction and Fertility* 61, 265–272.

Ramirez, D.V. and McCann, S.M. (1963) Comparison of the regulation of LH secretion in immature and adult rates. *Endocrinology* 72, 452–464.

Ricordeau, G., Thimonier, J., Poivey, J.P., Driancourt, M.A., Hochereau-De Reviers, M.T. and Tchamitchian, L. (1990) INRA research on the Romanov sheep breed in France: a review. *Livestock Production Science* 24, 305–310.

Ryan, K.D., Goodman, R.L., Karsh, F.J., Legan, S.J. and Foster, D.L. (1991) Patterns of circulating gonadotropins and ovarian steroids during the first periovulatory period in the developing sheep. *Biology of Reproduction* 45, 471–477.

Schlolaut, W. (1992) Effect of shearing on reproductive performance. *Kleinviehzuchter* 40, 856–857.

Smith, J.F., Frost, H., Fairclough, R.J., Peterson, A.J. and Tervit, H.R. (1977) Effect of age and peripheral levels of progesterone and oestradiol-17β and duration of oestrus in Romney Marsh ewes. *New Zealand Journal Agricultural Research* 19, 277–280.

Southam, E.R., Hulet, C.V. and Botkin, M.P. (1971) Factors influencing reproduction in ewe lambs. *Journal of Animal Science* 33, 1282–1287.

Stellflug, J.N., Rodriquez, F. and Fitzgerald, J.A. (1993) Influence of estrus induction with artificial insemination or natural mating on reproductive performance of fall-born ewe lambs during an out-of season breeding. *Sheep Research Journal* 9, 115–118.

Suttie, J.M., Kostyo, J.L., Ebling, J.P., Wood, R.I., Bucholtz, D. C., Skottner, A., Adel,

T.E., Towns, R.J. and Foster, D.L. (1991) Metabolic interfaces between growth and reproduction. IV. Chronic pulsatile administration of growth hormone and the timing of puberty in the female sheep. *Endocrinology* (Philadelphia) 129, 2024–2032.

Trounson, A., Willadsen, S.M. and Moor, R.M. (1977) Reproductive function in prepubertal lambs: ovulation, embryo development and ovarian steroidogenesis. *Journal of Reproduction and Fertility* 49, 69–75.

Watson, R.H. and Gamble, L.C. (1961) Puberty in the Merino ewe with special reference to the influence of season of birth on its occurrence. *Australian Journal of Agricultural Research* 12, 124–138.

Yellon, S.M. and Foster, D.L. (1985) Alternate photoperiods time puberty in the female lamb. *Endocrinology* 116, 2090–2097.

Zemdegs, I.Z., McMillen, I.C., Walker, D.W., Thorburn, G.D. and Nowak, R. (1988) Diurnal rhythms in plasma melatonin concentrations in the fetal sheep and pregnant ewe during late gestation. *Endocrinology* 123, 284–289.

Introduction to Controlled Breeding in Goats 12

12.1. Introduction

Goats were probably one of the first of the ruminants to be domesticated; they are valued for milk and meat production as well as for providing mohair and cashmere. Archaeological evidence indicates a long association between man and goats stretching back some 10,000 years. Several breeds of wild goat are thought to be the ancestors of the present day domesticated populations. Various authors have drawn attention to the fact that goats are often neglected in comparison with cattle and sheep and that part of this attitude towards them is probably due to a recognition of their capabilities, rather than any prejudice against them; it is said that farmers are aware that goats are intelligent, independent, agile, tolerant to many diseases and parasites and can look after themselves much better than some other forms of farm livestock. Goats thrive in a variety of climates but they are much more concentrated in the drier tropical and subtropical areas than are sheep. Goats are believed to have the widest ecological range of domestic livestock, ranging from extremes of tropical rain forests to dry deserts where sheep cannot exist.

Adaptability of goats

Goats that were developed in temperate countries, such as the Swiss Saanen goat, have been successfully acclimatized to the subtropical environment of countries such as Israel, producing 1.9 kids per birth, which exceeded the record for this breed in Switzerland (Epstein and Herz, 1964). Swiss breeds of goats have contributed more than most to milking herds throughout the world. Saanens have been exported from Switzerland to many countries, and their numbers have increased greatly through 'grading-up' breeding programmes based on the continuous introduction of Saanen bucks into the breeding herds of indigenous goats.

12.1.1. Common ancestry of goats and sheep?

It is believed that sheep and goats evolved from a common ancestor that carried 60 chromosomes, as in the present-day goat (*Capra hircus*). The sheep line underwent a series of translocations in their acrocentric autosomes resulting in a progressive reduction in the number from 60 to 54. According to Basrur (1986), breeders have attempted to cross domestic goats with their close relative, domestic sheep (*Ovis aries*) for centuries in the hope of developing a flock or herd that would provide milk, meat and wool as well as having the capability of thriving in hilly areas with poor grazing facilities. As discussed in some research reports (Roth *et al.*, 1989), using modern technology, it is now possible to go some way along that road.

Both goats and sheep belong to the family Bovidae (hollow-horned ruminants) and some varieties of goat are difficult to distinguish from sheep. Differences between the two species have been described by Terrill (1968). One common distinguishing characteristic is that the tail of the goat turns upwards while that of the sheep does not. Sheep have lacrimal face glands and interdigital foot glands while goats do not. Male goats generally have beards and a strong odour.

World population of goats

In 1994, the estimated world goat population was given as 574 million, with the vast majority of these (> 90%) being in the developing countries. In these countries, goats produce more milk than sheep, despite the fact that the sheep population is about 25% more than that of goats. Asia and Africa together account for 64% of the total output of goat milk (Devendra, 1991). However, milk is of secondary importance to goat meat production.

12.1.2. Goats for milk production

Goats in developing countries provide milk which is consumed fresh, or converted to cultured milk, butter and ghee, cheese or sweets. It is regarded as important to the health and nutrition of the landless and the rural poor. According to Devendra (1991), for socio-economic and nutritional reasons, increased investment in dairy goat schemes is necessary in the years ahead to support and expand household milk supply, and small-scale dairying at village level. The same author notes that development strategies need to be focused more thoroughly on the wider use of 'improver' indigenous dairy goat breeds.

Problems associated with improving goat production in India have been discussed by Misra (1993) who noted that goat production in that country was the primary occupation of poor and landless labourers. In the view of this author, research on genetic improvement, which had been carried out at institutional farms by crossing native breeds with élite high-yielding exotic goats, had invariably failed due to small population size and numerous operational constraints. A research strategy was advanced that involved organization of a breeders' cooperative and selection of high-yielding animals from individual herds to form a nucleus test herd; males born in this nucleus herd would be ranked on the basis of breeding value and used as improver sires for genetic improvement of the base herd.

Goat milk

Goat's milk differs from cow's milk in several respects: it has a higher percentage of non-protein nitrogen, a smaller proportion of coagulable protein and shows greater variability in its physical and chemical properties. The average composition of goat milk is 4.4–5.0% fat, 3.4–3.7% protein, 2.7–2.8% casein and 4.4–5.0% lactose (Kalantzopoulos, 1993). The discovery of the genetic polymorphism of alpha-casein indicates that there is not one single type of goat milk but several milks of different genotypes (Ricordeau, 1993). At the start of the 1990s, developed countries produced 25% of world goat milk production, while holding only 5% of the world goat population (Morand-Fehr and Jaouen, 1991). Suggestions for future developments in dairy goat farming in the developed countries have included: (i) improvement in monitoring systems for milk quality and in the consistency of production; (ii) application of artificial insemination (AI) techniques; (iii) goat health monitoring; and (iv) the development and marketing of new products.

12.1.3. Goat meat

Goat meat (chevon) is a red meat having a naturally occurring low fat content and for that reason constitutes a good source of lean in the preparation of low fat meat products. In the UK, Kirk and Austin (1992) reviewed data on the growth and carcass composition of goats reared in that country. Results reported by James and Berry (1995) in the USA indicated that it is quite feasible to use chevon either alone or in combination with beef in the development of low fat meat products; the same workers note that attention must be paid to the choice of cooking methods.

12.1.4. Goats in Europe

Ireland and the UK

Goats in the Republic of Ireland do not feature in official statistics and the population is small, relative to that of sheep. The size of the UK goat population has been variously estimated at about 75,000–100,000 with most of the dairy animals being in small herds averaging less than ten (Russel and Mowlem, 1988). Goats in the UK are mainly of the Saanen, Anglo-Nubian, Toggenburg and Alpine breeds, kept for milk production; goat meat (chevon) is generally a by-product from these milk breeds. The Swiss goat breeds (Toggenburg, Saanen, Alpine), like most indigenous British breeds of sheep, are restricted to breeding between August and February. The Nubian breed, which is of subtropical African origin, is believed to be less restricted in its breeding season.

Goat meat in the UK has been purchased mainly by ethnic communities of new Commonwealth origin, who have had a tradition of eating goat meat. Supplies come from kids, slaughtered at 8–12 weeks of age, young goat meat from animals up to 2 years of age and old goat meat. In terms of farming goats, research in Scotland, following on from work in New Zealand, has demonstrated that the grazing habits of the goat, with their selection of the fibrous parts of plants and a propensity to graze from the top of a sward, makes them suitable as a weed control species in sown grassland and as a complementary grazer (Russell *et al.*, 1986; Russel, 1988). Goats have the capability of being used over the whole spectrum of hill, upland and lowland farms.

Much valuable research in the reproduction and lactation of goats was carried out at the former National Institute of Dairying at Reading, which was an internationally recognized centre.

Goat farming in Greece

Greece is the main producer of goat milk in the countries of the European Union (EU), with an annual production output of 460,000 tonnes $year^{-1}$ and a goat population of about 6 million (85% of which are indigenous breeds). Goat production research in Greece is mainly focused on genetic improvement by way of selection and crossbreeding of indigenous goat breeds (Kalantzopoulos, 1993). Research in the processing of goat milk is concentrated on the manufacture of yoghurt from goat milk, the manufacture of Feta cheese from a mixture of ewe and goat milk and the development of new types of cheese.

Goats in France

Goat farming is an important part of the agricultural scene in France. In 1991, there were 20,540 goat herds and many of the animals were milk-recorded. An organization was established at that time, Caprigene France, which aimed at coordinating various aspects of goat breeding, with special emphasis on genetic improvement. A report at the time provided information of AI, performance and progeny tests and milking tests carried out under the auspicies of this organization (Chevre, 1992). According to Ricordeau (1993), the dairy goat population in France at that time was 960,000, with milk production totalling 450 million litres, of which 58% was processed commercially, mainly into cheese. New techniques such as membrane ultrafiltration and microfiltration and their adaptation for use with goat milk, have permitted new opportunities for industrial processing and storage. The average yield of milk recorded French goats was 681 kg, with a protein content of 29 g kg^{-1} and a fat content of 33.5 g kg^{-1}.

A paper by Jurkschat (1995) has tabulated data on the dairy performance of French Alpine and French Saanen goats as well as giving an account of nutrition, AI, milk recording and cheese production in France.

Research and development work with goats in France

According to Ricordeau (1993), research carried out over the past decade in France has led to important new technical and scientific developments. Research into reproduction, genetic improvement of goats and pathology has resulted in several advances: (i) the development of an earlier reproductive season, so that milk production can be extended throughout the year; (ii) a more efficient selection scheme; (iii) an increase in the true protein content of goat milk; (iv) the development of strategies against brucellosis, chlamydiosis, mastitis and Lentivirinae; and (v) strategies for improving milk quality. Such advances have been achieved by means of basic research, by cooperation between disciplines and between French research centres and industry.

Goats in Spain

The goat population of Spain, which numbers 3.6 million, is located mainly in Andalucia, which is the region that produces most of the goat milk used for cheese production. These goats are kept on poor quality land to which access is difficult. According to Analla *et al.* (1995), 60% of milk production is from goats of the Murcia-Granada and Malaga breeds; goats of these breeds have an average milk yield of 500kg at 5% fat and 3% protein and a litter size of 1.9. Kidding usually occurs in autumn and winter; the goats are grazed during the day and housed at night. Analla *et al.* (1995) note that a selection programme for milk yield was started in 1993; however, selection is now for protein percentage and yield.

Goats in Turkey

Data from a survey of Angora goats in Turkey have been discussed by Tuncel and Akman (1989). At that time, there were about two million Angora goats, most of them concentrated in the middle of the country. The sheep are shorn annually in spring, the yield of mohair averaging 1.5kg for the first shearing and 2.2–3.0kg for mature ewes. The authors mention that most Angora ewes do not show oestrus until 2.5 years of age, but with good management and nutrition, this can be reduced to 1–1.5 years. The Angora goat actually originated in Tibet, but Turkey is now recognized as having the most important Angora herds in the world.

12.1.5. Hill country goat farming in New Zealand

A 14-fold increase in the income of New Zealand farmers from goats kept for fibre was reported by Batten (1990) to have occurred over the previous decade; in 1988, 1,250,000 goats on 9437 farms contributed towards fibre production and it was expected that the goat population would continue to increase. The increase up to 1988 was ascribed to the use of highly productive goats, mains powered electric fences to control the grazing of goats, investment capital from outside the industry and the integration of the handling and marketing of the fibre by the producers. According to Batten (1990), each hill country farm could have a profitable goat flock integrated with sheep and cattle management.

12.1.6. Goats in North and South America

The USA goat population is estimated at two million head (Waldron, 1994). The leading goat producing states are Texas, Tennessee, Georgia and Alabama. Recognized dairy breeds in the country include Alpine, LaMancha, Nubian, Saanen and Toggenburg. According to Haenlein (1994) the USA dairy goat industry is on the threshold of being recognized as a necessary and legitimate industry. Much of the goat population, however, is kept as a low input enterprise and there would be scope for using imported breeds to improve their production efficiency. Boer goats have been imported from New Zealand because of their genetic potential to increase the amount of meat produced from goats. The Liaoning Cashmere goat of China has the potential to improve Cashmere production. However, as noted by Waldron (1994), the costly health regulations are severe barriers and producers would have to be certain of the genetic superiority of such imports before embarking on that approach to breed improvement.

Goats for meat in the USA

Farming goats in the USA for the primary purpose of producing meat is a relatively recent phenomena. Most goat meat has come from excess dairy goats, spent fibre goats and brush and briar goats (Pinkerton *et al.*, 1994). Goats kept for meat have been mainly of the Spanish breed type, which constitute less than 20% of the total goat population in the country (Glimp, 1995). However, there are major population centres on the US east coast that provide homes and work for a large segment of immigrants who regard the consumption of chevon as a natural and familiar practice. Gipson (1995) has dealt with the goat genetic resources in the USA and emphasizes the need for these to be maintained and evaluated for their possible contribution to the meat-goat industry.

The Boer goat

The introduction of the Boer goat, a meat breed, has increased interest in meat production. Until the arrival of this breed, the Spanish goat of Texas was considered the meat goat standard against which other breeds were compared. The Spanish goat is apparently well adapted to an extensive management system, but appears to be a seasonal breeder, which would limit accelerated kidding systems. The Myotonic or 'fainting' goat is considered to be a good candidate for meat production because of its muscular, meaty conformation; however, 'fainting' must clearly be a management nuisance and it is not known to what degree, if any, 'fainting' and muscularity are related.

Goats in Brazil

According to Machado *et al.* (1992), there were 11.3 million goats in Brazil in 1988, of which 90% were in the northeast of the country; they noted 19 indigenous and 12 imported breeds. Details are given by Ribeiro (1993) of milk yield, cheese manufacture and reproductive performance in goats in Brazil; the author draws attention to problems caused by the seasonal milk production of the European dairy breeds, resulting in peaks in the September–January period and extremely low yields in April to July.

12.1.7. Goats in China

A report by Cheng and Ma (1992) deals with achievements in the Chinese goat industry over the previous 40 years. They note a four-fold increase in the goat population in China between 1949 and 1989; goats numbered 16.1 million in 1949 and 98.1 million in 1989. Several new breeds had been developed by crossing with imported breeds, the improved goats being grouped as fur, cashmere, pelt and milk breeds. As a result of breed improvement during the decade 1979–89, goat milk production increased from 237,000 to 545,000 tonnes and cashmere production from 4000 to 5000 tonnes. According to Jian Ying (1995), the natural environments of China are well suited for goat production. Most of the country's harsher environments can be utilized efficiently only by goats. The native Chinese goat breeds mature early, with an all-year-round breeding capability; there would be advantages, however, in crossing the native goats with selected imported meat-goat breeds.

12.2. Areas of Controlled Breeding in Goats

The goat is a seasonally polyoestrous animal with a breeding season in Ireland and the UK beginning about September and ending in the early months of the new year. The length of the breeding season is a matter of some practical importance, since it is difficult to arrange breedings in dairy goats so that a herd will give a uniform yield of milk all the year round (Shelton, 1977). According to Asdell (1964), the usual number of young born at a kidding is two, but one to three are common and four and five are rare. In comparison with sheep, the amount of literature dealing with reproduction, and the control of reproduction in the goat, is limited.

In countries such as the USA and UK, goats have neither been a predominant livestock species nor a popular laboratory model for reproductive studies. As a consequence, goats have received little attention in professional/scientific journals. Various authors, nevertheless, have described the reproductive techniques that have been developed for use in the goat (BonDurant, 1981). Methods of semen collection, semen processing, AI, semen storage, oestrus synchronization, superovulation and embryo transfer have been covered by McKelvey (1990). A paper by Morand-Fehr (1993) examined goat research in Europe; according to this author, studies related to reproduction have been focused on the effects of photoperiodicity (in Norway and the UK), semen freezing (Germany), superovulation (Germany) and pregnancy diagnosis (in the UK and Germany).

12.3. Factors Affecting Fertility in the Female Goat

Many of the features of reproduction shown by sheep are common to the goat. Kadu and Kaikini (1987), for example, record greater ovulatory activity in the right than in the left ovary; transuterine migration of the early embryo was frequently observed. Goats, like sheep, are capable of responding to increased levels of nutrition ('flushing') in much the same way as sheep (Henniawati and Fletcher, 1986). At the same time, there are marked differences in some of the features of the reproductive system between sheep and goats, which need to be kept in mind.

12.3.1. *Seasonal breeding activity*

Work conducted more than 50 years ago by Bissonnette (1941) showed that in goats, as in sheep, the primary environmental cue used to regulate reproduction is daylength. However, many of the world's goats are located in tropical and subtropical regions and may show little response to photoperiod. In the USA, most breeds of goat show some seasonal restriction to mating. According to Shelton (1977), the Angora is the most restricted and seldom shows oestrus prior to September. Earlier authors in that country reviewed the breeding season of the dairy breeds and found that most matings occurred in the September to December period (Asdell, 1926; Turner, 1936). In more recent times, Australian workers have shown seasonality to be strongly present in female cashmere goats and the feral goats from which they are derived (Restall, 1992).

12.3.2. *Stress and reproduction in the doe*

Stress has been shown to prevent or terminate cyclic activity in goats raised under extensive conditions (Shelton, 1977). This type of stress would be described as psychological, rather than stress arising from nutrition, disease or climate. Instances are on record of does, raised under range conditions in the USA, when moved to unfamiliar surroundings, ceasing to show sexual activity, even though they were in the breeding season. Such an effect was observed by Shelton and Morrow (1965) in which distress communicated from an isolated male appeared to prevent cycling of Angora does. According to Shelton (1977), producers should be aware of the potential for stress-induced interference with reproduction in the goat.

Account must also be taken of the fact that, unlike sheep, anomalous sex differentiation leading to the intersex condition is relatively common in goats. According to Basrur and McKinnon (1986), it is more prevalent in dairy goats of the Saanen, Toggenburg and Alpine breeds than in any other species of domestic animal.

12.3.3. Intersexuality

Intersexes are classified according to the gonads that they are found to possess. The term 'true hermaphrodite' is employed to distinguish goats that carry both types of gonads from those that carry either one or the other types of gonads (pseudohermaphrodites). The term 'male hermaphrodite' is used in describing an intersex with testes and 'female hermaphrodite' refers to an intersex possessing ovaries. The gonads of intersex goats are usually testes, situated in the normal location of ovaries or they may be partially or fully descended. The intersexes are generally female-like at birth, but at puberty they start to behave like bucks and act aggressively towards goats and humans. According to Basrur and McKinnon (1986) a majority of intersexes showed marked male libido in the presence of a normal oestrous goat.

Inheritance of intersex condition

The intersex condition is believed to be inherited as a simple recessive, and is associated in some way with hornlessness, which is inherited as a dominant; horned hermaphrodites are rare. It appears that the homozygous state for the polled gene in goats is a severe disadvantage in both males and females. In the male, it causes poor differentiation of the duct system, leading to sterility, and in the genetic female it causes gonadal reversal leading to masculinization of the gonads and the genitalia. In India, Yadav *et al.* (1993) recorded the cytogenetic make-up of true hermaphrodites and male pseudohermaphrodites as 60,XX.

12.3.4. The freemartin syndrome

As in cattle, freemartinism in goats is caused by the fusion of fetal membranes (allantoida) and the subsequent vascular anastomoses that permit the passage of cells and other substances to pass from the male to the female fetus. However, unlike cattle, vascular anastomoses either do not occur in goats or occur after the critical period in gonadal differentiation. According to Basrur and McKinnon (1986), about 6% of intersexes may be expected to be freemartins; a prerequisite for the freemartin condition in goats is birth as a twin to a male kid or as one of heterosexual multiples. In France, Cribiu *et al.* (1991) recorded that of 54 female goats from births that included at least one male, three (5.5%) were freemartins (karotype 60,XX/60,XY); these authors suggest that the freemartin condition is more common in goats than indicated by the literature.

12.3.5. Mosaic Klinefelter's syndrome in goats

Klinefelter's syndrome is a frequent gonadal developmental disorder that has been found in approximately one in 500 newborn phenotypic human males. Although males with Klinefelter's syndrome are usually irreversibly sterile, men with mosaic Klinefelter's syndrome (XY/XXY) may have in their testes occasional seminiferous tubules with normal chromosomal constitution that produce sperm. Such sperm have been successfully employed by Harari *et al.* (1995) to fertilize human oocytes by intracytoplasmic sperm injection (ICSI). During routine cytogenetic screening in goats, Bhatia and Shanker (1992) confirmed the presence of XY/XXY cell lines in a fertile Saanen × Beetal goat which exhibited normal male morphology; drumstick appendages in polymorphonuclear leucocytes similar to those observed in the normal female animal indicated the presence of an additional X chromosome.

12.3.6. Pseudopregnancy in goats

Pseudopregnancy is characterized by an accumulation of fluid in the uterine lumen and the presence of a persistent corpus luteum. Mialot *et al.* (1991) examined 10,000 Alpine and Saanen goats on 71 farms in France by ultrasonics in 1989 and 1990 and found evidence of the condition on 55% of farms and in 2–3% of the goats; the incidence was greater than 5% on 11% of the farms. Goats of all ages were affected, though the condition was rare in young animals. The incidence was significantly higher in goats mated outside the normal breeding season and in does that had been subjected to oestrus synchronization treatment. Saanen goats showed a significantly higher incidence than French Alpine in one year but not in the next year (Duquesnel, 1991). Autopsy of animals diagnosed as pseudopregnant by Duquesnel *et al.* (1992) showed a distended uterus containing 1–7.2 litres of transparent sterile liquid; animals were in the luteal phase of the cycle, but their ovaries showed numerous follicles.

12.3.7. Pseudopregnancy and prostaglandin treatment

An experiment was carried out in the non-breeding-season (April to August) in Alpine and Saanen goats by Leboeuf *et al.* (1994). They diagnosed pseudopregnancy by real-time ultrasonics and administered 100 μg of the prostaglandin analogue cloprostenol 20 days ahead of inducing oestrus and ovulation by intravaginal progestagen–PMSG treatment. The authors recorded a fertility rate of 55% after AI and regarded this as being satisfactory. The fertility rate was comparable with that found in normal goats. It was concluded that prostaglandin treatment with only one injection of cloprostenol induced luteolysis in the pseudopregnant animals and resulted in normal reproductive performance.

12.4. Male Goat Fertility and Breeding Activity

The reproductive anatomy and physiology of the male goat have been described by Smith (1986). As in sheep, the testes are large in proportion to the animal's bodyweight in comparison with the testes of bulls. Normal scrotal circumference increases with age and liveweight and is larger during the breeding season than in the spring. The age at which puberty occurs is likely to vary with breed, liveweight and nutrition. In dairy breeds under conditions of good nutrition, fertile matings by young bucks are possible before they are six months old; buck and doe kids should be separated before five months of age to prevent breeding.

12.4.1. *Seasonal breeding activity*

In Australia, Walkden-Brown *et al.* (1994) reported the effects of season and nutrition on hormone concentrations and secondary sexual characteristics in mature cashmere bucks. It was evident that Australian cashmere bucks exhibited considerable reproductive seasonality, and that the expression of this seasonality was modified by the nutritional environment. Of the reproductive variables examined, testosterone level, sebaceous gland volume and odour score appeared to be ultimately dependent upon luteinizing hormone (LH) secretion and probably under photoperiodic control. However, the authors found that testicular size (which reflected sperm production capability) appeared to be influenced primarily by changes in feed intake and growth rather than by changing gonadotrophin concentrations. The study recorded a seasonal pattern of sebaceous gland volume which closely matched that of testosterone.

Sebaceous gland secretions

Changes in the secretion of sebum was believed to be responsible for seasonal changes in the odour of the bucks. Although the precise function of this seasonal pattern in sebum secretion was unknown, the odour may well contribute to the ovulatory response of does to the introduction of bucks (Claus *et al.*, 1990).

12.4.2. *Light manipulation to maintain high sperm production*

In Alpine and Saanen bucks, the breeding season is preceded by a progressive rise in LH secretion from June to September and a more rapid increase in testosterone secretion from August to September (Delgadillo and Chemineau, 1992). As a result of the increase in the secretion of these reproductive hormones, there is an increase in sexual behaviour, testis weight and sperm production during the breeding season (Delgadillo *et al.*, 1992b). Studies reported by Delgadillo *et al.* (1992a) employed two short photoperiodic cycles (monthly or two monthly) and showed that these were capable of increasing the number of insemination doses by 69% and 55%, respectively. The French workers have further shown that monthly alternation of long (16h light:8h darkness) and short days (8h light:16h darkness) can prevent seasonal changes in testosterone secretion and testis weight and increased sperm production during three consecutive years (Delgadillo *et al.*, 1993).

Writing of conditions in Australia, Ritar (1993) noted that bucks show strong seasonality in the quality and quantity of sperm produced, so that there is limited time in which the semen may be collected for storage and AI, but that this could be extended by optimizing nutrition and management.

Male sexual behaviour

The male goat shows clear courtship behaviour when in contact with oestrous goats. False mounting may occur during courtship. The buck frequently spills urine on its own head and forelegs or licks at the stream of its urine.

12.4.3. Preparation of teaser goats

An illustrated account is given by Pompermayer *et al.* (1993) in Brazil of a method of deviating the angle of the penis in teaser goats by surgically moving the preputial orifice to the lateral surface of the groin. The method was employed without postoperative complications and the bucks showed normal libido after the operation.

Attempts have been made to prepare female teaser goats using testosterone-impregnated intravaginal sponges. In Chile, Melandez and Campo (1989) administered testosterone (100mg) over a 10-day period and recorded that most female goats showed various types of male behaviour within 2–12 days of treatment and that this behaviour persisted for 2–21 days after sponge withdrawal; of 38 goats known to be in oestrus, 76% were marked by the treated females.

12.5. Artificial Insemination in Goats

The history of AI in goats in the UK has been traced by Clabburn (1992) and the status of AI compared with that in Norway. The same author also outlined the genetic improvement programme for dairy goats in France and stressed the need to set up a similar progeny testing scheme in the UK. The number of goats inseminated in France during the past decade has shown a steady increase (Montigny *et al.*, 1990; Montigny, 1992). Details are provided in Fig.

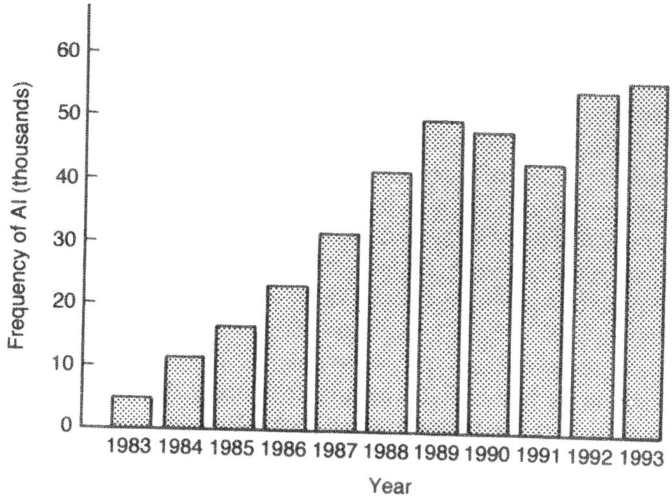

Fig. 12.1. Development of artificial insemination in goat breeding in France in the period 1983–93.

12.1 showing the increase in numbers in the ten-year period 1983–93. According to one report, of 41,139 milk-recorded French Alpine and Saanen goats inseminated in 1993, 62.4% conceived to first insemination and 87.9% to all inseminations (Chevre, 1994).

FAO training manual

A training manual on AI in goats and sheep has been prepared by Chemineau *et al.* (1991), covering the reproductive physiology of male and female goats and to various factors (season, environment and genetic) that influence reproductive traits. The manual also considers housing and management of animals in AI centres, control of oestrus and ovulation, AI methods and methods of pregnancy diagnosis. In Australia, Ritar (1993) noted that the use of AI under extensive grazing conditions in that country required control of ovulation to permit efficient and accurate timing of frozen-thawed semen inseminations.

12.5.1. Semen collection techniques

Ejaculation in the buck may be induced by electrical stimulation or semen may be collected by way of the artificial vagina (Fig. 12.2). In Australia, Carter *et al.* (1990) compared a sine-wave and a pulse-wave electro-ejaculator for the collection of semen from bucks. The authors recorded a significant difference in sperm concentration, which was higher with the sine-wave generator; this instrument was also considered less stressful to the animal. As in sheep, goat electro-ejaculated semen may differ physically and biochemically from that obtained with the artificial vagina. In a comparison of semen collection techniques reported by Memon *et al.* (1986), it was evident that ejaculates of greater volume but lower sperm concentration were obtained with electro-ejaculators than with the artificial vagina. According to Refsal (1986) bucks tend to respond more vigorously to stimulation than rams and often give a loud vocal response to electro-ejaculation that may be disturbing to owners and other observers.

Semen collection with the artificial vagina is obviously dependent on the libido of the buck and the skill and experience of the operator. Bucks are usually trained to mount a restrained oestrous teaser doe and to ejaculate into the artificial vagina. Training may take from two days to several weeks, but some bucks may prove impossible to train. Once trained, the bucks can be collected from once or twice daily during the breeding season and collections can be taken from some all year round.

Work in France has shown that libido may not be a limiting factor in semen collection in goats under appropriate management conditions (Corteel, 1977); training of the animals at an early age, maintaining relatively frequent semen collections at regular time intervals throughout the year and using an entire teaser doe appeared to be of greatest importance.

In Australia, Ritar and Salamon (1991) studied the effect of month of collection on the viability of Angora goat sperm and concluded from their results that the freezability of semen varied with the month of collection. It is now evident, as noted elsewhere, that rapid alternations of short and long days

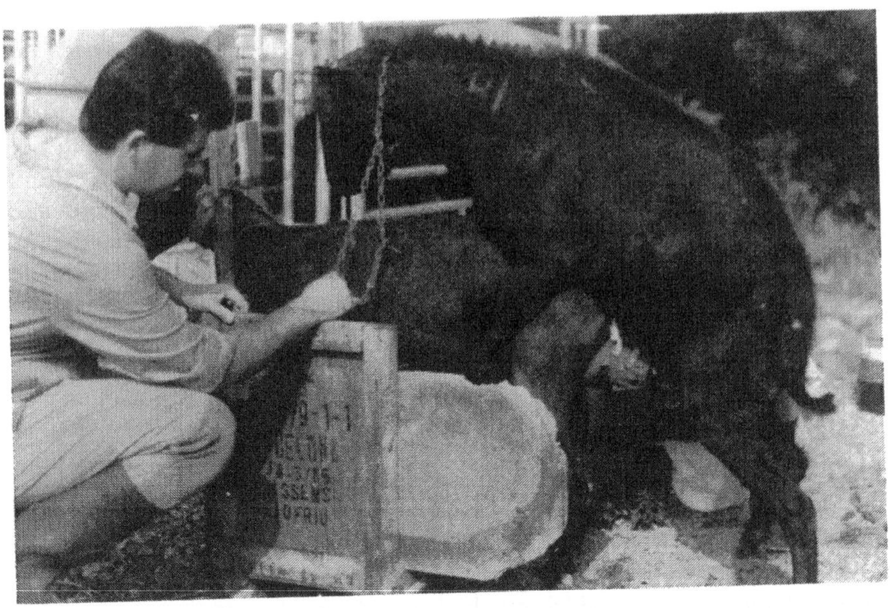

Fig. 12.2. Semen collection from a buck at the University of Barcelona for use in IVF research (photo courtesy of Teresa Mogas).

may be employed to prevent the regression of the Leydig cells in the testes and to permit sperm production at a high rate to occur throughout the year (Delgadillo *et al.*, 1995).

12.5.2. Semen examination

Semen is examined immediately after collection and its density and volume measured. It can be expected that the volume and sperm density of ejaculates will vary with season, age, breed, ambient temperature and nutritional status. According to Refsal (1986), most ejaculates should range in volume from 0.5 to 2.0ml and have 1500–4000 million sperm ml^{-1}. An optimal semen sample should show rapidly swirling wave motion and the percentage of sperm showing progressive motility should be greater than 75%. Coloured semen may be produced by some Angora rams. Mendoza *et al.* (1989) recorded yellow, light yellow and white ejaculates and suggested that the colour was due to riboflavin, since the concentration of riboflavin, which is probably produced by the vesicular glands, was correlated with these colour differences.

12.5.3. Processing semen

When raw (undiluted) semen is used, does can be inseminated with about 0.1ml semen immediately after semen collection; about 5–15 does may be inseminated per ejaculate. Semen can be diluted at 30°C with either a milk or a Tris-based diluent. The dilution is on either a volume basis (usually 1:2 to 1:4) or a constant sperm density (usually 250–600 million sperm ml^{-1}). The diluted semen is cooled to room temperature or 4°C and can be kept at these temperatures for up to 12 h. It is loaded into pipettes or straws for insemination; between 10 and 40 does can be inseminated per ejaculate.

12.5.4. Storage of semen

French workers believed that an important contribution to the choice of a storage medium for goat sperm was the discovery of an enzyme produced by the bulbo-urethral glands of the male; this enzyme catalyses the hydrolysis of lecithins in egg yolk to fatty acids and lysolecithins, which are apparently toxic to sperm. The presence of this phosphatidase in the seminal plasma of the buck indicated that media containing egg yolk should not be used for semen conservation. According to Corteel (1977), many workers ignored this information and continued to be perplexed by the variation in fertility of goats inseminated with sperm diluted in such media. A later report by Corteel *et al.* (1983) also noted that those who used egg yolk media to dilute goat semen should not be surprised at the disastrous results that may follow.

Egg-yolk media

In India, Chauhan and Anand (1990) reported studies to determine whether such hydrolysis by seminal plasma enzymes occurred; a quantitative estimate of phosphatidyl choline and phosphatidyl ethanolamine, the two major phospholipids in egg yolk, in diluted semen before and after freezing revealed that lipids in diluted semen were not hydrolysed by the enzymes in seminal plasma. The authors also recorded that of 26 goats inseminated with semen diluted and frozen in egg yolk–Tris diluent, 21 kidded. In Australia, Ritar (1993) found no improvement in the fertility of stored semen when seminal plasma was removed and a diluent containing a low dose of egg yolk was used for dilution. In a study by Maxwell *et al.* (1994), pooled goat semen was stored in a Tris-based diluent at 5°C under air and used to inseminate goats at 50–58h after sponge removal in progestagen–PMSG treated animals; fertility was maintained when semen stored for up to four days was deposited into the uterus, but declined after only two days storage when cervical insemination was used.

12.5.5. Timing of insemination

French workers have shown that there is much to support the view that insemination should be carried out between one and 12 h after the goat has first accepted mounting by a vasectomized or aproned entire male. The evidence suggests that intrauterine inseminations are more effective than intracervical inseminations and that vaginal inseminations should be avoided, particularly when frozen-thawed semen is employed. Two inseminations during the heat period are not advocated, one of the reasons being that vaginal distension during oestrus leads to the reflex release of oxytocin, which occasionally results in marked cervical and vaginal contractions; presumably, such contractions are regarded as having an adverse rather than a favourable affect on sperm transport.

12.5.6. Frozen goat AI

A paper by Corteel and Paquignon (1984) described freezing procedures which had been developed at that time. These procedures involved removal of seminal plasma before dilution and freezing by washing sperm as soon as they were collected. Earlier work by Corteel (1974) had shown that removal of seminal plasma from goat semen by washing the cells before dilution in a milk–glucose–glycerol medium significantly enhanced their ability to withstand freezing and thawing. In more recent times, an account is given by Corteel (1992) of the history of semen preservation in the goat and the effect of sperm washing and dilution with milk or egg yolk on the fertilizing ability of goat sperm. The author concluded that successful preservation of goat semen requires the removal of seminal plasma, and that dilution with heated cow's skimmed milk appears to produce a higher conception rate than dilution with egg-yolk buffers.

Seasonal effects

In Germany, Tuli and Holtz (1995) have dealt with the effect of season on the freezability of Boer goat semen in that country; the percentages of live sperm in semen frozen in winter was higher than semen frozen in spring (49 vs. 32%).

Sperm quality after frozen storage

In Australia, Ritar *et al.* (1990) examined methods of deep freezing of goat semen; optimal freezing was attained by holding straws in vapour 4 cm above liquid nitrogen for at least 30 s, followed by plunging into liquid nitrogen. The same workers found that goat semen was equally well stored in straws of 0.25 or 0.50 ml capacity. In other studies in that country, Ritar and Salamon (1991) found that the motility of Angora goat sperm deteriorated during six months of storage at −196°C, and that the rate of decline in viability was similar in the presence or absence of egg yolk, and regardless of whether seminal plasma was present (non-washed) or removed (washed) before dilution and freezing. Later studies by Ritar (1993) found that the post-thawing viability of goat sperm frozen in pellets on dry ice was higher than for semen frozen in straws in liquid nitrogen vapour, although the author noted that straws are preferred commercially.

A paper by Das and Rajkonwar (1994) reported morphological changes of the acrosome during equilibration and after freezing of goat semen in a raffinose egg-yolk–glycerol diluent; equilibrating for one hour in 7% glycerol resulted in the least sperm damage before freezing, but after freezing damage was least among sperm that had been equilibrated for 3 h.

Cryoprotectants

In India, Singh *et al.* (1995) undertook a study to determine the efficacy of various concentrations of DMSO and lactose as cryoprotective agents in diluted goat semen. Their results indicated that glycerol (8%) alone or glycerol (7%) plus DMSO (1%) provided better protection to goat sperm than any other concentration of DMSO either alone or in combination with glycerol. It was also observed that lactose in combination with glycerol acted as a better cryoprotectant than glycerol alone, and that the protective ability increased with increasing concentrations of lactose. The Indian workers recorded that the percentage of sperm with tail abnormalities was significantly lower in diluents containing lactose than in other diluents; one of the consequences of cold shock is a well-documented increase in tail abnormalities.

12.5.7. Intrauterine insemination by laparoscopy

The animal welfare considerations of laparoscopic AI in goats were raised by Conway (1987) in Australia. According to this author, minimal sedation/

analgesia techniques prior to laparoscopy are not acceptable from an animal welfare point of view. The use of intrauterine AI in goats in France was described and illustrated in an article by Fieri *et al.* (1991); the authors note that pregnancy rates were markedly higher for intrauterine insemination than for cervical insemination. In terms of sperm doses required for intrauterine AI, Ritar and Ball (1991) inseminated laparoscopically with 1 million, 5 million or 25 million motile sperm, which had been frozen in pellets or in straws; kidding rates did not differ among sperm dose treatments nor between the two freezing methods.

Cervical vs. intrauterine AI

Some results of field trials with goats in France were presented by Vallet *et al.* (1992); for 500 goats, fertility was 34.3% after cervical insemination and 44.3% after intrauterine insemination carried out via laparoscopy. According to Ritar (1993), for frozen-thawed semen, fertility is much higher for laparoscopic inseminations than for cervical inseminations, although deep placement of semen into the reproductive tract could decrease fertility. Females are best inseminated 5–10h before the expected time of ovulation. A dose of one million motile sperm may be used for laparoscopic insemination of thawed semen previously diluted at rates (semen:diluent) of 1:2 to 1:23. However, for cervical insemination, a low dilution rate of 1:2 is required to permit a sufficiently small, highly concentrated dose of at least 120 million motile frozen-thawed sperm to be deposited.

12.5.8. Transcervical intrauterine insemination

In developing countries, the use of surgical and laparoscopic methods of intrauterine insemination in goats has obvious economic and technical limitations. For such reasons, Garcia *et al.* (1994) conducted studies in Mexico to determine whether oxytocin could be used to dilate the cervix and permit transcervical intrauterine AI. Goats received a 12-day progestagen treatment, with oestradiol valerate at the start of treatment and prostaglandin $F_{2\alpha}$ at progestagen withdrawal. Oxytocin treatment at time of AI increased the number of intrauterine inseminated goats in comparison with controls (85% vs. 25%) without any apparent adverse effect on the fertility of does. The results led the authors to conclude that intrauterine AI was facilitated by the oxytocin treatment.

12.5.9. Artificial insemination after predicting ovulation time

Although the techniques of AI are well established and widely used for goats, the success rate (kidding rate) has shown considerable variation. In France, an LH kit was employed by Maurel *et al.* (1992) to examine the influence of the LH peak to AI interval on fertility results. Trials were conducted with Saanen and Alpine goats, which were treated before the breeding season with intravaginal sponges (45 mg FGA), prostaglandin (50 µg cloprostenol) and PMSG (400–500 IU). Results showed that LH peaks were spread out from 16 h after sponge removal until the time of AI (43–45 h post-treatment) and that no LH peak was evident in 20% of the goats. The authors concluded that the LH kit could be useful in investigating infertility problems and in determining the best range of times for AI in different goat breeds.

12.6. References

Analla, M., Munoz-Serrano, A. and Serradilla, J.M. (1995) Dairy goat breeding systems in the south of Spain. *Cahiers Options Mediterraneennes* 11, 143–154.

Asdell, S.A. (1926) Variation in the onset of the breeding year of the goat. *Journal of Agricultural Science, Cambridge* 16, 632.

Asdell, S.A. (1964) *Patterns of Mammalian Reproduction.* Cornell University Press, Ithaca, pp. 623–630.

Basrur, P.K. (1986) Goat-sheep hybrids. In: Morrow, D.A. (ed.) *Current Therapy in Theriogenology.* W.B. Saunders, Philadelphia, pp. 613–615.

Basrur, P.K. and McKinnon, A.O. (1986) Caprine intersexes and freemartins. In: Morrow, D.A. (ed.) *Current Therapy in Theriogenology.* W.B. Saunders, Philadelphia, pp. 596–600.

Batten, G.J. (1990) Hill country goat farming in perspective. *Proceedings of the New Zealand Grassland Association* 51, 61–64.

Bhatia, S. and Shanker, V. (1992) First report of a XX/XXY fertile goat buck. *Veterinary Record* 130, 271–272.

Bissonnette, T.H. (1941) Experimental modification of breeding cycles in goats. *Physiology and Zoology* 14, 379–383.

BonDurant, R.H. (1981) Reproductive physiology in the goat. *Modern Veterinary Practice* July issue, 525–529.

Carter, P.D., Hamilton, P.A. and Dufty, J.H. (1990) Electro-ejaculation in goats. *Australian Veterinary Journal* 67, 91–93.

Chauhan, M.S. and Anand, S.R. (1990) Effect of egg yolk lipids on the freezing of goat semen. *Theriogenology* 34, 1003–1013.

Chemineau, P., Cagnie, Y., Guerin, Y., Orgeur, P. and Vallet, J.C. (1991) Training manual on artificial insemination in sheep and goats. *FAO Animal Production and Health Paper* No. 83, 222 pp.

Cheng, G. and Ma, N. (1992) Achievements of Chinese sheep and goat raising industries over the last forty years. *Animal Genetics Resources Information* No. 9, 57–70.

Chevre. (1992) Caprigene France: balance sheet of a year of transition. *Chevre* No. 188, 20–23.

Chevre. (1994) Improving fertility. Suggestions for the AI campaign. *Chevre* No. 202, 21–22.

Clabburn, M.T.J. (1992) A.I. in goats – is there a future in the UK? *Goat Veterinary Society Journal* 13, 28–30.

Claus, R., Over, R. and Dehnhard, M. (1990) Effect of male odour on LH secretion and the induction of ovulation in seasonally anoestrous goats. *Animal Reproduction Science* 22, 27–38.

Conway, D.A. (1987) Animal welfare considerations of laparoscopic A.I. *New Zealand Veterinary Journal* 35, 117–118.
Corteel, J.M. (1974) Viabilite des spermatozoides de bouc conserves et congeles avec ou sans leur plasma seminal: effect du glucose. *Annales de Biologie animale Biochimie Biophysique* 14, 741–745.
Corteel, J.M. (1977) Production, storage and insemination of goat semen. In: *Management of Reproduction in Sheep and Goats*. Sheep Industry Development Program Symposium, pp. 41–57.
Corteel, J.M. (1992) Involvement of seminal plasma in goat sperm preservation. *Proceedings of the 5th International Conference on Goats* (New Delhi) 2(2), 290–297.
Corteel, J.M. and Paquignon, M. (1984) Preservation of the male gamete (ram, buck, boar). *Proceedings of the 10th International Congress Animal Reproduction and AI* (Illinois), II–20–27.
Corteel, J.M., Baril, G., Leboeuf, B. and Nunes, J.F. (1983) Goat semen technology. In: Courot M. (ed.) *The Male in Farm Animal Reproduction*. Seminar Proceedings, pp. 237–256.
Cribiu, E.P., Chauffaux, S. and Durand, V. (1991) The freemartinism syndrome in Alpine goats. *Recueil de Medecine Veterinaire* 167, 17–20.
Das, K.K. and Rajkonwar, C.K. (1994) Morphological changes of acrosome during equilibration and after freezing of buck semen with raffinose egg yolk glycerol extender. *Indian Veterinary Journal* 71, 1098–1102.
Delgadillo, J.A. and Chemineau, P. (1992) Abolition of the seasonal release of luteinizing hormone and testosterone in Alpine male goats (*Capra hircus*). *Journal of Reproduction and Fertility* 94, 45–55.
Delgadillo, J.A., Leboeuf, B. and Chemineau, P. (1992a) Abolition of seasonal variations in semen quality and maintenance of sperm fertilizing ability by photoperiodic cycles in goat bucks. *Small Ruminant Research* 9, 47–59.
Delgadillo, J.A., Leboeuf, B. and Chemineau, P. (1992b) Decrease in the seasonality of the sexual activity in bucks by short photoperiodic cycles. *Proceedings of the 5th International Conference on Goats* (New Delhi) 2, 279–289.
Delgadillo, J.A., Leboeuf, B. and Chemineau, P. (1993) Maintenance of sperm production in bucks during a third year of short photoperiodic cycles. *Reproduction, Nutrition and Development* 33, 609–617.
Delgadillo, J.A., Hochereau-de Reviers, M.T., Daveau, A. and Chemineau, P. (1995) Effect of short photoperiodic cycles on male genital tract and testicular parameters in male goats (*Capra hircus*). *Reproduction, Nutrition and Development* 35, 549–558.
Devendra, C.D. (1991) Milk and kid production from dairy goats in developing countries. *Proceedings of the 23rd International Dairy Congress* (Montreal) 1, pp. 327–351.
Duquesnel, R. (1991) Pseudopregnancy should be closely monitored. *Chevre* No. 187, 26–30.
Duquesnel, R., Parisot, D., Pirot, G., Mialot, J.P., Saboureau, L., Etienne, P., Delaval, J., Gueraud, J.M., Prengere, E., Etienne, P., Delaval, J., Gueraud, J.M., Pren gere, E., Montigny, G., Guerrault, P., Perrin, G., Humblot, P., Fontaubert, Y. and Chemineau, P. (1992) Pseudopregnancy in goats. *Annales de Zootechnie* 41, 407–415.
Epstein, H. and Herz, A. (1964) Fertility and birth weights of goats in a subtropical environment. *Journal of Agricultural Science, Cambridge* 62, 237–244.
Fieri, F., Buggin, M., Tainturier, D., Bruyas, J.F. and Mercier, A. (1991) Use of intrauterine artificial insemination in synchronized and superovulated goats. *Bulletin des G.T.V.* No. 4, 65–72.

Garcia, C.J., Padilla, R.G., De-Leon, T.M. and Martinez, C.A. (1994) Intrauterine artificial insemination through cervix in goats with synchronized oestrus. *Journal of Reproduction and Fertility* Abstract Series No. 13, p. 49.

Gipson, T.A. (1995) Goat genetic resources for meat production. *Journal of Animal Science* 73 (Suppl. 1), p. 123.

Glimp, H.A. (1995) Meat goat production and marketing. *Journal of Animal Science* 73, 291–295.

Haenlein, G.F.W. (1994) Status and prospects of the dairy goat industry in the U.S. *Journal of Animal Science* 72 (Suppl. 1)/*Journal of Dairy Science* 77 (Suppl. 1), p. 179.

Harari, O., Bourne, H., Baker, G., Gronow, M. and Johnston, I. (1995) High fertilization rate with intracytoplasmic sperm injection in mosaic Klinefelter's syndrome. *Fertility and Sterility* 63, 182–184.

Henniawati and Fletcher, I.C. (1986) Reproduction in Indonesian sheep and goats at two levels of nutrition. *Animal Reproduction Science* 12, 77–84.

James, N.A. and Berry, B.W. (1995) Use of Chevon in the development of low fat meat products. *Journal of Animal Science* 73 (Suppl. 1), p. 245.

Jian Ying (1995) Chinese meat goat breeds and their crosses. *Animal Genetic Resources Information* No. 15. FAO Publication, Rome, pp. 71–81.

Jurkschat, M. (1995) Diary goats in France. *Neue Landwirtschaft* No. 10, 87–88.

Kadu, M.S. and Kaikini, A.S. (1987) Functional activities of the ovaries and uterine horns of goats (*Capra hircus*). *Indian Veterinary Journal* 64, 945–949.

Kalantzopoulos, G. (1993) A review of current research on goat milk in Greece. *Lait* 73, 431–441.

Kirk, J.A. and Austin, S. (1992) Growth, carcass composition and sensory evaluation of goats reared in the United Kingdom. *Goat Veterinary Society Journal* 13, 31–39.

Leboeuf, B., Renaud, G., Broqua, C., Boue, P. and Terqui, M. (1994) Reproductive performance of pseudopregnant dairy goats after treatment with prostaglandins. *Proceedings of the 10th Meeting European Embryo Transfer Association* (Lyon), p. 204.

Machado, T.M., Lauvergne, J.J. and Souvenir Zafindrajoana, P. (1992) The goat population in Brazil since its discovery. *Archivos de Zootecnia* 41, 455–466.

Maurel, M.C., Leboeuf, B., Baril, G. and Bernelas, D. (1992) Determination of the preovulatory LH peak in dairy goats using an ELISA kit on farm. *Proceedings of the 8th Meeting of the European Embryo Transfer Association* (Lyon), p. 186.

Maxwell, W.M.C., Pomares, C.C. and Eppleston, J. (1994) Fertility of liquid-stored goat sperm. *Journal of Reproduction and Fertility* Abstract Series No. 13, p. 32.

McKelvey, W.A.C. (1990) Reproductive techniques in the goat. *Goat Veterinary Society Journal* 11, 33–40.

Melendez, M.F. and Campo, C.H. Del (1989) Preparation of female teaser goats using testosterone-impregnated vaginal sponges. *Archivos de Medicina Veterinaria, Chile* 21, 13–21.

Memon, M.A., Bretzlaff, K.N. and Ott, R.S. (1986) Comparison of semen collection in goats. *Theriogenology* 26, 823–827.

Mendoza, G., White, I.G. and Chow, P. (1989) Studies of chemical components of Angora goat seminal plasma. *Theriogenology* 32, 455–466.

Mialot, J.P., Saboureau, L., Gueraud, J.M., Prengfre, E., Parizot, D., Pirot, G., Duquesnei, R., Petat, M. and Chemiau, P. (1991) Pseudopregnancy in goats. *Recueil de Medecine Veterinaire* 167, 383–390.

Misra, R.K. (1993) Improvement of goat production in farmers' flocks – network approach. *Indian Dairyman* 45, 312–315.

Montigny, G. De. (1992) AI in 1991: a good year. *Chevre* No. 189, 16–18.

Montigny, G. de., Boue, P., Lahaye, P. and Sigwald, J.P. (1990) Milk recording and artificial insemination: results for 1989. *Chevre* No. 178, 14–17.

Morand-Fehr, P. (1993) The situation with regard to goat research, and particularly with regard to goat milk, in the non-Mediterranean European countries. *Lait* 73, 455–464.

Morand-Fehr, P. and Jaouen, J.C. Le (1991) The production of goat milk and kids in dairy goat farming in developed countries. *Proceedings of the 23rd International Dairy Congress* (Montreal) 1, 352–364.

Pinkerton, F., Escobar, E.N., Harwell, L. and Drinkwater, W. (1994) Marketing channels for meat goats in southern United States. *Journal of Animal Science* 72 (Suppl. 1)/*Journal of Dairy Science* 77 (Suppl. 1), p. 111.

Pompermayer, L.G., Borges, A.P.B., Espeschit, C.J.B. and Neves, M.T.D. (1993) Preparation of teaser goats using a technique of transplanting the preputial orifice to the inguinal region. *Arquivo Brasileiro de Medicina Veterinaria e Zootecnia* 45, 305–313.

Refsal, K.R. (1986) Collection and evaluation of caprine semen. In Morrow D.A. (ed.) *Current Therapy in Theriogeneology*. W.B. Saunders, Philadelphia, 619–621.

Restall, B.J. (1992) Seasonal variation in reproductive activity in Australian goats. *Animal Reproduction Science* 27, 305–318.

Ribeiro, S.D.A. (1993) Intensive production of goats. *Proceedings of the 10th Brazilian Congress on Animal Reproduction* 1(4), pp. 143–149.

Ricordeau, G. (1993) Current situation as regards research into goat milk in France. *Lait* 73, 443–453.

Ritar, A.J. (1993) Control of ovulation, storage of semen and artificial insemination of fibre-producing goats in Australia: a review. *Australian Journal of Experimental Agriculture* 33, 807–820.

Ritar, A.J. and Ball, P.D. (1991) Fertility of young cashmere goats after laparoscopic insemination. *Journal of Agricultural Science Cambridge* 117, 271–273.

Ritar, A.J. and Salamon, S. (1991) Effects of month of collection, method of processing, concentration of egg yolk and duration of frozen storage on viability of Angora goat spermatozoa. *Small Ruminant Research* 4, 29–37.

Ritar, A.J., Ball, P.D. and O'May, P.J. (1990) Examination of methods for the deep freezing of goat semen. *Reproduction, Fertility and Development* 2, 27–34.

Roth, T.L., Anderson, G.B., Bon Durant, R.H. and Pashen, R.L. (1989) Survival of sheep × goat hybrid inner cell masses after injection into ovine embryos. *Biology of Reproduction* 41, 675–682.

Russel, A.J.F. (1988) Recent developments in goat production from hill land. *Goat Veterinary Society Journal* 9, 1–4.

Russel, A.J.F. and Mowlem, A. (1988) Goats. In: *Management and Welfare of Farm Animals* 3rd edn. Universities Federation for Animal Welfare Handbook, Baillière Tindall, London, pp. 125–141.

Russell, A.J.F., Lippert, M., Ryder, M.L. and Grant, S.A. (1986) Goat production in the hills and uplands. *Hill Farming Research Organization, Biennial Report* 1984–5, pp. 135–141.

Shelton, M. (1977) Management of reproduction of the goat. In: *Management of Reproduction in Sheep and Goats* Sheep Industry Development Program Symposium, pp. 134–139.

Shelton, M. and Morrow, T. (1965) A study of the mechanism of male stimulation in Angora does. *Texas Agricultural Experiment Station Report* p. 2340.

Singh, M.P., Sinha, A.K. and Singh, B.K. (1995) Effect of cryoprotectants on certain seminal attributes and on the fertility of buck spermatozoa. *Theriogenology* 43, 1047–1053.
Smith, M.C. (1986) The reproductive anatomy and physiology of the male goat. In: Morrow D.A. (ed.) *Current Therapy in Theriogenology*. W.B. Saunders, Philadelphia, pp. 616–618.
Terrill, C.E. (1968) Adaptation of sheep and goats. In: Hafez, E.S.E. (ed.) *Adaptation of Domestic Animals*. Lea and Febiger, Philadelphia, pp. 246–263.
Tuli, R.K. and Holtz, W. (1995) Effect of season on the freezability of Boer goat semen in the northern temperature zone. *Theriogenology* 43, 1359–1363.
Tuncel, E. and Akman, N. (1989) Breed characteristics of Angora goat in Turkey. *EUR Publication* No. 11893, 518–532.
Turner, C.W. (1936) Seasonal variation in the birth rate of the milking goat in the United States. *Journal of Dairy Science* 19, 619.
Vallet, J.C., Baril, G., Leboeuf, B. and Perrin, J. (1992) Intrauterine insemination by laparoscopy in domestic small ruminants. *Annales de Zootechnie* 41, 305–309.
Walkden-Brown, S.W., Restall, B.J., Norton, B.W., Scaramuzzi, R.J. and Martin, G.B. (1994) Effect of nutrition on seasonal patterns of LH, FSH and testosterone concentration, testicular mass, sebaceous gland volume and odour in Australia cashmere goats. *Journal of Reproduction and Fertility* 102, 351–360.
Waldron, D.F. (1994) Global existence and availability of genetic variation in goats. *Journal of Animal Science* (72 (Suppl. 1)/*Journal of Dairy Science* 77 (Suppl. 1), p. 179.
Yadav, B.R., Singh, C., Kumar, P., Tomer, O.S. and Yadav, J.S. (1993) Morphological anatomical and cytogenetical investigations in sexually anomalous goats. *Small Ruminant Research* 11, 331–342.

13 Artificial Control of Oestrus and Breeding Activity in Goats

13.1. Introduction

Oestrus control measures are likely to be of good practical interest as the means of facilitating the application of artificial insemination (AI) in goats. The development of methods that permit inseminations to be carried out at a predetermined time may also be of value in goat herds. In some countries, due to the seasonality of reproduction in dairy goats, considerable difficulty may be experienced in goat dairies in maintaining an adequate volume of milk to supply established markets during the winter months. Such considerations have been responsible for interest in the induction of oestrus in does outside the normal breeding season. It is now well established that various forms of hormonal treatment or light manipulation can be effective in reducing such seasonal effects on reproduction in goats. The effectiveness of such measures in practice is likely to vary according to age, breed and a variety of environmental factors.

13.2. Oestrus and the Oestrous Cycle in Goats

The goat is a spontaneous ovulator with an oestrous cycle interval of 20–21 days. The duration of oestrus is somewhat longer than that of the ewe, with ovulation occurring some 30–36h after the start of sexual receptivity. Such characteristics appear to be common to goats both in the temperate and tropical regions. Dealing with West African Dwarf goats in the humid tropics, Chibooka *et al.* (1988) recorded the duration of oestrus and length of the oestrous cycle to be 33h and 20.4 days, respectively.

Knowledge about the duration of oestrus in the goat and the factors which may influence it is important for those who are to use AI in the breeding of their animals. The usual recommendation is that goats are inseminated 12h after the onset of oestrus and inseminated again the following day if they are still in oestrus. Reports in the literature indicate breed differences in the duration of oestrus (Jarosz *et al.*, 1971; Van Rensburg, 1971) and breed differences have been recorded in the interval from the onset of oestrus to ovulation (Harrison, 1948; Salama, 1972).

13.2.1. Oestrous symptoms

Oestrus is evident in the doe by an increase in the animal's activity and attentiveness. It is also found that many does are more vocal during oestrus and may walk around with their tails raised, displaying an oedematous and reddened vulva (BonDurant, 1981). Homosexual activity (does mounting other does) is not uncommon; there is the observation by Katz and McDonald (1992) that the goat represents an excellent model for the cow in terms of its reproductive behaviour. According to these authors, the sexual behaviour of oestrous goats and cows are very similar, with females typically mounting other females.

Other features noted in the oestrous goat include the fact that urination may increase in frequency; milk production and appetite in dairy goats may decrease. According to Smith (1986a), a common method of heat detection for small herds is to rub a rag on the buck's scent glands and store it in a tightly closed container. The buck jar is opened and presented warm to the doe each day; if the doe is in oestrus, then she will show great interest in the jar. Although obviously not examples of advanced technology, such strategies may have a place in some areas of goat-keeping.

Behavioural traits

A study of behavioural traits exhibited by the British White goat showed that females sought out the buck and displayed tail wagging, bleating and restlessness from −60 to +35h relative to the onset of oestrus (Llewelyn *et al.*, 1993); the incidence of these activities rose at −12h and peaked at the onset of oestrus. Tail wagging also increased in intensity at the onset of oestrus, as did the intensity with which the female actively sought out the buck. Symptoms such as vulva redness and a clear vaginal discharge were observed 1–2 days before oestrus and were most evident in oestrus. It was concluded that onset of frequent tail wagging was the most useful trait for detecting the onset of oestrus.

Male behavioural patterns

The various behavioural symptoms displayed by the buck in the detection of oestrus include the flehmen reaction, tongue-lapping, vocalization, fore-leg striking and the mounting of the oestrous doe.

13.2.2. Effect of mating on oestrus duration

A study of Romano (1993) with Nubian goats in Uruguay determined the effect of mating with a teaser buck on oestrus duration in multiparous and nulliparous goats; results showed that oestrus was significantly shortened by service in both groups. Further work reported by Romano (1994a) showed that one service reduced the duration of oestrus by 45%; whether the goat was serviced once, twice or three times did not affect the response. It was evident in other studies that mounting accompanied by penile intromission was necessary to stimulate the mechanisms involved in oestrus shortening (Romano, 1994b). In a study of oestrus in West African dwarf goats, Akusu and Egbunike (1990) also recorded a significant reduction in the duration of the heat period when goats were serviced by bucks. It should also be noted that some workers have reported that oestrus duration shows seasonal changes, being shorter at the beginning and at the end of the breeding season than in mid-season (Chemineau *et al.*, 1991).

As observed by Romano (1994a), information on this male effect may be of practical value; goats detected in oestrus by sterile teaser bucks for breeding by AI may be separated from the teaser immediately after service, thereby decreasing the number of AI doses required and leading to more rational use of the teaser.

Onset of oestrus to ovulation

It is evident from the literature that there is variation in the interval from oestrus onset to ovulation in goats (Harrison, 1948; Salama, 1972). In the West African dwarf goat, Akusu *et al.* (1986) recorded ovulation occurring 20–48h after the onset of oestrus.

13.3. Physiology and Endocrinology of the Oestrous Cycle

According to Asdell (1964), the length of the caprine oestrous cycle is variable and he quoted an average figure of about 21 days. The oestrous cycle in the French Alpine goat has been recorded as about 20 days in length on average (Chemineau *et al.*, 1991). Looked at more closely, about 77% of the cycles can be regarded as normal (17–25 days), 14% as short (< 17 days) and 9% as long (> 25 days). The distribution of normal and extended cycles was found to be significantly influenced by season in studies reported by Llewelyn *et al.* (1993) from Zimbabwe. The proportion of normal cycles was highest in the cool, dry winter months (June to August) and was low during the hot, rainy months (September to February).

According to Ott (1986), short oestrous cycles should be regarded as a natural phenomenon, especially early or late in the breeding season. Studies by Cerbito *et al.* (1995) with Philippine goats in the tropics showed that short oestrous periods were associated with short oestrous cycles, which was in agreement with earlier French studies in the temperate zone. The same authors recorded that short oestrous cycle in goats was associated with a lower ovulation rate, and its occurrence in the tropics related to rainfall.

13.3.1. Endocrine events

Concentrations of oestradiol and progesterone in plasma during the oestrous cycle in goats were measured by Abeyawardene and Pope (1990) in studies in the UK. Their results led them to suggest that in goats, as in cattle, ovarian follicular oestradiol secretion approaching the preovulatory level is restored by four days after oestrus and its rapid decline after this time may be due to the inhibitory influence of the rapidly rising plasma progesterone concentrations. In India, Sureshkumar and Janakiraman (1992) measured pituitary luteinizing hormone (LH) levels at different stages of the goat's oestrous cycle and in agreement with previous studies (Mgongo *et al.*, 1984) recorded high levels of LH in the pituitary 3–22h before ovulation with simultaneous increase in blood LH; the pituitary LH showed marked changes during the cycle, and was highest in the follicular phase.

Circulating levels of follicle stimulating hormone (FSH) during the oestrous cycle of goats have received little attention, although there was one study which described a peak in FSH concentration during oestrus (Chemineau *et al.*, 1982); according to later studies by Ginther and Kot (1994) this peak could be related to their data showing the emergence of the first follicular wave of the cycle. According to Chemineau and Delgadillo (1994), enhancement of the negative feedback of oestradiol on the hypothalamic–pituitary axis is responsible for the low gonadotrophin release evident during anoestrus.

13.3.2. Follicular dynamics in the goat

Studies reported by Ginther and Kot (1994) attempted to use ultrasound to determine the nature of follicular dynamics in goats during the breeding season. Previous reports in which ovarian follicular activity was examined by way of laparotomy (Camp *et al.*, 1983), exteriorization of the ovaries (Akusu *et al.*, 1986) and examination of slaughterhouse ovaries (Greyling and Van Niekerk, 1990; Al-Baggal *et al.*, 1993; Sureshkumar and Janakiraman, 1993) had not clarified the nature of follicular dynamics in this species. According to the result obtained by Ginther and Kot (1994), however, there are four follicular waves during the caprine oestrous cycle, with ovulation occurring during wave 4. The phenomenon of follicular dominance was, however, difficult to assess in the goat because of its apparent presence during some waves and absence during others and because two dominant follicles per wave was a common occurrence. The authors concluded that follicular dominance was expressed more weakly and less frequently in goats than had been reported in cattle.

Table 13.1. Average number of kids per birth in goats (from Asdell, 1964).

Breed	Number of kids
Toggenburg	1.8
Saanen	1.9
Anglo-Nubian	2.1
Alpine	1.8
Anglo-Nubian	2.1
Angora	1.2
Saanen	1.8
Toggenburg	1.8
Jamnapari	1.1
Black Bengal	2.1

13.3.3. *Factors affecting ovulation rate*

The usual number of young born at kidding is two, but one to three are common and four and five are rare. As noted by Asdell (1964) there are indications that breed differences exist in the species (Table 13.1). On an age basis, some authors have shown that the number of kids born to does below 18 months of age average 1.5, whereas in does above that age the average is 2.1.

13.3.4. *Seasonal variation in reproductive activity*

In temperate regions the goat can be regarded as an autumnal breeder, like the sheep, with sexual activity occurring in northern latitudes between September and January. In Egypt, working with Egyptian-Nubian goats, Aboul-Ela *et al.* (1988) recorded that the non-breeding season started in late January and had a duration of more than four months. In the southern hemisphere, Ritar and Salamon (1991) recorded the goat breeding season in Australia as extending from March to August. In the same country, Restall (1992) studied seasonal variation in the reproductive activity of goats, recording that females did not begin to ovulate spontaneously until April, with the peak incidence in June (90%); no ovulations were recorded between September and February.

There are breeds of goats in which, like the sheep, breeding activity never stops completely when kept in a suitable environment. This is true for breeds in tropical and subtropical regions, but may also be evident in other breeds which are located further from the equator. In South Africa, Greyling and Van Niekerk (1987) observed that the Boer goat showed a peak in breeding activity during autumn and low activity in late spring to mid-summer; however, periods of complete anoestrus were never observed. In Japan, Sawada *et al.* (1995) recorded that Shiba goats, which are continuous breeders, did not show seasonal variations in fertility.

13.3.5. Onset of breeding season

Although some reviews of goat reproduction in the 1970s suggested otherwise, the first oestrus of the breeding season apparently occurs without the 'silent heat' which is characteristic of the onset of the breeding season in sheep. According to Billings and Katz (1995), in the female goat, progesterone facilitates oestradiol-induced sexual behaviour during anoestrus but not during the breeding season; they produced evidence indicating that photoperiod may contribute to such a seasonal difference. A review of the neuroendocrinology of reproduction in goats by Chemineau and Delgadillo (1994) notes that progesterone priming is not necessary for the induction of oestrous behaviour by oestradiol. The same authors note that the negative feedback effects of oestradiol during anoestrus is mediated by the photoperiod, which acts on the central nervous system by modification of the duration of melatonin secretion at night.

13.4. Artificial Control of Oestrus and Ovulation

There are several reasons for wishing to control the time of oestrus in the goat. In small herds, it may be a question of not having a buck available to detect oestrus in does that have to be transported to the male for breeding. In larger herds, AI can be applied on a fixed-time basis when an accurate method of controlling oestrus and ovulation are employed. The management of goats before breeding, methods of oestrus control, oestrus detection procedures and timing of AI have been discussed, among others, by Bowen (1988) in New Zealand and by Chemineau *et al.* (1991) in France.

13.4.1. Use of prostaglandins

It is clear, from many reports that during the breeding season, when goats are actively cycling, oestrus can be effectively synchronized with prostaglandin (PG) $F_{2\alpha}$ or one of its analogues, such as cloprostenol (Fig. 13.1). An injection of PG will induce luteal regression and does can be expected to exhibit oestrus after an average interval of about 50h (Bretzlaff *et al.*, 1980, 1983). The treatment is effective as early as day 4 following oestrus; according to BonDurant (1981) and Ott (1986), this is earlier than either in sheep or cattle. In terms of dosage, work by Bretzlaff *et al.* (1980) indicated that as little as 1.25mg $PGF_{2\alpha}$ was effective for oestrus control in the dairy goat. According to Ott (1986), 2.5mg PG should be effective in does weighing up to 65kg. The same author notes that goat owners should keep in mind that some does may experience a short cycle, usually within 10 days of the controlled oestrus.

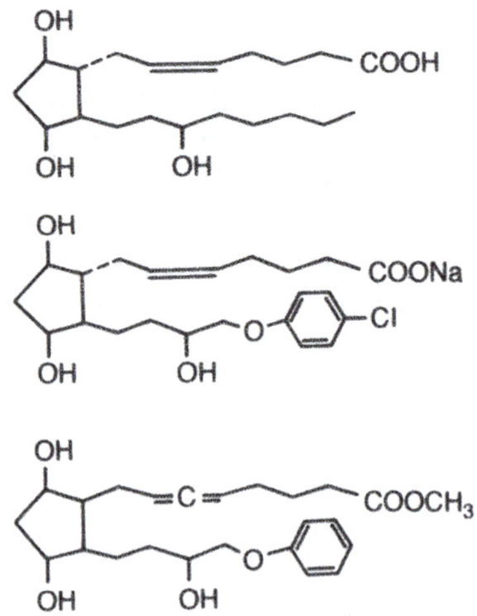

Fig. 13.1. Molecular structure of prostaglandin $F_{2\alpha}$ and some analogues.

Interval between PG doses

There is ample evidence to show that oestrus synchronization using two PG treatments, 11 days apart, has no adverse effect on pregnancy rate in goats (Ott *et al.*, 1980). In Egypt, El-Amrawi *et al.* (1993a) injected cycling Saanen goats twice with 8mg $PGF_{2\alpha}$ with an interval of 11 days between injections; 100% exhibited oestrus within 48h and 80% conceived and kidded. Oestrus control by two injections of $PGF_{2\alpha}$ at an 11-day interval was reported by Kumar and Thomas (1994) in India. After the first and second injections, 75% and 100% of goats came in oestrus, respectively; the mean interval between PG injection and oestrus was 46h for the first treatment and 48h for the second. The same authors found no adverse effect on fertility. In Germany, Meinecke-Tillmann (1988) used two injections, each of 150μg of the prostaglandin analogue, cloprostenol, with an interval of 10–11 days between doses in attempts to control oestrus in various breeds of goats.

The fertility of dairy goats injected twice with 100μg cloprostenol, at an interval of 10 days, and which were inseminated at a predetermined time after the treatment, was reported by Simplicio and Machado (1991a) in Brazil; for goats inseminated once at 60, 72 or 84h after the second PG dose, the kidding rate was 10%, 44.7% and 21.4%, respectively.

13.4.2. *Use of progesterone and progestagens*

The use of progestagens in association with PMSG in controlling oestrus in goats has been reported in many papers. There are also those who have concluded that high fertility may be achieved using a progestagen-impregnated sponge inserted for 18–21 days without subsequent treatment with pregnant mare serum gonadotrophin (PMSG), provided the ovaries are active at the start of treatment (Llewelyn and Kadzere, 1992). Results of French applications in the breeding season were summarized 20 years ago by Corteel (1975). He employed vaginal sponges impregnated with 45mg fluorogestone acetate (FGA) left in place for 19–21 days, with 400IU PMSG being injected at sponge removal; data showed that 95% of the goats exhibited oestrus, most of these coming in oestrus within a 24h period starting 12h after sponge removal (see Fig. 13.2). Breeding by AI, using either fresh or frozen semen, resulted in kidding rates of approximately 56%. Fixed-time AI was employed; with fresh semen the sponge was removed in the morning and the first insemination performed in the afternoon of the following day, and a second insemination 24h later.

Sponge treatment prior to insertion

According to Smith (1986b), the mucopurulent vaginal discharge that is evident when the intravaginal sponge is removed does not interfere with fertility but can be partially controlled by spraying the sponge with an antibiotic preparation (e.g. 2% chlorotetracycline hydrochloride) at the time of insertion. The same author notes that sponges may be lost if the withdrawal strings are not trimmed short to prevent extraction by other goats in the pen. In Egypt, El-Amrawi *et al.* (1993b) treated goats during the breeding season with an FGA-impregnated vaginal sponge for 17 days, followed by 400 or 500IU PMSG on sponge withdrawal; 100% of goats showed oestrus with satisfactory fertility and litter size.

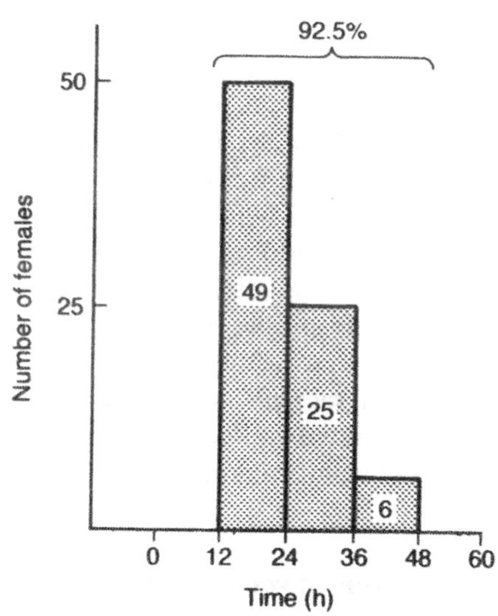

Fig. 13.2. Oestrus synchronization during the breeding season: interval between sponge withdrawal and onset of oestrus (from Corteel, 1975).

Implant treatments

Age and gonadotrophin effects on the oestrous and ovulation response to norgestomet synchronization in Spanish does was examined by Wildeus and Moore (1995) in Virginia, USA. They found that gonadotrophin injection (400 IU PMSG/200 IU human chorionic gonadotrophin (hCG)) after a 12-day implant treatment produced a superovulatory response (ovulation rate > 3) in 50% of young (11 months) and older (22 months) does. Other data suggested that age affected oestrous response to synchronization in cycling does at the end of the breeding season whereas gonadotrophin treatment affected the ovulatory response.

13.4.3. Progestagen–hCG and PG combinations

Some reports have attempted oestrus control in goats destined for breeding by AI, including the use of frozen-thawed semen, by way of a treatment combining progestagen with hCG and prostaglandin. In Brazil, for example, treatment involved medroxyprogesterone acetate (MAP) sponges for 10 days, with 100 μg cloprostenol and 300 IU hCG given 48 h before pessary withdrawal (Simplicio and Machado, 1991b); the authors recorded the best kidding rates in goats inseminated with frozen-thawed semen during a spontaneous oestrus (73%) or inseminated at 50 h after sponge removal (38%).

13.4.4. Use of controlled internal drug release device

In Australia, Ritar *et al.* (1989) showed that kidding rates for Cashmere goats after laparoscopic AI with frozen-thawed semen were not significantly different for CIDR and FGA sponge treatments when PMSG was injected at their removal or 48 h before removal. They also recorded that kidding rates for dairy goats after cervical insemination with fresh, diluted semen were not different when PMSG was administered 48 h before or at removal of an intravaginal sponge; the authors reported that kidding percentage was significantly higher after uterine (via cervix) than cervical inseminations. In New Zealand, Moore *et al.* (1988, 1989) used CIDRs in studies of AI in farmed feral goats bred with fresh or frozen-thawed semen. In a further report from Australia, Ritar *et al.* (1990) found that intravaginal sponges containing FGA and CIDR devices containing progesterone were equally effective for the control of ovulation in Cashmere goats when combined with an injection of 200 IU PMSG.

13.4.5. Hormonal induction of ovulation in young goats

The ovulatory activity of goats aged 6–19 months was examined by laparoscopy in a study reported by Ritar *et al.* (1994) in Australia. It was found that Cashmere goats commenced ovulation as young as 7 months of age (12 kg bodyweight) and that almost all Cashmeres were ovulating by 8–10 months, at a liveweight of at least 18kg. These authors employed CIDRs with PMSG (200–400IU) given at withdrawal of the device to induce ovulation, reporting kidding rates (to AI plus follow-up natural mating) of 75% and 84% for 8- and 20-month old Cashmeres, respectively. The same authors mention that ovulatory activity and fertility in the young goats outside the normal breeding season was not improved by melatonin treatment.

13.4.6. Intrauterine insemination

Fresh semen

Various reports on the insemination of goats with fresh semen, either by cervical or intrauterine AI, are in the literature. In New Zealand, Moore *et al.* (1988) reported on cervical versus laparoscopic AI after PMSG injection at, or 48h before, CIDR removal. They recorded a kidding rate of 24% for cervical and 50% for laparoscopic insemination; no effect of PMSG was evident in the kidding percentage.

Frozen semen

Reports from various countries have shown that intrauterine AI with frozen-semen can result in acceptable conception rates in goats synchronized by progestagen–PMSG and prostaglandin. In Rwanda, Leboeuf *et al.* (1994) used an 11-day FGA sponge treatment in 605 indigenous Rwandan goats, with cloprostenol 48h before pessary removal and PMSG at removal; pregnancy rates ranged from 49% to 73% after a single insemination of frozen semen 41–45h after the end of treatment. In Argentina, Mareco and Arosteguy (1994) used an 11-day MAP sponge treatment, with prostaglandin administered on day 9 and PMSG at pessary removal; AI at 55–58h after the end of treatment resulted in 60% of goats kidding.

Transcervical AI in goats

A study by Garcia *et al.* (1994) in Mexico examined the use of oxytocin (100 USP (IU) at insemination) in dilating the cervix of the goats so as to permit intrauterine AI; they recorded that such treatment significantly increased the percentage of goats inseminated through the cervix (85% vs. 25%) without affecting the conception rate.

13.4.7. Oxytocin-controlled oestrus in goats

It was shown by Cooke and Kniffton (1981) that oxytocin is luteolytic in the goat. It is believed that the effect is probably mediated by the release of uterine prostaglandin $F_{2\alpha}$; there is evidence showing that prostaglandin concentrations are elevated shortly after oxytocin administration (Cooke and Homeida, 1982) and the luteolytic effect can be prevented by a PG synthetase inhibitor (Cooke and Homeida, 1983). Studies by Homeida and Cooke (1989) examined hormonal events accompanying oxytocin-induced oestrus and compared them with those occurring at natural oestrus. Treatment involved daily injections of 50IU oxytocin on days 3–6 of the oestrous cycle; results indicated that the induced oestrus was accompanied by hormonal events similar to those in a natural oestrus, and also ovulation. In Japan, Sawada *et al.* (1994) recorded that daily injections of oxytocin on days 3–6 of the cycle decreased the progesterone concentration, increased the oestradiol concentration to a level similar to that found at oestrus and shortened the cycle.

13.4.8. Methods of increasing litter size in goats

Fecundin treatment

Methods of influencing the ovulation rate in sheep can be applied to goats, whether this takes the form of PMSG administration, PMSG in conjunction with progestagen or the employment of the immunological approach. In Greece, Driancourt *et al.* (1990) and Avdi *et al.* (1991) have reported on the effect of active immunization against the androgen, androstendione, on the reproductive performance of dairy goats. In this, goats were immunized against androstendione (two injections of Fecundin given 3 weeks apart) at the start of the breeding season; conception rate at first mating was unaffected by treatment but litter size increased significantly from 1.25 to 1.63. The authors also note that total milk yield between 5 and 19 weeks after kidding was increased in the immunized goats, presumably a reflection of the increased litter size.

13.5. Controlling the Goat's Breeding Season

There are two non-hormonal methods of influencing the breeding season of does: these are (i) the use of an artificially altered photoperiod; and (ii) the sudden introduction of the buck. Altering the photoperiod can enable the breeding of does in the period usually considered to be deep anoestrus; introducing the buck is only effective in stimulating oestrous activity if the goats in question are in late anoestrus.

13.5.1. Manipulation of the light environment

Various light treatments have been employed to permit goats to breed out of season. Some methods have involved the preliminary use of artificially extended daylength during the winter. Ashbrook (1982) recommended an extended daylength regime in which the does were exposed to 20h of light per day for two months during the January to March period; oestrus was expected to occur some 3–4 months after the goats were returned to natural daylength.

The implications of light manipulations on the subsequent performance of the offspring may be something to keep in mind. In the UK, for example, Deveson *et al.* (1992) advanced the breeding season of Saanen goats by four months using a treatment of long days (20h light; 4h darkness) during the winter followed by three months of melatonin treatment in the spring. In this work, evidence was found that the goat fetus detected photoperiod via the maternal melatonin signal and that this significantly delayed puberty onset in female kids (16.5 vs. 12.8 weeks) and resulted in a significant delay in testes development in the male kids.

An article by BonDurant (1986) in the USA has described a lighting system whereby yearling does could be induced into oestrus at least 60–80 days early by providing the feeding areas in a barn with 19 hours of artificial light daily, starting in mid- to late-winter. An inexpensive timer was employed to turn on lights (two 8-foot, 40-watt fluorescent tubes/36–40 m^2 of pen space) at about 05.00 and to turn them off 19 h later. The goats were fed in the evening to encourage them to expose themselves to the augmented light.

Light pulse treatment

Workers in the UK have examined the sensitivity of goats to a light pulse (1 h duration) during the night, as assessed by suppression of plasma melatonin concentrations (Deveson *et al.*, 1990); their results showed that a light intensity of only 150 lux suppressed melatonin levels by more than 80%, which they believed was enough to advance the breeding season.

13.5.2. Photoperiod and male goat performance

Although BonDurant (1986) concluded that there may be seasonal changes in testicular function in the buck which could be reversed by light manipulation, there was no direct evidence from studies in this area in the mid-1980s. However, it later became clear from the results of work in France (Delgadillo *et al.*, 1991, 1992, 1993) that rapid (monthly) alternation of long and short days would prevent seasonality of testosterone secretion and could be used to increase sperm production. The photoperiodic treatment in this instance consisted of 16h light and 8h darkness (long-day) and 8h light and 16h darkness (short-day). It is apparent from such work that bucks are adequately able to interpret rapid changes in daylength.

13.5.3. The 'female effect' on bucks

Although recognition of the 'male effect' (male introduction stimulating breeding activity in females) goes back more than 50 years in the literature of small ruminants (Underwood *et al.*, 1944), a much more recent finding was an analogous 'female effect' on males of sheep and goats, with exposure to oestrous females resulting in rapid increases in LH pulse frequency in goat bucks (Howland *et al.*, 1985). Studies reported by Walkden-Brown *et al.* (1994) in Australia have confirmed earlier findings that there is a seasonal cycle in LH and testosterone secretion in mature Cashmere bucks, and that nutrition and oestrous females are powerful modulators of the secretion of these hormones in a seasonally dependent way. Although the functional significance of the 'female effect' is not entirely clear, it is possible that it contributes in a positive way towards successful matings. In an earlier paper, Walkden-Brown *et al.* (1993c) suggested that the 'female effect' may be one component of a self-reinforcing cycle of stimulation that may be initiated by either does or bucks.

13.6. Progesterone and Progestagens in the Induction of Pregnancy in Anoestrus

The treatment of anoestrous goats with intravaginal progestagen (e.g. FGA) alone is known to be inadequate to induce either a behavioural oestrus or an LH surge with a consequent ovulation (Tamanini *et al.*, 1985); FGA treatment must be followed by gonadotrophin administration to stimulate oestradiol secretion by the ovary.

13.6.1. Progestagen–PMSG treatment in France

Details of treatments for inducing oestrus and ovulation in goats in the non-breeding season in France were provided by Corteel (1975). On the basis of earlier French findings, the time of administering PMSG was modified to 48h prior to sponge removal. French studies indicated that post-treatment fertility in the goat increased as the breeding season approached (see Fig. 13.3); the evidence also suggested that seasonal and lactational effects influenced fertility. The depressive effects of such seasonal and lactational effects could be corrected by increasing the number of motile sperm used in the inseminations, which were carried out at the same time (post-treatment) as in the breeding season.

13.6.2. Studies in California

Authors in other countries drew attention to the fact that there are goat farming systems in which it may be economically desirable to change the annual distribution of kidding from spring to early winter. For example, in California, goat dairies receive a premium price for winter milk and/or have their summer milk production quota increased on their ability to supply winter milk to the creamery. For such reasons, East and Rowe (1989) treated dairy goats (Anglo-Nubian, Saanen, Toggenburg and Alpine) during the transition from anoestrus to the breeding season either (i) with a 6mg norgestomet implant (Synchro-Mate B implanted under the ventral surface of the tail) for 9 days or (ii) an intravaginal sponge containing 30mg FGA for 16 days, plus an injection of 250IU PMSG two days before implant or sponge removal. When compared with control does, acceptable levels of oestrus (53% vs. 93 and 95%) and fertile matings (34 vs. 64 and 59%) resulted in the goats by using norgestomet implants or FGA sponges, respectively, as a progestagen source. Although the authors note that the administration of PMSG at progestagen withdrawal would save time and labour, their protocol followed the recommendation of Corteel *et al.* (1982) who found a superior oestrous response when PMSG was given two days in advance of FGA sponge removal.

Practical merit of treatment

In North American dairy goat farming terms, the progestagen–PMSG treatments used by East and Rowe (1989) increased the proportion of lactating mature does during January and February at a time when less than 30% of does would normally have kidded, when milk production was declining to the seasonal low and when many does had ceased their lactation prior to the next kidding. However, at the time of the work (1989), neither sponge nor implant treatment had Food and Drug Administration approval for use in dairy goats in the USA.

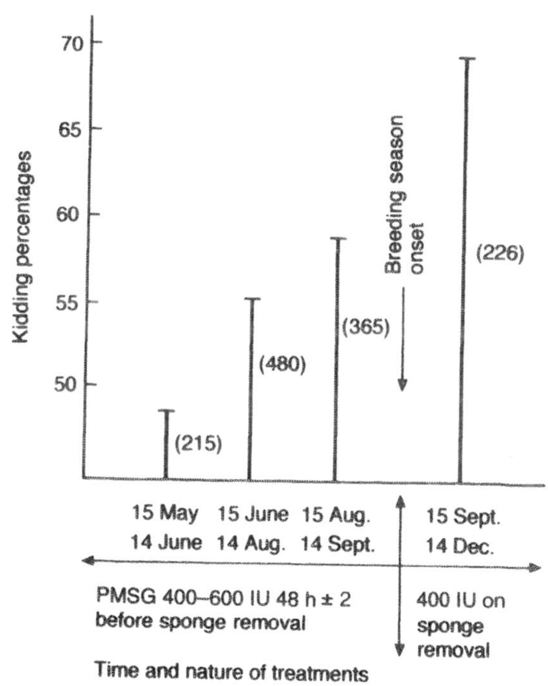

Fig. 13.3. Kidding percentage in synchronized goats before and after the onset of the breeding season (from Corteel, 1975) Values in parentheses indicate numbers of animals.

13.6.3. Long vs. short progestagen treatments

Further studies by French workers compared a long (21 days) with a short (11 days) treatment of progestagen; in the short treatment, prostaglandin (200μg cloprostenol) was given in addition to PMSG two days before sponge removal; the goats were bred using frozen semen, 30 and 48h after sponge removal, on each occasion using doses of 150–200 million breeding season sperm. The French research went on to show that satisfactory levels of fertility could be achieved with only one insemination performed on a fixed-time basis after short-term FGA treatment, PMSG and cloprostenol. In terms of the quality of sperm used in breeding, it had become evident that, under French conditions, it was necessary to set up an efficient long-term storage technique for goat sperm collected in the breeding season but inseminated in the forthcoming non-breeding-season.

13.6.4. Set-time AI in French goat breeding

The development of the progestagen–PG–PMSG protocol has enabled the French to develop their AI-based genetic improvement programme in goats. The average kidding rate of about 60% is acceptable but it has been observed that in some 25% of flocks the kidding rate is below 50%. In studies reported by Baril *et al.* (1993), dairy goats were treated with progestagen-sponges and received PG and PMSG nine days after sponge insertion; sponges were removed 11 days after insertion and does inseminated with frozen-thawed sperm (60 million for Saanen; 100 million for Alpine) 43 or 45h after progestagen withdrawal. Oestrus occurred in 98% of the goats between 24 and 72h after sponge withdrawal; fertility was significantly lower in does that came in heat later than 30h after progestagen withdrawal (33.3 vs. 65%). The occurrence of late oestrus was not affected by age, but increased with the number of treatments that an individual animal had previously received; it may be that anti-PMSG antibody is responsible for some of the observed delay in oestrus onset. An article by Leboeuf (1993) also drew attention to the relationship between conception rate and time of oestrus onset; in this instance, the author showed that does exhibiting oestrus within 24h of sponge withdrawal had a higher conception rate (66.5%) than those exhibiting oestrus later.

13.6.5. Progestagen–PMSG trials elsewhere

In Spain, Forcada *et al.* (1989) compared different methods involving sponges, PG and PMSG for the induction of oestrus in dairy goats in spring; no particular treatment was clearly superior in the study. In South Africa, Greyling and Van Nieker (1991) similarly examined different synchronization techniques in Boer goats during the non-breeding season; they concluded that the intravaginal sponge plus PG and PMSG was an effective treatment.

A study by Alvarez *et al.* (1995) in Mexico sought to determine the effect of two norgestomet doses (3 and 6mg) used in an 11-day treatment period in conjunction with PMSG on fertility in anoestrous goats; both treated groups received 500IU PMSG 24h prior to implant removal. The authors concluded from their results that a 3mg norgestomet implant (half the normal implant) combined with PMSG was sufficient for oestrus control in dairy goats.

A study by Figueiredo Freitas *et al.* (1994) tested the hypothesis that an inadequate progestagen level on the day of sponge removal may be responsible for the variability in the occurrence of oestrus after hormonal treatment of the anoestrous dairy goat. The basic treatment employed was an 11-day treatment using a 45mg FGA sponge, with 50μg cloprostenol and 400–500IU PMSG administered 48h prior to sponge withdrawal. They found, however, that a higher level of progestagen at the end of treatment did not improve oestrus synchronization and apparently led to some loss of fertility.

13.6.6. Repeated use of gonadotrophins

Papers by Baril *et al.* (1991, 1992a,b) in France reported that for goats treated twice, there was an increase in the level of antibodies against PMSG which might explain the decrease in the efficiency of the gonadotrophin in controlling oestrus; the authors suggested that it would be preferable to avoid treating goats with PMSG more than once a year.

13.6.7. Effect of the male

The sudden introduction of a buck during late anoestrus can hasten the onset of the breeding season in the goat (Shelton, 1960, 1980; Chemineau, 1983; Corteel *et al.*, 1983; BonDurant, 1986; Belibasaki *et al.*, 1993; Walkden-Brown *et al.*, 1993a,b,c). There are several reports showing that most does are likely to have ovulated by the 10th day after the introduction of the male. A study of reproduction in the Australian feral goat reported by Restall (1987) showed that the introduction of bucks stimulated oestrus in female goats after 2–3 days, and again after 7–9 days. The conception rate to insemination during the first oestrus was low but increased to over 80% for the second heat period. The author recorded the incidence of ovulation, measured by endoscopic examination, was 10% before introduction of the buck, 90% after 3 days and 100% after 5–10 days.

In a later paper, the same author reviewed the sequence of physiological and endocrinological events occurring in anovular female goats after exposure to bucks (Restall, 1992). It was noted that rapid changes in LH pulse frequency were generally followed by ovulation within three days of exposure, a short luteal phase and a second ovulation 5–6 days later. The male stimulus was apparently multisensorial, with odour as a major component.

Breed effects

In France, Chemineau (1987) showed that female goats responded to the introduction of the buck by a rapid increase in LH plasma levels and by ovulation, not always associated with oestrus and often followed by a short luteal cycle. The quality of the response was found to depend on the intensity of the stimulation and on the depth of anoestrus at the introduction of the bucks. The French worker found the buck effect was more efficient at inducing sexual activity in goat breeds of low seasonality, such as breeds from tropical and subtropical regions, than in breeds originating in mid and high latitudes. In Germany, Claus *et al.* (1990) studied the effect of buck odour on LH secretion and the induction of ovulation in seasonally anoestrous goats; LH concentration increased after exposure of the females to hair from bucks and led to oestrus and ovulation in some of the animals. Evidence reviewed by Chemineau and Delgadillo (1994) has also noted that the presence of the buck can increase the LH pulse frequency in does.

13.6.8. Progestagen treatment and the buck effect

It is evident that the re-establishment of normal ovarian function, at the second occurring ovulation, enables goats to conceive much more readily than they do at the first ovulations of the season. French studies suggested making the buck effect more effective by employing a single progestagen treatment (sponge or injection) before the introduction of males to the anovulatory goats. This treatment simulated the effect of progesterone action during an initial short oestrous cycle. According to Chemineau (1985), such progestagen treatment resulted in 5% of FGA-treated Creole goats experiencing a short cycle, compared with 80% of control does; oestrus was exhibited by 100% of treated females compared with 55% of untreated goats. The eventual outcome was that the conception rate of FGA-treated does at the first induced oestrus was markedly higher than in control animals (78% vs. 15%).

13.7. Use of Melatonin

Studies reported by Deveson *et al.* (1992), in the UK, showed that the administration of melatonin (3mg day^{-1} orally at 16.00h) or artificially short days of 8h light:16h darkness from mid-summer failed to induce early oestrus onset in maiden British Saanen dairy goats. However, long-days (20L:4D) for a period of two months in early spring, followed by 2–3 months of melatonin treatment (implant or daily oral doses), induced oestrus onset and kidding 2–3 months earlier than in untreated controls. Some reports have employed melatonin treatments (Regulin) alone or in conjunction with CIDR and PMSG treatments (Huang *et al.*, 1993).

13.7.1. Melatonin and ram introduction

The effect of melatonin on the response of lactating goats to the introduction of the buck in late anoestrus was examined in studies reported by Belibasaki *et al.* (1993) in Greece. In this work, the introduction of males initiated short cycles but only melatonin-treated does continued to cycle normally; untreated does returned to anoestrus. Other studies reported from that country by Zygoyiannis *et al.* (1993) examined the effect of melatonin implants, administered during anoestrus in indigenous (*Capra prisca*) and crossbred dairy goats on commercial farms. There were some indications that the onset of breeding was advanced by such treatment; the only significant effect recorded was a higher litter size in the melatonin-treated animals.

13.7.2. Melatonin and moulting in Cashmere goats

In Germany, Brackel-Bodenhausen *et al.* (1994) examined the effects of photoperiod and a slow-releasing preparation of melatonin on reproductive activity in Boer × German Improved Fawn goats; it was concluded that photoperiodic control of seasonal cyclicity was modulated by the melatonin treatment. The use of melatonin in goats may also be related to the induction of a moult as well as in influencing reproduction. In Aberdeen, Dicks *et al.* (1995) have reported on the effect of melatonin implants administered from December until April to Cashmere goats. They recorded that adult goats treated with melatonin began their spring moult of cashmere and subsequent growth of guard hairs and cashmere significantly earlier than untreated animals.

13.8. References

Abeyawardene, S.A. and Pope, G.S. (1990) Concentrations of oestradiol-17B in plasma and milk and progesterone in plasma during the oestrous cycle and in early pregnancy in goats. *British Veterinary Journal* 146, 101–105.

Aboul-Ela, M.B., Aboul-Naga, A.M., El-Nakhla, S.M. and Mousa, M.R. (1988) *Proceedings of the 11th International Congress of Animal Reproduction and AI* (Dublin) 4, Paper No. 545.

Akusu, M.O. and Egbunike, G.N. (1990) Effects on oestrus duration of West African dwarf goats. *Small Ruminant Research* 3, 413–418.

Akusu, M.O., Osuagwuh, A.I.A., Akpokodje, J.U. and Egbunike, G.N. (1986) Ovarian activities of the West African dwarf goat (*Capra hircus*) during oestrus. *Journal of Reproduction and Fertility* 78, 459–462.

Al-Baggal, H.A.R., Al-Dahash, S.Y.A. and Alwan, A.F. (1993) Macroscopic study of the female genital system in Iraqi goats. *Small Ruminant Research* 9, 341–346.

Alvarez, F.D., Avendano, L. and Correa, A. (1995) Induction of estrous in anestrous goats after a treatment combining Norgestomet and pregnancy mare serum gonadotropin (PMSG). *Journal of Animal Science* 73 (Suppl. 1), p. 253.

Asdell, S.A. (1964) *Patterns of Mammalian Reproduction*, 2nd edn. Cornell University Press, Ithaca, pp. 623–630.

Ashbrook, P.F. (1982) Year-round breeding for uniform milk production. *Proceedings of the 3rd International Conference on Goat Production*, Scottsdale, Arizona, pp. 153–154.

Avdi, M., Alifakiotis, T. and Driancourt, M.A. (1991) Effect of active immunization against androstendione on reproductive and productive performance of dairy goats. *Livestock Production Science* 29, 87–92.

Baril, G., Vallet, J.C., Beckers, J.F. and Remy, B. (1991) Repeated hormone treatments – caution needed. *Chevre* No. 183, 34–35.

Baril, G., Remy, B., Vallet, J.C. and Beckers, J.F. (1992a) Observations on the repeated use of gonadotropin treatment in dairy goats. *Annales de Zootechnie* 41, 191–296.

Baril, G., Remy, B., Vallet, J.C. and Beckers, J.F. (1992b) Effect of repeated use of progestagen-PMSG treatment for estrus control in dairy goats out of breeding season. *Reproduction in Domestic Animals* 27, 161–168.

Baril, G., Leboeuf, B. and Saumande, J. (1993) Synchronization of estrus in goats: the relationship between time of occurrence of estrus and fertility following artificial insemination. *Theriogenology* 40, 621–628.

Belibasaki, S., Zygoyiannis, D., Davies, P. and Doney, J.M. (1993) Milk progesterone profiles during anoestrus through to pregnancy in Greek dairy goats (*Capra prisca*): the effect of melatonin treatment and male introduction. *Animal Production* 56, 333–339.

Billings, H.J. and Katz, L. S. (1995) The effect of progesterone on the ability of estradiol to stimulate sexual behaviour in female goats in controlled photoperiods. *Biology of Reproduction* 52 (Suppl. 1), p. 130.

BonDurant, R.H. (1981) Reproductive physiology in the goat. *Modern Veterinary Practice* July issue, 525–529.

BonDurant, R.H. (1986) Induction of estrus in does by introduction of buck or photoperiod manipulation. In: Morrow, D.A. (ed.) *Current Therapy in Theriogenology*. W.B. Saunders, Philadelphia, pp. 579–581.

Bowen, G.M. (1988) Experiences with artificial insemination in goats. *Proceedings of the New Zealand Society of Animal Production* 48, 65–67.

Brackel-Bodenhausen, A. Von., Wuttke, W. and Holtz, W. (1994) Effects of photoperiod and slow-release preparations of bromocryptine and melatonin on reproductive activity and prolactin secretion in female goats. *Journal of Animal Science* 72, 955–962.

Bretzlaff, K.N., Ott, R.S., Weston, P.G. and Hixon, J.E. (1980) Doses of prostaglandin $F_{2\alpha}$ effective for induction of estrus in goats. *Theriogenology* 16, 587.

Bretzlaff, K.N., Hill, A. and Ott. R.S. (1983) Induction of estrus in goats with prostaglandin $F_{2\alpha}$. *American Journal Veterinary Research* 44, 1162.

Camp, J.C., Wildt, D.E., Howard, P.K., Stuart, L.D. and Chakraborty, P.K. (1983) Ovarian activity during normal and abnormal length estrous cycles in the goat. *Biology of Reproduction* 28, 673–681.

Cerbito, W.A., Natural, N.G., Aglibut, F.B. and Sato, K. (1995) Evidence of ovulation in goats (*Capra hircus*) with short estrous cycle and its occurrence in the tropics. *Theriogenology* 43, 803–812.

Chemineau, P. (1983) Effect on oestrus and ovulation of exposing Creole goats to the male at three times of the year. *Journal of Reproduction and Fertility* 67, 65–72.

Chemineau, P. (1985) Effects of a progestagen on buck-induced short ovarian cycles in the Creole meat goat. *Animal Reproduction Science* 9, 87–94.

Chemineau, P. (1987) Possibilities for using bucks to stimulate ovarian and oestrous cycles in an ovulatory goats – a review. *Livestock Production Science* 17, 135–147.

Chemineau, P. and Delgadillo, J.A. (1994) Neuroendocrinology of reproduction in goats. *Productions Animales* 7, 315–326.

Chemineau, P., Gauthier, D., Poirier, J.C. and Saumande, J. (1982) Plasma levels of LH, FSH, prolactin, oestradiol-17B and progesterone during natural and induced oestrus in the dairy goat. *Theriogenology* 17, 313–323.

Chemineau, P., Cognie, Y., Guerin, Y., Orgeur, P. and Vallet, J.C. (1991) Training manual on artificial insemination in sheep and goats. *FAO Animal Production and Health Paper*, No. 83, 222 pp.

Chibooka, O., Somade, B. and Montsma, G. (1988) Reproduction of West African Dwarf goats – a summary of research work at Ile-Ife, Nigeria. In: Smith, O.B. and Bosman, H.G. (eds) *Goat Production in the Humid Tropics* pp. 125–136.

Claus, R., Over, R. and Dehnhard, M. (1990) Effect of male odour on LH secretion and the induction of ovulation in seasonally anoestrous goats. *Animal Reproduction Science* 22, 27–38.

Cooke, R.G. and Kniffton, A. (1981) Oxytocin-induced oestrus in the goat. *Theriogenology* 16, 95–97.

Cooke, R.G. and Homeida, A.M. (1982) Plasma concentrations of 13, 14-dihydro-15-keto-prostaglandin f2 and progesterone during oxytocin-induced oestrus in the goat. *Theriogenology* 18, 453–460.

Cooke, R.G. and Homeida, A.M. (1983) Prevention of the luteolytic action of oxytocin in the goat by inhibition of prostaglandin synthesis. *Theriogenology* 20, 363--365.

Corteel, J.M. (1975) The use of progestagens to control the oestrous cycle of the dairy goat. *Annales de Biologie animale Biochimie Biophysique* 15, 353–363.

Corteel, J.M., Gonzalez, C. and Nunes, J.F. (1982) Research and development in the control of reproduction. *Proceedings of the 3rd International Conference on Goat Production and Diseases*, pp. 584–601.

Corteel, J.M., Baril, G., Leboeuf, B. and Nunes, J.F. (1983) Goat semen technology. In: Courot, M. (ed.) *The Male in Farm Animal Reproduction* Seminar Proceedings, pp. 237–256.

Delgadillo, J.A., Leboeuf, B. and Chemineau, P. (1991) Decrease in the seasonality of sexual behaviour and sperm production in bucks by exposure to short photoperiodic cycles. *Theriogenology* 36, 755–770.

Delgadillo, J.A., Leboeuf, B. and Chemineau, P. (1992) Abolition of seasonal variations in semen quality and maintenance of sperm fertilizing ability by photoperiodic cycles in goat bucks. *Small Ruminant Research* 9, 47–59.

Delgadillo, J.A., Leboeuf, B. and Chemineau, P. (1993) Maintenance of sperm production in bucks during a third year of short photoperiodic cycles. *Reproduction, Nutrition and Development* 33, 609–617.

Deveson, S.L., Arendt, J. and Forsyth, I.A. (1990) Sensitivity of goats to a light pulse during the night as assessed by suppression of melatonin concentrations in the plasma. *Journal of Pineal Research* 8, 169–177.

Deveson, S.L., Forsyth, I.A. and Arendt, J. (1992) Induced out-of-season breeding in British Saanen dairy goats: use of artificial photoperiods and/or melatonin administration. *Animal Reproduction Science* 29, 1–15.

Dicks, P., Russell, A.J.F. and Lincoln, G.A. (1995) The effect of melatonin implants administered from December until April, on plasma prolactin, tri-iodothyronine and thyroxine concentrations and on the timing of the spring moult in cashmere. *Animal Science* 60, 239–247.

Driancourt, M.A., Philipon, P., Terqui, M., Molenat, G., Mirman, B., Louault, C., Avdi, M., Folch, J. and Cognie, Y. (1990) Possibilities for immunization against steroids to improve ovulation rate and litter size in sheep and goats. *Production Animales* 3, 31–37.

East, N.E. and Rowe, J.D. (1989) Subcutaneous progestin implants versus intravaginal sponges for dairy goat estrus synchronization during the transitional period. *Theriogenology* 32, 921–928.

El-Amrawi, G.A., Hussein, F.M. and El-Bawab, I.E. (1993a) Fertility of Saanen goats following induction of oestrus using PGF 2 ×. *Assiut Veterinary Medical Journal* 29, 241–248.

El-Amrawi, G.A., Hussein, F.M. and El-Bawab, I.E. (1993b) Oestrus synchronization and kidding rate in does treated with a vaginal sponge. *Assiut Veterinary Medical Journal* 29, 249–259.

Figueiredo Freitas, V.J., Baril, G. and Saumande, J. (1994) The effect of progestagen level at the end of estrus induction treatment on estrus synchronization, fertility and fecundity in the anoestrous dairy goat. *Proceedings of the 10th Meeting of the European Embryo Transfer Association* (Lyon), 170.

Forcada, M.F., Sierra, A.I. and Callen, M.A. (1989) A comparison of the efficiency of different methods for the induction and synchronization of oestrus in dairy goats in spring. *Archivos de Zooteccnia* 38, 211–222.

Garcia, C.J., Padilla, R.G., De-Leon, T.M. and Martinez, C.A. (1994) Intrauterine artificial insemination through cervix in goats with synchronized oestrus. *Journal of Reproduction and Fertility* Abstract Series No. 13, p. 49.

Ginther, O.J. and Kot, K. (1994) Follicular dynamics during the ovulatory season in goats. *Theriogenology* 42, 987–1001.

Greyling, J.P.C. and Van Niekerk, C.H. (1987) Occurrence of oestrus in the Boer goat doe. *South African Journal of Animal Science* 17(3), 147–149.

Greyling, J.P.C. and Van Niekerk, C.H. (1990) Ovulation in the Boer goat doe. *Small Ruminant Research* 3, 457–464.

Greyling, J.P.C. and Nieker, C.H. Van (1991) Different synchronization techniques in Boer goat does outside the normal breeding season. *Small Ruminant Research* 5, 233–243.

Harrison, R.J. (1948) The changes occurring in the ovary of the goat during the oestrous cycle and in early pregnancy. *Journal of Anatomy* 82, 21–48.

Homeida, A.M. and Cooke, R.G. (1989) Hormonal events at oxytocin-induced estrus in the goat. *Theriogenology* 32, 1007–1010.

Howland, B.E., Sanford, L.M. and Palmer, W.M. (1985) Changes in the serum levels of LH, FSH, prolactin, testosterone and cortisol associated with season and mating in male pygmy goats. *Journal of Andrology* 6, 89–96.

Huang, J.C., Lin, J.H. Yuan, H.H. and Tseng, J.L. (1993) Induction of oestrus in dairy goats during the anoestrous season with melatonin and progesterone plus PMSG. *Journal of Taiwan Livestock Research* 26, 189–202.

Jarosz, S.J., Deans, R.J. and Dukelow, W.R. (1971) The reproductive cycle of the African Pygmy and Toggenburg goat. *Journal of Reproduction and Fertility* 24, 119–123.

Katz, L.S. and McDonald, T.J. (1992) Sexual behaviour of farm animals. *Theriogenology* 38, 239–253.

Kumar, S.S. and Thomas, C.K. (1994) Synchronization of oestrus in goats. I. Effect on reproductive performance and man hour requirement. *Indian Journal of Animal Production and Management* 10, 74–80.

Leboeuf, B. (1993) Recommendations for successful hormone treatment and AI in goats. *Chevre* No. 195, 13–16.

Leboeuf, B., Nercy, C. and Ruyter, T.De (1994) Artificial insemination of goats in Rwanda. Adaptation to Rwandan goats of the method used for European dairy breeds. *Revue d'Elevage et de Medecine Veterinaire des Pays Tropicaux* 47, 240–243.

Llewelyn, C.A. and Kadzere, C.J. (1992) Oestrous synchronization and fertility following treatment with fluorogestone acetate (Chronolone) impregnated intra-vaginal sponges in indigenous goats of Zimbabwe. *Zimbabwe Veterinary Journal* 23, 159–165.

Llewelyn, C.A., Perrie, J., Luckins, A.G. and Munro, C.D. (1993) Oestrus in the British White goat: timing of plasma luteinizing hormone surge and changes in behavioural and vaginal traits in relationships to onset of oestrus. *British Veterinary Journal* 149, 171–182.

Mareco, G. and Arosteguy, R.P. (1994) Intra-uterine insemination with frozen semen by laparoscopy in goats. *Veterinaria Argentine* 11, 676–678.

Meinecke-Tillmann, S. (1988) Problems of controlling the cycle in goats. *Deutsche Veterinarmedizinische Gesellschaft* 75–77.

Mgongo, F.O.K., Gombe, S. and Ogua, J.S. (1984) Progesterone, estrogen, LH and corticosteroids in goats. *Indian Journal Animal Reproduction* 4, 1–5.

Moore, R.W., Miller, C.M., Hall, D.R.H. and Dow, B.W. (1988) Cervical versus laparoscopic AI of goats after PMSG injection at or 48 hours before CIDR removal. *Proceedings of the New Zealand Society of Animal Production* 48, 69–70.

Moore, R.W., Dow, B.W. and Staples, L.D. (1989) Artificial insemination of farmed feral goats with frozen-thawed semen. *Proceedings of the New Zealand Society of Animal Production* 49, pp. 171–173.

Ott, R.S. (1986) Prostaglandins for induction of estrus, estrus synchronization, abortion and induction of parturition. In: Morrow, D.A. (ed.) *Current Therapy in Theriogenology.* W.B. Saunders, Philadelphia, 583–585.

Ott, R.S., Nelson, D.R. and Hixon, J.E. (1980) Fertility of goats following synchronization of estrus with prostaglandin $F_{2\alpha}$. *Theriogenology* 13, 341.

Restall, B.J. (1987) Reproduction in the Australian feral goat. In: *Biannual Research Report* (1984–86), North Coast Agricultural Institute Wollongbar, pp. 24–26.

Restall, B.J. (1992) The male effect in goats. *Proceedings of the 5th International Conference on Goats* (New Delhi) 2(2), pp. 322–331.

Restall, B.J. (1992) Seasonal variation in reproductive activity in Australian goats. *Animal Reproduction Science* 27, 305–318.

Ritar, A.J. and Salamon, S. (1991) Effects of month of collection, method of processing, concentration of egg yolk and duration of frozen storage on viability of Angora goat spermatozoa. *Small Ruminant Research* 4, 29–37.

Ritar, A.J., Salamon, S., Ball, P.D. and O'May, P.J. (1989) Ovulation and fertility in goats after intravaginal device-PMSG treatment. *Small Ruminant Research* 2, 323–331.

Ritar, A.J., Ball, P.D. and O'May, P.J. (1990) Artificial insemination of cashmere goats: effects on fertility and fecundity of intravaginal treatment, method and time of insemination, semen freezing process, number of motile spermatozoa and age of females. *Reproduction, Fertility and Development* 2, 377–384.

Ritar, A.J., Robertson, J.A. and Evans, G. (1994) Ovulatory activity, hormonal induction of ovulation and fertility of young cashmere and Angora female goats in a temperate environment. *Reproduction, Fertility and Development* 6, 737–747.

Romano, J.E. (1993) Effect of service on estrus duration in dairy goats. *Theriogenology* 40, 77–84.

Romano, J.E. (1994a) Effects of service number on estrus duration in dairy goats. *Theriogenology* 41, 1273–1277.
Romano, J.E. (1994b) Effects of different stimuli of service on estrus duration in dairy goats. *Theriogenology* 42, 875–879.
Salama, A. (1972) Ovarian changes in goats during oestrus. *Indian Journal of Animal Science* 42, 436–438.
Sawada, T., Fujikawa, Y., Sato, S. and Mori, J. (1994) Effect of oxytocin and indomethacin on the oestrous cycle of goats. *Prostaglandins* 48, 91–98.
Sawada, T., Takahara, Y. and Mori, J. (1995) Secretion of progesterone during long and short days of the oestrous cycle in goats that are continuous breeders. *Theriogenology* 43, 789–795.
Shelton, M. (1960) The influence of the presence of the male goat on the initiation of oestrous cycling and ovulation in Angora goats. *Journal of Animal Science* 19, 368–375.
Shelton, M. (1980) Influence of various exteroceptive factors on initiation of oestrus and ovulation. *International Goat and Sheep Research* 1, 156–162.
Simplicio, A.A. and Machado, R. (1991a) Fertility of goats inseminated with frozen semen during spontaneous oestrus or synchronized with MGA, hCG and cloprostenol. *Proceedings of the Congress Animal Reproduction* (Brazil) 2, p. 363.
Simplicio, A.A. and Machado, R. (1991b) Fertility of dairy goats subjected to oestrus synchronization and insemination at a predetermined time. *Proceedings of the 9th Congress Animal Reproduction* (Brazil), 2, 351.
Smith, M.C. (1986a) The reproductive anatomy and physiology of the female goat. In: Morrow, D.A. (ed.) *Current Therapy in Theriogenology*. W.B. Saunders, Philadelphia, pp. 577–579.
Smith, M.C. (1986b) Synchronization of estrus and the use of implants and vaginal sponges. In: Morrow, D.A. (ed.) *Current Therapy in Theriogenology*. W.B. Saunders, Philadelphia, pp. 582–583.
Sureshkumar, P.K. and Janakiraman, K. (1992) Changes in concentrations of serum progesterone, prolactin and LH in relation to pituitary LH during the caprine estrous cycle. *Journal of Reproduction and Development* 38, 303–307.
Sureshkumar, P.K. and Janakiraman, K. (1993) Histomorphological changes in the caprine ovary relative to the stages of the estrous cycle. *Small Ruminant Research* 12, 287–300.
Tamanini, C., Bono, G., Cairoli, F. and Chiesa, F. (1985) Endocrine responses induced in anestrous goats by the administration of different hormones after a fluorogestone acetate treatment. *Animal Reproduction Science* 9, 357–364.
Underwood, E.J., Shier, F.L. and Davenport, N. (1944) Studies in sheep husbandry in Western Australia. V. The breeding season of Merino crossbred and British breed ewes in the Agricultural districts. *Journal Department Agriculture of Western Australia* 11, 135–143.
Van Rensburg, S.J. (1971) Reproductive physiology and endocrinology of normal and habitually aborting Angora goats. *Onderstepoort Journal of Veterinary Research* 38, 1–62.
Walkden-Brown, S.W., Restal, B.J. and Henniawati (1993a) The male effect in the Australian cashmere goat. 1. Ovarian and behavioural response of seasonally anovulatory does following the introduction of bucks. *Animal Reproduction Science* 32(1–2), 41–53.
Walkden-Brown, S.W., Restall, B.J. and Henniawati (1993b) The male effect of the Australian cashmere goat. 2. Role of olfactory cues from the male. *Animal Reproduction Science* 32, 55–67.

Walkden-Brown, S.W., Restall, B.J. and Henniawati (1993c) The male effect in the Australian cashmere goat. 3. Enhancement with buck nutrition and use of oestrous females. *Animal Reproduction Science* 32(1–2), 69–84.

Walkden-Brown, S.W., Restall, B.J., Norton, B.W. and Scaramuzzi, R.J. (1994) The 'female effect' in Australian cashmere goats: effect of season and quality of diet on the LH and testosterone response of bucks to oestrous does. *Journal of Reproduction and Fertility* 100, 521–531.

Wildeus, S. and Moore, G.A. (1995) Age and gonadotropin effects on the estrus and ovulation response to norgestomet synchronization in Spanish does. *Journal of Animal Science* 73 (Suppl. 1), p. 252.

Zygoyiannis, D., Davies, P.H. and Doney, J.M. (1993) The effect of melatonin on seasonal reproduction of indigenous and crossbred dairy goats in Greece. *Animal Production* 57, 273–279.

14 Pregnancy Testing and the Control of Parturition in Goats

14.1. Introduction

In dairy goat herds, where the commercial success of the enterprise depends on the ability of the producer to meet market requirements, there is likely to be ample justification for considering the pregnancy diagnostic techniques that may be employed in this species. The choice of such methods will usually depend on the stage of gestation. Some methods give a higher degree of accuracy at the early stage of gestation, whereas others are more appropriate at later stages of gestation (Bernard, 1994). Diagnosis of pregnancy based on hormonal assays of blood plasma, serum or milk, and estimation of pregnancy-specific antigens or proteins, can provide a higher degree of accuracy than other methods for diagnosing early pregnancy. Ultrasonic pregnancy detectors can detect pregnancy at 50–70 days after mating with > 90% accuracy. With the advent of real-time ultrasonic scanners, it is possible to diagnose pregnancy at 40–45 days after mating with 90% accuracy. Other methods that have been used for pregnancy diagnosis in goats include abdominal palpation, radiography and vaginal cytology.

In terms of the commercial advantages of induced parturition in goats, these include: (i) reducing the incidence of neonatal mortality by allowing closer attention by the stockperson and earlier intervention in instances of dystocia; and (ii) arranging to have kiddings at the most appropriate times for the available labour resources.

14.2. Physiology and Endocrinology of Early Pregnancy in the Goat

14.2.1. Maintenance of pregnancy in the goat

As noted by Currie (1977), despite the many similarities during pregnancy in sheep and goats, one important difference lies in the regulation of pregnancy and parturition. Although both species depend on the luteal production of progesterone in the first trimester of pregnancy, the sheep subsequently recruits placental production of this hormone thereby rendering luteal function dispensable. The goat, on the other hand, requires a functional corpus luteum (CL) at all times during gestation as this body is the major source of progesterone; interference with luteal function at any stage of pregnancy will provoke abortion (Meites *et al.*, 1951).

14.2.2. Nutrition and progesterone levels in the goat

A high concentration of circulating progesterone is known to be essential for the establishment and maintenance of pregnancy in the goat. In sheep, an association has been demonstrated between progesterone profiles and embryo survival and it is evident that the effects of nutrition may be mediated by way of progesterone concentrations. A study reported by Mani *et al.* (1995) aimed to determine the effects of undernutrition on progesterone concentration during the early luteal phase and mid-gestation in the goat. Results showed that undernutrition had no effect on plasma progesterone concentration during the early luteal phase but by mid-gestation there was an inverse relationship between level of nutrition and progesterone level. This appears to be the first report that nutrition can influence progesterone concentration in the goat.

The higher progesterone concentration in feed-restricted goats during mid-gestation is in agreement with studies in pregnant sheep which have shown an inverse relationship and the concentration of plasma progesterone. It must be remembered, however, that the source of progesterone in mid-pregnancy is quite different in these two species due to the reliance of the ewe on placental production of the steroid.

14.2.3. Maintenance of corpus luteum

Mechanisms controlling progesterone production by the corpus luteum of pregnancy in the goat and in other CL-dependent species, such as the pig, have not been fully elucidated. However, there have been those who believed that luteal secretion of progesterone in the goat is controlled in large part by maternal pituitary gonadotrophins (Denamur, 1974) and it has been shown that removal of the pituitary will result in abortion. At later stages of gestation, however, the sheep tolerates hypophysectomy without pregnancy being disrupted whereas the goat still responds by aborting (Cowie *et al.*, 1963). Although the placentas of the ewe and doe are very similar in gross and fine morphological and structural detail, there is apparently no evidence of significant progesterone production at this site (Thorburn and Schneider, 1972); even if some progesterone synthesis does occur, it is clearly insufficient to maintain pregnancy in the absence of corpora lutea.

According to Currie (1977), it seemed likely, but not proven at that time, that goat placental lactogen contributes to the regulation of progesterone secretion by the CL. Studies reported by Malecki *et al.* (1987) in Australia suggested that the goat does not require maternal pituitary LH to maintain pregnancy between days 50 and 130 of gestation; these authors concluded that CL function may be maintained by luteotrophins or an antiluteolytic agent produced by the uterine contents. They also noted the possibility that the CL of the goat may be capable of producing progesterone independently of luteotrophins (Rothschild, 1981).

14.3. Physiology and Endocrinology of Late Pregnancy and Kidding in the Goat

14.3.1. Factors affecting length of the gestation period

The duration of gestation in the goat is similar to that in the sheep. A large series of goats of several breeds quoted by Asdell (1964) gave a mean of 150.8 days; the modal distribution was 151 days, and 86% of gestation periods fell between 147 and 155 days. The same author drew attention to some of the breed differences which had been recorded in the literature at that time (see Table 14.1) and to the fact that the month of conception had some influence; it averaged 151.3 days for August conceptions and 149.8 days for conceptions occurring in February. Age is believed to have an effect, being least (150.1 days) when the goat conceives in her first year and gradually rises to a maximum (151.3 days) at 6 years (Asdell, 1964). There is an effect of litter size on gestation length in goats as in sheep; according to Peaker (1978), quadruplets are born three days earlier than singletons. In Germany, Sambraus and Wittmann (1993) recorded an average of 151.4 and 150.3 days for adult goats giving birth to twins and triplets, respectively.

Table 14.1. Breed differences in the duration of gestation in goats (from Asdell, 1964).

Breed	Duration of gestation (days)
Bar Bari	146
East African Dwarf Goat	146.5
Angora	148.08 ± 0.09
Philippine goats	148.1 ± 0.07
Jumna Pari	150
Anglo-Nubian	150.0 ± 0.1
Schwartzwald	150.8 ± 0.2

14.3.2. Endocrine events

The plasma level of progesterone, which is produced almost entirely by the corpora lutea of the pregnant doe, remains high until about four days before parturition. Oestrone sulphate starts to increase at 40–50 days of gestation, and can be employed as a means of pregnancy diagnosis after that time; peak oestrogen levels are found at parturition and a very low concentration is evident one day after parturition (Challis and Linzell, 1971). It is believed that placental lactogen, which is also produced by the placenta, follows a similar pattern; placental lactogen is first detectable at about two months of gestation (Thomas *et al.*, 1977).

14.3.3. Fetal signals in late pregnancy

The way in which the developing goat fetus signals the timing of labour onset differs in detail from that found in the sheep because the critical event for activating the caprine uterus is prepartum luteal regression; this maternal event occurs abruptly about 24 h before delivery (Currie and Thorburn, 1977) and results in a near complete removal of progesterone from the maternal circulation by the time uterine contractions are initiated.

14.3.4. Prepartum management

Dairy goats, like milking cows, require a 6–8-week dry period to ensure good milk production during the subsequent lactation. Although the pregnant goat should not be allowed to become fat, attention to the diet, in terms of supplying high-quality roughage and appropriate amounts of meals during the final trimester of pregnancy, is essential to avoid problems from pregnancy toxaemia (ketosis). Periparturient care of the doe is dealt with in an article by Smith (1986), the author covering disease control and the appropriate treatment for pregnancy toxaemia.

14.3.5. Maternal behaviour in goats

Maternal and neonatal behaviour in goats is similar to that of sheep (Blauvellt, 1954; Klopfer, 1971). However, according to Alexander *et al.* (1974), goats appear less likely than sheep to reject one of twins. It is not certain whether the goat is a 'follower' or a 'hider' species. In feral goats, animals show a hiding phase of up to four days, at least in mountainous regions (Rudge, 1970); this is associated with a clear preference for maternal isolation at the time of parturition.

The first sign that kidding is imminent is the increased engorgement of the udder. About a day before parturition, the relaxation of the pelvis is evident as hollows appear on either side of the tail. The doe may be restless and paw the ground as well as alternately lying down and standing up. Normal parturition usually starts with a period of uterine contractions of 1–10 h. This is followed by rupture of the placental membranes and presentation of the kid. Most goats kid lying down and once labour commences, regular checks of progress should be made. Usually, kids are born within three hours and the fetal membranes are passed within two hours of the last kid. According to the behaviour studies of Sambraus and Wittmann (1993), the interval between the birth of the first and second kids in triplet-bearing does, averaged 15 min and that between the birth of the second and third kid 6 min. They also recorded that 98% of kids did not attempt to suck any doe other than their own mother and that the average time from parturition to first sucking was 35 min for first-born kids and 51 min for subsequent kids.

If the goat has been observed straining for more than one hour without apparent progress in the birth process, it is time to provide assistance; the most common and easily corrected dystocia results from two or more kids in the birth canal simultaneously (Franklin, 1986). Assistance should only be attempted by a veterinarian or a stockperson who is thoroughly conversant with dealing with difficult births. On no account should an inexperienced person attempt to give assistance.

Once the kids are delivered, it is important to ensure that the dam has the opportunity to lick them dry. As well as providing essential stimulation for the kids, it helps in establishing the mother–kid bond; licking for 5–10 min is usually adequate for acceptance. According to Smith (1986), there is a critical period of about two hours after birth during which the doe must be exposed to her kid if she is to accept it. It is also important that the kids feed within an hour or two of birth. As with sheep, newborn goats that are too weak to suck from their mothers should be fed by stomach-tube until they have gained enough strength to suck on their own. In dairy herds, most kids, whether they are to be reared as herd replacements or for meat, are taken from their dams after receiving colostrum for 24h. Goats kept for fibre production may be run as a suckler herd, with the kids remaining with their mothers until weaned at about 12–14 weeks of age.

14.4. Methods Used in Pregnancy Testing and Detection of Multiples in the Goat

A number of authors have reviewed the growing range of techniques that are available for determining pregnancy in the goat (BonDurant, 1981; Williams, 1986; Goel and Agrawal, 1992; Bernard, 1994).

14.4.1. Non-return of oestrus

Although the absence of a heat period may be considered a useful indication of the doe's pregnancy status, in practice the non-return of oestrus may not be reliable. Clearly, if the doe has been bred out of season, the absence of a heat period may simply mean the animal has returned to its anoestrous condition. According to Williams (1986), many goats have a tendency to show some signs of heat during pregnancy. Restall *et al.* (1990) recorded that diagnoses of pregnancy based on returns to service were not accurate in Thai goats because 36.5% of pregnant animals were recorded as returning to service.

14.4.2. Use of ultrasonics

Doppler ultrasound technique

The Doppler ultrasound technique has been reported to work satisfactorily in the hands of skilled operators. According to Williams (1986), it is possible to test goats as early as 25–30 days, but in the author's experience false-negatives were a problem. She found it more satisfactory performing the test at 35–40 days of pregnancy; at this stage, it was easy to detect the sound of the fetal blood flow, which was much faster than that of the mother.

Echo-pulse ultrasonics

Using amplitude/depth ultrasonics (A-mode sound), it is possible to make accurate diagnoses at 65–100 days after mating. This technique involves detection of a sound wave reflected from each wall of a cross-section of the uterus; when the distance between echoes is extended, this is interpreted as reflecting the distension characteristic of a gravid uterus. As described by BonDurant (1981), the transducer probe is placed in the right lower flank just above the udder and directed in an arc from the left stifle to the last left rib (see Fig. 14.1). Neither fasting nor exceptional restraint is required in applying the technique.

14.4.3. Real-time ultrasonic scanning

It has been evident from the early 1980s that real-time ultrasonic scanning is a very useful method for use with goats and is one of the means of identifying the false pregnancies which can be troublesome in this species. In Canada, Baronet and Vaillancourt (1989) demonstrated that accurate pregnancy diagnosis in goats was rapidly obtained using real-time ultrasonic scanning with a 5MHz rectal probe; the earliest day that the presence of the embryos could be reliably observed in early pregnancy was day 25. If the abdomen of the doe was raised, the techniques could be employed from day 25 to 100 after the breeding date. By using the probe rectally, the authors noted that no preparation of the goat was required and the minimal stress was created.

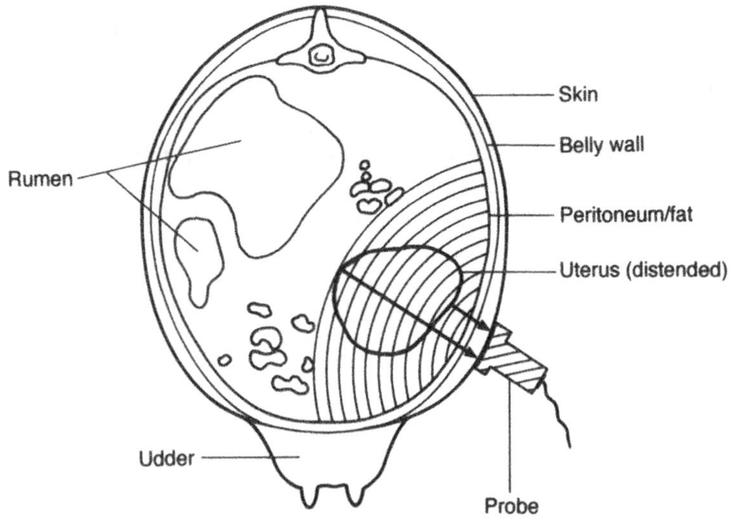

Fig. 14.1. Diagrammatic representation of the use of A mode ultrasound in pregnancy diagnosis in the goat (from BonDurant, 1981).

Distinguishing true from false pregnancies

According to Russel (1990), writing about goats in the UK, real-time ultrasonic scanning can be used to distinguish between true and false pregnancies in does as well as to count the number of kids carried. The author also provides information on the stage of gestation at which scanning is possible, the accuracy of the diagnosis and the speed of scanning. Elsewhere, pregnancy diagnosis using real-time ultrasonics was used by Restall *et al.* (1990) in Thai goats; they recorded an accuracy of 100% in diagnosing pregnancy but only a 25% accuracy in detecting twin fetuses. In France, Tainturier *et al.* (1993) recorded that real-time ultrasonics, with a human prostate probe, was capable of visualizing embryonic vesicles after 16–17 days of gestation; embryos and heart-beats were detected at day 23 and the number of embryos could be estimated from day 25.

In the USA, Dawson *et al.* (1994) reported on the determination of fetal numbers in Alpine does by real-time ultrasonic scanning by the trans-abdominal route. The does were restrained while standing and the transducer probe was placed on the hairless caudal ventral abdominal wall cranial to the udder; non-pregnant females were easily recognized 5 and 7 weeks after breeding with 100% accuracy. The authors recorded the accuracy for determining singles, twins and triplets at 5 weeks of gestation as 44, 73 and 67%, respectively; at 7 weeks, it was 83, 89 and 100%, respectively. In Spain, Santiago Moreno *et al.* (1995) examined goats daily by way of a transrectal 7.5 MHz ultrasonic probe over the period 10–40 days after mating; they record 100% accuracy in detecting pregnancy at day 24 and 100% accuracy in detecting multiples at day 30.

14.5. Progesterone and Oestrogen Assays

14.5.1. Progesterone tests

According to BonDurant (1981), from 19 to 22 days after breeding, the plasma progesterone concentrations of pregnant does will usually be greater than 1.4ng ml^{-1}; goats that are non-pregnant will be returning to oestrus at this time and can be expected to have minimal progesterone concentrations, certainly less than 1ng ml^{-1}. Milk progesterone concentrations in the goat closely follow those of plasma progesterone, apart from the fact that because progesterone is a fat-soluble hormone, the concentrations are higher in milk. Does with milk progesterone levels less than 1.5ng ml^{-1} at 19–22 days following breeding are generally considered to be non-pregnant.

Progesterone testing in the UK and France

The Milk Marketing Board (England and Wales) in the UK offered a progesterone assay service to goat breeders in 1980, based on the earlier work of Holdsworth and Davies (1979), using a single milk sample obtained 24 days after service. However, a high incidence of false positive results, which were obtained when the anti-serum used in this procedure was changed, prevented the service from being developed.

In France, Thibier *et al.* (1982) observed that early pregnancy diagnosis by progesterone measurement from blood samples 21–22 days after artificial insemination (AI) in the goat, although useful, was a cumbersome and expensive technique. They investigated the accuracy of progesterone testing in milk samples and concluded that two major problems prevented the launch of a routine early pregnancy diagnosis service, using milk samples: (i) a low ratio of the progesterone levels between pregnant and non-pregnant does; and (ii) large between-flocks variations in the mean concentrations of progesterone in pregnant and non-pregnant animals.

Variability in results

In the UK, Holdsworth *et al.* (1983) suggested reasons for the substantial differences observed between the concentrations of progesterone in goat's milk and plasma. An analysis of their data showed that the type of progesterone antiserum employed, the presence of other progesterone metabolites and the physiological function of the mammary gland, especially in pregnancy, could all influence the progesterone concentration recorded in milk. In practical terms, plasma gave a more precise low progesterone value for predicting non-pregnancy in the goat. A report of work by Murray and Newstead (1988) using commercially available milk and plasma ELISA diagnostic kits (Ovucheck; Cambridge), recorded the difficulties they encountered in interpreting progesterone concentrations in goat's milk; the wide range of values obtained indicated that prediction of pregnancy with such kits is very uncertain.

Sampling methods employed

The use of progesterone tests for diagnosing pregnancy in goats takes various forms. Where the mating dates of goats is not accurately known, it may be attempted by taking several consecutive samples. A paper by Restall *et al.* (1990) reported pregnancy diagnosis in Thai native goats, which was attempted by measuring plasma progesterone concentrations at four 7-day intervals after mating; 96.2% of goats with all four progesterone concentrations >2 ng ml^{-1} were pregnant. In Cuba, Carmenate *et al.* (1992) monitored blood progesterone concentration 8 and 21 days after insemination in goats; at 21 days, pregnancy was diagnosed accurately in 77.4% of does and non-pregnancy in 95.7%.

Problem of false positives

Although as a test for non-pregnancy, the progesterone assay, used in both milk and serum samples, proved to be satisfactory (Holdsworth and Davies, 1979), high levels of the steroid can be indicative of several conditions other than pregnancy. As noted by Williams (1986), the goat in question may have a shorter or longer cycle than 21 days and may have a functional corpus luteum; hydrometra, pseudopregnancy and persistent corpus luteum are also mentioned as capable of giving false-positives.

In the USA, Fleming *et al.* (1990) predicted pregnancy status in goats from serum samples collected 21 days after the last breeding and analysed using a commercial bovine milk progesterone enzyme immunoassay (EIA) test and a radioimmunossay (RIA) test. Both tests detected non-pregnancy (EIA, 100%; RIA, 80%) more accurately than pregnancy (EIA, 66%; RIA, 75%). The authors concluded that commercial cowside progesterone tests had potential as a rapid, inexpensive screening test for non-pregnant does bred out of season. In Australia, Dionysius (1991), working with dairy goats, tested the accuracy of two qualitative on-farm assay kits and two quantitative assay kits, all designed for use in the dairy cow; accuracy of a positive diagnosis ranged from 83 to 88% and of non-pregnancy from 80 to 100%.

14.5.2. Oestrogen tests

Data on the concentration of unconjugated total oestrogens and of oestrone and oestradiol-17-alpha in the peripheral plasma of pregnant goats were reported by various workers in the early 1970s (Thornburn *et al.*, 1972; Challis and Linzell, 1973). It also became clear during the 1970s that the increase in plasma oestrone concentration in pregnant goats was accompanied by an increase in its conjugated form which rose gradually during gestation (Heap and Hamon, 1979). Because of its transport by the mammary gland (Heap *et al.*, 1984), oestrone sulphate is detectable in milk as well as in blood samples. A typical oestrone sulphate profile of a goat during pregnancy is provided in Fig. 14.2.

According to a paper by Chaplin and Holdsworth (1982) and Davies and

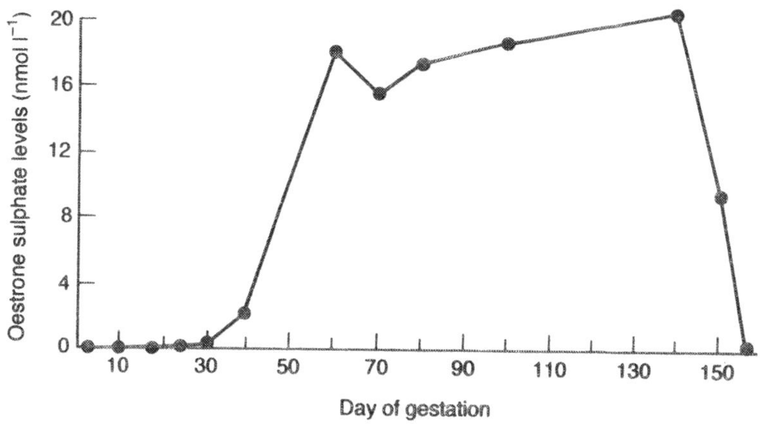

Fig. 14.2. Oestrone sulphate profile of a goat during pregnancy (from McArthur and Geary, 1986).

Chaplin (1983) in the UK, oestrone sulphate radioimmunossay of goat's milk at 50+ days of gestation is capable of giving an accurate diagnosis of pregnancy; the test did not give false-positives with hydrometra or persistent corpus luteum. The oestrone sulphate test was also reported by Booth (1984) to be highly accurate and capable of distinguishing accurately between a true and false pregnancy in the goat. According to Williams (1986), a comparable oestrone sulphate urine test can be employed in goats that are dry or in their first pregnancy.

True pregnancy and pseudopregnancy

The need by dairy goat owners for an accurate means of pregnancy detection to aid in management, where false heat and pseudopregnancy may be a problem, resulted in the development in the USA of a direct RIA for oestrone sulphate, similar to that described by Holdsworth *et al.* (1982) in the UK. A report by McArthur and Geary (1986) dealt with a field test of this system which was carried out on a large commercial dairy goat farm. This study confirmed that the measurement of oestrone sulphate in milk was an extremely accurate means of pregnancy diagnosis, and that it was highly effective in enabling true and pseudopregnancy to be distinguished. It was believed that this was because oestrone sulphate is produced by the fetus and the continued production of this steroid is indicative of fetal viability. However, studies reported by Sawada *et al.* (1995) in Japan have indicated that the fetus may not take part in oestrogen production and that the main site of oestrone in the pregnant goat may be the placenta.

Accuracy of results

In the UK, a paper by Murray and Newstead (1988) dealt with the use of an oestrone sulphate ELISA milk assay, the authors recording an 80% accuracy in determining both pregnancy and non-pregnancy when compared with actual kidding data; the authors discuss reasons for these errors. It is apparent that assay difficulties in the measurement of oestrone sulphate need to be kept in mind in assessing the reliability of this test.

14.5.3. *Oestrogen in faeces*

Faecal sampling from week 5 to week 13 of gestation in goats revealed that the oestrogen concentration of faeces was significantly higher in pregnant than in non-pregnant animals from day 42 of gestation and that the concentrations of non-cycling goats were lower than those in cycling does (Sindermann, 1991). The overall accuracy of pregnancy diagnosis carried out on day 50 of gestation was 100% for blood samples and 91.3% for faeces; no significant difference was evident, however, in the oestrone sulphate level between does carrying twins and those with singles.

14.6. Other Pregnancy Detection Methods

14.6.1. *Detection of pregnancy-specific protein B*

Studies in the USA in the 1980s led to the detection of pregnancy by RIA of a novel pregnancy specific protein (PSPB) in the serum of cows (Sasser *et al.*, 1986); the PSPB test was effective from day 24 of pregnancy until parturition. This assay was more accurate than the progesterone test in detecting pregnant and non-pregnant cows 30 days after AI because PSPB is pregnancy-specific, unlike the progesterone test. In France, Humblot *et al.* (1990) found that the PSPB concentration in the blood of pregnant goats was significantly different from that in non-pregnant goats 24 days after AI; the authors concluded that PSPB profiles in goats are similar to those found in cows throughout pregnancy, and that RIA may be useful for the diagnosis of pregnancy or late embryo mortality.

14.6.2. *Abdominal palpation/ballottement*

According to Williams (1986a), abdominal palpation in late pregnancy is possible in slab-sided, thin, relaxed goats, but big-bodied, strong-willed goats resisted and tightened the abdominal muscles, making palpation very difficult. In a doe that is pregnant for 120 days or more, BonDurant (1981) noted that an experienced examiner can palpate the fetuses in either flank by using a gentle closed-fist technique. The same author mentions that giving the doe a cold drink of water often induces abrupt fetal movements that can be felt in the flank. In India, Goel and Agrawal (1990) found that diagnosis by abdominal palpation was not possible at 51–60 days but that the accuracy increased to 70% at 61–70 days, 90.3% at 71–80 days and 95.4% at 80 days and later.

14.6.3. Diagnosis by laparotomy

Under the conditions prevailing in Brazil, Espechit *et al.* (1990) suggested that laparotomy was more suitable than some other methods of pregnancy diagnosis. These authors recorded data for 111 goats laparotomized 40–49 days after mating, showing that 91.9% were diagnosed pregnant and that 100% of these does kidded. Of the total number of fetuses, 86.3% were detected by laparotomy; use of the technique was held to have no adverse effect on the goats or the fetuses.

14.7. Hormones in the Detection of Multiples

14.7.1. Progesterone

In Illinois, Jarrell and Dziuk (1991) showed that goats with multiple corpora lutea had a significantly higher progesterone concentration from days 7–30 than goats with one corpus luteum; of goats with two corpora lutea, those with two fetuses at day 45 of gestation had a significantly higher progesterone concentration on day 13 than those with a single fetus. The same authors noted that the number of corpora lutea or fetuses did not influence the progesterone concentration after day 30.

14.7.2. Oestrogens

According to Thimonier *et al.* (1977), total oestrogen levels in the goat during pregnancy did not permit the exact determination of the number of fetuses. However, they recorded that at day 120, 93% of does with one kid in their studies had less than 6ng ml^{-1} of total oestrogens in their peripheral plasma, whereas 63% of goats with two kids or more showed a level higher than 6ng ml^{-1}. Elsewhere, Dhindsa *et al.* (1981) examined oestrogens in blood and recorded that the concentrations of such steroids were significantly higher in pregnant does carrying multiple fetuses than in those with a single fetus.

A paper by Refsal *et al.* (1984) observed a difference in oestrone sulphate concentrations between goats carrying small or large litters. In Italy, however, studies reported by Tamanini *et al.* (1986) failed to find any such relationship. The authors suggest that this may have been due to the fact that the number of fetuses per pregnant goats was very high (up to five) in the work of Refsal *et al.* (1984) whereas in their study litter size never exceeded two kids. A later paper by Refsal *et al.* (1991) confirmed that serum oestrone sulphate concentrations at 50–55 days and after 90 days of gestation in pregnant goats were positively and significantly correlated with the number of kids born. The same authors recorded that serum oestrone sulphate levels increased up to 65–75 days of gestation, remained at high levels until 120 days, and then gradually increased for the remainder of the pregnancy period. They noted that a goat that had a marked decrease in its oestrone level on days 94 and 108 of gestation delivered a stillborn kid and a decomposed fetus.

In Saudi Arabia, Salah (1994) studied the pre- and postpartum concentrations of oestradiol-17β in Aardi goats and recorded that does carrying twin fetuses had significantly higher oestradiol concentrations than those carrying a single fetus.

14.7.3. Pregnancy-specific protein B

In France, Humblot *et al.* (1990) found that the PSPB concentration in the blood of goats pregnant with twins was significantly higher from day 25 and throughout gestation than those with a single fetus.

14.8. Induction of Kidding in the Goat

Prostaglandin $F_{2\alpha}$ and its analogues have been successfully employed to control parturition in goats, dating back to the early 1970s (Currie and Thorburn, 1973; Umo and Fitzpatrick, 1976; Ott *et al.*, 1980). Williams (1986b) described her practice of injecting goats with prostaglandin (5–10mg $PGF_{2\alpha}$ or 62.5–125μg cloprostenol) at 144 days of gestation at 07.00–08.00 and expecting the kids to be born during the afternoon of the following day. According to this author, rarely did a doe deliver earlier than 27 h after injection; a peak of deliveries occurred between 30 and 35 h; if the goat had not given birth by 55 h, she would probably go to term and deliver spontaneously.

Small-scale studies in the induction of parturition in goats in India, using either prostaglandin or dexamethasone, were described by Jain and Madan (1989). Pregnant goats were injected with 20mg $PGF_{2\alpha}$ or 20mg dexamethasone 10 days before the expected date of kidding; does injected with PG and dexamethasone delivered their kids 35.8 and 54.7h respectively, after treatment. In other work in that country, Thakur and Verma (1990a) treated goats with PG at 135–140 days of pregnancy; they recorded parturition occurring after an average interval of 29h and noted that PG treatment did not affect placental expulsion, birth weight, kid survival or postpartum fertility. In a further report by the same authors (Thakur and Verma, 1990b), the injection of 10mg dexamethasone at 135–140 days of gestation resulted in parturition after about 48h; they found no adverse effect on subsequent fertility or kid survival.

14.9. Treatment of Hydrometra in Goats

Hydrometra is an accumulation of aseptic fluid in the uterine lumen in the presence of a persistent corpus luteum and the terms hydrometra and pseudopregnancy are regarded as synonymous. Pseudopregnancy is an important cause of infertility in dairy goats with an incidence varying between farms from 0 to 20% (Mialot *et al.*, 1991). In the treatment of hydrometra, the first consideration is in inducing regression of the corpus luteum, after which the goats will show oestrus and the fluid will be discharged from the uterus. The term 'cloudburst' is employed when the uterine fluid is discharged spontaneously around the expected time of parturition in goats which have been mated and have developed a hydrometra. It is also known that a hydrometra can develop after a period of anoestrus.

In The Netherlands, Hesselink (1993) investigated the use of prostaglandin in the treatment of goats with hydrometra under field conditions. Treatment with 5 mg prostaglandin $F_{2\alpha}$ caused a cloudburst in all the goats treated and it was evident that the PG effectively induced luteolysis; however, some fluid remained in the uterus after the cloudburst and was believed to be responsible for the recurrence of the hydrometra by preventing luteolysis. A second administration of PG 12 days after the cloudburst improved the reproductive performance of the does.

14.10. Use of Prostaglandin in Terminating Pregnancy

As noted by Memon *et al.* (1986), there may be occasions when the veterinary practitioner may be called upon to terminate pregnancy in goats which have been bred with an undesirable buck or a young doe accidentally bred by a buck. The same workers sought to determine the effectiveness of prostaglandin as an abortion-inducing agent in 41 goats with unknown gestation lengths. The treatment involved two doses of 5 mg $PGF_{2\alpha}$, 24 h apart and this was found to be effective in inducing abortion at what was estimated to be about 3 months of pregnancy. In this study, 31 of the 41 does retained their fetal membranes for 12–72 h. The authors note that although goats and cows have a cotyledonary type of placenta, it is not known why cows retain their membranes and goats do not when parturition is induced.

14.11. References

Alexander, G., Signoret, J.P. and Hafez, E.S.E. (1974) Sexual and maternal behaviour. In: (ed.) Hafez, E.S.E. *Reproduction in Farm Animals* 3rd edn. Lea and Febiger, Philadelphia, pp. 222–254.

Asdell, S.A. (1964) *Patterns of Mammalian Reproduction* 2nd edn. Cornell University Press, Ithaca, pp. 623–630.

Baronet, D. and Vaillancourt, D. (1989) Pregnancy diagnosis in goats by echotomography. *Medecin Veterinaire du Quebec* 19, 67–73.

Bernard, A. (1994) Pregnancy diagnosis in goats. *Kleinviehzuchter* 42, 1135–1138.

Blauvelt, H. (1954) Dynamics of the mother-newborn relationship in goats. In: Schaffner B. (ed.) *Group Processes; Transactions of the First Conference.* Macy Foundation, New York, pp. 221–258.

BonDurant, R.H. (1981) Reproductive physiology in the goat. *Modern Veterinary Practice* (July issue), 525–529.

Booth, J.M. (1984) Pregnancy testing goats. *Veterinary Record* 115, 668.

Carmenate, C., Pedroso, R., Gonzalez, N., Arencibia, J. and Alvarez, T. (1992) Pregnancy diagnosis in goats on the basis of blood progesterone concentration, using radioimmunoassay. *Revista Cubana de Reproduction Animal* 17–18(1–2), 93–100.

Challis, J.R.G. and Linzell, J.L. (1971) The concentration of total unconjugated oestrogens in the plasma of pregnant goats. *Journal of Reproduction and Fertility* 26, 401–405.

Challis, J.R.G. and Linzell, J.L. (1973) Oestrone metabolism in pregnant and lactating goats. *Journal of Endocrinology* 57, 451–457.

Chaplin, V.M. and Holdsworth, R.J. (1982) Oestrone sulphate in goat's milk. *Veterinary Record* 111, 224–226.

Cowie, A.T., Daniel, P.M., Prichard, M.M.L. and Tindal, J.S. (1963) Hypophysectomy in pregnant goats, and section of the pituitary stalk in pregnant goats and sheep. *Journal of Endocrinology* 28, 93–98.

Currie, W.B. (1977) Endocrinology of pregnancy and parturition in sheep and goats. In: *Management of Reproduction in Sheep and Goats.* Sheep Industry Development Program Symposium pp. 72–78.

Currie, W.B. and Thorburn, G.D. (1973) Induction of premature parturition in goats by prostaglandin $F_{2\alpha}$ administered into the uterine vein. *Prostaglandins* 2, 201–214.

Currie, W.B. and Thorburn, G.D. (1977) The fetal role in timing the initiation of parturition in the goat. In: Wolstenholme, G.E.W. and Knight, J. (eds) *The Fetus and Birth* Ciba Foundation Symposium No. 47, Associated Scientific Publishers, Amsterdam.

Davies, J. and Chaplin, V.M. (1983) A caprine pregnancy test based upon measurement of oestrone sulphate in milk. *Proceedings of the 3rd International Symposium World Association of Veterinary Laboratory Diagnosticians* Ames, Vol. 1, pp. 119–125.

Dawson, L.J., Sahlu, T., Hart, S.P., Detweiler, G., Gipson, T.A., Teh, T.H., Henry, G.A. and Bahr, R.J. (1994) Determination of fetal numbers in Alpine does by real-time ultrasonography. *Small Ruminant Research* 14, 225–231.

Denamur, R. (1974) Luteotrophic factors in the sheep. *Journal of Reproduction and Fertility* 38, 251.

Dhindsa, D.S., Metcalfe, J. and Resko, J.A. (1981) Oestrogen concentrations in systemic plasma of pregnancy pygmy goats. *Journal of Reproduction and Fertility* 62, 99–103.

Dionysius, D.A. (1991) Pregnancy diagnosis in dairy goats and cows using progesterone assay kits. *Australian Veterinary Journal* 68, 14–16.

Espechit, C.J.B., Fandino, B.A.R., Rodrigues, M.T. and Fonseca, F.A. (1990) Pregnancy diagnosis and determination of the number of foetuses in goats by means of laparotomy. *Revista da Sociedade Brasileira de Zootecnia* 19, 72–76.

Fleming, S.A., Camp, S.D. Van and Chapin, H.M. (1990) Serum progesterone determination as an aid for pregnancy diagnosis in goats bred out of season. *Canadian Veterinary Journal* 31, 104–107.

Goel, A.K. and Agrawal, K.P. (1990) Pregnancy diagnosis in goats. *Indian Veterinary Medical Journal* 14, 77–78.
Goel, A.K. and Agrawal, K.P. (1992) A review of pregnancy diagnosis techniques in sheep and goats. *Small Ruminant Research* 9, 255–264.
Heap, R.B. and Hamon, M. (1979) Oestrone sulphate in milk and its association in pregnancy. *British Veterinary Journal* 135, 462–463.
Heap, R.B., Hamon, M. and Fleet, I.R. (1984) Transport of oestrone sulphate by the mammary gland in the goat. *Journal of Endocrinology* 101, 221–230.
Hesselink, J.W. (1993) Hydrometra in dairy goats: reproductive performance after treatment with prostaglandins. *Veterinary Record* 133, 186–187.
Holdsworth, R.J. and Davies, J. (1979) Measurement of progesterone in goat's milk: an early pregnancy test. *Veterinary Record* 105, 535.
Holdsworth, R.J., Heap, R.B., Booth, J.M. and Hamon, M. (1982) A rapid direct radioimmunoassay for the measurement of oestrone sulphate in the milk of dairy cows and its in pregnancy diagnosis. *Journal of Endocrinology* 95, 7–12.
Holdsworth, R.J., Heap, R.B., Goode, J., Peaker, M. and Walters, D.E. (1983) Progesterone concentrations in the milk and plasma of the goat. *Journal of Endocrinology* 98, 263.
Humblot, P., Montigny, G. De., Jeanguyot, N., Tetedoie, F., Payen, B., Thibier, M. and Sasser, R.G. (1990) Pregnancy-specific protein B and progesterone concentrations in French Alpine goats throughout gestation. *Journal of Reproduction and Fertility* 89, 205–212.
Jain, G.C. and Madan, M.L. (1989) Plasma prostaglandin F, oestradiol-17B, cortisol and progesterone in induced parturient goats. *International Journal of Animal Sciences* 4(2), 152–156.
Jarrell, V.L. and Dziuk, P.J. (1991) Effect of number of corpora lutea and fetuses on concentrations of progesterone in blood of goats. *Journal of Animal Science* 69, 770–773.
Klopfer, P.H. (1971) Mother love: what turns it on? *Scientific American* 59, 404–407.
Malecki, J., Jenkin, G. and Thorburn, G.D. (1987) Passive immunization of pregnant goats against ovine LH. *Journal of Endocrinology* 114, 413–436.
Mani, A.U., Watson, E.D. and McKelvey, W.A.C. (1995) Effect of undernutrition on progesterone concentration during the early luteal phase and mid-gestation in goats. *Veterinary Record* 136, 518–519.
McArthur, C.P. and Geary, A. (1986) Field evaluation of a pregnancy immunoassay for the detection of oestrone sulphate in goats. *Journal of Endocrinology* 110, 133–136.
Meites, J., Webster, H.D., Young, F.W., Thorp, F. and Hatch, R.N. (1951) Effects of corpora lutea removal and replacement with progesterone on pregnancy in goats. *Journal of Animal Science* 10, 411–416.
Memon, M.A., Archbald, L.F., Olcott, B.M., Memon, H.S., Oz, H.H., Chandler, J.E. and Ingraham, R.H. (1986) Observations on the use of prostaglandin $F_{2\alpha}$ as an abortifacient and effect of gonadotrophin-releasing hormone on ovarian activity after induced abortion during the breeding season in goats. *Theriogenology* 25, 653–658.
Mialot, J.P., Saboureau, L., Gueraud, J.M., Pregere, E., Parizot, D., Pirot, G., Duquesnel, R., Petat, M. and Chemiau, P. (1991) Pseudopregnancy in goats. *Recuil de Medecine Veterinaire* 167, 383–390.
Murray, R.D. and Newstead, R. (1988) Determination of steroid hormones in goats' milk and plasma as an aid to pregnancy diagnosis using an ELISA. *Veterinary Record* 1222, 158–161

Ott, R.S., Nelson, D.R., Memon, M.A., Lock, T.F. and Hixon, J.E. (1980) Dexamethasone and prostaglandin $F_{2\alpha}$ for induction of parturition in goats. *Proceedings of the 9th International Congress of Animal Reproduction and AI* (Madrid) 4, 492–495.

Peaker, M. (1978) Gestation and litter size in the goat. *British Veterinary Journal* 134, 379.

Refsal, K.P.R., Marteniuk, J.V., Nachreiner, R.F. and Williams, C.S.F. (1984) Effects of gestation length and fetal number on serum estrone sulphate concentrations in pregnant goats. *Proceedings of the 10th International Congress Animal Reproduction and AI* (Illinois) 2, pp. 96–98.

Refsal, K.R., Marteniuk, J.V., Williams, C.S.F. and Nachreiner, R.F. (1991) Concentrations of estrone sulphate in peripheral serum of pregnant goats: relationships with gestation length, fetal number and the occurrence of fetal death in utero. *Theriogenology* 36, 449–461.

Restall, B.J., Milton, J.T.B., Klong-Yutti, P. and Kochapakdee, S. (1990) Pregnancy diagnosis in Thai native goats. *Theriogenology* 34, 313–317.

Rothschild, I. (1981) The regulation of the mammalian corpus luteum. *Recent Progress in Hormone Research* 37, 183–298.

Rudge, M.R. (1970) Mother and kid behaviour in feral goats (*Capra hircus*). *Zeitschrift für Tierpsychology* 27, 687–692.

Russel, A.J.F. (1990) The application of real-time ultrasonic scanning to the diagnosis of pregnancy and determination of foetal numbers in goats. *Goat Veterinary Society Journal* 11(1), 9–14.

Salah, M.S. (1994) Pre- and post-partum levels of serum progesterone and oestradiol-17 B in Aardi goat. *Tropenlandwirt* 95, 77–86.

Sambraus, H.H. and Wittmann, M. (1993) Parturition and sucking behaviour in goats. *Kleinviehzuchter* 41, 582–584.

Santiago Moreno, J., Gonzales de Bulnes, A., Garcia Lopez, M. and Lopez-Sebastian, A. (1995) Early pregnancy diagnosis and determination of the number of embryos by transrectal ultrasonics in goats. *ITEA Producion Animal* 91, 37–43.

Sasser, R.G., Ruder, C.A., Ivani, K.A., Butler, J.E. and Hamilton, W.C. (1986) Detection of pregnancy by radioimmunoassay of a novel pregnancy-specific protein in serum of cows and a profile of serum concentrations during gestation. *Biology of Reproduction* 35, 936–942.

Sawada, T., Nakatani, T., Tamada, H. and Mori, J. (1995) Secretion of unconjugated estrone during pregnancy and around parturition in goats. *Theriogenology* 44, 281–286.

Sindermann, B. (1991) Pregnancy diagnosis in goats by means of the determination of oestrone sulphate in blood or oestrogen in faeces. Thesis, Justus-Liebig-Universitat Giessen, 91 pp.

Smith, M.C. (1986) Perparturient care of the doe. In: Morrow, D.A. (ed.) *Current Therapy in Theriogenology*. W.B. Saunders, Philadelphia, 589–590.

Tainturier, D., Fieni, F., Bruyas, J.F., Chemli, J., Allaire, F. and Zaiem, I. (1993) Diagnosis of pregnancy in goats using an ultrasound transrectal scanner with a human prostate probe. *Sciences Veterinaires Medecine Comparee* 95, 81–87.

Tamanini, C., Chiesa, F., Prandi, A. and Galeati, G. (1986) Estrone and estrone conjugate plasma levels throughout pregnancy in the goat: their determination as a pregnancy diagnosis test. *Animal Reproduction Science* 11, 35–42.

Thakur, M.S. and Verma, S.K. (1990a) Induction of parturition in goats with prostaglandin $F_{2\alpha}$. *Indian Journal of Animal Reproduction* 11(2), 148–151.

Thakur, M.S. and Verma, S.K. (1990b) Use of dexamethasone for induction of parturition in goats. *Archiv fur Experimentelle Veterinarmedizin* 44, 459–463.

Thibier, M., Jeanguyot, N. and De Montigny, G. (1982) Accuracy of early pregnancy diagnosis in goats based on plasma and milk progesterone concentrations. *International Goat and Sheep Research* 2, 1–6.

Thimonier, J., Bosc, M., Djiane, J., Martal, J. and Terqui, M. (1977) Hormonal diagnosis of pregnancy and number of fetuses in sheep and goats. In: *Management of Reproduction in Sheep and Goats* Sheep Industry Development Program Symposium (Madison), pp. 79–88.

Thomas, C.R., Forsyth, I.A. and Hart, I.C. (1977) Goat placental lactogen: levels through pregnancy and variation over 24 hour periods. *Journal of Endocrinology* 75, 51P.

Thorburn, G.D. and Schneider, W. (1972) The progesterone concentration in the plasma of the goat during the oestrous cycle and pregnancy. *Journal of Endocrinology* 52, 23–28.

Thorburn, G.D., Nichol, D.H., Bassett, J.M., Shutt, D.A. and Cox, R.I. (1972) Parturition in the goat and sheep: changes in corticosteroids, progesterone, oestrogens and prostaglandin F. *Journal of Reproduction and Fertility* Suppl. 16, 61–84.

Umo, I. and Fitzpatrick, R.J. (1976) Introduction of parturition in goats with prostaglandin $F_{2\alpha}$. *Proceedings of the 8th International Congress of Animal Reproduction and AI* (Krakow) 3, 411–413.

Williams, C.S.F. (1986a) Pregnancy diagnosis. In: Morrow, D.A. (ed.) *Current Therapy in Theriogenology*. W.B. Saunders, Philadelphia, pp. 587–588.

Williams, C.S.F. (1986b) Practical management of induced parturition. In: Morrow, D.A. (ed.) *Current Therapy in Theriogenology*. W.B. Saunders, Philadelphia, pp. 588–589.

15 Embryo Transfer and Associated Techniques in Goats

15.1. Introduction

Although embryo transfer (ET) has been used more frequently in sheep than in goats, the procedures that have been employed are essentially similar (Moore, 1982). The first report of the birth of a kid after embryo transfer goes back more than 60 years to a paper by Warwick *et al.* (1934). Forty years and more were to pass before serious attempts to apply ET technology in goats were reported, first in Australia (Moore, 1974; Bilton and Moore, 1976; Moore and Eppleston, 1979a) and subsequently in many other countries including New Zealand (Tervit *et al.*, 1983) France (Chemineau *et al.*, 1986; Baril *et al.*, 1993) and India (Agrawal and Goel, 1991).

Test-tube goat embryos

In the production of goat embryos by *in vitro* procedures, a paper by Hanada (1985) reported the birth of a kid after *in vitro* fertilization (IVF) of ovulated goat oocytes more than ten years ago; the first goat offspring produced after *in vitro* maturation and fertilization (IVM/IVF) of the caprine oocyte is much more recent (Crozet *et al.*, 1993); these were French studies in which the sheep oviduct was employed in the early culture of the caprine embryos. The first kids to be born in Europe after transfer of *in vitro* matured, fertilized and *in vitro* cultured goat embryos appears to be that reported by Pereira *et al.* (1995) in Germany.

15.1.1. Advantages of embryo transfer

Among the various advantages of ET is its use in expanding the populations of scarce and expensive animals. In France, for example, work began in 1984 on goat ET in order to increase more rapidly the number of Angora goats from imported animals. In 1985, Creole goats were introduced into France by way of ET (Chemineau *et al.*, 1986). In the same country, Deguet *et al.* (1989)

subsequently used the technology in expanding the populations of Angora goats imported from New Zealand, Texas and Australia.

In the Czech Republic, Riha *et al.* (1994) note that the economic problems associated with restructuring animal production in that country has led farmers to seek new sources of income; the breeding of Cashmere goats ranks as one of the most promising of new animal enterprises. One of the most rapid ways of establishing high quality herds is to use imported frozen embryos. Workers in the UK have also drawn attention to the fact that demand for high quality pure-bred Angora and Cashmere goats in Europe and Australasia has resulted in extensive use of the ET technique. It would not be unreasonable to expect, with one flushing of embryos and the remating of the doe shortly afterwards for a normal pregnancy, that the rate of production would be 6–7 times faster than normal.

15.2. Superovulation Techniques

In goats, a major and unresolved problem of superovulation is the extreme variability in ovulatory response that is evident among animals treated with similar agents. It is known that some part of this variability arises from genetic factors (Nuti *et al.*, 1987), age (Moore, 1982; Krogh, 1991; Mahmood *et al.*, 1991), stage of the cycle at which treatment is applied (Wani *et al.*, 1990) and the type of gonadotrophin employed in stimulating an ovarian response.

15.2.1. PMSG vs. FSH preparations

As in cattle and sheep, many comparisons have been made between pregnant mare serum gonadotrophin (PMSG) and follicle stimulating hormone (FSH)-preparation in the induction of superovulation in the goat (Armstrong *et al.*, 1983a; Tsunoda and Sugie, 1989; Pampoukidou *et al.*, 1992; Pendleton *et al.*, 1992); as with the other ruminants, the evidence favours the use of the FSH in the goat, in terms of ovulatory response and embryo yield.

In [illegible]d, for example, Eiamvitayakorn *et al.* (1988) recorded the average number of ovulations after FSH was 9.6 as compared with 5.7 after PMSG treatment. In Japan, Tsunoda and Sugie (1989) also reported that the average number of normal embryos recovered was significantly higher in FSH-treated goats (9.4) than in those treated with PMSG (5.7). In India, Mahmood *et al.* (1991) recorded an average of 16.55 corpora lutea with FSH and 11.70 with PMSG. Similarly, in Greece, Pampoukidou *et al.* (1992) recorded an average of 12.5 ovulations after treatment with 21 mg FSH-P and 3.88 with 1200 IU PMSG. In Malaysia, Rosnina *et al.* (1992) recorded a difference in favour of FSH in their comparison with PMSG (6.8 vs. 3.0 embryos per doe recovered). In the USA, Pendleton *et al.* (1992) also recorded evidence suggesting that FSH was superior to PMSG for superovulation of non-lactating dairy goats during the breeding season. Despite the failings now well-recognized in using PMSG in superovulation, reports continue on its use in the goat (Pargaonkar *et al.*, 1994); presumably this is often a question of cost and availability rather than the technical merits of the agent.

15.2.2. Progestagen–FSH treatments

Progestagen–gonadotrophin treatment is effective for the induction of superovulation in the doe both during the breeding season and during anoestrus. Progesterone has been administered daily by intramuscular injections (10–12mg day^{-1}) as in the protocol described by Tervit *et al.* (1984); synchronizing injections of progesterone were given daily and, beginning at the time of the 16th injection, donors were injected twice daily with gradually decreasing doses of FSH, so that 21 mg were administered over a 4-day period (see Fig. 15.1).

An alternative was administering progestagens by way of the intravaginal sponge. A typical protocol employed in the breeding season was one in which FGA sponges were used over an 11-day period (see Fig. 15.2), with FSH being administered over 3–4 days, starting two days before sponge withdrawal, and around the time at which prostaglandin is administered (Baril *et al.*, 1988). Deguet *et al.* (1989) recorded data for 17 collections made during the breeding season and 12 during the non-breeding-season, with the number of embryos collected per doe averaging 11 and 10.4, respectively. In Scotland, Mani *et al.* (1994) achieved superovulation in Angora goats by the injection of 22mg porcine FSH, divided into four decreasing twice-daily, commencing one day before sponge withdrawal after a 17-day progestagen treatment.

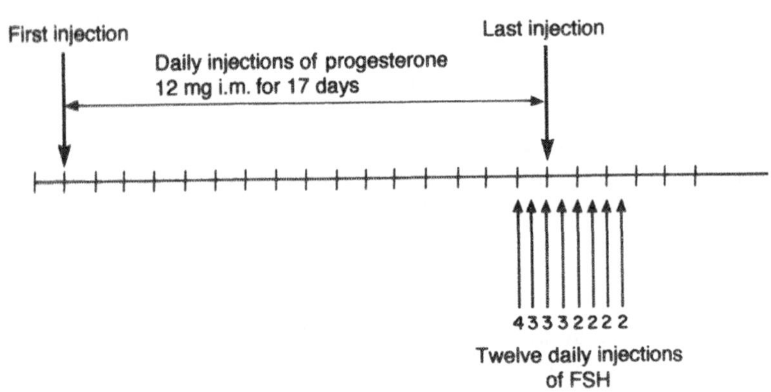

Fig. 15.1. Hormone treatment of donors in goat studies in New Zealand (from Tervit *et al.*, 1984).

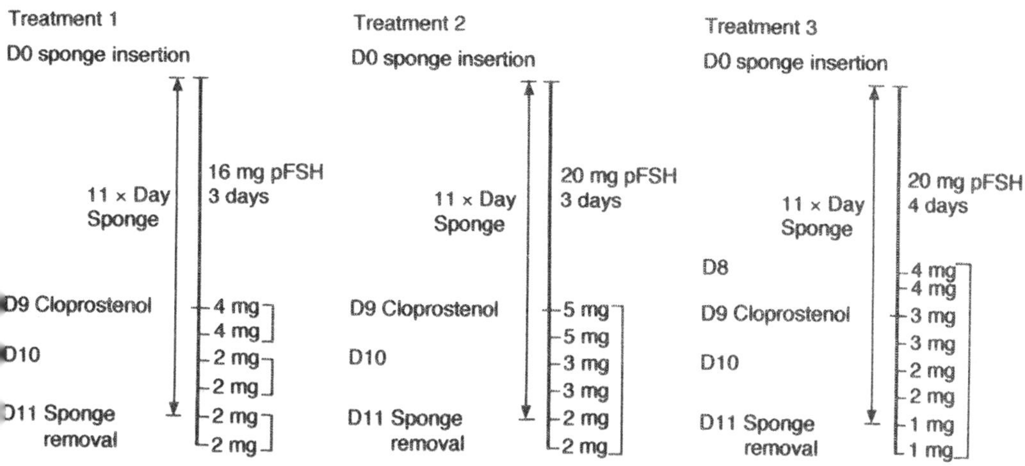

Fig. 15.2. Progestagen (sponge)–FSH–prostaglandin treatments for superovulation in the goat (from Baril *et al.* 1988).

Norgestomet implants

The feasibility of using norgestomet ear implants and FSH in a goat ET programme was investigated by Pendleton *et al.* (1986); implants were employed in does during late anoestrus over a period of 16–17 days, the implants being replaced with new implants after 10 days to ensure adequate progestagen treatment. Superovulation was induced by twice-daily injections of descending doses of FSH, starting two days prior to implant removal; results demonstrated that such treatment was effective. Elsewhere, progestagen treatment in the form of a 6mg norgestomet implant was employed in work reported by Akinlosotu and Wilder (1993); they treated mature goats during anoestrus with the implant for 9 days and injected FSH in four doses starting 24h before implant removal.

Origin of FSH preparation

No difference was evident in the average ovulation rate between two commercial forms of FSH (FSH-P, Burns-Biotec; FSH, Sanofi) used in the superovulation of goats (Baril *et al.*, 1988).

FSH:LH ratios

A study by Puls-Kleingeld *et al.* (1991) compared purified FSH (pFSH) with either 40 or 80% purified LH (pLH) in goat superovulation; results indicated that 40% was superior, bringing about higher ovulation rates and increase in the yield of transferable embryos. A subsequent study by Nowhsari *et al.* (1995) compared the superovulatory response of does to purified FSH supplemented with 30, 40 or 50% pLH; results confirmed the earlier finding of a good response of goats to pFSH preparations with a high FSH:LH ratio, and suggested that supplementation with approximately 40% pLH may be close to the optimum.

FSH–PMSG combinations

Some authors have examined the use of superovulation protocols based on a single injection of FSH combined with PMSG on the understanding that this may give similar results to regimens based on multiple doses of FSH. In Australia, Batt *et al.* (1993) employed single doses of FSH with single doses of PMSG and concluded that, at certain dose levels, such combinations could be of value.

15.2.3. Repeated superovulation

There is some evidence that certain FSH-preparations are able to induce a superovulatory response in goats over a period of several treatments to a greater extent than others (Table 15.1). In France, Baril *et al.* (1992) tested the possibility that superovulation could be induced repeatedly in donors by employing gonadotrophins which had either no (caprine FSH) or a low (ovine FSH) potential to induce the production of antigonadotrophin antibodies. These authors concluded that the production of embryos in permanent goat donors would be more efficient if they are induced to superovulate with the caprine or ovine forms of FSH.

15.2.4. Immunization against inhibin

Results from studies reported by Dietrich *et al.* (1995) in Germany indicate that immunization against recombinant human inhibin alpha-subunit may be a useful alternative to the conventional methods of superovulation. In this, the immune response of Boer goats to the antigen given during the breeding season varied greatly among animals but a booster dose administered during anoestrus elicited a prompt and fairly uniform response. The authors recorded that embryo yield from immunized goats was more than three times the number of embryos recovered from untreated controls.

Table 15.1 Ovulation rate in superovulated goats after repeated treatment in relation to source of FSH (from Baril *et al.*, 1992).

	FSH porcine ($n=9$)	FSH caprine ($n=9$)	FSH ovine ($n=10$)
1st Treatment	17.0 ± 3.0	11.8 ± 3.7	12.3 ± 2.1
2nd Treatment	12.1 ± 3.8	10.8 ± 3.9	12.6 ± 3.1
3rd Treatment	10.7 ± 4.0	16.0 ± 5.6	11.0 ± 2.8
4th Treatment	3.6 ± 3.0	17.4 ± 5.9	14.6 ± 3.2
5th Treatment	1.6 ± 1.4	11.9 ± 5.6	11.7 ± 4.3

15.2.5. Seasonal effects

Some authors have drawn attention to marked differences in the super-ovulatory response of goats according to the stage of the breeding season. In the USA, Senn and Richardson (1992) superovulated Anglo-Nubian goats, using norgestomet ear-implants and FSH; they recorded that does treated early in the season average 15.1 viable embryos per goat compared with 3.33 for those treated late.

15.2.6. Superovulation in goats in the tropics

Temperate breeds of goat are 'short-day' breeders and the results of super-ovulation treatments may be influenced by whether they are applied in the breeding season or during anoestrus. Goat breeds in the tropics, on the other hand, are capable of breeding throughout the year. A study by Rosnina *et al.* (1992) was undertaken to determine whether the indigenous goat (*Kambing kacang*) would respond to gonadotrophins throughout the year in Malaysia; the authors record that these goats could be superovulated throughout the year although their response was depressed during the dry months. Such results apparently agreed with a previous report that bodyweight loss during severe drought depressed the ovulatory response of Angora goats (Tervit *et al.*, 1983).

15.3. Breeding the Donor Goat

Breeding may be by natural service or by artificial insemination (AI). According to Moore (1982), when bucks producing high quality semen are available, then natural mating should be employed. In France, Baril *et al.* (1988), working with Angora goats, arranged for matings at intervals of four hours throughout oestrus; with dairy goats, matings were at 12 and 24h after the onset of oestrus or three times after the end of sponge treatment (30, 48 and 54h). In Scotland, Mani *et al.* (1994) used vasectomized bucks in the detection of oestrus in donor does and then hand-mated the donors at bucks at 6-hourly intervals throughout the duration of the ensuing oestrus.

It is believed that fertilization failure after AI is more common in the ewe than in the doe and that this may be associated with species differences in the patency of the cervix; Moore and Eppleston (1979b) showed that the cervix of a high proportion of does could be penetrated with an insemination pipette. When frozen-thawed semen is to be used, then intrauterine insemination is warranted.

15.3.1. *Controlling the time of ovulation*

In breeding donor goats, AI at a predetermined time after the end of a superovulation treatment can only be efficient if ovulations are closely synchronized. As noted by Baril *et al.* (1994), this is usually not true in superovulated goats, and for a good fertilization rate the does have to be inseminated according to oestrus or the preovulatory LH peak; this technique is efficient but tedious. Earlier studies by Baril and Vallet (1990) led them to conclude that although the interval between the LH peak and ovulation is quite constant (about 24h), the variability of the intervals between sponge removal and the onset of oestrus, and between the onset of oestrus and the LH peak, precluded the determination of an optimum time for AI after sponge removal or onset of oestrus.

15.3.2. *Use of a GnRH antagonist*

For such reasons, Baril *et al.* (1994) developed a method of controlling the time of the LH peak so that AI could be carried out at a predetermined time. They did this by using a gonadotrophin releasing hormone (GnRH) antagonist (Antarelix), which was given 12h after progestagen-sponge removal, and inducing ovulation by way of 3mg purified LH administered after a further 24h; the authors recorded that the quality of embryos was significantly higher in the doses treated with Antarelix.

15.3.3. *Use of GnRH*

In studies where goats were bred by natural service, Akinlosotu and Wilder (1993) used GnRH, administered 24 or 48h after removal of a progestagen implant, as part of their superovulation treatment; they recorded a significantly higher ovulation rate in the GnRH treated does. They also found that embryos obtained from the GnRH treated goats were at a more uniform developmental stage compared with those of controls.

15.4. Embryo Recovery and Storage

15.4.1. *Surgical recovery of embryos*

Embryo recovery by surgical intervention is carried out under general anaesthesia, the does having been starved during the previous 24h. As described by Baril *et al.* (1988), general anaesthesia may be induced by intravenous injection of sodium thiopental (8mg per kg bodyweight) or penthobarbital (13mg per kg bodyweight). An endotracheal catheter is introduced to permit the maintenance of anaesthesia under halothane and oxygen for the duration of the recovery procedure. For the collection of tubal embryos, Nuti *et al.* (1987) cannulated the oviducts at the fimbriated end with a plastic catheter and flushed fluid gently through from the uterotubal junction.

Six to eight days after the onset of oestrus, the embryos are collected after the uterus and ovaries have been exteriorized by way of laparotomy; the uterine horn is punctured at its base and a Foley catheter introduced and its cuff inflated. A second catheter is introduced at the uterotubal junction and 50ml of warm (37°C) flushing medium used to flush the embryos into the Foley catheter. A suitable medium flushing medium is phosphate-buffered saline (PBS) to which 2–3% of bovine serum albumin (BSA) or 10% of fetal calf serum has been added; it is also possible to enrich PBS with homologous goat serum (Amoah and Gelaye, 1991).

Laparotomy vs. endoscopy

The advantages and disadvantages of surgical vs. endoscopic recovery of embryos in the goat are discussed by Baril *et al.* (1988); whichever technique is employed, the length of the recovery occupies 20–30 min. The same authors noted that the number of oocytes and embryos recovered (relative to the number of corpora lutea) by endoscopy was lower than by way of surgery (62% vs. 85%). Against this was the fact that recovery under endoscopy could be repeated successfully on the same goat (see Table 15.2) whereas with surgery, a decrease in embryo recovery rate was evident as soon as the second collection due to the problem of adhesions.

15.4.2. Non-surgical recovery techniques

Investigations on dilation of the cervix and the transcervical recovery of goat embryos were reported from Germany by Grupp (1991); intracervical treatment with PGE_2, with or without oestradiol benzoate, made such embryo

Table 15.2 Effect of repeated attempts at laparoscopic recovery on the recovery rate in goats (from Baril *et al.*, 1988).

	Recovery no.		
	1st	2nd	3rd–7th
Recovery rate	63.2%	62.3%	60.7%
	(709–50) N.S.	(470–34) N.S.	(491–38) N.S.

N.S.: $P > 0.05$.
( ) Number of CL – Number of collected does.

recovery possible in each of ten goats used in the study. The author suggested that hormonal cervical dilation, followed by transcervical embryo recovery, provided a rapid and non-traumatic method for use in the goat. In India, Agrawal *et al.* (1991) also described a procedure which was successfully employed in the non-surgical recovery of goat blastocysts six days after breeding from a superovulated Jamnapari goat.

15.4.3. Factors affecting embryo recovery

Premature luteal regression

Premature luteal regression/or failure of normal luteal development has been reported to occur more frequently in goats superovulated with PMSG than with FSH (Armstrong *et al.*, 1983a; Pendleton *et al.*, 1992); it has also been reported that a high incidence of early return to oestrus occurred in goats treated with PMSG in the postpartum period and at either end of the breeding season in cyclic goats (Armstrong *et al.*, 1983b). In Australia, Moore (1982) noted that in Angora goats, which show a well-defined period of non-breeding activity during the spring and early summer, some 30% or more of does showed evidence of luteal regression within four or five days of multiple ovulation, with subsequent loss of viability of embryos.

Action of prostaglandin

It is believed that premature release of prostaglandin $F_{2\alpha}$ is implicated in this early luteal regression in the superovulated goat (Battye *et al.*, 1988). Noting instances of premature regression of corpora lutea in their goat studies in Malaysia, Rosnina *et al.* (1992) suggested that this may have contributed to some of the low embryo recoveries in their work; they also mention that such luteal regression may have been related to high circulating levels of oestrogen during the early luteal phase.

Exogenous progesterone as a possible treatment?

Studies reported by Borque *et al.* (1993) in Spain with superovulated Murciana goats recorded that does which failed to yield viable embryos on day 6 (postovulation) had shown a significant decrease in progesterone concentrations as a consequence of premature luteal regression. They not only found that regression of the corpora lutea was evident on day 4, but that day-3 progesterone levels were lower than they should be if the doe possessed normally functioning CL. The Spanish workers suggested that determination of progesterone levels on day 3 might be used as a diagnostic tool, avoiding loss of embryos if they were recovered on day 4; they also suggested that administraton of exogenous progesterone might be a possible treatment for early luteal regression.

15.4.4. Development of the early goat embryo

The development of the early embryo in the goat differs in detail from that of the sheep. Six days after the onset of oestrus, most goat embryos are at the morula stage and it is not until the eighth day that all embryos reach the blastocyst stage (see Table 15.3). A study of the early goat embryo by Yang *et al.* (1991) showed that the first cleavage occurred about 24h after ovulation; two, four-, eight-, 16-cell embryos, morulae and blastocysts were found about 32–42, 48–52, 62, 72–80, 96h and 7–8 days after ovulation, respectively.

The same authors recorded that embryos up to the 16-cell stage were recovered from the oviduct and morulae/blastocysts from the uterus. According to Baril *et al.* (1988) there can be a great variation in embryo developmental stages among goats collected at the same interval after oestrus as well as within a batch of embryos collected from the individual doe; such variation increases the difficulty of selecting embryos for transfer or for freezing. In China, Yang *et al.* (1991) also recorded that embryos at several stages of development were often obtained from the same goat.

15.4.5. Short-term storage

Preservation of goat embryos by cooling them to refrigerator temperature and storing them over a period of several days has been reported by some workers. In India, Ryot and Vadnere (1989) cooled 8–16 cell embryos to 5°C and stored them for up to four days; after two days, slowly cooled embryos developed after incubation and gave rise to pregnancies after transfer to recipients.

15.4.6. Freezing and vitrification of goat embryos

Conventional freezing procedures

In the freezing procedure described by Baril *et al.* (1988), embryos of suitable

Table 15.3 Developmental stages of goat embryos recovered at different intervals after oestrus (from Baril *et al.*, 1993).

Day of collection	Developmental stage
2	1–2 cells
3	4–8 cells
4	8–20 cells
5	20 cells – early morula
6	Morula
7	Compact morula – expanded blastocyst
8	Expanded blastocyst – hatched blastocyst

quality were washed and immersed in PBS to which a suitable cryoprotectant was added (glycerol: 5% (0.7 M) for 10 min then 10% (1.4 M) for 10 min). The embryos were then loaded into 0.25ml straws and cooled 1°C per minute to −7°C; at this point, seeding was induced and the temperature decreased at the rate of 0.3°C per minute down to −35°C. After a 15-min holding period at −35°C, the straws were plunged into liquid nitrogen. Such straws were thawed by immersion in a water-bath held at 37°C and passed through four decreasing concentrations of glycerol (7.5%–5%–2.5–0%) to permit progressive rehydration of the embryonic cells. A report from Australia by Li *et al.* (1990) suggested that maximum survival of goat embryos after freezing was attained at the expanded, hatching and hatched-blastocysts stages.

Sensitivity of goat morulae to low temperature

Studies by Puls-Kleingeld *et al.* (1992) clearly indicated that, compared with blastocysts, goat morulae were not suited for freeze-thawing; the authors noted that only 11% of the recipients of frozen-thawed morulae became pregnant in contrast to 90% of those receiving blastocysts. Further work in Germany, reported by Nowshari and Holtz (1995), also showed a high pregnancy rate (83%) after the transfer of blastocysts; these workers also presented results suggesting that it was justifiable to culture caprine morulae *in vitro* to the blastocyst stage if they are to be cryopreserved.

Since the report of Bilton and Moore (1976) of the first kid born after transfer of a frozen-thawed embryo, variable survival rates have been evident in embryos transferred after freezing in glycerol (Chemineau *et al.*, 1986; Rao *et al.*, 1988; Wang *et al.*, 1988; Baril *et al.*, 1989; Li *et al.*, 1990; Le Gall *et al.*, 1993), ethylene glycol (Le Gall *et al.*, 1993; Riha *et al.*, 1994) and dimethylsulphoxide (Bilton and Moore, 1976; Rao *et al.*, 1988; Li *et al.*, 1990). Data reported by Fieni *et al.* (1990) showed that, of three cryoprotectants which they examined, ethylene glycol was the most suitable for freezing goat embryos (see Table 15.4).

Table 15.4 Comparison of cryoprotectants in the freezing of goat embryos (from Fieni *et al.*, 1990).

	Morula		Blastocysts		Morula and blastocysts	
	No.	%	No.	%	No.	%
Glycerol	2/84	2.4[a]	4/43	9.3[bc]	6/127	4.7[de]
Ethylene glycol	15/93	16.1[ax]	19/35	54.3[bx]	34/128	26.6[e]
DMSO	9/104	8.6[y]	21/51	41.2[cy]	30/155	19.3[d]
Total	26/281	9.2[z]	44/129	34.1[z]	70/410	17

a: DS.: $P < 0.05$.
b, c, d, e, x, y: D.S.: $P < 0.01$.
z: D.S.: $P < 0.001$.

In this, freezing was carried out using a conventional protocol (seeding at −7°C; cooling at 0.4°C min^{-1} until −35°C; immersion in liquid nitrogen at −196°C); embryo viability after freeze-thawing was evaluated by culturing for 48h in B2 Menezo medium. A further report by Fieni *et al.* (1995) led them to conclude that ethylene glycol and DMSO (1.5 M in PBS) are convenient for freezing goat embryos and that under the same conditions, glycerol appeared to give lower survival rates. The data provided also indicated that the introduction of sucrose (0.25 M) in the dilution solution facilitated the removal of the cryoprotectant and increased the survival rate of the embryos.

Vitrification

The successful transfer of vitrified goat embryos was reported by Yuswiati and Holtz (1990) in Germany; 16 good quality embryos were transferred to nine recipients and two births recorded. In the vitrification procedure, embryos were placed in a solution consisting of modified PBS, 10% glycerol and 20% propanediol for 10 min and then drawn individually into 0.25ml straws containing 0.04ml of a solution consisting of PBS, 25% glycerol and 25% propanediol, this solution being separated from a 0.2ml solution of 1.0 M sucrose by an air bubble.

15.5. Embryo Transfer and Recipient Management

Although the first surgical goat embryo transfers date back to the early 1930s, the increasing importance of goat ET in breeding improvement programmes around the world in recent decades has led to the use of laparoscopy as an alternative transfer procedure. A report by Vallet *et al.* (1989) showed that comparable results could be achieved by laparoscopy in comparison with surgery; the authors listed several advantages to laparoscopy under field conditions, including speed of transfer (5 vs. 15 min) and freedom from genital tract adhesions.

15.5.1. Recipient surgical procedures

Recipients are subjected to the same anaesthetic procedures as already described earlier for donors. After the reproductive tract has been exteriorized and the presence of at least one functional corpus luteum established, the recipient usually receives either two embryos (transferred either singly to each uterine horn or both embryos transferred to the uterine horn ipsilateral to a corpus luteum); if three embryos are transferred, this usually takes the form of two to the ipsilateral and one to the contralateral horn. Embryo survival rates of 60% may be obtained.

15.5.2. Tubal transfer using endoscopy

The successful laparoscopic transfer of early stage embryos to the oviducts of goats was reported by Besenfelder *et al.* (1994); according to the authors, used routinely, the technique should take less than 5 min. After manipulation of the reproductive organs, the authors recorded no visible alterations or injuries.

15.5.3. Factors affecting pregnancy rate after embryo transfer

Current reports suggest that pregnancy rates after transfer of goat embryos range from 45 to 80% depending on the quality of embryos, the nutritional status of does and transfer experience (Godke *et al.*, 1985). The effect of undernutrition either before and/or after embryo transfer on pregnancy rate and embryo survival was studied by Mani *et al.* (1994); they concluded that pregnancy rate and embryo survival were low in does underfed both before and after embryo transfer when compared with does on a maintenance diet; adequate feeding of recipient goats is necessary for an effective superovulation programme. It has to be admitted that the mechanism by which nutrition affects embryo survival in the goat is not fully understood.

15.5.4. Pregnancies after transfer of fresh and frozen embryos

In Denmark, Krogh (1991) recorded a pregnancy rate of 88% and a kidding rate of 70% after the transfer of fresh Angora embryos and rates of 75% and 55%, respectively, for the transfer of 270 imported frozen embryos.

15.6. Laboratory Production of Goat Embryos

During the past decade, considerable effort has been devoted to the development of *in vitro* maturation (IVM) and *in vitro* fertilization (IVF) of oocytes from farm animals (Gordon, 1994). In most of these species, young have been born after the transfer of embryos produced by *in vitro* procedures. The use of ovaries collected from animals at the abattoir as a source of oocytes for IVM–IVF allows for the large-scale production of embryos that can be used in the development of new biotechnologies such as cloning and genetic engineering.

In goats, unlike cattle, however, one of the limiting factors in deriving goat oocytes from abattoir sources is likely to be the lack of commercial enterprises capable of providing consistent supplies of ovaries. Research in the area of *in vitro* maturation and fertilization of goat oocytes has been reported from several countries, including India (Chauhan and Anand, 1991; Agrawal, 1992; Pawshe *et al.*, 1993), the USA (Younis *et al.*, 1991), Spain (Martino *et al.*, 1994a,b; Mogas *et al.*, 1995a,b,c,), France (De Smedt *et al.*, 1992; Crozet *et al.*, 1995a,b) and Germany (Pereira *et al.*, 1995). The development of IVF technology in species such as the goat holds out the promise of useful advances in certain animal biotechnology programmes, such as those involving gene transfer. The goat would permit more rapid progress in the production of transgenic animals than would be possible in cattle and other livestock with longer generation intervals.

15.6.1. Recovery and maturation of goat oocytes

Recovery methods

A study by Mogas *et al.* (1992) in Spain examined ovary slicing in comparison with aspiration and follicle dissection in recovering oocytes from prepubertal goats. In that country, there are commercial abattoirs where goats are slaughtered regularly for meat, but these animals are normally only two months old. Although it is evident that such prepubertal goat oocytes are capable of undergoing nuclear maturation, they may not have the same capacity for fertilization and normal development as oocytes from older goats. Mogas *et al.* (1992) were able to recover higher numbers of oocytes by slicing than by the other procedures. The effect of these recovery methods on the IVM of the goat oocyte was dealt with by Martino *et al.* (1994b); they concluded that the slicing technique yielded more oocytes per ovary than dissection and aspiration, but the IVF capacity of oocytes obtained by slicing was lower than that obtained by dissection. It is known that the oocyte maturation period is, as with sheep, about 24h.

Maturation media

Other work in Spain (Mogas *et al.*, 1993) demonstrated a beneficial effect on cleavage rate when hormones (FSH/LH/oestradiol) were employed in the artificial maturation medium; in Germany, Pereira *et al.* (1995) similarly employed FSH, LH and oestradiol in their IVM medium. The effect of granulosa cell source on IVM, IVF and IVC of prepubertal goat oocytes was examined by Mogas *et al.* (1995a); their results indicated that granulosa cells from gonadotrophin-stimulated ovaries, either from adult or prepubertal animals, may confer developmental competence on oocytes recovered from prepubertal ovaries. Other studies reported in the same year by Mogas *et al.* (1995b) led them to conclude that oestrous goat serum was better than fetal calf serum in terms of its effect on IVF and early embryo development of prepubertal goat oocytes.

Effect of follicle size

In France, Crozet *et al.* (1995a,b) investigated the effect of the size of follicles from which goat oocytes originate on their subsequent ability to be fertilized and to undergo early embryonic development *in vitro*. Their results indicated that developmental competence of goat oocytes is acquired progressively during final follicular growth and that only a small proportion of oocytes (those isolated from large antral follicles) have the capacity to progress to the blastocyst stage after IVM/IVF/IVC. In Malaysia, Rajikin *et al.* (1994) have described the ultrastructural changes that occur in goat oocytes at different stages of development *in vitro*.

15.6.2. Oocyte recovery from live goats

Studies in the USA attempted to recover goat oocytes by way of either a laparoscopic aspiration procedure or by a transvaginal ultrasound-guided aspiration technique (Graff *et al.*, 1995). The number of follicles aspirated (6.3 vs. 16.1) and oocytes harvested (4.3 vs. 11.5) using the transvaginal approach was less than that obtained by the laparoscopic method. The authors note that this was due to decreased follicle visibility at collection using the transvaginal procedure. In contrast to does subjected to repeated laparoscopic recoveries, there was no evidence of adhesions in donor does subjected to the transvaginal aspiration procedure; the authors note that the transvaginal approach should not be overlooked in an effort to minimize the problem of adhesions in valuable donor goats.

Further work in the USA has led to the birth of kids from the transfer of blastocysts after *in vitro* embryo production based on oocytes recovered by transvaginal ultrasound-guided aspiration of follicles (Han *et al.*, 1996); the authors conclude that the recovery of oocytes by non-invasive transvaginal ultrasound-guided aspiration for *in vitro* embryo production may be a viable alternative to conventional embryo transfer in the goat (Table 15.5).

Table 15.5 Kids born after the intrauterine transfer of IVM/IVF/IVC produced goat embryos (from Han *et al.*, 1996).

Oocyte source	No. embryos/ recipient	Total no. recipients	Total no. embryos transferred	No. w/fetal heart beats	No. pregnant recipients	No. kids born
TVA*	1	2	2	1	1	0
TVA	2	5	10	3	2	1
TVA	3	1	3	1	1	1
Abattoir	1	9	9	3	3	2‡
Abattoir	2	14	28	6	5†	4
Total		31	52	14	12	8

*Transvaginal ultrasound-guided aspiration.
†Two single pregnancies with normal IVF offspring were lost due to prepartum pregnancy toxaemia.
‡One of these offspring (a male) resulted from a frozen-thawed IVM/IVF/IVC blastocyst.

15.6.3. *Capacitation of goat sperm*

In the capacitation of sperm and its use in IVF in goats, a report by Younis *et al.* (1991) in the USA was one of the first to document the establishment of a pregnancy after IVM and IVF of goat oocytes; they attempted to establish an IVF system whereby goat oocytes matured *in vivo* or *in vitro* could be fertilized and cultured to stages suitable for transfer. The use of a modified defined medium (Brackett and Oliphant, 1975) was found to give the best results in these studies; this method has also been employed by other workers as a capacitation procedure (Crozet *et al.*, 1993; Keskintepe *et al.*, 1994; Mogas *et al.*, 1995c). Other studies have examined the use of heparin in the capacitation of sperm.

In Spain, Palomo *et al.* (1995) investigated the effect of heparin and sperm concentration on the IVF of prepubertal goat oocytes; they concluded that 100μg ml^{-1} heparin with a final concentration of 5.01 million sperm ml^{-1} resulted in higher cleavage rates and a lower incidence of polyspermy than other dose levels tested. In Chile, Cox *et al.* (1994) conducted a study to assess whether capacitated goat sperm are able to penetrate cattle and sheep oocytes in a way comparable with that observed in homologous fertilization; their results suggested that heterologous fertilization could be used to study sperm physiology and fertilization in goats.

15.6.4. In vitro *culture of goat embryos*

Results reported by Sakkas *et al.* (1989) demonstrated that co-culture of goat embryos with oviductal cells (in various media) enabled a high percentage of goat embryos to develop through the period of cleavage arrest; they observed, however, that the co-culture system provided a less adequate environment to the blastocyst stage. In contrast, the studies of Richardson *et al.* (1992) indicated that goat embryos may not exhibit the stage-specific developmental block characteristic of cattle and sheep. However, the studies of Pivko *et al.* (1995) in the Czech Republic led them to conclude that, in comparison with other farm ruminants, there is probably no great difference in goats in the timing of the shift in the control of gene transcription from the maternal to the embryonic genome. It is evident that the nutrient uptake of the goat is very similar to that of the sheep embryo (Gardner *et al.*, 1994); low levels of glucose uptake prior to compaction are consistent with low levels of glucose utilization.

In the USA, Pritchard *et al.* (1992) demonstrated that co-culture with oviductal cells resulted in a greater rate of goat embryo development than medium alone (0 vs. 70% blastocysts). The same workers also showed that transfer of early stage goat embryos (2–4-cell) from oviductal to uterine cell monolayers during culture did not enhance development in comparison with embryos maintained on oviductal cells alone; such findings are in accord with similar data for cattle, sheep and pig embryos.

15.7. Embryo Splitting and Cloning

15.7.1. *Splitting goat embryos*

Embryos were collected from superovulated donor goats 6–8 days after oestrus and bisected symmetrically by way of a microblade, in Chinese studies reported by Wang *et al.* (1990); demi-embryos with normal morphology were transferred to the uterine horn ipsilateral to the corpus luteum of recipients. The authors record transfers to five recipients, two of which gave birth to live kids. Identical goat twins were also produced in China by the efforts of Zhang *et al.* (1991a, 1992), working with frozen embryos. In this work, hatched blastocysts were embedded in agar and frozen; after thawing, five embryos were bisected and transferred to five recipients, four of which became pregnant, with two producing monozygotic twins.

The adverse effect of freeze-thawing on the viability of goat demi-embryos was clearly evident in results from Germany, where Nowshari and Holtz (1993) transferred goat demi-embryos either fresh or after freezing. Transfer of 11 pairs of fresh demi-embryos resulted in the birth of five twin and three single kids; transfer of 11 pairs of thawed demi-embryos resulted in two single kids.

Genetically identical goats could be very useful in certain lines of research. In France, for example, Chesne *et al.* (1995) described their attempts to produce identical twin goats for use in the development of a vaccine against *Cowdria ruminantium*, a major pathogen of ruminants in the tropics. Embryos recovered at the hatched blastocyst stage were recovered at day 10, split and transferred immediately to recipient does. Although the survival rate of demi-embryos was low, the authors did succeed in producing enough twin sets to be used in the pathology research (see Table 15.6).

Table 15.6 Results of the transfer of split-goat embryos (from Chesne *et al.*, 1995).

	Number	Percentage*
Pairs demi-embryos transferred	48	
Recipient goats	48	
Presumed pregnant	35	72.9
Goats kidding	18	37.5
Kids born	23	
Set of twins	5	10.5

*Percentages are relative to the number of recipients.

15.7.2. Cloning by nuclear transfer

The amount of information currently available on cloning by nuclear transfer in goats is much less than that in sheep, but there is little reason to believe that this species should present problems that cannot be solved using procedures based on sheep and cattle work. Already, normal kids have been born in China after nuclear transfer (Yong *et al.*, 1991; Zhang *et al.*, 1991b); they employed *in vivo* produced oocytes and blastomeres from 4- to 32-cell embryos in their work (Fig. 15.3).

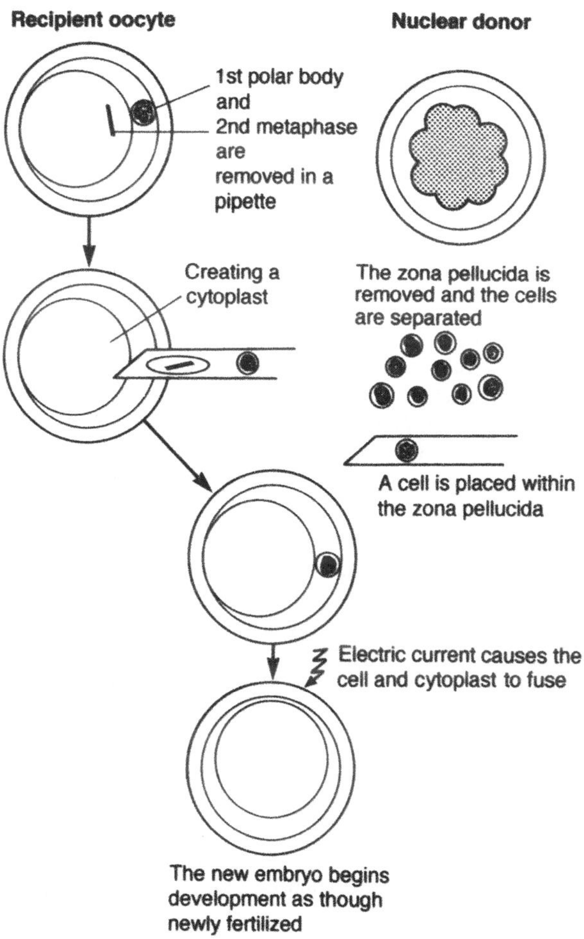

Fig. 15.3. Steps in cloning by nuclear transfer in farm animals. Studies in China have led to the birth of kids after the transfer of a blastomere from 4- to 32-cell embryos into an enucleated mature oocyte. As with sheep and cattle, commercial cloning of goats may be a reality by the end of the century (from Woolliams and Wilmut, 1989).

15.8. Production of Transgenic Goats

Commercial production of human pharmaceutical proteins in the milk of dairy goats may eventually be feasible as a result of studies being carried out in various countries, mainly the USA, the UK and Australia (Moffat, 1991). Already, in the USA, researchers have developed transgenic goats carrying the gene for a longer-acting form of tissue plasminogen activator (tPA), with some goats capable of producing this clot-dispersing factor at the rate of about 3 g per litre of milk (Ebert *et al.*, 1991). A later report by Ebert and Schindler (1993) referred to one of their transgenic does expressing tPA at an initial concentration of 9 mg/ml, leveling off to 3 mg through the remainder of the lactation; these authors concluded that the potential for transgenic technology to create beneficial changes in farm animal species remained speculative but that their recent investigations with goats were encouraging.

15.9. References

Agrawal, K.P. (1992) In vitro maturation of caprine oocytes. *Indian Journal of Animal Reproduction* 13, 195–297.

Agrawal, K.P. and Goel, A.K. (1991) Production of elite Jamunapari kids by embryo transfer technology. *Indian Journal of Animal Reproduction* 12, 78–80.

Agrawal, K.P., Goel, A.K. and Tyagi, S. (1991) Successful non-surgical embryo recovery from a goat. *Indian Journal of Experimental Biology* 29, 1144.

Akinlosotu, B.A. and Wilder, C.D. (1993) Fertility and blood progesterone levels following LHRH-induced superovulation in FSH-treated anestrous goats. *Theriogenology* 40, 895–904.

Amoah, E.A. and Gelaye, S. (1991). Embryo recovery, evaluation, storage and transfer in goats. *Small Ruminant Research* 6, 119–129.

Armstrong, D.T., Pfitzner, A.P., Warnes, G.M., Ralph, M.M. and Seamark, R.F. (1983a) Endocrine responses of goats to superovulation with PMSG and FSH. *Journal of Reproduction and Fertility* 67, 395–401.

Armstrong, D.T., Pfitzner, A.P., Warnes, G.M. and Seamark, R.F. (1983b) Superovulation treatment and embryo transfer in Angora goats. *Journal of Reproduction and Fertility* 67, 403–410.

Baril, G. and Vallett, J.C. (1990) Time of ovulations in dairy goats induced to superovulate with porcine follicle stimulating hormone during and out of the breeding season. *Theriogenology* 34, 303–311.

Baril, G., Casamitjana, P., Perrin, J. and Vallet, J.C. (1988) Embryo production, freezing and transfer in Angora, Alpine and Saanen goats. *Proceedings of the 4th Meeting European Embryo Transfer Association* (Lyon), pp. 67–93.

Baril, G., Casamitjana, P., Perrin, J. and Vallet, J.C. (1989) Embryo production, freezing and transfer in Angora, Alpine and Saanen goats. *Zuchtungskunde* 24, 101–115.

Baril, G., Remy, B., Leboeuf, B., Vallet, J.C., Beckers, J.F. and Saumande, J. (1992) Comparison of porcine FSH, caprine FSH and ovine FSH to induce repeated superovulation in goats. *Proceedings of the 8th Meeting of the European Embryo Transfer Association* (Lyon), p. 126.

Baril, G., Brebion, P. and Chesne, P. (1993) Manuel de formation pratique pour la transplantation embryonnaire chez la brebis et la chevre. *FAO Publication* 115, Rome, 183 pp.

Baril, G., Leboeuf, B., Figueiredo Freitas, V.J., Pougnard, J.L. and Saumande, J. (1994) Control of the LH preovulatory surge by a gonadotropin releasing hormone antagonist in superovulated goats. *Proceedings of the 10th Meeting European Embryo Transfer Association* (Lyon), p. 148.

Batt, P.A., Kilben, J.D. and Cameron, A.W.N. (1993) Use of a single or multiple injections of FSH in embryo collection programmes in goats. *Reproduction, Fertility and Development* 5, 49–56.

Battye, K.M., Fairclough, R.J., Cameron, A.W.N. and Trounson, A.O. (1988) Evidence of prostaglandin involvement in early luteal regression of the superovulated nanny goat (*Capra hircus*). *Journal of Reproduction and Fertility* 84, 425–430.

Besenfelder, U., Zinovieva, N., Dietrich, E., Sohnrey, B., Holtz, W. and Brem, G. (1994) Tubal transfer of goat embryos using endoscopy. *Veterinary Record* 135, 480–481.

Bilton, R.J. and Moore, N.W. (1976) In vitro culture, storage and transfer of goat embryos. *Australian Journal of Biological Science* 29, 125–129.

Borque, C., Pintado, B., Perez, B., Gutierrez, A., Munoz, I. and Mateos, E. (1993) Progesterone levels in superovulated Murciana goats with or without successful embryo collection. *Theriogenology* 39, 192.

Brackett, B.G. and Oliphant, G. (1975) Capacitation of rabbit spermatozoa in vitro. *Biology of Reproduction* 12, 260–274.

Chauhan, M.S. and Anand, S.R. (1991) In vitro maturation and fertilization of goat oocytes. *Indian Journal of Experimental Biology* 29, 105–110.

Chemineau, P., Procureur, R., Cognie, Y., Lefevre, P.C., Locatelli, A. and Chupin, D. (1986) Production, freezing and transfer of embryos from a bluetongue-infected goat herd without bluetongue transmission. *Theriogenology* 26, 279–289.

Chesne, P., Cognie, Y., Martinez, D., Depres, E. and Aumont, G. (1995) Production of identical tropical goat twins for pathology research. *Proceedings of the 11th Meeting of the European Embryo Transfer Association* (Hanover), p. 144.

Cox, J.F., Avila, J., Saravia, F. and Santa Maria, A. (1994) Assessment of fertilizing ability of goat spermatozoa by in vitro fertilization of cattle and sheep intact oocytes. *Theriogenology* 41, 1621–1629.

Crozet, N., Smedt, De V., Ahmed-Ali, M. and Sevellec, C. (1993) Normal development following in vitro oocyte maturation and fertilization in the goat. *Theriogenology* 39, 206.

Crozet, N., Ahmed-Ali, M. and Dubos, M.P. (1995a) Developmental competence of goat oocytes from follicles of different size categories following maturation, fertilization and culture in vitro. *Journal of Reproduction and Fertility* 103, 293–298.

Crozet, N., Ahmed-Ali, M., Cognie, Y. and Dubos, M.P. (1995b) Effect of follicle size on goat oocytes developmental competence in vitro. *Proceedings of the 11th Meeting European Embryo Transfer Association* (Hanover), p. 150.

Deguet, M., Joisel, F., Boender, G. and Dobbelaere, T. (1989) Embryo transfer in the goat. Practical application in 3 herds of Angora goats. *Recueil de Medecine Veterinaire* 165, 807–813.

De Smedt, V., Crozet, N., Ahmed-Ali, M., Martino, A. and Cognie, Y. (1992) In vitro maturation and fertilization of goat oocytes. *Theriogenology* 37, 1049–1060.

Dietrich, E., Hennies M., Holtz, W. and Voglmayr, J.K. (1995) Immunization of goats against recombinant human inhibin alpha-subunit: effects on inhibin binding, mating behaviour, ovarian activity and embryo yield. *Animal Reproduction Science* 39, 119–128.

Ebert, K.M. and Schindler, J.E.S. (1993) Transgenic farm animals: progress report. *Theriogenology* 39, 121–135.

Ebert, K.M., Selgrath, J.P., DiTullio, P., Denman, J., Smith, T.E., Memon, M.A., Schindler, J.E., Monastersky, G.M., Vitale, J.A. and Gordon, K. (1991) Transgenic production of a variant of human tissue-type plasminogen activator in goat milk; generation of transgenic goats and analyses of expression. *Biotechnology* 9, 835–838.

Eiamvitayakorn, J., Natural, N.G. and Apelo, C.L. (1988) Superovulatory treatment in goats (*Capra hircus*). *Thai Journal of Veterinary Medicine* 18, 251–258.

Fieni, F., Buggin, M., Tainturier, D., Bruyas, J.F., Perrin, J., Dumont, P., Beckers, J.P., Chupin, D. and Daubie, M. (1990) Comparison of the efficiency of three cryoprotectants for freezing goat embryos. *Proceedings of the 6th Meeting of the European Embryo Transfer Association* (Lyon), p. 144.

Fieni, F., Beckers, J.P., Buggin, M., Bruyas, J.F., Perrin, J., Daubie, M. and Tainturier, D. (1995) Evaluation of cryopreservation techniques for goat embryos. *Reproduction, Nutrition, Development* 35, 367–373.

Gardner, D.K., Lane, M. and Batt, P.A. (1994) Nutrient uptake and enzyme activity of the pre-attachment goat embryo developed in vivo. *Theriogenology* 41, 204.

Godke, R.A., Overskei, T.L. and Voelkel, S.A. (1985) The potential of micromanipulation and embryo transfer in breeding goats. *Dairy Goat Journal* 63, 154–157.

Gordon, I. (1994) *Laboratory Production of Cattle Embryos.* CAB International, Wallingford, 640 pp.

Graff, K.J., Meintjes, M., Paul, J.B., Dyer, V.W., Denniston, R.S., Ziomek, C. and Godke, R.A. (1995) Ultra-sound guided transvaginal oocyte recovery from FSH-treated goats for IVF. *Theriogenology* 43, 223.

Grupp, T. (1991) Investigations on dilation of the cervix and transcervical embryo recovery in ewes and goats. Thesis, Ludwig-Maximilians-Universitat Munchen, Germany, 144 pp.

Han, Y., Meintjes, M., Graff, K.J., Denniston, R.S., Ebert, K.M., Ziomek, C. and Godke, R.A. (1996) Offspring born from the transfer of caprine blastocysts after IVM/IVF and IVC of transvaginal ultrasound-guided aspirated oocytes. *Theriogenology* 45, 357.

Hanada, A. (1985) In vitro fertilization in goats. *Japanese Journal of Animal Reproduction* 31, 21–26.

Keskintepe, L., Luvoni, G.C., Bassiony, M.M. and Brackett, B.G. (1994) In vitro caprine blastocyst development from immature oocytes. *Biology of Reproduction* 50 (Suppl. 1), p. 189.

Krogh, K. (1991) Ova transfer in sheep and goats. *Dansk Fareavl* 56(12), 4–6.

Le Gall, F., Baril, G., Vallet, J.C. and Leboeuf, B. (1993) In vivo and in vitro survival of goat embryos after freezing with ethylene glycol or glycerol. *Theriogenology* 40, 771–777.

Li, R., Cameron, A.W.N., Batt, P.A. and Trounson, A.O. (1990) Maximum survival of frozen goat embryos is attained at the expanded, hatching and hatched blastocyst stages of development. *Reproduction, Fertility and Development* 2, 345–350.

Mahmood, S., Koul, G.L. and Biswas, J.C. (1991) Comparative efficacy of FSH-P and

PMSG on superovulation in Pashmina goats. *Theriogenology* 35, 1191–1196.

Mani, A.U., Watson, E.D. and McKelvey, W.A.C. (1994) The effects of subnutrition before or after embryo transfer on pregnancy rate and embryo survival in does. *Theriogenology* 41, 1673–1678.

Martino, A., Mogas, T., Palomo, M. J. and Paramio, M.T. (1994a) Meiotic competence of prepubertal goat oocytes. *Theriogenology* 41, 969–980.

Martino, A., Palomo, M.J., Mogas, T. and Paramio, M.T. (1994b) Influence of the collection technique of prepubertal goat oocytes on in vitro maturation and fertilization. *Theriogenology* 42, 859–873.

Moffat, A.S. (1991) Transgenic animals may be down on the pharm. *Science* 254, 35–36.

Mogas, T., Martino, A., Palomo, M.J. and Paramio, M.T. (1992) Effect of method of recovery on the number and type of oocytes obtained for IVM. *Journal of Reproduction and Fertility* Abstract Series No. 9, p. 52.

Mogas, T., Palomo, M.J., Izquierdo, D., Martino, A. and Paramio, M.T. (1993) Effect of hormones on the in vitro maturation, fertilization and early cleavage of prepubertal goat oocytes. *Journal of Reproduction and Fertility* Abstract Series No. 11, p. 67.

Mogas, T., Izquierdo, M.D., Palomo, M.J. and Paramio, M.T. (1995a) Effect of hormones, serum source and culture system on the IVM and IVF of prepubertal goat oocytes and subsequent embryo development. *Theriogenology* 43, 284.

Mogas, T., Palomo, M.J., Izquierdo, D. and Paramio, M.T. (1995b) Effect of granulosa cell source on in vitro maturation, fertilization and embryo development of prepubertal goat oocytes. *Journal of Reproduction and Fertility* Abstract Series No. 15, pp. 68–69.

Mogas, T., Palomo, M.J., Izquierdo, M.D. and Paramio, M.T. (1995c) The effect of estrus goat serum (EGS) and FCS on the in vitro fertilization and early cleavage of prepubertal goat oocytes. *Proceedings of the 11th Meeting of the European Embryo Transfer Association* (Hanover), p. 212.

Moore, N.W. (1974) Multiple ovulation and ovum transfer in the goat. *Proceedings of the Australian Society of Animal Production* 10, 246–249.

Moore, N.W. (1982) Egg transfer in the sheep and goat. In: Adams, C.F. (ed.) *Mammalian Egg Transfer*. CRC Press, Boca Raton, Florida, pp. 119–133.

Moore, N.W. and Eppleston, J. (1979a) Embryo transfer in the Angora goat. *Australian Journal of Agricultural Research* 30, 973–981.

Moore, N.W. and Eppleston, J. (1979b) Control of oestrus, ovulation and fertility in relation to artificial insemination in the Angora goat. *Australian Journal of Agricultural Research* 30, 965–972.

Nowshari, M.A. and Holtz, W. (1993) Transfer of split goat embryos without zonae pellucidae either fresh or after freezing. *Journal of Animal Science* 71, 3403–3408.

Nowshari, M.A. and Holtz, W. (1995) In vitro culture of goat morulae to blastocysts before freezing. *Theriogenology* 44, 983–988.

Nowshari, M.A., Beckers, J.F. and Holtz, W. (1995) Superovulation of goats with purified pFSH supplemented with defined amounts of pLH. *Theriogenology* 43, 797–802.

Nuti, L.C., Minhas, B.S., Baker, W.C., Capehart, J.S. and Marrack, P. (1987) Superovulation and recovery of zygotes from Nubian and Alpine dairy goats. *Theriogenology* 28, 481–488.

Palomo, M.J., Mogas, T., Izquierdo, M.D. and Paramio, M.T. (1995) Effect of heparin and sperm concentration on IVF of prepubertal goat oocytes. *Theriogenology* 43, 292.

Pampoukidou, A., Alifakiotis, T., Avdi, M. and Magras, I. (1992) Superovulation and embryo transfer in goats by using PMSG or FSH. *Proceedings of the 8th Meeting European Embryo Transfer Association* (Lyon), p. 198.

Pargaonkar, M.D., Bakshi, S.A., Pargaonkar, D.R., Tandle, M.K. and Doijode, S.V. (1994) Studies on superovulation response of goats treated with PMSG. *Indian Journal of Dairy Science* 47, 149–150.

Pawshe, C.H., Jain, S.K. and Totey, S.M. (1993) Effect of commercially available follicle stimulating hormone on in vitro maturation of goat oocytes. *Indian Journal of Animal Reproduction* 14, 69–71.

Pendleton, R.J., Youngs, C.R., Rorie, R.W., Memon, M.A. and Godke, R.A. (1986) The use of Norgestomet implants and FSH for estrus synchronization and superovulation in goats. *Theriogenology* 25, 180.

Pendleton, R.J., Youngs, C.R., Rorie, R.W., Pool, S.H., Memon, M.A. and Godke, R.A. (1992) Follicle stimulating hormone versus pregnant mare serum gonadotropin for super-ovulation of dairy goats. *Small Ruminant Research* 8, 217–224.

Pereira, R.J.T.A., Ajoub, M. and Holtz, W. (1995) Birth of live kids after transfer of in vitro matured and fertilized goat embryos. *Proceedings of the 11th Meeting European Embryo Transfer Association* (Hanover), p. 224.

Pivko, J., Grafenau, P. and Kopecny, V. (1995) Nuclear fine structure and transcription in early goat embryos. *Theriogenology* 44, 661–671.

Pritchard, J.F., Thibodeaux, J.K., Pool, S.H., Blakewood, E.G., Menezo, Y. and Godke, R.A. (1992) In vitro co-culture of early stage caprine embryos with oviduct and uterine epithelial cells. *Human Reproduction* 7, 553–557.

Puls-Kleingeld, M., Yuswiati, E., Nowshari, M.A., Beckers, J.F. and Holtz, W. (1991) The effect of FSH/LH ratio and treatment schedule on the superovulatory response in goats. *Journal of Reproduction and Fertility* Suppl. 43, 308.

Puls-Kleingeld, M., Nowhsari, M.A. and Holtz, W. (1992) Cryopreservation of goat embryos by the one-step or three-step equilibration procedure. In: Lokeshwar, R.R. (ed.) *Recent Advances in Goat Production*. Nutan Printers, New Delhi, pp. 1388–1391.

Rajikin, M.H., Yusoff, M. and Abdullah, R.B. (1994) Ultra-structural studies of developing goat oocytes in vitro. *Theriogenology* 42, 1003–1016.

Rao, V.H., Sarmah, B.C., Agrawal, K.P., Ansari, M.R. and Bhattacharyya, N.K. (1988) Survival of goats embryos frozen and thawed rapidly. *Animal Reproduction Science* 16, 261–264.

Richardson, M.E., Peltz, J.M., Long, J.A. and Dickey, J.F. (1992) In vitro culture of goat (*Capra hircus*) embryos. *Biology of Reproduction* 46 (Suppl. 1), p. 121.

Riha, J., Cunat, L., McKelvey, W.A.C., Millar, P. and Bernatsky, C. (1994) Transfer of imported frozen cashmere goat embryos. *Zivocisna Vyroba* 39, 881–888.

Rosnina, Y., Jainudeen, M.R. and Nihayah, M. (1992) Super-ovulation and egg recovery in goats in the tropics. *Veterinary Record* 130, 97–99.

Ryot, K.D. and Vadnere, S.V. (1989) Preservation of goat embryos by slow and rapid methods of cooling. *Indian Veterinary Journal* 66, 1086–1087.

Sakkas, D., Batt, P.A. and Cameron, A.W.N. (1989) Development of preimplantation goat (*Capra hircus*) embryos in vivo and in vitro. *Journal of Reproduction and Fertility* 87, 359–365.

Senn, B.J. and Richardson, M.E. (1992) Seasonal effects on caprine response to synchronization of estrus and superovulatory treatment. *Theriogenology* 37, 579–585.

Tervit, H.R., Goold, P.G., McKenzie, R.D. and Clarkson, D.T. (1983) Techniques and success of embryo transfer in Angora goats. *New Zealand Veterinary Journal* 31, 67–70.

Tervit, H.R., Goold, P.G. McKenzie, R.D., Clarkson, D.J. and Drummonds, J. (1984) Embryo transfer in Angora and Saanen goats. *New Zealand Veterinary Journal* 33, 78–80.

Tsunoda, Y. and Sugie, T. (1989) Superovulation in nonseasonal Japanese native goats, with special reference to the developmental progression of embryos. *Theriogenology* 31, 991–996.

Vallet, J.C., Baril, G. and Loysel, C. (1989) Efficiency of laparoscopic embryo transfer in goats. *Proceedings of the 5th Meeting of the European Embryo Transfer Association* (Lyon), p. 186.

Wang, G., Baohua, M., Wang, J., Qian, J. and Zhang, Y. (1988) Embryo freezing and transfer in milk goats. *Theriogenology* 29, 322.

Wang, B., Ziong, H.Q. and Fan, B.Q. (1990) A study of embryo bisection in goats. *Jiangsu Journal of Agricultural Science* 9 (Suppl.), 66–70.

Wani, G.M., Geldermann, H. and Hahn, J. (1990) Superovulations during early luteal phase in goats. *World Review of Animal Production* 25(2), 41–43.

Warwick, B.L., Berry, R.O. and Horlacher, W.R. (1934) Results of mating rams to Angora female goats. *Proceedings of the American Society of Animal Production* p. 225.

Woolliams, J.A. and Wilmut, I. (1989) Embryo manipulation in cattle breeding and production. *Animal Production* 48, 3–30.

Yang, Z.M., Tan, J.H. and Qin, P.C. (1991) A preliminary study on the preimplantation development of goats. *Acta Veterinaria et Zootechnica Sinica* 22(1), 32–37.

Yong, Z., Jianchen, W., Jufen, Q. and Zhiming, H. (1991) Nuclear transplantation in goats. *Theriogenology* 35, 299.

Younis, A.I., Zuelke, K.A., Harper, K.M., Oliveira, M.A.L. and Brackett, B.G. (1991) In vitro fertilization of goat embryos. *Biology of Reproduction* 44, 1177–1182.

Yuswiati, E. and Holtz, W. (1990) Work in progress: successful transfer of vitrified goat embryos. *Theriogenology* 34, 629–632.

Zhang, Y., Wang, J.C., Xu, J. and Qian, J.F. (1991a) Freezing of goat half-embryos. *Scientia Agricultura Sinica* 24(1), 11–15.

Zhang, Y., Wang, J.C., Quian, J.F. and Hao, Z.M. (1991b) Nuclear transplantation in goat embryos. *Scientia Agricultura Sinica* 24(5), 1–6.

Zhang, Y., Wang, J.C. and Qian, J.F. (1992) Bisection of frozen-thawed goat embryos. *Acta Veterinaria et Zootechnica Sinica* 23, 193–197.

Index

Printed and bound by CPI Group (UK) Ltd, Croydon, CR0 4YY

16/09/2026

14963433-0003